Elements of Computer Algebra
with Applications

518.562.A31

Elements of Computer Algebra
with Applications

Alkiviadis G. Akritas
University of Kansas

A WILEY-INTERSCIENCE PUBLICATION
JOHN WILEY & SONS
New York • Chichester • Brisbane • Toronto • Singapore

Copyright © 1989 by John Wiley & Sons, Inc.

All rights reserved. Published simultaneously in Canada.

Reproduction or translation of any part of this work beyond that permitted by Section 107 or 108 of the 1976 United States Copyright Act without the permission of the copyright owner is unlawful. Requests for permission or further information should be addressed to the Permissions Department, John Wiley & Sons, Inc.

Library of Congress Cataloging in Publication Data:

Akritas, Alkiviadis G., 1949–
    Elements of computer algebra with applications / Alkiviadis G. Akritas.
      p. cm.
    "A Wiley-Interscience publication."
    Includes bibliographies and index.
    ISBN 0-471-61163-8
      1. Algebra—Data processing. I. Title.
QA155.7.E4A37 1988
512—dc19

                                                              88-14870
                                                                   CIP

Printed in the United States of America

10 9 8 7 6 5 4 3 2 1

To the memory of my father

Preface

A computer is basically a device for processing information, and the kind of information does not matter much. Computers have to be "taught" to do arithmetic so that they can be *number*-crunchers. The business world has generated the word processor that crunches *words*. And there now exists the capability to crunch *symbols*: Computer Algebra. By dealing mainly with *exact* numbers, and algebraic expressions in terms of their symbolic representations, computer algebra systems can help scientists gain more *insight* into the various physical phenomena under examination.

Computer Algebra differs from Numerical Analysis, where the emphasis is on the errors that might arise in computations while executing a certain algorithm. These errors, truncation and round-off, occur because of the use of floating-point and single- or double-precision arithmetic. In general, the smaller the overall error, the better the algorithm.

Since 1960, many software systems devoted to various classes of symbolic computations have been developed; these systems are continuously increasing in efficiency and capability, and an expansion of their use can be expected in the future. Polynomial and rational function operations form the fundamental basis of any symbolic manipulation system, and, therefore, research in this area has included the development and analysis of efficient algorithms for polynomial greatest common divisors, polynomial factorization, and polynomial real root isolation.

Computer algebra includes a wide variety of topics and, since it is still in a developmental phase, new ones are constantly added to the list. Only those topics the author considers "classical" are covered in this book, which can be used both as a reference source for the researcher and as a text in Numerical Methods in the Computer Science or Mathematics curriculum.

Since there is little, if any, overlap in the *treatment* of topics between Numerical Analysis and Computer Algebra a course, complementary or alternative to Numerical Analysis, can be based on this book.

For the Mathematics curriculum, on one hand, this book provides an excellent vehicle for teaching about both theory and applications of algebra and effectively joins traditional Algebra with Computer Science. On the other hand, the Computer Science student taking Computer Algebra utilizes a great number of concepts learned in previous courses and is made aware of the beautiful works of some of the giants of Mathematics of the previous centuries (Galois, Hensel, Lagrange, Sturm, Sylvester, and Vincent to name a few) whose basic approach to computation *cannot* be dealt with in Numerical Analysis, but closely resembles what researchers are now trying to accomplish using computer algebra.

The main firsts of the book are

1. The best subresultant method for computing polynomial remainder sequences, developed in 1986 by the author based on a paper by Sylvester (1853).
2. Budan's theorem showing its importance and its relation to a theorem by Fourier.
3. The fastest existing method for isolating the real roots of a polynomial equation, developed in 1978 by the author based on a theorem by Vincent (1836).

What this book is not about

This book is not about proofs; proofs are included if they contribute to a better understanding of the material or if they are to be found only in scientific journals. Also, this book is not about data structures; when used as a text, the implementation of the various algorithms described in this book can be done in various ways and is left to the instructor and the ingenuity of the students.

This book has grown out of the senior/graduate course "Computer Algebra" that I have been teaching both at the University of Kansas, U.S.A. and at the National Technical University of Athens, Greece. Students majoring in Computer Science, Mathematics, and Electrical and Computer Engineering have successfully taken this course; regarding prerequisites, a good course in data structures as well as a course in modern algebra and/or in linear algebra are *desirable* but not required.

The material in the book is divided in three parts:

- Part I is the introduction with Chapter 1 explaining what computer algebra is.

- Part II contains the mathematical foundations and basic algorithms. Since computer algebra deals mainly with integers and polynomials with integer coefficients, Chapter 2 covers the basic properties of integers, whereas Chapter 3 covers those of the polynomials.
- In Part III, we find applications of the ideas developed in the previous parts and advanced topics; namely, Chapter 4 is devoted to error correcting codes and cryptography, Chapter 5 to the computation of polynomial greatest common divisors and polynomial remainder sequences, Chapter 6 to the factorization of polynomials with integer coefficients, and Chapter 7 to the isolation and approximation of the real roots of polynomial equations.

Selection by the instructor of topics of interest will be very important, because I have found that I cannot cover anywhere near all of this material in a semester. A reasonable choice is to cover in detail Parts I (Introduction) and II (Mathematical Foundations and Basic Algorithms) and then various chapters from Part III (Applications and Advanced Topics) as time permits, bearing in mind that Section 7.2 depends on Section 5.2. Finally, the use of the computer algebra system **maple** is *strongly* recommended in such a course.

I am grateful to Maria Taylor, my editor, and all the staff at John Wiley for our wonderful collaboration.

I thank Zamir Bavel, my colleague and friend at the University of Kansas, for his valuable advice, Manolis Protonotarios, Chairman of the Electrical and Computer Engineering Department of the National Technical University of Athens, for giving me the opportunity to teach Computer Algebra in Greece, and my Mother for taking care of everything while I was working on the book in Greece.

ALKIVIADIS G. AKRITAS

University of Kansas

Contents

Part I. Introduction   1

1. What Is Computer Algebra?   3

    1.1. Computer Algebra versus Numerical Analysis   4
    1.2. Computer Algebra: Exact Integer and Polynomial
           Arithmetic   6
    1.3. Computer Algebra Systems   16
           Programming Exercises   18
           References   20

Part II. Mathematical Foundations and Basic Algorithms   21

2. Integers   23

    2.1. Fundamental Concepts   24
           2.1.1. Sets   24
           2.1.2. Equivalence Relations   26
           2.1.3. Functions and Algebraic Systems   28
    2.2. Greatest Common Divisors of Integers   34
           2.2.1. Divisibility of Integers   34
           2.2.2. Euclid's Algorithm and Lamé's Theorem   38
           2.2.3. Extended Euclidean Algorithm   42
           2.2.4. Euclid's Algorithm and Continued Fractions   44

xii CONTENTS

    2.3. Integer Factorization   52
        2.3.1. Prime Numbers and the Sieve of Eratosthenes   52
        2.3.2. Integers Modulo m and the Greek–Chinese Remainder Algorithm   59
        2.3.3. Primality Testing   77
        2.3.4. Factorization of Large Integers   82
    2.4. Exact Computations Using Modular Arithmetic   83
    Exercises   91
    Programming Exercises   94
    Historical Notes and References   97

3. Polynomials   101

    3.1. Fundamental Concepts   102
        3.1.1. Basic Facts of Polynomials   102
        3.1.2. The Ruffini–Horner Method   107
        3.1.3. Interpolation over a Field   110
        3.1.4. Computations Using the Evaluation–Interpolation Scheme   114
    3.2. Greatest Common Divisors of Polynomials over a Field   120
        3.2.1. Divisibility of Polynomials   121
        3.2.2. Euclid's Algorithm for Polynomials over a Field   123
        3.2.3. Irreducible Factors of Polynomials   126
        3.2.4. Squarefree Factorization of Polynomials   132
    3.3. Galois Fields $GF(p^r)$ 135
        3.3.1. Basic Facts of Finite Fields   135
        3.3.2. Construction of Galois Fields $GF(2^r)$ 144
        3.3.3. Circuits for Polynomial Arithmetic in $GF(2^r)$ 146
    Exercises   151
    References   154

Part III. Applications and Advanced Topics   155

4. Error-Correcting Codes and Cryptography   157

    4.1. Error-Correcting Codes—General Concepts   158
        4.1.1. Hamming Codes   164
        4.1.2. BCH Codes   174
    4.2. Cryptography—General Concepts   188
        4.2.1. Symmetric (Single-Key) Cryptosystems   190
        4.2.2. Asymmetric (Public-Key) Cryptosystems   201

	Exercises	209
	Programming Exercises	212
	References	213

5. Greatest Common Divisors of Polynomials over the Integers and Polynomial Remainder Sequences — 215

    5.1. Introduction and Motivation   216
        5.1.1. A General Overview of the Classical PRS Algorithms   218
        5.1.2. A Nonclassical Approach: The Modular Greatest Common Divisor Algorithm   224
    5.2. The Sylvester–Habicht Pseudodivisions Subresultant PRS Method   227
        5.2.1. Sylvester's Reduced (Subresultant) PRS Algorithm   227
        5.2.2. The Resultant of Two Polynomials   230
        5.2.3. Habicht's Subresultant PRS Algorithm   243
    5.3. The Matrix-Triangularization Subresultant PRS Method   253
        5.3.1. Gaussian Elimination and Polynomial Division   254
        5.3.2. Gaussian Elimination and the Bruno, Trudi Forms of the Resultant   256
        5.3.3. Gaussian Elimination + Sylvester's Form of the Resultant = the Matrix-Triangularization Subresultant PRS Method   260
    5.4. Empirical Comparisons Between the Two Subresultant PRS Methods   267
    Exercises   279
    Programming Exercises   281
    Historical Notes and References   282

6. Factorization of Polynomials over the Integers — 285

    6.1. Introduction and Motivation   286
        6.1.1. The Schubert–Kronecker Factorization Method over the Integers   286
        6.1.2. A General Overview of the Roundabout Factorization Method over the Integers   288
    6.2. Factorization of Polynomials over a Finite Field   292
        6.2.1. Squarefree Factorization over Finite Fields   293
        6.2.2. Counting Irreducible Polynomials over Finite Fields   296

              6.2.3. Distinct Degree (Partial) Factorization over Finite Fields — 299
              6.2.4. Berlekamp's Factorization Algorithm over Finite Fields — 304
       6.3. Lifting a (mod p)-Factorization to a Factorization over the Integers — 316
              6.3.1. Linear and Quadratic Lifting — 320
              6.3.2. Finding the True Factors over the Integers — 326
       Exercises — 327
       Historical Notes and References — 329

7. Isolation and Approximation of the Real Roots of Polynomial Equations — 333

   7.1. Introduction and Motivation — 335
   7.2. Fourier's Theorem and Sturm's Bisection Method for Isolation of the Real Roots — 336
          7.2.1. Fourier's Theorem — 337
          7.2.2. Sturm's Theorem and Sturm's Bisection Method for Isolation of the Real Roots — 341
          7.2.3. Computation of an Upper (Lower) Bound on the Values of the Positive Roots of a Polynomial Equation — 349
          7.2.4. Computation of a Lower Bound on the Distance Between Any Two Roots of a Polynomial Equation — 353
   7.3. Budan's Theorem and the Two Continued Fractions Methods for Isolation of the Real Roots — 357
          7.3.1. Budan's Theorem — 357
          7.3.2. Moebius Substitutions and their Effect on the Roots of an Equation — 358
          7.3.3. Vincent's Theorem: Extension and Application — 363
          7.3.4. The Two Continued Fractions Methods for Isolation of the Real Roots — 373
   7.4. Empirical Comparisons Between the Two Real Root Isolation Methods — 380
   7.5. Approximation of the Real Roots of a Polynomial Equation — 381
          7.5.1. Real Root Approximation Using Bisection — 382
          7.5.2. Real Root Approximation Using Continued Fractions — 383

7.6. Empirical Comparisons Between the Two Real Root Approximation Methods	388
Exercises	391
Programming Exercises	393
Historical Notes and References	394

Appendix: Linear Algebra   **401**

Index   **413**

Part I
Introduction

This part of the book is about computer algebra and computer algebra systems; basic notions of algorithms and their complexity are also presented along with concepts of data structures. The difference between computer algebra and numerical analysis is made clear.

In numerical analysis the real numbers are approximated by floating-point numbers because most computers cannot internally store numbers having more than 10 decimal digits. This results in nonexact computations that are executed very quickly because the arithmetic operations are hardware implemented.

Computer algebra deals mainly with integers of arbitrary precision, employing appropriate data structures; this results in error-free computations that are executed somewhat slowly because the arithmetic operations have to be software implemented.

What is Computer Algebra?

1.1 Computer Algebra versus Numerical Analysis

1.2 Computer Algebra: Exact Integer and Polynomial Arithmetic

1.3 Computer Algebra Systems

Programming Exercises

References

The term *computer algebra* (or *symbolic and algebraic computations*) refers to the capability of computers to manipulate mathematical expressions in a symbolic rather than numerical way, much as one does algebra with pencil and paper. By dealing mainly with exact numbers (infinite precision integers and rational numbers) and algebraic expressions in terms of their symbolic representations, computer algebra systems can free scientists from the painstaking concern for numerical errors (truncation and round-off) and thereby help them to gain more insight into the various physical phenomena under examination—to quote R. W. Hamming, "the purpose of computing is insight, not numbers." Insight is sometimes obtained by evaluating a mathematical expression, but in many cases the relations of the quantities are made clearer by algebraic means.

Computer algebra has been applied to a variety of problems. Consider, for example, the theory of gravitation, where possible variants of Einstein's general theory of relativity are investigated. In order to agree with experimental findings, these variants must satisfy G. Birkoff's theoretical criterion; computer algebra is well suited for applying this criterion. Another example is from neurology, where a system of equations models the propagation of a signal along a nerve. Under some conditions these equations can produce a repeated series of signals, known as a *wamn*. In order to test the stability of the wave train, it is sufficient to calculate the sign of a particular mathematical expression. To produce the expression itself is an enormous task by hand, but with computer algebra it becomes a routine calculation (Pavelle et al., 1981).

1.1
Computer Algebra Versus Numerical Analysis

Before discussing exact arithmetic we look more closely at the inherent limitations of numerical computations using computers. The reader should keep in mind that a computer is a machine with finite memory that is composed of words having finite length; 16 or 32 bits are common lengths of computer words, in which case the maximum integer that can be stored in them is $2^{16} - 1$ or $2^{32} - 1$, corresponding to a 5- or 10-decimal digit number, respectively.

In performing numerical computations on a computer, one is therefore faced with the problem of representing the *infinite* set of real numbers within a computer of finite memory and of given word length. The most widely implemented solution in numerical analysis is to approximate the real numbers using the *finite* set of floating-point numbers. A set F of floating-point numbers is characterized by a number base β, a precision t, and an exponent range $[L, U]$, where the parameters β, t, L, and U clearly depend on the computer. Each floating-point number f in F can be represented as

$$f = \pm \left(\frac{d_1}{\beta} + \frac{d_2}{\beta^2} + \cdots + \frac{d_t}{\beta^t} \right) \beta^e$$

where the integers d_i, $i = 1, 2, \ldots, t$ satisfy the inequality $0 \le d_i \le \beta - 1$ and $L \le e \le U$; if we require $d_1 \ne 0$ for all f in F, $f \ne 0$, we have the *normalized* floating-point numbers.

It should be pointed out that with the use of floating-point numbers (or integers that can be stored in one computer word) the arithmetic operations $+$, $*$, and so on are executed very fast. This is because the computer circuitry performs these operations instead of having a software routine do the job; therefore, we say that the arithmetic operators $+$, $*$, and so forth are *hardware* implemented.

The careful reader should suspect, by now, the kind of problems that arise from this approximation of the reals by the floating-point numbers. First, the set F is not continuous or even an infinite set. There are exactly $2(\beta - 1)\beta^{t-1}(U - L + 1) + 1$ normalized floating-point numbers (including zero) in F; moreover, these numbers are not equally spaced throughout their range but only between successive powers of β. As an example, consider the 33-point set F with $\beta = 2$, $t = 3$, $L = -1$, and $U = 2$, which is shown in Figure 1.1.1.

Because of the above, it is possible that given f_1 and f_2 in F, their sum (or product) will *not* be in F and will have to be approximated by the closest floating-point number. This difference between the true and the approximated sum (or product) is the *round-off* error. It should also be noted that the operations of addition and multiplication in F are not associative and the distributive law also fails. Consider, for example, in our toy 33-point set F, the expression $5/4 + (3/8 + 3/8) = 2$, where $5/4$, $3/8$, and 2 belong to F. In this expression, though, $(5/4 + 3/8) + 3/8 \ne 2$ because the sum $(5/4 + 3/8)$ does not belong to F and has to be approximated by either $3/2$ or $7/4$. Round-off errors do not occur only while using floating-point numbers; they may also appear when one is dealing with integers, such as in the case when one wants to calculate the product, say, of two s-digit numbers in a computer that cannot handle numbers having more than s digits.

We see, therefore, that in numerical analysis one has to carefully estimate

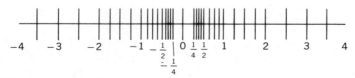

Figure 1.1.1. The floating-point number system for $\beta = 2$, $t = 3$, $L = -1$, and $U = 2$. (Forsythe, G. E., M. A. Malcolm, C. B. Moler: *Computer methods for mathematical computations*, © 1977, p. 12. Reprinted by permission of Prentice-Hall, Inc., Englewood Cliffs, New Jersey.)

(and compute bounds of) the round-off errors performed by each algorithm instead of only focusing attention on the algorithms themselves and their efficiency. This, combined with the fact that mathematical expressions can be stored and manipulated numerically only, points to the need for software systems capable of manipulating expressions symbolically and of performing error-free computations (and this is how computer algebra was "born"). As we will immediately see in Section 1.2, these systems avoid floating-point numbers and use integers of arbitrary precision employing appropriate data structures. In Section 1.3 we mention some of the various systems that are available and briefly describe the capabilities of two of them.

1.2
Computer Algebra: Exact Integer and Polynomial Arithmetic

As mentioned above, in order to be able to represent integers of arbitrary precision and to do exact arithmetic, we have to introduce (whenever not available) appropriate data structures (or use modular arithmetic, as we will see in Chapter 2) and avoid floating-point numbers; this is done in computer algebra. In this section we introduce a computational model for integer and polynomial representation and analyze certain "pencil-and-paper" algorithms for doing arithmetic with them. Obviously, one can use arrays to represent integers and polynomials, but these are not "dynamic" data structures; we will discuss only their list representations (Horowitz et al., 1976).

Lists and Basic List Operations We begin by recursively defining a *list* over an arbitrary set S to be a finite sequence $(a_1, a_2, \ldots, a_n)$, $n \geq 0$, such that each a_i is either an element of S or a list over S; the empty list is represented by 0 and corresponds to $n = 0$. When we write $a = (a_1, a_2, \ldots, a_n)$ we interpret it in two ways: (1) a is considered to be a pointer to the beginning of the list and (2) a represents the entire list, so that when we write $a \oplus x$, where $\oplus$ is any one of the binary operators on scalars, we mean that the operation $\oplus$ has been performed between each element of a and the scalar x. (The meaning will always be made clear from the context.)

Given the list $a = (a_1, a_2, \ldots, a_n)$, where now a is a pointer to the beginning of the list, one can define various operations on it. Of interest to us are the following: length$(a) = n$; first$(a) = a_1$; last$(a) = a_n$; tail$(a) = (a_2, a_3, \ldots, a_n)$; invert$(a) = (a_n, \ldots, a_1)$; prefix $b_1, \ldots, b_k$ to a, $k \geq 1$, yields the list $(b_1, \ldots, b_k, a_1, \ldots, a_n)$; advance $b_1, \ldots, b_k$ in a, $k \leq n$, results in b_i pointing to a_i, $1 \leq i \leq k$ and $a = (a_{k+1}, \ldots, a_n)$. If $a = 0$, the empty list, we define prefix a_1 to a to mean $a = (a_1)$. In what follows, lists and list elements of a list can easily be distinguished from elements of S.

List Representation of Integers In order to be able to store inside a computer integers of arbitrary precision and to do exact integer arithmetic, we have to represent them as lists. In this way, though, we loose the ability to perform on these integers the hardware-implemented operators +, *, and so on; instead, as we will see below, we have to develop special software routines that will do the job.

We distinguish two types of integer—those that are represented by lists, called *multiple precision* or *long* integers, and those that are not, which are called *single precision* or *small* integers. An integer of the first type is represented as $i = (i_0, i_1, \ldots, i_n)$, $n \geq 1$, where these i_j values $0 \leq i_j \leq \beta - 1$, are the coefficients of β^j values in the expression $i = \Sigma_{0 \leq j \leq n} i_j \beta^j$ and are all nonnegative or all nonpositive, according to whether $i > 0$ or $i < 0$, respectively; $\beta - 1 = 2^\mu - 1$ is the largest value stored in a computer word.

Each i_j is stored in a separate computer word and, except for i_n, takes up μ bits. The list may be formed in one of two ways; namely, i_n—the most significant β-digit—can be either at the beginning or at the end of the list. For the most part we follow the latter approach, whose order is the reverse of the natural representation of an integer but is chosen since most arithmetic operations begin with the low-order digit; however, when we study long-integer division below, the most significant β-digit will be at the beginning of the list. Sign$(i) = \pm 1$ depending on whether $i_n > 0$ or $i_n < 0$; that is, the sign of i is "stored" only in i_n.

For example, suppose that $\beta = 10^3$, that is, a computer word can hold only three decimal digits, and that we want to store the number $i = +23456789$ in memory. We can store it as shown in Figure 1.2.1.

The arrows in Figure 1.2.1 indicate that a *node*, or *cell* (which is made up of one or more computer words), is linked to the next one. The whole structure is addressed by the variable i. Similarly, a rational number n/d is represented by the list $r = (\mathbf{n}, \mathbf{d})$, where $\mathbf{n}$ and $\mathbf{d}$ are the integers n and d represented by lists.

Data Structures The first thing that needs to be taken care of when dealing with linked structures is how to construct a node; that is, how many computer words per node do we use, how many data fields per node are we going to have, and what will the size of these fields be. These node characteristics were omitted in Figure 1.2.1; however, a reasonable choice could be to use two words per node (or cell) divided into three fields whose functions are explained below: the type field T, the element field E, and the successor field S, as shown in Figure 1.2.2.

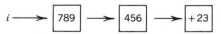

Figure 1.2.1. Internal representation of the integer $i = +23456789$ in a computer whose word can store only three decimal digits.

Figure 1.2.2. Internal representation of a *cell* or *node*.

Consider the general list $(x_1, x_2, \ldots, x_n)$. The computer representation of this list consists of a set of n cells linked via their successor fields, together with representations, assumed already given, of any of the x_i values that are themselves lists. The successor field of the ith cell contains the address of the $(i+1)$th cell $(1 \leq i < n)$, whereas the successor field of the nth cell contains 0. (It is assumed, of course, that 0 is not the address of any cell.) The type field of the ith cell contains the value zero or one according as x_i is an atom or a list. If x_i is an atom, then the element field of the ith cell contains x_i; if x_i is a list, then the element field of the ith cell contains the location of some representation of the list x_i. The location of the list $(x_1, x_2, \ldots, x_n)$ is taken to be the address of the first cell in its representation. In this connection the internal representation of the integer i in Figure 1.2.1 is as shown in Figure 1.2.3.

Once these decisions have been taken, then, during computations, all nodes not being used are linked together in arbitrary order and form the *available space list*. When a new node is needed, the first node is unlinked from the available space list and linked to the appropriate data list. When a data list is no longer needed in a computation, its nodes are linked to the head of the available space list. In Programming Exercises 1 and 2 for this section (at the end of the chapter, before the References) the reader is asked to write some of these routines. Of interest to both the novice and the experienced programmer is the technique with which we can check the value of a given field and/or update it in a high-level language without destroying any existing information in the given word. We will explain it with the following example using the decimal system.

Example. Suppose that the number $n = 123456789$ is stored in a computer word with *three* fields of length 1, 3, and 5 decimal digits respectively; that is, the number can be regarded as 1–234–56789 although the computer sees the nine-digit number as a unit. In order to update the value of the central

Figure 1.2.3. The complete internal list representation of the integer in Figure 1.2.1.

three-digit field from 234 to, say, 432, we do the following:

Divide the number $n = 123456789$ by 10^5, obtaining the quotient $q_1 = 1234$ and the remainder $r_1 = 56789$.
Then, divide $q_1 = 1234$ by 10^3, obtaining $q_2 = 1$ and $r_2 = 234$ (at this point r_2 is disregarded).
Next, multiply $q_2 = 1$ times 1000 and add the quantity 432 (which was supposed to replace the quantity 234), to obtain the partial number $n_p = 1432$.
Finally, multiply $n_p = 1432$ times 10^5 and add $r_1 = 56789$ to obtain the number 143256789 updated as desired.

The reader should do some more examples to thoroughly understand this technique.

Algorithms and Their Complexity An algorithm is a method for solving a class of problems. The cost of using the algorithm to solve one of these problems is called the *complexity* of the algorithm and is measured in whatever units are relevant (running time, or storage). In this book we are interested only in the *time* complexity of a computation, and we will measure it by expressing the running time of the computation as a function of some measure (to be defined below) of the amount of data that was presented as input.

Definition 1.2.1. Let f, g be real-valued functions defined on a set S. Then we say that (1) f is *dominated* by g, and write $f = 0(g)$, in case there exists a positive real number c_1 and x_1 in S, such that $|f(x)| \le c_1 |g(x)|$ for all x in S, $x > x_1$; (2) f *dominates* g, and write $f = \Omega(g)$, if g is dominated by f; and (3) f and g are *codominant*, and write $f \sim g$, if $f = 0(g)$ and $f = \Omega(g)$.

Some properties of dominance and codominance, which will be used subsequently, follow easily from the definition.

Definition 1.2.2. By the β-*length* of an integer i we mean the number of β-digits in its representation, and we write $L_\beta(i)$. If $\lceil x \rceil$ is the *ceiling function*, the least integer greater than or equal to x, and $\lfloor x \rfloor$ is the *floor function*, the greatest integer less than or equal to x, then

$$L_\beta(i) = \begin{cases} 1, & \text{if } i = 0 \\ \lceil \log_\beta(|i|+1) \rceil = \lfloor \log_\beta |i| \rfloor + 1, & \text{if } i \ne 0 \end{cases}$$

Example. The integer $i = +23456789$ of Figure 1.2.1, with $\beta = 10^3$, has β-length, $L_\beta(i) = 3$; i.e., the length of an integer is the number of computer cells needed for its representation. In what follows the subscript β will be omitted since for any other base γ we have $L_\beta \sim L_\gamma$ (if we think of L_β and L_γ as functions defined on the set of integers).

Definition 1.2.3. Let **A** be any algorithm and S the set of all valid inputs to **A**. The integer $t_A(n)$, for n in S, is the number of basic operations performed by the algorithm **A** when presented with the input n and is called the *computing-time function* associated with **A** and defined on S. Basic operations consist of such things as single precision additions and multiplications, replacements, unconditional transfers, and subroutine calls.

Polynomials are formally defined in Chapter 3; for the theorem below, we appeal to the reader's intuitive understanding of them.

Theorem 1.2.4. If $p(n) = p_m n^m + \cdots + p_1 n + p_0$ is a polynomial of degree m, then $p(n) = 0(n^m)$.

Proof. Taking $n \geq 1$, we have

$$|p(n)| \leq |p_m|n^m + \cdots + |p_1|n + |p_0| \leq \left(|p_m| + \cdots + \frac{|p_1|}{n^{m-1}} + \frac{|p_0|}{n^m}\right)n^m$$

$$\leq (|p_m| + \cdots + |p_1| + |p_0|)n^m$$

The theorem is now proved if we set $c = |p_m| + \cdots + |p_1| + |p_0|$. □

As an application of Theorem 1.2.4, consider an algorithm with k instructions (or steps), each of which is executed in time $c_i n^{m_i}$, $1 \leq i \leq k$. Then, the whole algorithm is executed in time $0(n^m)$, where m is the maximum of the m_i, $1 \leq i \leq k$.

If the computing time function of an algorithm **A** is of the form $t_A(n) = 0(n^3)$, or $t_A(n) = 0(n^{13})$, and so on where n is the input size (i.e., if the computing time function is at most a polynomial function of the amount of the input data), then **A** is called a *polynomial-time* algorithm. There also exist *exponential-time* algorithms, whose computing time functions are exponential functions of the amount of the input data, that is, of the form $t_A(n) = 0(2^n)$. Obviously,

$$0(\log n) < 0(n) < 0(n \log n) < 0(n^2) < 0(2^n)$$

and the general rule is that the calculation is "easy" if we are dealing with a polynomial-time algorithm and "hard" if we are dealing with an exponential-time algorithm.

In this book we will be concerned mostly with the *worst-case* bound, which is the maximum running time needed to execute the algorithm. Another kind is the *average-case* bound; it simply states the average amount of computing time that is obtained if the performance of the algorithm is averaged over all possible inputs.

Classical Integer Arithmetic Algorithms and Their Complexity Let us now examine the classical algorithms for performing long-integer arithmetic along with their complexity. As we have already indicated, we have to develop software routines for integer arithmetic because the computer is unable to perform the hardware-implemented operators $+$, $*$, and so on on long integers; therefore, we say that now the arithmetic operators $+$, $*$, and so forth are *software* implemented, and obviously they are slower than their hardware-implemented counterparts. Below, in calculating the computing time function of an algorithm, the reader should think of the grade-school algorithms for integer arithmetic; a node, or cell, is then analogous to a decimal digit, and we simply count digit additions and/or multiplications. The various operations will be performed on the two long integers i_1 and i_2, represented by the lists.

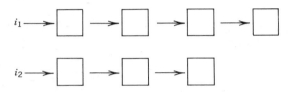

Suppose, at first, that we want to compute the sum of i_1 and i_2. From the programming point of view we can follow one of two approaches:

1. Write a procedure, call it **ISUM** for *i*nteger *sum*mation, which will accept i_1 and i_2 as inputs and will return their sum s as output.
2. "Overload" the $+$ operator; that is, when the $+$ operator is encountered, the type of the surrounding variables is checked, and when long integers are discovered, there is a branch to the procedure **ISUM**. (We call this a "user-friendlier" approach because, in this case, the user does not have to memorize the name of any procedure that needs to be invoked.) *These two approaches are implied for any other integer operation and will not be explicitly mentioned.*

As far as addition is concerned, the output $s = i_1 + i_2$ is another list that is derived by simultaneously scanning (advancing through) the lists i_1 and i_2 and adding the small integers of the corresponding nodes (using the hardware implemented $+$), propagating, of course, any carry. (Recall that, by our conventions, for addition and multiplication the least significant node appears first in the list representation of an integer.) The computing time function is $t_{\text{ISUM}}(i_1, i_2) = 0\{\max[L(i_1), L(i_2)]\}$; this is easily seen if we consider the fact that we are counting the number of single precision additions and that there are at most $\max[L(i_1), L(i_2)]$ of them (this number also bounds the carries that take place) (see Definition 1.2.3). Also, the number of cells in the new list that is created is at most $\max[L(i_1), L(i_2)] + 1$.

We leave it as an exercise for the reader to verify that employing the grade-school algorithm for integer multiplication to compute the product $i_1 \cdot i_2$ (of the two long-integers mentioned above), we have $t_{\text{IMULT}}(i_1, i_2) = 0[L(i_1)L(i_2)]$.

Division is somewhat more involved. When we divide i_1 by i_2, we are actually looking for integers q and r satisfying the division property $i_1 = i_2 q + r$, $0 \le r < i_2$. A moment's reflection about the grade-school process of long division shows that it can be performed if we can repeatedly divide an $(k+1)$-place number $m = (m_0, m_1, \ldots, m_k)$ by the k-place divisor $n = (n_1, n_2, \ldots, n_k)$, $n \le m < bn$, where b is the base of the number system (in computer applications $b = \beta$, where β was used above). In a typical computer application, $b = 2^{32}$ or some other power of 2; in this discussion, we consider the most significant digits first, and so $m = m_0 b^k + \cdots + m_k$ and $n = n_1 b^{k-1} + \cdots + n_k$. For example, if we are asked to divide 1234 by 23, we first divide 123 (our starting m) by 23 (our n), obtaining 5 and a remainder of 8; then we divide 84 (our new m) by 23, obtaining 3 and a remainder of 15. Clearly, this same idea works in general. The most obvious approach to this problem is to make a guess about the quotient q based on the most significant digits of m and n; the quotient obtained in this way is called *trial quotient* and is denoted by q_t. The standard guess is to divide n_1, the most significant digit of n, into the two-digit number $m_0 b + m_1$, and use the quotient as q_t. So we define

$$q_t = \begin{cases} b - 1, & \text{if } n_1 = m_0 \\ \left\lfloor \dfrac{m_0 b + m_1}{n_1} \right\rfloor, & \text{if } n_1 > m_0 \end{cases}$$

where in either case, $q_t \le b - 1$ and $q_t n_1 \le m_0 b + m_1$. (In this discussion, why can we not have the case $n_1 < m_0$?)

Example. Denote by q the correct quotient; then for $b = 10$, we have the following:

If $n = 69$ and $m = 600$, $n_1 = m_0$, and so $q_t = 9$. In this case $q = 8$.
If $n = 69$ and $m = 480$, then $q_t = 48/6 = 8$. In this case $q = 6$.
If $n = 29$ and $m = 200$, $n_1 = m_0$, and so $q_t = 9$. In this case $q = 6$.

For each case above verify the validity of $n \le m < bn$, which does not permit us to consider cases like $n = 59$ and $m = 600$ or even $n = 60$ and $m = 600$; is this a restriction to the generality? (To answer this question, see Theorem

1.2.5 and the comments following it.) Moreover, observe that in all the above cases q_t is too big; however, for $n = 69$ the guess is not as bad as it is for $n = 29$. The reason why this is true is explained by the following theorem.

Theorem 1.2.5. Let b be the base of a number system and consider the numbers $m = m_0 b^k + \cdots + m_k$ and $n = n_1 b^{k-1} + \cdots + n_k$, $n \leq m < bn$. If, on dividing m by n, we denote by q_t and q (both integers) the trial quotient and quotient, respectively, we have $q_t \geq q$; moreover, if $n_1 \geq b/2$, then $q_t - 2 \leq q \leq q_t$, which means that q_t is equal to q, or to $q + 1$ or to $q + 2$.

Proof. Throughout this proof it should be kept in mind that $m = m_0 b^k + m_1 b^{k-1} + m_2 b^{k-2} + \cdots + m_k$ and $n = n_1 b^{k-1} + n_2 b^{k-2} + \cdots + n_k$, $n \leq m < bn$.

From $qn \leq m$ and using the fact that $m_2 b^{k-2} + \cdots + m_k < b^{k-1}$, we easily see that $qn_1 b^{k-1} < (m_0 b + m_1 + 1) b^{k-1}$ and hence $qn_1 \leq m_0 b + m_1$, where $q \leq b - 1$. By definition, however, q_t is either $b - 1$ or, if $n_1 > m_0$, the largest multiple of n_1, which is $\leq m_0 b + m_1$. Clearly, then, $q_t \geq q$.

To prove the second part of the theorem, suppose now that $n_1 \geq b/2$; it will suffice to show that $(q_t - 2)n \leq m$. Using the fact that $n_2 b^{k-2} + \cdots + n_k < b^{k-1}$, we obtain

$$(q_t - 2)n < (q_t - 2)(n_1 + 1) b^{k-1} = [q_t n_1 + (q_t - 2 - 2n_1)] b^{k-1}$$
$$\leq (m_0 b + m_1) b^{k-1} + (q_t - 2 - 2n_1) b^{k-1}$$

by the definition of q_t. Since $n_1 \geq b/2$ and $q_t \leq b - 1$ we have $q_t - 2 - 2n_1 < 0$ and the right-hand side of this relation is $\leq (m_0 b + m_1) b^{k-1} \leq (m_0 b + m_1) b^{k-1} + m_2 b^{k-2} + \cdots + m_k = m$, thus proving the second part of the theorem. □

To ensure that the leading digit of the divisor is $\geq b/2$, we have to *normalize*, that is, multiply m and n by 2^e, where 2^e is the largest power of 2 such that $2^e \cdot n < b^{k+1}$. Then divide $2^e \cdot n$ into $2^e \cdot m$. As a demonstration, consider the last case of the last example where we had $b = 10$, $n = 29$, and $m = 200$. Then we compute the largest e such that $2^e \cdot 29 < 1000$; this turns out to be $e = 5$ and the normalized n and m are 928 ($=32 \cdot 29$) and 6400 ($=32 \cdot 200$), respectively. Normalizing does not affect the quotient; however, we do have to divide out 2^e from the remainder.

Note that in Theorem 1.2.5 it does not matter what the base b is. Also, we can easily adjust q_t by 1 or 2 if necessary to obtain the correct quotient at each stage of the long-integer division. Moreover, as we see in the following example, we can change our guessing strategy and use more leading digits both from m and n.

Example. Dividing 3242 into 272828282, we have

$$
\begin{array}{r}
84462 \\
3242\overline{)272828282} \\
25936 \\
\overline{13468} \\
12968 \\
\overline{15002} \\
12968 \\
\overline{20348} \\
19452 \\
\overline{8962} \\
6484 \\
\overline{2478} \\
\end{array}
$$

$q_t = \lfloor 272/32 \rfloor = 8$

$q_t = \lfloor 134/32 \rfloor = 4$

$q_t = \lfloor 150/32 \rfloor = 4$

$q_t = \lfloor 203/32 \rfloor = 6$

$q_t = \lfloor 89/32 \rfloor = 2$

remainder

Note that we use the leading two or three digits of the divisor.

This guessing method will always give us either the correct quotient digit or a digit that is one greater than the correct digit. The proof of this fact, as well as implementations details, are left as a programming exercise for the reader. [*Hint*: See Knuth (1981), pp. 235–238 and p. 246 for Exercises 19–21; also see Flanders (1984), pp. 342–357 for complete Pascal programs for long-integer arithmetic.)

We leave it as an exercise for the reader to prove that using the preceding pencil-and-paper algorithm we have $t_{\text{IDIV}}(i_1, i_2) = 0[L(i_2)\{L(i_1) - L(i_2) + 1\}]$; that is, the time to divide i_1 by i_2, $(i_1 \geq i_2)$, which means that computing q and r, satisfying the division property, is essentially the time it takes to compute the product $i_2 \cdot q$; see also Programming Exercise 4 for this section.

List Representation of Polynomials We next turn our attention to polynomials with integer coefficients. There are several ways to represent an univariate polynomial $p(x)$ of degree n [and the equation $p(x) = 0$] inside the computer; we will represent it by the ordered list $p = (x, c_r, e_r, c_{r-1}, e_{r-1}, \ldots, c_1, e_1)$, $r \geq 1$, where each integer coefficient c_i is $\neq 0$ and is represented by the list $c_i = (c_{i1}, c_{i2}, \ldots, c_{im_i})$, $m_i \geq 1$; the exponents e_i are in decreasing order $e_r > e_{r-1} > \cdots > e_1$. The degree of $p(x)$ is $n = e_r$, and we consider the sign of $p(x)$ to be the sign of c_r. [Another way to represent the univariate polynomial $p(x)$ of degree $n \geq 0$ is to use the list $p = (x, n, c_n, c_{n-1}, \ldots, c_0)$; in this case zero coefficients are included. We use the first representation.]

As in the case of the list representation of integers, our cells (or nodes) will again consist of two computer words with the same three fields. For example, the polynomial $p(x) = x^3 - 7$ can be represented as shown in Figure 1.2.4.

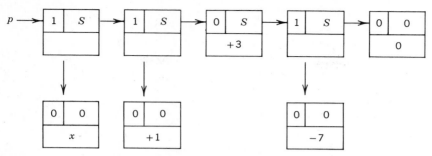

Figure 1.2.4. Internal list representation of the univariate polynomial $x^3 - 7$; x denotes the numerical value used by the computer for the internal representation of the variable x. Note that the coefficients can be of arbitrary length, whereas the exponents are single-precision integers.

The empty list represents the polynomial $p(x) = 0$. Multivariate polynomials over the integers can be represented in recursive canonical form; that is, a polynomial in v variables $x_1, x_2, \ldots, x_v$ is considered to be a polynomial in one variable, x_v, with coefficients c_i that are polynomials in $v - 1$ variables $x_1, x_2, \ldots, x_{v-1}$. We leave it as an exercise for the reader to draw the internal represention of multivariate polynomials.

Classical Polynomial Arithmetic Algorithms and Their Complexity
Computing-time bounds for operations on polynomials are typically given as functions of the degrees and the lengths of the norms of the polynomials.

Definition 1.2.6. Let $p(x) = \Sigma_{0 \leq i \leq n} c_i x^i$ be a univariate polynomial equation with integer coefficients. (If the coefficients are rational, we first turn them into integers by multiplying each one of them by the least common multiple of the denominators.) The *max-norm* (or subinfinity norm) or $p(x)$ is $|p(x)|_\infty = \max_{0 \leq i \leq n}(|c_i|)$, the *sum-norm* (or sub-one norm) is $|p(x)|_1 = \Sigma_{0 \leq i \leq n} |c_i|$, and the *Euclidean norm* is $|p(x)|_2 = (\Sigma_{0 \leq i \leq n} |c_i|^2)^{1/2}$.

Traditionally, only the first two norms are used in the computing-time analysis of algorithms having polynomials as inputs, whereas the Euclidean norm is used in the proofs of general theorems. From the definition it follows that $|p(x)|_\infty \leq |p(x)|_1 \leq (n+1)|p(x)|_\infty$, where n is the degree of $p(x)$; therefore, $L[|p(x)|_\infty] \sim L[|p(x)|_1]$.

Let $p_1(x) = \Sigma_{0 \leq i \leq m} c_i x^i$ and $p_2(x) = \Sigma_{0 \leq i \leq n} d_i x^i$ be two polynomials with integer coefficients of degrees m and n, respectively. We are interested in deriving bounds as functions of the degrees m and n and the lengths $L[|p_1(x)|_\infty], L[|p_2(x)|_\infty]$, on the times required to compute $p_1(x) \pm p_2(x)$, $p_1(x) \cdot p_2(x)$, and $q(x)$, $r(x)$ satisfying the division property $p_1(x) =$

16   WHAT IS COMPUTER ALGEBRA?

$p_2(x)q(x) + r(x)$, where the degree of $r(x)$ is $<n$ (assuming, of course, that $m \geq n$).

Using the representation of polynomials mentioned above, let us derive the computing time function for the routine **PSUM**, which stands for *p*olynomial *sum*mation; namely, this routine takes as input the polynomials $p_1(x)$ and $p_2(x)$ and returns their sum $p_1(x) + p_2(x)$, itself a new list. [More precisely, $2[\max(m, n) + 1]$ is the length of the new list representing the sum $p_1(x) + p_2(x)$.] At first we see that at most $\max(m, n) + 1$ additions of coefficients will be performed. Next, recall that for any two long integers i_1 and i_2, we have $t_{\text{ISUM}}(i_1, i_2) = 0\{\max[L(i_1), L(i_2)]\}$. In our case, and in the worst possible scenario, the coefficients of $p_1(x)$ will all be equal to $|p_1(x)|_\infty$ [the maximum coefficient of $p_1(x)$], whereas the coefficients of $p_2(x)$ will all be equal to $|p_2(x)|_\infty$ [the maximum coefficient of $p_2(x)$]; therefore, one addition of coefficients is performed in time $0(\max\{L[|p_1(x)|_\infty], L[|p_2(x)|_\infty]\})$. Since there are at most $\max(m, n) + 1$ additions of coefficients we easily see that the computing time function of this routine is

$$t_{\text{PSUM}}[p_1(x), p_2(x)] = 0([\max(m, n) + 1]\max\{L[|p_1(x)|_\infty], L[|p_2(x)|_\infty]\}).$$

We leave it as an exercise for the reader to prove that the time needed to compute the product $p_1(x) \cdot p_2(x)$ is $t_{\text{PMULT}}[p_1(x), p_2(x)] = 0\{(m + 1)(n + 1)L[|p_1(x)|_\infty]L[|p_2(x)|_\infty]\}$. Moreover, if we let $p_1(x)$ be the divisor and $p_2(x)$ be the quotient, the last expression bounds the time needed to perform the usual division algorithm for polynomials with integer coefficients. (Algorithms for polynomial division will be presented in later chapters.)

1.3
Computer Algebra Systems

There are now several computer algebra systems available, most of which were developed in the United States; in the Soviet Union there is one called **analitik** which is implemented in hardware. Excellent sources of information on these systems are periodic conferences, symposia, and so on, on symbolic and algebraic manipulation, SYMSAC in the United States and EUROCAL or EUROSAM in Europe. The proceedings of these conferences provide both surveys of current developments and further references; see, for example, the proceedings of 1971 edited by Petricle. Moreover, the *S*pecial *I*nterest *G*roup on *S*ymbolic and *A*lgebraic *M*anipulation (SIGSAM) of the *A*ssociation for *C*omputing *M*achinery (ACM) publishes a quarterly bulletin. The *Journal of Symbolic Computation* is also published quarterly.

A standard test for all computer algebra systems is Delaunay's calculation of the motion of the moon. Finding the exact position of the moon at any given moment is extremely important for navigation and astronomy and is surprisingly complicated; this calculation was started in 1847 by the French astronomer Charles Delaunay and took him 10 *years* to finish, *plus* another 10 years to check. The formula occupies an entire book. In 1970 three researchers at the Boeing Laboratories in Seattle checked Delaunay's work on a computer, taking only 20 *hours*. They found only three minor mistakes, which is almost incredible.

Computer algebra systems exhibit great sophistication and diversity of design and can be classified into two main groups that reflect their development.

The systems in the first group can be considered as *special-purpose* programs; that is, they were designed to solve specific problems in various fields such as mathematics, theoretical physics, and chemistry. All these systems can operate at relatively high speeds because a special-purpose program can be tuned for the kind of input expected. The following list is a sample of such systems: **camal**, a British system for lunar theory and general relativity; **schoonship** for high-energy physics; and **altran**, **sac-1**, and **sac-2** for polynomial arithmetic.

For purely sentimental reasons, we briefly present the capabilities of **sac-1** (*s*ymbolic and *a*lgebraic *c*alculations—version *1*), which is a highly portable system.

Sac-1 is a FORTRAN-based portable system for performing a variety of tasks on many different algebraic structures. It is actually a hierarchy of subsystems or modules, each consisting of a number of subroutines that perform related tasks. Each module depends directly on one or more (and perhaps indirectly on many) other preceding modules of the system, except for the list processing system that stands alone. The following list indicates the relationship of the various subsystems (i.e., from the list below we see that, e.g., the polynomial system depends both on the list processing and on the integer arithmetic system): (1) list processing; (2) integer arithmetic; (3) polynomial; (4) modular arithmetic; (5) greatest common divisor and resultant; (6) linear algebra, polynomial factorization, rational function, gaussian polynomial; (7) rational function integration, real zero; (8) real algebraic number, complex zero.

To use **sac-1**, one writes a FORTRAN program that makes the usual FORTRAN subroutine and function calls to the desired **sac-1** routines; hence no new syntax must be learned if that of FORTRAN is familiar to the user. The names of the **sac-1** algorithms indicate their function; for instance, **PSUM** stands for *p*olynomial *sum*mation, whereas **ISUM** stands for *i*nteger *sum*mation.

The systems in the second group are the *general-purpose* programs, which provide the user with as many mathematical capabilities as possible. Most of them are accessible through the various computer networks. Major representatives of this group are **macsyma**, **reduce**, **schratchpad**, and **maple** (a Canadian system).

For example, **reduce** is a LISP-based system, and the following are some of its capabilities: (1) expansion and ordering of polynomials and rational functions; (2) symbolic differentiation; (3) symbolic integration; (4) substitution and pattern matching; (5) calculation of the greatest common divisor of two polynomials; (6) automatic and user-controlled simplification of expressions; and (7) a complete language for symbolic calculations in which the **reduce** program itself is written.

The newest and by far the best computer algebra system to use is **maple**. It was developed in the 1980s, and incorporates the best features of the other systems that were developed in the late 1960s. For users there is a high-level language with modern syntax more suitable for describing algebraic algorithms.

General-purpose computer algebra systems have been developed for microcomputers but generally are slower and less comprehensive than their counterparts designed for large computers; the most widely available such system is μ-**math** (mu-math).

In the past, computer algebra systems, in general, lacked widespread usage because of (1) slowness, as the basic arithmetic operations must be explicitly programmed, rather than being performed by the hardware and (2) rapid exhaustion of the allocated data storage space for symbolic expressions, because of the growth of the final or intermediate expressions of a computation. Today, however, as inexpensive computers are improved, computer algebra systems become more readily available for teaching, study, and research. At the time of writing this book (early in 1987), Hewlett-Packard released the HP-28c hand-held symbolic algebra calculator, which retails for about $200.

Programming Exercises

Section 1.2

This set of exercises is for experienced programmers only.

1. Suppose that your design a linked allocation system to represent and manipulate univariate polynomials with integer coefficients, and that you have decided to construct a node using two successive computer words with one word for the exponent and successor field and the other word for the coefficient field:

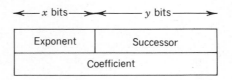

Write procedures in a high-level language (without using logical operations, or record structures) that will store a value in, and will read a value from, any field without destroying the contents of the other fields.

2. Having established the format of the nodes, write procedures that will (a) initialize the available space, that is, construct a singly linked list (linked through the successor field) and return a pointer pointing to the first node in this list (300 nodes should be enough for our project); (b) get a node from the available space for use in the computations; (c) return a node to the available space when it is no longer needed; (d) determine the length of any list, that is, the number of nodes.

3. Implement the classical algorithms for doing long-integer arithmetic.

4. If $i_1 = (a_0, \ldots, a_{n-1})$ and $i_2 = (b_0, \ldots, b_{m-1})$ are two long integers with $L(i_1) = n$, $L(i_2) = m$, show that $t_{\text{IDIV}}(i_1, i_2) = 0[m(n - m + 1)]$. [*Hint*: Show that $L(q) \leq n - m + 1$ so $q = q_{n-m}, \ldots, q_0$, and that determining each q_i involves at most m multiplications.]

5. Assume that the *external* representation of an univariate polynomial with integer coefficients is of the same form as its *internal* representation described in this section; that is, $(x, c_r, e_r, c_{r-1}, e_{r-1}, \ldots, c_1, e_1)$, $r \geq 1$, where the c_i terms represent the coefficients and the e_i terms, the exponents. Using procedures a–d of Programming Exercise 2, assume user-friendly integer addition and multiplication and write new routines that will (a) read in an input polynomial and convert it to its internal linked representation (a pointer to the beginning of this list is returned); (b) write out an output polynomial; (c) compute the sum $p_1(x) + p_2(x)$ for any two given polynomials; and (d) compute the product $p_1(x) \cdot p_2(x)$ for any two given polynomials.

6. Let $p(x_1, x_2, \ldots, x_v) = \Sigma_{0 \leq j \leq n} p_j(x_1, x_2, \ldots, x_{v-1}) \cdot (x_v)^j$ be a polynomial in v variables, over the integers and denote by $\vartheta_i[p(x)]$ its degree in x_i. Below we denote $p(x_1, x_2, \ldots, x_v)$ by $p(\mathbf{x})$ for any value of v. By induction on v, we define two norms, $|p(\mathbf{x})|_\infty$ and $|p(\mathbf{x})|_1$, as follows: if $v = 0$, then $|p(\mathbf{x})|_\infty = |p(\mathbf{x})|_1 = |p(\mathbf{x})|$ since $p(\mathbf{x})$ is an integer; for $v > 0$, we define $|p(\mathbf{x})|_\infty = \max_{0 \leq j \leq n} |p_j(\mathbf{x})|_\infty$ (largest coefficient) and $|p(\mathbf{x})|_1 = \Sigma_{0 \leq j \leq n} |p_j(\mathbf{x})|_1$ (sum of all coefficients).

(a) Show that $L[|p(\mathbf{x})|_\infty] \sim L[|p(\mathbf{x})|_1]$.

Design a pencil-and-paper algorithm that, given two multivariate polynomials $p_1(\mathbf{x})$ and $p_2(\mathbf{x})$ with integer coefficients, will:

(b) Compute their sum $p_1(\mathbf{x}) + p_2(\mathbf{x})$ in time

$$0[\max\{L[|p_1(\mathbf{x})|_\infty], L[|p_2(\mathbf{x})|_\infty]\} \cdot \prod_{1 \leq i \leq v} (\max\{\vartheta_i[p_1(\mathbf{x})], \vartheta_i[p_2(\mathbf{x})]\} + 1)]$$

(c) Compute their product $p_1(\mathbf{x}) \cdot p_2(\mathbf{x})$ in time

$$O(L[|p_1(\mathbf{x})|_\infty] \cdot L[|p_2(\mathbf{x})|_\infty] \cdot \prod_{1 \leq i \leq v} \{\vartheta_i[p_1(\mathbf{x})]\} + 1)\{\vartheta_i[p_2(\mathbf{x})] + 1\}$$

References

Flanders, H.: *Scientific Pascal*. Reston, Reston, VA, 1984.

Forsythe, G. E., M. A. Malcolm and C. B. Moler: *Computer methods for mathematical computations*. Prentice-Hall, Englewood Cliffs, NJ, 1977.

Horowitz, E., and S. Sahni: *Fundamentals of data structures*. Computer Science Press, Rockville, MD, 1976.

Knuth, D.: *The art of computer programming*. Vol. 2: *Seminumerical algorithms*. Addison-Wesley, Reading, MA, 1981.

Pavelle, R., M. Rothstein, and J. Fitch: Computer algebra. *Scientific American*, 136–152, December 1981.

Petricle, S. R. (ed.): *Proceedings of the 2nd symposium on symbolic and algebraic manipulation*. Association for Computing Machinery, New York, NY, 1971.

Part II
Mathematical Foundations and Basic Algorithms

Fundamental to any system of computation is the underlying algebraic structure. It took mathematicians a long time to realize that the important idea behind computing is the operation itself rather than the objects being computed. Although many important examples of algebraic systems were well known in the nineteenth century, it was not until the 1920s that abstract algebra came into its own, giving rise to a unified and simplified treatment of many topics that had previously seemed quite unrelated. In this part of the book we introduce some of the important algebraic systems—groups, rings, and fields—and develop some basic algorithms for computing in these systems.

In Chapter 2 we examine some basic properties of the integers along with some important algorithms for computing with them, and in Chapter 3 we do the same for polynomials. The material presented is essential for a thorough understanding of Part III of this book.

2
Integers

2.1 Fundamental Concepts
2.2 Greatest Common Divisors of Integers
2.3 Integer Factorization
2.4 Exact Computations Using Modular Arithmetic
Exercises
Programming Exercises
Historical Notes and References

One of the oldest branches of mathematics is number theory, the study of the properties of the (positive) integers, parts of which are extremely relevant to current research in algorithms. Many famous problems of number theory, when viewed from an algorithmic viewpoint, present deep and attractive unsolved problems; they are unsolved if we regard the question as not just how to do these problems computationally, but how to do them as fast as possible. Moreover, integers are encountered frequently in many branches of mathematics, and, in the study of algebra, we repeatedly come across concepts whose significance was first observed in the context of the integers; these same concepts were later found useful in much more general situations. In this chapter we will examine those properties, and algorithms, of the integers that are useful in the development of our topics.

2.1
Fundamental Concepts

In this section we present material with which the reader is no doubt familiar. Our treatment will be rather informal (Sims, 1984).

2.1.1 Sets

An algebraic system is a set of objects together with some operations that combine the objects. The objects we study and the operations on them can be effectively characterized in the language of set theory; see historical note 1.

We can consider a set simply as a collection of some specified elements. If A is a set and x an element of the set, we write $x \in A$ and read "x belongs to A" or "x is an element A"; $x \notin A$ reads "x is not an element of A." A set A is *subset* of a set B, or A is *contained* in B, written $A \subset B$, if every element of A is an element of B; in other words, $x \in A$ implies $x \in B$. Two sets A and B are *equal*, $A = B$, if every element of one is an element of the other and conversely, in other words, $A \subset B$ and $B \subset A$. A set A is a *proper subset* of a set B if $A \subset B$ and $A \neq B$ (not equal). There is a special set called the *empty set*, denoted by $\emptyset$, and it is the set that has no elements; thus $x \notin \emptyset$ for all x. Furthermore, $\emptyset$ is a subset of every set, that is, $\emptyset \subset A$, for all A.

In our informal set theory we suppose that once we specify the elements we have defined a set. The notation we use is $\{x : P(x)\}$ for the set of elements x with property P. Occasionally we use just the bracket notation $\{\cdots\}$ to mean the set consisting just of the elements between the brackets. $\{x\}$ is a *singleton*, the set containing the element x.

We have special notation for sets of numbers that we frequently encounter. The *natural numbers* form the set $\mathbf{N} = \{0, 1, 2, \ldots\}$. The *integers* form the set $\mathbf{Z} = \{\ldots, -2, -1, 0, 1, 2, \ldots\}$. The *rational numbers* form the set

FUNDAMENTAL CONCEPTS   25

$\mathbf{Q} = \{x/y : x, y \in \mathbf{Z}, y \neq 0\}$. If $\mathbf{R}$ denotes the set of *real numbers*, then the *complex numbers* form the set $\mathbf{C} = \{x + iy : x, y \in \mathbf{R}\}$, where $i^2 = -1$. We have the chain of subsets $\mathbf{N} \subset \mathbf{Z} \subset \mathbf{Q} \subset \mathbf{R} \subset \mathbf{C}$, and $\mathbf{Z}^*$, $\mathbf{Q}^*$, $\mathbf{R}^*$, and $\mathbf{C}^*$ denote the nonzero elements in the corresponding sets. The real numbers are divided into positive, negative, or zero; $\mathbf{Z}^+$, $\mathbf{Q}^+$, and $\mathbf{R}^+$ denote the positive elements in the corresponding sets.

There are several operations we perform on sets to obtain new ones. Let A and B denote sets. The *union*, written $A \cup B$, is the set $\{x : x \in A$ or $x \in B\}$. The union is the set of elements that are in either A, B, or both. The *intersection*, written $A \cap B$, is the set $\{x : x \in A$ and $x \in B\}$. The intersection is the set of elements that are common to both A and B. We say that A and B are *disjoint* if $A \cap B = \emptyset$, that is, they have no element in common. The *difference* is the set $A - B = \{x : x \in A, x \notin B\}$. The (cartesian) *product* is the set $A \times B = \{(x, y) : x \in A, y \in B\}$, where the term cartesian is derived from the name of the French mathematician R. Descartes (1596–1650). The elements of the product are the pairs (x, y), which are called *two-tuples*. Finally, from the set A we form the set of all subsets of A, called the *power set* of A, $\wp(A) = \{B : B \subset A\}$. Note that A and $\emptyset$ always belong to $\wp(A)$.

A set I is an *index set* for a family of sets $F = \{A_i\}$ if for every $i \in I$ there is a set A_i in the family F. If I is a set of n elements, we usually denote a family of distinct sets as $F = \{A_1, \ldots, A_n\}$. As in the case of two sets, we can form unions and intersections over families of sets, $\cup_{i \in I} A_i = \{x : x \in A_i, \text{some } i\}$ and $\cap_{i \in I} A_i = \{x : x \in A_i, \text{all } i\}$. For a family of n sets the product is $\Pi_{1 \leq i \leq n} A_i = A_1 \times \cdots \times A_n = \{(x_1, \ldots, x_n) : x_i \in A_i\}$. The elements of the product are n-tuples. If $A = A_1 = \cdots = A_n$, then the n-fold product is simply denoted by A^n.

Definition 2.1.1. A *partition* of a set S is a set π of subsets of S, such that:

a. If $A \in \pi$, then $A \neq \emptyset$.
b. If $A \in \pi$ and $B \in \pi$, then either $A = B$ or $A \cap B = \emptyset$.
c. Every element of S is in some element of π.

In other words, we can say that a partition of a set S is a family of nonempty subsets of S such that each element of S belongs to exactly one member of the family. The elements of the partition are called *blocks*. For example, $\{\{1, 2, 3\}, \{4, 5, 6\}\}$ is a partition of the set $\{1, 2, 3, 4, 5, 6\}$ into two blocks. A subset R of S is called a set of *representatives* for a partition π of S if R contains exactly one element from each block of π.

We now give a proposition that says that intersection distributes over union. The exercises contain other results relating unions and intersections.

Proposition 2.1.2. For sets A, B, C the following holds: $A \cap (B \cup C) = (A \cap B) \cup (A \cap C)$.

Proof. We will prove half of the result by showing the inclusion $A \cap (B \cup C) \subset (A \cap B) \cup (A \cap C)$. Suppose an element $x \in A \cap (B \cup C)$. Then $x \in A$ and ($x \in B$ or $x \in C$). If $x \in B$, then $x \in A \cap B$. If $x \in C$, then $x \in A \cap C$. In either case x belongs to the union $(A \cap B) \cup (A \cap C)$. We leave it as an exercise to show the reverse inclusion $(A \cap B) \cup (A \cap C) \subset A \cap (B \cup C)$. □

2.1.2 Equivalence Relations

A *binary relation* R on a nonempty set A is a subset of $A \times A$, that is, $R \subset A \times A$. For example, the "less than" relation, $x < y$, on $\mathbf{Z}$ is given by $\mathbf{Z} \times \mathbf{Z} \supset LT = \{(x, y): y - x \text{ positive}\}$. More generally, an *n*-ary relation R on a nonempty set A is a subset of A^n, that is, $R \subset A^n$. The set $R^{-1} = \{(y, x):(x, y) \in R\}$ is called the *inverse* of R. Clearly $(R^{-1})^{-1} = R$. For example, the inverse of $>$ is $<$.

Definition 2.1.3. An *equivalence relation* E on a set A is a binary relation $E \subset A \times A$ that satisfies three properties:

a. $(x, x) \in E$ for all $x \in A$ (reflexivity).
b. $(x, y) \in E$ implies $(y, x) \in E$ (symmetry).
c. $(x, y) \in E$ and $(y, z) \in E$ implies $(x, z) \in E$ (transitivity).

We write $x \equiv_E y$ (or simply $x \equiv y$ if E is understood) for $(x, y) \in E$ and read "x is equivalent to y." For each element $x \in A$ we form the set of elements that are equivalent to it, $\mathbf{x} = \{y \in A: y \equiv x\}$, denoted by $\mathbf{x}$ and called the *equivalence class* of x. By reflexivity, $x \in \mathbf{x}$. An element $y \in \mathbf{x}$ is called a *representative* of the equivalence class. If $y \in \mathbf{x}$, then by symmetry and transitivity it follows that $\mathbf{y} = \mathbf{x}$, so any representative of an equivalence class determines the same equivalence class. We form the set of all (distinct) equivalence classes, $A/\equiv = \{\mathbf{x}: x \in A\}$. $A/\equiv$ is called the *factor set* or the *quotient set* of the equivalence relation and, as we will see below, it is a partition of A.

Example. On $\mathbf{Z} \times \mathbf{Z}^*$ we define the equivalence relation $(x, y) \equiv (x', y')$ if $xy' = x'y$. For example, $(-3, -9) \equiv (2, 6)$ and $(-1, 2) \equiv (3, -6)$. Verify that this defines an equivalence relation. We claim that the quotient or factor set $\mathbf{Z} \times \mathbf{Z}^*/\equiv$ can be identified with the rationals $\mathbf{Q}$. Namely, elements of $\mathbf{Q}$ are represented by quotients x/y, $x \in \mathbf{Z}$, $y \in \mathbf{Z}^*$, and $x/y = x'/y'$ if $xy' = x'y$. Thus an element of $\mathbf{Q}$ is precisely an equivalence class.

Below we show that there is a close relation between equivalence relations on a set A and partitions of A.

Proposition 2.1.4. If E is an equivalence relation on a set A, then the factor set $A/E = \{\mathbf{a}: a \in A\}$ is a partition of A. Conversely, if π is a partition of A, then π is expressible as A/E for an equivalence relation E on A.

Proof. To prove the first part of this proposition, we must show that the equivalence classes are disjoint sets and that their union is A.

Since E is reflexive, we have $a \in \mathbf{a}$ for all $a \in A$. It then follows that each $\mathbf{a}$ is nonempty and that their union is A.

To prove disjointness, suppose that for some $x \in A$, $x \in \mathbf{a}$ and $x \in \mathbf{b}$; we will show that $\mathbf{a} = \mathbf{b}$. By the definitions of $\mathbf{a}$ and $\mathbf{b}$ we have $(x, a) \in E$ and $(x, b) \in E$; however, $(x, a) \in E$ and $(x, b) \in E$ imply, by symmetry, $(a, x) \in E$ and $(x, b) \in E$, which imply $(a, b) \in E$ and $\mathbf{a} = \mathbf{b}$.

To prove the second part of the theorem, define the relation E_π on A, such that $(a, b) \in E_\pi$ if and only if a and b belong in the same block of π. (Show that this is an equivalence relation.) We will show that $A/E_\pi = \pi$; below π_a denotes the unique block of π containing a for some $a \in A$.

To show $\pi \subset A/E_\pi$, let $\pi_a \in \pi$; then $a \in \pi_a$, since π_a is not empty. We claim that $\pi_a = \mathbf{a} \in A/E_\pi$. To prove the claim, let $x \in \mathbf{a}$; then $(x, a) \in E_\pi$ by the definition of $\mathbf{a}$, from which we obtain $x \in \pi_a$ by the definition of E_π, proving that $\mathbf{a} \subset \pi_a$. For the reverse inclusion let $x \in \pi_a$; then $(x, a) \in E_\pi$ by the definition of E_π, from which we obtain $x \in \mathbf{a}$ by the definition of $\mathbf{a}$, thus proving that $\pi_a \subset \mathbf{a}$. Therefore, we have that each block π_a of π is equal to an equivalence class $\mathbf{a}$ of A/E_π.

The proof of the reverse inclusion $A/E_\pi \subset \pi$ is left as an exercise to the reader. □

In practice, the importance of equivalence relations is tied up with the idea of *suppression of detail*; that is, when we deal with the equivalence classes of $A/\equiv$, we agree to ignore differences among equivalent elements of A.

Example. Fix an integer $m > 1$. We define an equivalence relation on $\mathbf{Z}$ by $b \equiv_m a$ if $b - a = mq$ for some $q \in \mathbf{Z}$; that is, $b \equiv_m a$ if their difference is an integer multiple of m. Some examples are $-4 \equiv_5 16$, $91 \equiv_7 0$, and $1087 \equiv_2 1$. The equivalence class of a is $\mathbf{a} = \{a + mq : q \in \mathbf{Z}\}$, which we also write as $\mathbf{a} = a + m\mathbf{Z}$. For $m = 5$, $\mathbf{0} = \{\ldots, -10, -5, 0, 5, 10, \ldots\}$, $\mathbf{1} = \{\ldots, -9, -4, 1, 6, 11, \ldots\}$, and so forth; therefore, we obtain a decomposition of $\mathbf{Z}$ into the mutually disjoint subsets $\mathbf{0}, \mathbf{1}, \mathbf{2}, \mathbf{3}, \mathbf{4}$ and the factor set is $\mathbf{Z}/\equiv_5 = \{\mathbf{0}, \mathbf{1}, \mathbf{2}, \mathbf{3}, \mathbf{4}\}$. Note that while each of the equivalence classes, contains an *infinite* number of elements, the set of equivalence classes, $\mathbf{Z}/\equiv_5$, contains only *five* elements. In general, $\mathbf{Z}/\equiv_m = \{\mathbf{0}, \ldots, \mathbf{m} - \mathbf{1}\}$ contains m elements.

2.1.3 Functions and Algebraic Systems

We begin with the following.

Definition 2.1.5. For nonempty sets A and B, a *function* or *mapping* (or *map*) from A to B is a rule that assigns to each element $x \in A$ a unique element $y \in B$.

Set-theoretically, we can characterize a function f from A to B as a subset of $A \times B$, that is, $f \subset A \times B$, with the property that for every $x \in A$ there is a unique $y \in B$ with $(x, y) \in f$. For $x \in A$ $f(x)$, the *image* of x or the *value* of f at x, is the unique element $y \in B$ corresponding to x. We use the notation $f: A \to B$ or $A \xrightarrow{f} B$ for the function f from A to B. The set $f(A) = \{f(x) : x \in A\}$ is a subset of B, caled the *image* of A under the function f. For $y \in B$ the set $f^{-1}(y) = \{x \in A : f(x) = y\}$ is a subset of A called the *inverse image* of y. It is the set of all the elements in A whose image is y. (*Note*: As we will see below, the inverse f^{-1} of f is a relation from B to A, but not a function.) When the sets have an algebraic structure (i.e., if we are dealing with groups, rings or fields—to be defined below), then the functions that preserve (the binary operations of) the algebraic structure are called *homomorphisms* (homo = same, morphism = form). More precisely, if $f: A \to B$, where A, B are sets having an algebraic structure, then f is called a homomorphism if:

  i. $f(a + b) = f(a) + f(b)$.
  ii. $f(a \cdot b) = f(a) \cdot f(b)$.
  iii. $f(1_A) = 1_B$.

Here 1_A and 1_B are the multiplicative identities (see Definition 2.1.13) of A, B respectively. (Note that condition iii does not hold in general, but will always be true in this book.)

A function $f: A \to B$ is *onto* or *surjective* if $f(A) = B$; in other words, every element in B is the image of some element in A. f is *one-to-one* or *injective* if whenever $x \ne x'$, then $f(x) \ne f(x')$; in other words, distinct elements in A have distinct images in B. A function that is one-to-one and onto is called a *bijective* function or simply a *bijection*. A bijective homomorphism is called an *isomorphism*, and the sets A, B are called *isomorphic*, denoted by $A \cong B$.

Examples of Functions Let A, B denote sets.

  1. For every set A the identity function $id: A \to A$ is given by $id(x) = x$ for all $x \in A$. Clearly, id is a bijection.

2. For a product of sets we have the projection onto the first factor, $p_1: A \times B \to A$, given by $p_1[(x, y)] = x$. p_1 is surjective. What is $p_1^{-1}(x)$ for $x \in A$? Similarly, we define the projection onto the second factor B.

3. If A has an equivalence relation $\equiv$ defined on it, then there is the canonical surjective quotient function $s: A \to A/\equiv$, which maps each element x to its equivalence class, $s(x) = \mathbf{x}$, in the quotient set $A/\equiv$. We use the same notation for the equivalence class $\mathbf{x}$ whether we consider it as a subset of A or as an element of the quotient set $A/\equiv$; while this may sometimes be confusing, the meaning is usually clear from the context. For example, in the case $s_5: \mathbf{Z} \to \mathbf{Z}/\equiv_5$ the inverse image of the equivalence class $\mathbf{4} = \{\ldots, -6, -1, 4, 9, \ldots\}$ considered as an *element* of the quotient set is the set $s_5^{-1}(\mathbf{4}) = \mathbf{4}$, now considered as a *subset* of $\mathbf{Z}$.

4. The successor function $\mathrm{succ}: \mathbf{N} \to \mathbf{N}$ is defined by $\mathrm{succ}(n) = n + 1$. S is injective. Why?

5. The addition function $+: \mathbf{Z} \times \mathbf{Z} \to \mathbf{Z}$ where $+(x, y) = x + y$ is an example of a binary operation. What is $+^{-1}(0)$? Is $+$ surjective or injective?

As stated above, addition is an example of a binary operation. There are two important properties a binary operation $\oplus$ on a set S *may* possess. If for every pair of elements x, y in S we have $x \oplus y = y \oplus x$, we say that $\oplus$ is *commutative*. If for every triple of elements x, y, z in S we have $x \oplus (y \oplus z) = (x \oplus y) \oplus z$, we say that $\oplus$ is *associative*. On $\mathbf{Z}$, for example, the operations $+$ and $\cdot$ are commutative and associative, but $-$ is neither.

Let us now precisely define a finite set. We say that a set S is *finite* if, whenever a function $f: S \to S$ is injective, then f is surjective. As we will see in Theorem 2.1.7, the notions of injectivity and surjectivity coincide for functions from a finite set S to S; this is the most important property of finite sets. The successor function $\mathrm{succ}: \mathbf{N} \to \mathbf{N}$ defined by $\mathrm{succ}(n) = n + 1$, is injective but not surjective and, therefore, by our definition, $\mathbf{N}$ is not finite. We have the following theorems.

Theorem 2.1.6. A set S is finite if and only if there exists unique nonnegative $n \in \mathbf{N}$ and a bijection from $\{1, 2, \ldots, n\}$ to S.

Proof. This theorem is intuitively obvious. □

Theorem 2.1.7. Let S be a finite set and suppose that $f: S \to S$ is a surjective function. Then f is injective.

Proof. Since f is surjective, for each element $s \in S$ we can choose $t \in S$ such that $f(t) = s$; moreover, define $g: S \to S$ by $g(s) = t$. If $g(s_1) = g(s_2)$, then

$s_1 = f[g(s_1)] = f[g(s_2)] = s_2$, and so g is injective; moreover, since S is finite, g must also be surjective. Now, if f were not injective, then there would be $s_1 \neq s_2$ in S such that $f(s_1) = f(s_2)$. However, since g is surjective, there exist t_1 and t_2 in S such that $g(t_1) = s_1$ and $g(t_2) = s_2$. Then we obtain $t_1 = f[g(t_1)] = f(s_1) = f(s_2) = f[g(t_2)] = t_2$, which is the contradiction, since $g(t_1) \neq g(t_2)$. □

Two functions $f, g: A \to B$ are *equal*, $f = g$, if $f(x) = g(x)$ for all $x \in A$. That is, they are equal if they have the same image at each x. If $f: A \to B$ and $g: B \to C$ are functions, then for $x \in A$, it follows that $f(x)$ belongs to B, and so $g[f(x)]$ makes sense. We define the *composition* of f and g to be the function $g \circ f: A \to C$ given by $g \circ f(x) = g[f(x)]$ for $x \in A$. $g \circ f$ says first perform f and then apply g.

Example. Addition modulo 5 is the function $+_5: \mathbf{Z} \times \mathbf{Z} \to \mathbf{Z}/\equiv_5$ given by $+_5(x, y) = s_5(x + y)$. This function is the composition of the addition function $+$ followed by the canonical quotient function s_5 of the previous example; that is, $+_5 = s_5 \circ +$.

Whenever the composition of functions makes sense, the associative law holds.

Proposition 2.1.8 (Associative Law for Composition). *If* $f: A \to B$, $g: B \to C$, *and* $h: C \to D$ *are functions, then* $h \circ (g \circ f) = (h \circ g) \circ f$.

Proof. For $x \in A$, $h \circ (g \circ f)(x) = h[g \circ f(x)] = h\{g[f(x)]\} = h \circ g[f(x)] = (h \circ g) \circ f(x)$. □

If $f: A \to B$ is a bijection, that is, f is one-to-one and onto, then for each $y \in B$ the inverse image set $f^{-1}(y)$ consists of exactly one element $\{x\}$, where $f(x) = y$. We can define the *inverse function* $f^{-1}: B \to A$ by letting $f^{-1}(y)$ equal to unique element x such that $f(x) = y$, where $y \in B$. Composing f and f^{-1}, we obtain the functions $f^{-1} \circ f: A \to A$ and $f \circ f^{-1}: B \to B$; if $x \in A$ with $f(x) = y$, $y \in B$, then $f^{-1} \circ f(x) = f^{-1}[f(x)] = f^{-1}(y) = x$, and a similar computation shows that $f \circ f^{-1}(y) = y$. Thus $f^{-1} \circ f = id_A$ and $f \circ f^{-1} = id_B$, where id_A and id_B are the identity functions on the respective sets.

We summarize some properties of functions which hold under composition.

Proposition 2.1.9. *Let* $f: A \to B$, $g: B \to C$; *then*:

  a. *If* f, g *are surjective, then* $g \circ f$ *is surjective.*
  b. *If* f, g *are injective, then* $g \circ f$ *is injective.*
  c. *If* f, g *are bijective, then* $g \circ f$ *is bijective and* $(g \circ f)^{-1} = f^{-1} \circ g^{-1}$; *in other words, the inverse of the composition is the composition of the inverses in reverse order.*

Proof. We prove only part a, leaving the rest to the reader. Let $c \in C$; then, since g maps B onto C, there is an element $b \in B$ such that $g(b) = c$. Moreover, since f is also surjective, there is an element $a \in A$ such that $f(a) = b$. Then $g \circ f(a) = g[f(a)] = g(b) = c$, and $g \circ f$ is surjective. □

Let $f: A \to B$ be a function. We define an equivalence relation on A as follows; for $x, x' \in A$, let $x \equiv x'$ if $f(x) = f(x')$. Verify that this is an equivalence relation. We now define a function on the set of equivalence classes, $i: A/\equiv \to B$, by $i(\mathbf{x}) = f(x)$ for an equivalence class $\mathbf{x}$; that is, we have defined the image of the class $\mathbf{x}$ under i in terms of the representative x. We must check to see that if we choose another representative $x' \in \mathbf{x}$, so that $\mathbf{x}' = \mathbf{x}$, we obtain the same image. By our definition of i, $i(\mathbf{x}') = f(x')$. But $x' \equiv x$ means that $f(x') = f(x)$, and so the image $i(\mathbf{x}') = i(\mathbf{x})$ is independent of the choice of representative and i is well defined on the set of equivalence classes $A/\equiv$. By construction, i is injective. Let $s: A \to A/\equiv$ be the canonical function onto the quotient set. Using this notation, we have shown the following.

Theorem 2.1.10 (Factor Theorem for Functions). Let $f: A \to B$ be a function. Then in the diagram

f decomposes into the surjection s followed by the injection i; that is, $f = i \circ s$.

Example. Consider the function $f: \mathbf{Z} \to \mathbf{N}$ where, for x in $\mathbf{Z}$, $f(x) = x^2$. Clearly, $f = i \circ s$ where the surjection is $s: \mathbf{Z} \to \mathbf{Z}/\equiv_{()}2 = \{\mathbf{x} = \{x, -x\} : x \in \mathbf{Z}^*\}$, and the injection is $i: \mathbf{Z}/\equiv_{()}2 \to \mathbf{N}$.

In Theorem 2.1.10, we say that f factors through the quotient $A/\equiv$. When our sets have an algebraic structure, such as a group or ring structure (to be defined below), the quotient set also has the same structure and we can use the factor theorem to show that every homomorphism decomposes into a surjective homomorphism followed by an injective homomorphism. An immediate conclusion is Corollary 2.1.11.

Corollary 2.1.11. The function $i: A/\equiv \to f(A)$, $f(A)$, a subset of B, is a bijection onto the image $f(A)$.

Let us now consider an important case that will lead us to groups, the first algebraic system to be presented.

If A is a set, then the set of bijections from A to A is denoted by $Bij(A) = \{A \xrightarrow{f} A, f \text{ bijection}\}$. By Proposition 2.1.8 we know that the composition of functions is associative, and from Proposition 2.1.9 we know that the composition of two bijections is again a bijection. Furthermore, the identity function id is in $Bij(A)$, and if f is in $Bij(A)$, then the inverse function f^{-1} is also in $Bij(A)$. Putting these results together, we have the following.

Theorem 2.1.12 (Definition of a Group). Let A be a set and consider the set of bijections on A, $Bij(A) = \{A \xrightarrow{f} A, f \text{ bijection}\}$. Then:

a. $f, g \in Bij(A)$ implies $f \circ g \in Bij(A)$.
b. $h \circ (f \circ g) = (h \circ f) \circ g$ for $f, g, h \in Bij(A)$.
c. $id \circ f = f \circ id = f$ for $f \in Bij(A)$.
d. If $f \in Bij(A)$, then $f^{-1} \in Bij(A)$ and $f^{-1} \circ f = f \circ f^{-1} = id$.

In Theorem 2.1.12, part a says that $Bij(A)$ is closed under the binary operation $\circ$, part b says that $\circ$ is associative, part c says the identity function is an identity element of $Bij(A)$, and part d says elements have inverses in $Bij(A)$. In general, a set that is closed under an associative, binary operation and has an identity element and inverses is called a *group*. Thus $Bij(A)$ is a group under composition $\circ$. In general, if the operation in the group is addition (multiplication), then the group is called *additive* (*multiplicative*). If $\tilde{n} = \{1, \ldots, n\}$ denotes the set with n elements, then $S_n = Bij(\tilde{n})$ denotes the set of permutations on n elements and is called the *symmetric group of degree n*. S_n contains $n!$ elements. Obviously, the *nonzero* rational numbers with multiplication form a group, with 1 as the identity and $1/x$ as the inverse of each nonzero rational number; the entire set of rationals with multiplication is *not* a group.

We now present algebraic systems having two binary operations—addition and multiplication.

Definition 2.1.13. A *ring* $(R, +, \cdot)$ is an algebraic system that satisfies:

a. R is a commutative group under $+$; in other words, for $x, y \in R$, $x + y = y + x$. The identity element with respect to addition is called the *zero* element and is denoted by 0_R or simply 0 (the additive identity element).
b. R may *not* be a group under multiplication because (some of) the inverses of the elements may not exist; however, multiplication is associative and there does exist an identity element with respect to it, denoted by 1_R or simply 1 (the multiplicative identity element).
c. Multiplication distributes over addition; that is, for $x, y, z \in R$, $x(y + z) = xy + xz$ and $(y + z)x = yx + zx$.

FUNDAMENTAL CONCEPTS 33

The sets $\mathbf{Z}, \mathbf{Q}, \mathbf{R}, \mathbf{C}$ are all rings under the usual operations. Some examples of sets that are not rings are the following: the set of natural numbers $\mathbf{N}$; the set of positive real numbers $\mathbf{R}^+$, with the usual $+$ and $\cdot$ operations; and the set $\mathbf{Z} - \{5\}$ of all integers except 5.

The set $\mathbf{Z} - \{5\}$ is not a ring with respect to the usual addition and multiplication in $\mathbf{Z}$ because if a, b are in $\mathbf{Z} - \{5\}$, then $a + b$ need not be in $\mathbf{Z} - \{5\}$; for example, $2 + 3 = 5$. In this case we say that $\mathbf{Z} - \{5\}$ is not *closed under addition*, and we mean that $\mathbf{Z} - \{5\}$ is a subset of a bigger set $\mathbf{Z}$, on which addition and multiplication always make sense. The notion of not *closed under multiplication* is similar.

A ring is called *commutative* if its multiplication is commutative; in other words, $ab = ba$. Let R be a commutative ring, where $0, 1$ represent the identity elements. Then, two possible situations can arise. One is that $0 \neq 1 + 1 + \cdots + 1$ (n times) for any $n > 0$. In this case R is said to have *characteristic* 0; examples are $\mathbf{Z}, \mathbf{Q}, \mathbf{R}$, and $\mathbf{C}$. The other situation is that there is some $n > 0$ such that $0 = 1 + 1 + \cdots + 1$ (n times), and this case is examined below.

We now focus our attention on two special kinds of commutative rings that are of extreme importance.

Definition 2.1.14. An element $a \neq 0$ in a commutative ring R is said to be a *zerodivisor* if $ab = 0$ for some $b \neq 0$ in R (likewise, b is also zerodivisor). An element $u \neq 0$ in R is said to be a *unit* if u is invertible; that is, $uv = 1$ for some $v \in R (v = u^{-1})$.

Example. Consider the algebraic system $(\mathbf{Z}/\equiv_m, +, \cdot)$; verify that this is a commutative ring. As we will see again later, when we study integers modulo m, we define the product of two equivalence classes in $\mathbf{Z}/\equiv_m$ to be the equivalence class determined by the product of the representatives. So for $m = 8$, and $\mathbf{2}, \mathbf{4} \in \mathbf{Z}/\equiv_8$, we have $\mathbf{2} \cdot \mathbf{4} = \mathbf{0}$; that is, both $\mathbf{2}$ and $\mathbf{4}$ are zerodivisors. Note that this would not have happened had we picked $m = 5$. Indeed, every nonzero element of $\mathbf{Z}/\equiv_5$ is a unit; the inverses of 2, 3, and 4 are 3, 2, and 4 respectively.

From the preceding definition it easily follows that a ring element cannot simultaneously be a zerodivisor and a unit, and we are thus lead to two important classes of nontrivial commutative rings, where *nontrivial* means that the ring has more than one element.

Definition 2.1.15. An *integral domain* is a nontrivial commutative ring with no zerodivisors.

The classical example of an integral domain (from which also the name is derived) is the ring $\mathbf{Z}$ of integers; its only units are 1 and -1.

Definition 2.1.16. A *field* is a nontrivial commutative ring in which every nonzero element is a unit; or, equivalently, a *field* $(F, +, \cdot)$ is an algebraic system that satisfies:

   a. F is a commutative group under $+$.
   b. The set of nonzero elements in F is a commutative group under $\cdot$.
   c. Multiplication distributes over addition.

Note that a field is an integral domain. We can easily verify that the set **R** of real numbers is a field under real-number addition and multiplication; **Q** and **C** are also fields. These fields have an infinite number of elements. However, there do exist fields with finite numbers of elements; these are called *Galois fields* (GF). For example, the set $\{0, 1\}$ together with modulo 2 addition and multiplication is a field denoted by $GF(2)$; verify this.

Consider a finite field of q elements $GF(q)$, and form the sequence of sums

$$\Sigma_{1 \leq i \leq k} 1 = 1 + 1 + \cdots + 1 \qquad (k \text{ times})$$

$k = 1, 2, 3, \ldots$. Since the field is closed under addition, these sums must be elements in the field; moreover, since the field has a finite number of elements, there must be a repetition in the sequence of sums. Therefore, there must exist two positive integers k' and k'', $k' < k''$ such that

$$\Sigma_{1 \leq i \leq k'} 1 = \Sigma_{1 \leq i \leq k''} 1$$

However, this implies that $\Sigma_{1 \leq i \leq k'' - k'} 1 = 0$ and, hence, there must exist a *smallest* positive integer λ such that $\Sigma_{1 \leq i \leq \lambda} 1 = 0$. This integer λ is called the *characteristic* of the field $GF(q)$. For example, the characteristic of the field $GF(2)$ is 2, since $1 + 1 = 0$.

2.2
Greatest Common Divisors of Integers

Computing the grestest common divisor of integers is one of the oldest topics investigated by mathematicians, and the method presented by Euclid for achieving it, is the oldest existing numerical algorithm. In this section we present Euclid's algorithm and related topics.

2.2.1 Divisibility of Integers

We begin with an extremely important principle, which will be used in proofs (Childs, 1979).

GREATEST COMMON DIVISORS OF INTEGERS 35

Well-Ordering Principle Let k_0 be any fixed integer. Then any nonempty set of integers $\geq k_0$ has a least element.

Proof. We use induction to prove that if there is a set of integers $\geq k_0$, with no least element, then it must be empty. Let S be this set of integers $\geq k_0$ with no least element, and let $p(k)$ be the proposition: "no number $\leq k$ is in S." If we show that $p(k)$ is true for all $k \geq k_0$, we will have shown that S is empty, because if j is in S, then $p(j)$ is false. Obviously, $p(k_0)$ is true, because otherwise S would have a least element since k_0 would be in S, and all integers in S are $\geq k_0$. Suppose now that $p(n)$ is true for some $n \geq k_0$; we will show that $p(n+1)$ is also true completing the induction. If $p(n+1)$ is false, then some number $\leq n+1$ is in S. However, since $p(n)$ is true, no such number $\leq n$ is in S, which implies that $n+1$ would be in S and would be the least element of S, a contradiction. So $p(n+1)$ is true, and S is empty. □

It runs out that the well-ordering principle is equivalent to induction (see Exercise 3 for Section 2.1.1), and its dual also holds; that is, if S is a set of integers all elements of which are $\leq k_0$, then S has a maximal element. We will use this principle many times in this book.

We say that the nonzero integer a *divides* b or a is a *divisor* of b, written $a|b$, if there exists c such that $b = a \cdot c$. For example, $\pm 7|28$ since $28 = 7 \cdot 4$ and $28 = (-7)(-4)$. For any nonzero a we have $a|0$, $\pm 1|a$, and $\pm a|a$. The divisors ± 1 and $\pm a$ are called the *trivial divisors* of a. The concept of divisors is very important in our study of the integers.

One of the basic properties of integers is the *division* or *Euclidean property*, which is well known from arithmetic.

Theorem 2.2.1 (Euclidean Property). For any a and nonzero b there exists a unique *(integral) quotient* q and *remainder* r such that $a = b \cdot q + r$, $0 \leq r < |b|$.

Proof. Consider the set of integers of the form $a - kb$, where k runs over all integers, positive and nonpositive; that is, consider the progression

$$\ldots, a - 3b, a - 2b, a - b, a, a + b, a + 2b, a + 3b, \ldots$$

In this sequence, select the smallest nonnegative number and denote it by r, whereas by q denote the corresponding value of k. (Such an r exists because the set $\{a - kb\}$ contains negative and nonnegative values, and by *well-ordering*, any nonempty set of nonnegative integers contains a least element.) Then by definition

$$r = a - qb \geq 0$$

To show uniqueness, suppose that also

$$a = b \cdot q_1 + r_1, \qquad 0 \leq r_1 < |b|$$

and that $r_1 \neq r$. Let us assume that $r_1 < r$ so that $0 < r - r_1 < |b|$; then clearly we have

$$r - r_1 = (q_1 - q)b$$

and $b|(r - r_1)$, a contradiction to the fact that $0 < r - r_1 < |b|$. □

Definition 2.2.2. If a and b are not both zero, then $d > 0$ is the *greatest common divisor* of a and b if:

  a. $d|a$ and $d|b$.
  b. Whenever $c|a$ and $c|b$, then $c|d$.

The greatest common divisor of a and b is written $gcd(a, b)$ or (a, b). The latter is not to be confused with the two-tuple, and the meaning should be clear from the context. The uniqueness of the greatest common divisor follows from property b and the fact that it is positive: if d' were another greatest common divisor, then by property b $d|d'$ and $d'|d$, and $d' = d$ since both are positive. For example, we have $(12, 30) = (12, -30) = (-12, 30) = (-12, -30) = 6$. The greatest common divisor of two integers not both zero always exists and can be expressed in the following manner.

Theorem 2.2.3 (Existence of the *gcd*). If a and b are not both zero, then there exist integers x and y such that $(a, b) = ax + by$.

Proof. Let d be the smallest positive integer of the form $ax + by$, say, $d = ax_0 + by_0$, where x_0, y_0 might not be unique. (As in the proof of Theorem 2.2.1, such a d exists because of *well-ordering*.) Clearly, $d > 0$ and it satisfies property b of Definition 2.2.2. By contradiction we will show that it also satisfies property a. Assume that property a is false and suppose, without loss of generality, that d does not divide b. Then, $b = d \cdot q + r$, $0 < r < d$, and hence, $r = b - dq = b - (ax_0 + by_0)q = a(-qx_0) + b(1 - qy_0)$, contradicting the minimality of d. □

The theorem does not assert that x and y are unique, which they are not, but only that the greatest common divisor can be expressed in this form. For instance, $6 = (12, -30) = 12(3) + (-30)(1) = 12(-2) + (-30)(-1)$. Below, based on the Euclidean property, we will present an algorithm for expressing the greatest common divisor in the form of Theorem 2.2.3.

With our understanding of greatest common divisors we can characterize the integer solutions to linear equations in two variables (linear Diophantine equations).

Theorem 2.2.4. Suppose that $ax + by = c$ is given with a and b not both zero and $d = (a, b)$; then:

a. The equation is solvable for x and y if and only if $d|c$.
b. If x_0, y_0 is a particular solution, then all solutions are of the form $x_0 - n(b/d)$, $y_0 + n(a/d)$ for all n.

Proof. We will prove only part a, and leave part b as an exercise for the reader. Suppose that x and y are integers such that $ax + by = c$; then, c is a multiple of d and, therefore, $d|c$. Suppose now that $d|c$, so that $c = dk$, for some integer k. From Theorem 2.2.3 we know that there exist integers s, t such that $d = as + bt$. Multiplying by k, we obtain $c = dk = a(sk) + b(tk)$, which implies that $x = sk$ and $y = tk$ satisfy $ax + by = c$. □

As an example, the equation $12x - 30y = 84$ is solvable since $(12, -30) = 6|84$. $x = 2$, $y = -2$ is a solution and all solutions are of the form $x = 2 + 5n$, $y = -2 + 2n$.

Two integers a and b are *relatively prime* if $(a, b) = 1$. By Theorem 2.2.3, this is equivalent to requiring that there exist integers s, t such that $as + bt = 1$. We have the following theorems.

Theorem 2.2.5. Let a and b be integers not both zero, and let $d = (a, b)$. Then a/d and b/d are relatively prime.

Proof. By Theorem 2.2.3 there exist integers s, t such that $d = as + bt$. Dividing through by d, we obtain $1 = (a/d)s + (b/d)t$, which implies that $(a/d, b/d) = 1$. □

Theorem 2.2.6. Let a, b, and c be integers, and let $d = (a, b)$. If a divides bc, then a/d divides c.

Proof. If $a|bc$, then $(a/d)|(b/d)c$. However, by Theorem 2.2.5 $(a/d, b/d) = 1$, and by Theorem 2.2.3 there exist integers s, t such that $(a/d)s + (b/d)t = 1$; multiplying through by c in the last equality, we have $c(a/d)s + c(b/d)t = c$. Since (a/d) divides $c(a/d)s$ and $c(b/d)t$, (a/d) divides c. □

Definition 2.2.7. If a and b are both nonzero, then $m > 0$ is the *least common multiple* of a and b if:

a. $a|m$ and $b|m$.
b. Whenever $a|c$ and $b|c$, then $m|c$.

The least common multiple of a and b is written $lcm(a, b)$ or $[a, b]$. Here also uniqueness follows from part b and the fact that m is positive.

Theorem 2.2.8 (Existence of *lcm*). If a and b are both nonzero, then the least common multiple of a and b exists and, in fact, $[a, b] = |ab|/(a, b)$.

Proof. Since a and b are both nonzero, their greatest common divisor, $d = (a, b)$, is also nonzero. From $(ab)/d = a(b/d) = (a/d)b$, and the fact that b/d and a/d are integers, it follows that $m = |ab|/d$ is a positive common multiple of a and b. Let n be any other common multiple of a and b; then, $n = as = bt$ for some integers s, t. Since $a|bt$, then, from Theorem 2.2.6 it follows that $(a/d)|t$, and $t = u(a/d)$ for some integer u. However, from $n = bt = u(ab)/d$ we see that $m|n$, and so m is a least common multiple of a and b. As stated above, uniqueness follows from part b of Definition 2.2.7 and the fact that m is positive. □

2.2.2 Euclid's Algorithm and Lamé's Theorem

We now present the classical *Euclidean algorithm* for computing the greatest common divisor of two integers. The Euclidean property says that for a and nonzero b there exists a unique quotient q and remainder r such that $a = bq + r$ with $0 \le r < |b|$. In Programming Exercise 1 for Section 2.2.1, the reader was asked to write procedures **QUO**(a, b) and **MOD**(a, b), which return the quotient and nonnegative remainder, respectively, on dividing a by b; these procedures will be used in propositions and in describing other algorithms throughout this book.

The key to Euclid's algorithm is the observation that if $a = bq + r$ and d divides a and b, then $d|r = a - bq$ (Exercise 3 for Section 2.2.1). Since this is true for any divisor, it is true for $d = gcd(a, b)$, and we have that $gcd(a, b) = gcd[b, \textbf{MOD}(a, b)]$. Also, $(a, 0) = |a|$ for any a, and for convenience we define $gcd(0, 0)$ to be 0. Therefore, given a and nonzero b, we perform a sequence of divisions as follows.

Let $a_0 = a$ and $a_1 = b$; then

$$a_0 = a_1 q_1 + a_2, \qquad 0 < a_2 < |a_1|$$
$$a_1 = a_2 q_2 + a_3, \qquad 0 < a_3 < a_2$$
$$\cdots$$
$$a_{k-2} = a_{k-1} q_{k-1} + a_k, \qquad 0 < a_k < a_{k-1}$$
$$a_{k-1} = a_k q_k + 0$$

The process eventually stops since the remainders $|a_1| > a_2 > a_3 > \cdots > 0$ form a strictly decreasing sequence of nonnegative integers and a_k is the greatest common divisor.

We have just observed that $(a_0, a_1) = (a_1, a_2) = \cdots = (a_k, 0) = a_k$, so that we have computed the greatest common divisor of a and b and verified that the following algorithm works. (In our notation $x := y$ means assign the value of y to x; $(x, y) := (x_1, y_1)$ means $x := x_1, y := y_1$.)

EA (Euclidean Algorithm)

Input: a and $b \neq 0$.
Output: $d = \gcd(a, b)$.

1. [Initialize.] $(a_0, a_1) := (a, b)$.
2. [Main loop.] While $a_1 \neq 0$, do $(a_0, a_1) := [a_1, \mathbf{MOD}(a_0, a_1)]$.
3. [Exit.] Return $d := a_0$.

Consider the example $a = 342$ and $b = 612$. The sequence we have is $(342, 612) = (612, 342) = (342, 270) = (270, 72) = (72, 54) = (54, 18) = (18, 0)$, and so $d = 18$.

Computing-Time Analysis of EA. Without loss of generality, we assume below that $a \geq b$.

Since steps 1 and 3 are executed in time ~ 1, it is obvious that the execution time of the Euclidean algorithm is dominated by the execution time of step 2.

In step 2 a number n of integer divisions is performed and we need to compute an upper bound on n; however, before doing this, note that the first division involves the integers a, b and from Section 1.2 we know that $t_{\mathbf{IDIV}}(a, b) = 0\{L(b)[L(a) - L(b) + 1]\}$. Since the integers involved in the subsequent divisions are smaller than a and b, it follows that the expression for $t_{\mathbf{IDIV}}(a, b)$ is also an upper bound on the execution time of all these other divisions. Therefore, in the worst case, we conclude that *each* integer division is executed in time

$$t_{\mathbf{IDIV}}(a, b) = 0\{L(b)[L(a) - L(b) + 1]\} \ .$$

Next, to compute an upper bound on the number n (of integer divisions that have to be performed in order to find the greatest common divisor of a and b) we use Lamé's theorem stated below; according to this theorem the quantity $5 \cdot$ (the number of digits in the smaller number), is an upper bound on n, and in our case this becomes $n \leq 5 \cdot L(b)$.

Therefore, $t_{\mathbf{EA}}(a, b) =$ (number of divisions) $\cdot$ (time for each division) $\leq$ $[5 \cdot L(b)] \cdot \{L(b)[L(a) - L(b) + 1]\}$, which results in

$$t_{\mathbf{EA}}(a, b) = 0\{L(b)^2[L(a) - L(b) + 1]\}$$

We next present Lamé's theorem, which is probably the first theorem ever proved in what is now known as "computational complexity."

Lamé's Theorem (1844) (A Worst-Case Bound for Euclid's Algorithm). The number of divisions that need to be performed in order to find the greatest common divisor of two integers is always less than five times the number of digits in the smaller integer.

Proof. Consider the *Fibonacci* sequence of numbers (see historical note 2)

$$1, 1, 2, 3, 5, 8; 13, 21, 34, 55, 89; 144, 233, 377, 610, 987; 1597, \ldots$$

where each term is the sum of the preceding two. (Note that 1 is the only number that appears twice; for the rest of this proof the sequence $\{1, 1, 2, 3, 5, 8\}$ is equivalent to $\{1, 2, 3, 5, 8\}$.)

It is easy to show that the number of terms of the Fibonacci sequence, which have the same number of digits, is at least four and at most five. Indeed, if we represent by t_1 the first term with $k + 1$ digits, then we clearly have

$$10^k < t_1 < 2 \cdot 10^k$$

because t_1 is the sume of two k-digit terms. Likewise, since

$$(1/2)10^k < t_0 < 10^k$$

and $t_2 = t_1 + t_0$, we have

$$(3.2)10^k < t_2 < 3 \cdot 10^k$$

and then, proceeding in the same way, we obtain the inequalities

$$(5/2)10^k < t_3 < 5 \cdot 10^k$$
$$4 \cdot 10^k < t_4 < 8 \cdot 10^k$$
$$(13/2)10^k < t_5 < 13 \cdot 10^k$$
$$(21/2)10^k < t_6 < 21 \cdot 10^k$$

Therefore, the group of $(k + 1)$-digit terms has at least four members and at most five.

If we denote by $f_0, f_1, f_2, \ldots$, the terms of the Fibonacci sequence, then the number n of the terms preceding f_n will be at most equal to $5 \cdot$ (the number of digits of f_n minus one). Therefore, in order to find the greatest common divisor of the two consecutive terms f_n, f_{n+1}, we will perform a number of divisions smaller than five times the number of digits of f_n. Suppose now that we have two integers $a, b (a > b)$ whose greatest common divisor we want to find. If we denote by $r_{n'}, r_{n'-1}, \ldots, r_2, r_1, r_0$ the decreasing sequence of remainders obtained by applying Euclid's algorithm, we have $r_i = qr_{i-1} + r_{i-2}$, $q \geq 1$; moreover, suppose that the integer b is between f_{n+1} and f_n. Then the remainders $r_{n'}, r_{n'-1}, \ldots, r_0$ are going to be in the various intervals formed by the members of the sequence $f_{n+1}, f_n, f_{n-1}, \ldots, f_2, f_1, f_0$.

First consider the case $q = 1$; that is, the quotient is one in all the divisions. If there are two remainders r_h and r_{h-1} in the same interval (f_k, f_{k-1}), such that $f_k > r_h > r_{h-1} > f_{k-1}$, then (since $f_k = f_{k-1} + f_{k-2}$ and f_{k-1} appears in this summation only once), we are going to have $r_h = r_{h-1} + r_{h-2}$ and $f_{k-2} > r_{h-2}$. That is, there will be no remainders in the interval (f_{k-1}, f_{k-2}). The same conclusion holds also for the cases $f_k = r_h > r_{h-1} > f_{k-1}$ and $f_k > r_h > r_{h-1} = f_{k-1}$. Therefore, if all the quotients of the Euclidean algorithm equal 1, then the remainders will be distributed in such a way (among the decreasing sequence of the Fibonacci numbers) that there will not be more than two in an interval, and every interval with two remainders will be followed by an interval with no remainders.

Next, consider the case $q > 1$; that is, in one division of the Euclidean algorithm we have $r_i = 2 \cdot r_{i-1} + r_{i-2}$ (the smallest $q > 1$). Let f_{j+1} and f_j be the two consecutive Fibonacci terms within which lies r_i; then we have

$$r_i - 2r_{i-1} > 0, \qquad 2f_j - f_{j+1} > 0$$

and

$$2(f_j - r_{i-1}) - (f_{j+1} - r_i) > 0$$

from which we conclude that $f_j > r_{i-1}$. If r_{i-1} is also less than f_{j-1}, then the interval (f_j, f_{j-1}) will be empty. On the other hand, if $r_{i-1} > f_{j-1}$, then, since we have $f_{j+1} = 2f_{j-1} + f_{j-2}$, $r_i = 2r_{i-1} + r_{i-2}$, and $r_i < f_{j+1}$, it must be the case that $f_{j-2} > r_{i-2}$; that is, the interval (f_{j-1}, f_{j-2}) will be empty. Therefore, if a quotient of the Euclidean algorithm is other than one, there will exist at least one interval of the Fibonacci sequence that will not contain a remainder, and this will not be made up by an interval containing two remainders.

Therefore, in order for the remainder sequence $r_{n'}, r_{n'-1}, \ldots, r_1, r_0$, to have the same number of terms as the sequence $f_n, f_{n-1}, \ldots, f_1, f_0$, the quotients of all the divisions must be one, and so must be r_0. So just like the Fibonacci sequence where $f_0 = 1$, $f_1 = 2$, the remainder sequence must have $r_0 = 1$; however, r_1 cannot be 2 because then the two sequences would be identical and b would be equal to f_{n+1}, which is not the case. Hence, r_1 will be at least 3 and the remainder sequence will have, at the end, one or more terms less than the Fibonacci sequence. $\square$

Below we present another worst case bound for Euclid's algorithm (Wilf, 1986). We first need the following.

Lemma 2.2.9. If $a \geq b \geq 1$, then $\mathrm{MOD}(a, b) \leq (a-1)/2$.

Proof. By definition, $\mathrm{MOD}(a, b) = a - \lfloor a/b \rfloor \cdot b \leq a - b$, and obviously, $\mathrm{MOD}(a, b) \leq b - 1$. Therefore, $\mathrm{MOD}(a, b) \leq \min(a - b, b - 1)$, and to prove our lemma we distinguish the following two cases:

i. $b \le (a+1)/2$. Then $b - 1 \le a - b$ and $\mathbf{MOD}(a, b) \le b - 1 \le (a + 1)/2 - 1 = (a-1)/2$.

ii. $b > (a+1)/2$. Then $a - b \le b - 1$ and $\mathbf{MOD}(a, b) \le a - b < a - (a + 1)/2 = (a-1)/2$. $\square$

Theorem 2.2.10 (Another Worst-Case Bound for Euclid's Algorithm). Let a and b be two positive integers, and let $M = \max(a, b)$. The number of divisions that need to be performed in order to find the greatest common divisor of a and b is at most $\lfloor 2 \log_2 M \rfloor + 1$.

Proof. Without loss of generality we assume $a \ge b$. We know that Euclid's algorithm generates the decreasing sequence of nonnegative integers $a_0, a_1, \ldots, a_k$, where $a_0 = a$, $a_1 = b$, and $a_i = \mathbf{MOD}(a_{i-2}, a_{i-1})$, $i \ge 2$. By Lemma 2.2.9 we have $a_i \le (a_{i-2} - 1)/2 \le a_{i-2}/2$. By induction on i we obtain $a_{2i} \le a_0/2^i$ and $a_{2i+1} \le a_1/2^i$, $i \ge 0$, and therefore, $a_k \le 2^{-\lfloor k/2 \rfloor} M$. The algorithm terminates when $a_k < 1$; this happens when $2^{-\lfloor k/2 \rfloor} M < 1$, or when $k > 2 \log_2 M$. $\square$

Example. Let us estimate the number of divisions needed to compute $(144, 89)$ and $(21, 13)$. Using Lamé's theorem we see that, for both cases, the number of divisions is less than $5 \cdot 2 = 10$, whereas using Theorem 2.2.10 we find that the number of divisions is at most 15, for the first case, and 9 for the second; in fact, $(144, 89)$ is obtained after 9 divisions, and $(21, 13)$ after 5.

Finally, an interesting result by Dirichlet (1849) states that if a and b are two randomly chosen integers, then the probability that $\gcd(a, b) = 1$ is $6/\pi^2 \approx .60793$. (Additional results on this topic can be found in Bradley, 1970; Collins, 1974; Knuth, 1969; Lipson, 1981; Motzkin, 1949; and Schroeder, 1986.)

2.2.3 Extended Euclidean Algorithm

Given a and b, it is important for any applications in later sections to express $\gcd(a, b)$ in the form $ax + by$ (Theorem 2.2.3). Obviously, one way to achieve this is to apply Euclid's algorithm and then work backward. That is, for $a = 612$ and $b = 342$ we have

$$612 = 341 \cdot 1 + 270, \quad \text{or} \quad 270 = 612 - 342$$
$$342 = 270 \cdot 1 + 72, \quad \text{or} \quad 72 = 342 - 270$$
$$270 = 72 \cdot 3 + 54, \quad \text{or} \quad 54 = 270 - 72 \cdot 3$$
$$72 = 54 \cdot 1 + 18, \quad \text{or} \quad 18 = 72 - 54$$
$$54 = 18 \cdot 3 + 0,$$

and we know that 18 is the greatest common divisor of 612 and 342.

Working backward, we have

$$18 = 72 - 54 = (342 - 270) - (270 - 3 \cdot 72)$$
$$= [342 - (612 - 342)] - [(612 - 342) - 3 \cdot (342 - 270)]$$
$$= [342 - (612 - 342)] - \{(612 - 342) - 3 \cdot [342 - (612 - 342)]\}$$
$$= 9 \cdot 342 + (-5) \cdot 612$$

that is, $18 = 9 \cdot 342 + (-5) \cdot 612$, and we have solved the problem.

Another approach to the same problem, with many applications, as we will see later in this book, is the *extended Euclidean algorithm*. The values of x and y are calculated in a sequence of computations, by expressing *each* a_i (computed in the process of the Euclidean algorithm, Section 2.2.2) in the form $ax_i + by_i$; that is, consider the sequence

$$a_0 = a \qquad\qquad a_0 = ax_0 + by_0$$
$$a_1 = b \qquad\qquad a_1 = ax_1 + by_1$$
$$a_2 = a_0 - a_1 q_1 \qquad\qquad a_2 = ax_2 + by_2$$
$$a_3 = a_1 - a_2 q_2 \qquad\qquad a_3 = ax_3 + by_3$$
$$\cdots \qquad\qquad \cdots$$
$$a_i = a_{i-2} - a_{i-1} q_{i-1} \qquad\qquad a_i = ax_i + by_i$$
$$\cdots \qquad\qquad \cdots$$
$$a_k = a_{k-2} - a_{k-1} q_{k-1} \qquad\qquad a_k = ax_k + by_k$$
$$0 = a_{k-1} - a_k q_k \qquad\qquad 0 = ax_{k+1} + by_{k+1}$$

The left column is the division sequence, which we have seen before, where now we have solved for the remainders. The right column expresses each remainder in the form $ax_i + by_i$; we want to find x_i and y_i. Obviously, we have $x_0 = 1$, $y_0 = 0$, and $x_1 = 0$, $y_1 = 1$. Comparing both sides at the ith step, we have $a_i = ax_i + by_i = a_{i-2} - a_{i-1} q_{i-1} = (ax_{i-2} + by_{i-2}) - (ax_{i-1} + by_{i-1}) q_{i-1} = a(x_{i-2} - x_{i-1} q_{i-1}) + b(y_{i-2} - y_{i-1} q_{i-1})$, from which we obtain the following inductive procedure for computing the x_i and y_i values:

$$q_{i-1} = \mathbf{QUO}(a_{i-2}, a_{i-1})$$
$$a_i = a_{i-2} - a_{i-1} q_{i-1}$$
$$x_i = x_{i-2} - x_{i-1} q_{i-1}$$
$$y_i = y_{i-2} - y_{i-1} q_{i-1}$$

Of course, we can also compute a_i as $\mathbf{MOD}(a_{i-2}, a_{i-1})$, but the above expression emphasizes that a_i is computed in a manner similar to that for x_i and y_i. We have the following algorithm:

XEA (Extended Euclidean Algorithm)

Input: a and $b \neq 0$.
Output: d, x, y such that $d = \gcd(a, b) = ax + by$.

1. [Initialize.] $(a_0, a_1) := (a, b)$; $(x_0, x_1) := (1, 0)$; $(y_0, y_1) := (0, 1)$.
2. [Main loop.] While $a_1 \neq 0$, do $\{q := \mathbf{QUO}(a_0, a_1); (a_0, a_1) := (a_1, a_0 - a_1 q); (x_0, x_1) := (x_1, x_0 - x_1 q); (y_0, y_1) := (y_1, y_0 - y_1 q)\}$.
3. [Exit.] Return $(d, x, y) := (a_0, x_0, y_0)$.

The computing-time analysis of **XEA** is similar to that of **EA**, and details are left to the reader. Applying the extended Euclidean algorithm to our example $a = 342$, $b = 612$, we obtain:

Iteration	q	a_0	a_1	x_0	x_1	y_0	y_1
0	—	342	612	1	0	0	1
1	0	612	342	0	1	1	0
2	1	342	270	1	−1	0	1
3	1	270	72	−1	2	1	−1
4	3	72	54	2	−7	−1	4
5	1	54	18	−7	9	4	−5
6	3	18	0	9	−34	−5	19

Note that $a_0 = ax_0 + by_0$ holds at each iteration step. The procedure returns $d = 18$, $x = 9$, and $y = -5$, which checks: $18 = 342 \cdot (9) + 612 \cdot (-5)$.

2.2.4 Euclid's Algorithm and Continued Fractions

Continued fractions play an important role in many areas of mathematics. In Chapter 7 of this book we will see how they are used to develop very efficient algorithms for isolating and approximating the real roots of polynomial equations with integer coefficients. In this section we introduce continued fractions with the help of Euclid's algorithm (Olds, 1963; Richards, 1981).

Given any rational fraction a_0/a_1, in lowest terms so that $(a_0, a_1) = 1$ and $a_1 > 0$, we apply the Euclidean algorithm as formulated previously to obtain

$$a_0 = a_1 c_0 + a_2, \qquad 0 < a_2 < a_1$$
$$a_1 = a_2 c_1 + a_3, \qquad 0 < a_3 < a_2$$
$$a_2 = a_3 c_2 + a_4, \qquad 0 < a_4 < a_3$$
$$\ldots$$
$$a_{k-2} = a_{k-1} c_{k-2} + a_k, \qquad 0 < a_k < a_{k-1}$$
$$a_{k-1} = a_k c_{k-1}$$

GREATEST COMMON DIVISORS OF INTEGERS

This notation is a slightly altered variation of the one used in Section 2.2.2; namely, we have replaced $q_1, \ldots, q_k$ by $c_0, \ldots, c_{k-1}$. If we write ξ_i in place of a_i/a_{i+1} for all values of i in the range $0 \le i \le k-1$, then the above equations become

$$\xi_i = c_i + \frac{1}{\xi_{i+1}}, \quad 0 \le i \le k-2, \quad \xi_{k-1} = c_{k-1}$$

If in $\xi_0 = c_0 + 1/\xi_1$ we replace ξ_1 by its value $c_1 + 1/\xi_2$, we have $\xi_0 = c_0 + 1/(c_1 + 1/\xi_1)$. Continuing in this way, we obtain

$$\frac{a_0}{a_1} = \xi_0 = c_0 + \cfrac{1}{c_1 + \cfrac{1}{\ddots + \cfrac{1}{c_{k-2} + \cfrac{1}{c_{k-1}}}}}$$

This is the *continued fractions expansion* of a_0/a_1. The integers c_i are called the *partial quotients*. Since, in general, a_0 will not be positive (by assumption $a_1 > 0$), c_0 may be positive, negative, or zero. However, since in the Euclidean algorithm $0 < a_2 < |a_1|$, we note that the quotient c_1 is positive, and likewise, $c_2, \ldots, c_{k-1}$ are positive. We shall use the notation $(c_0; c_1, \ldots, c_{k-1})$ to represent the continued fraction above.

Example. Consider the rational fraction $8/5$. It is easily seen that $8/5 = (1; 1, 1, 2)$. Moreover, it can be verified that $8/5 = (1; 1, 1, 1, 1)$. But as it turns out, these are the only two continued fraction expansions of a rational number. In general, we have

$$\frac{a_0}{a_1} = (c_0; c_1, \ldots, c_{k-2}, c_{k-1}) = (c_0; c_1, \ldots, c_{k-2}, c_{k-1} - 1, 1)$$

Note: Our condition that the quotients $c_1, c_2, \ldots, c_{k-1}$ be positive is not universally accepted. In that case, the fraction $-8/5$ can also be represented by the continued fractions $(-2; 2, 2)$ and $(-1; -1, -1, -2)$.

According to their expansion, continued fractions can be finite or infinite; for example, the continued fraction expansion of $8/5$ is finite. The following two theorems establish some results about finite continued fractions. We first need the following definition.

Definition 2.2.11. Given the number c, its *integral part* $[c]$ is defined by

$$[c] = \begin{cases} \lfloor c \rfloor, & \text{if } c \ge 0 \\ \lceil c \rceil, & \text{if } c < 0 \end{cases}$$

Theorem 2.2.12 (Uniqueness). If $(c_0; c_1, \ldots, c_m) = (d_0; d_1, \ldots, d_n)$, and if $c_m > 1$ and $d_n > 1$, then $m = n$ and $c_i = d_i$ for $i = 0, 1, \ldots, n$.

Proof. We will use induction. Define the quantities $s_i = (c_i; c_{i+1}, \ldots, c_m)$, and $t_i = (d_i; d_{i+1}, \ldots, d_n)$. Clearly, we have $s_i = (c_i; c_{i+1}, \ldots, c_m) = c_i + 1/s_{i+1}$, and likewise $t_i = (d_i; d_{i+1}, \ldots, d_n) = d_i + 1/t_{i+1}$. Note that $s_i > c_i$, $s_i > 1$ for $i = 1, 2, \ldots, m-1$, and $s_m = c_m > 1$, and $t_i > d_i$, $t_i > 1$ for $i = 1, 2, \ldots, n-1$, and $t_n = d_n > 1$; moreover, $c_i = [s_i]$, and $d_i = [t_i]$ for all i in their respective ranges. The hypothesis of the theorem is $s_0 = t_0$, and taking integral parts, we have $c_0 = [s_0] = [t_0] = d_0$. By definition, though, we have $1/s_1 = s_0 - c_0 = t_0 - d_0 = 1/t_1$, which implies $s_1 = t_1$, and $c_1 = [s_1] = [t_1] = d_1$. We leave it as an exercise for the reader to complete the induction step; that is, $s_i = t_i$ and $c_i = d_i$ imply $s_{i+1} = t_{i+1}$ and $c_{i+1} = d_{i+1}$. In addition, m must equal n. To see this suppose, without loss of generality, that $m < n$. Then, from the preceding discussion we know that $s_m = t_m$ and $c_m = d_m$. However, we have $s_m = c_m$ and $t_m > d_m$, a contradiction. □

Theorem 2.2.13. Any finite continued fraction represents a rational number, and conversely, any rational number can be expressed as a finite continued fraction, and in exactly two ways.

Proof. The first part follows using induction on the number of terms in the continued fraction and the formula $(c_0; c_1, \ldots, c_m) = c_0 + 1/(c_1; c_2, \ldots, c_m)$. The second part follows by expressing any rational number, say, a_0/a_1, as a continued fraction and using Theorem 2.2.12. □

So far we dealt with rational numbers and finite continued fractions, but what about irrational numbers and their expansion? Some very important properties of the continued fraction expansion of irrational numbers have been grouped together in Theorem 2.2.14, which is presented without proof.

Theorem 2.2.14. Any irrational number ξ is uniquely expressible as an infinite continued fraction $(c_0; c_1, \ldots, c_n, \ldots)$, where the c_i values are computed by the algorithm

Set $\xi_0 = \xi$ and define inductively $c_i = [\xi_i]$
and $\xi_{i+1} = 1/(\xi_i - c_i)$, $i \geq 0$

Conversely, any infinite continued fraction determined by the integers c_i, $c_i > 0$ for all i, represents an irrational number ξ. Moreover, if we define

$$p_{-2} = 0, \quad p_{-1} = 1, \quad p_i = c_i p_{i-1} + p_{i-2}, \quad i \geq 0$$
$$q_{-2} = 1, \quad q_{-1} = 0, \quad q_i = c_i q_{i-1} + q_{i-2}, \quad i \geq 0$$

then the finite continued fraction $(c_0; c_1, \ldots, c_n)$ has the rational value $r_n = p_n/q_n$, $(p_n, q_n) = 1$, and it is called the nth *convergent* to ξ. The denominators q_n of the convergents form an increasing sequence of positive integers for $n > 0$, and the following relations hold:

a. If $\xi = (c_0; c_1, \ldots, c_{n-1}, \xi_n)$, where $\xi_n = (c_n; c_{n+1}, \ldots)$, $n \geq 0$, then we have

$$\xi = \frac{p_{n-1}\xi_n + p_{n-2}}{q_{n-1}\xi_n + q_{n-2}}$$

b.
$$\xi = c_0 + \cfrac{1}{c_1 + \cfrac{1}{\ddots + \cfrac{1}{c_{n+1} + \cfrac{1}{\xi_n}}}}, \quad n \geq 1$$

c.
$$\frac{p_n}{q_n} = c_0 + \cfrac{1}{c_1 + \cfrac{1}{\ddots + \cfrac{1}{c_{n-1} + \cfrac{1}{c_n}}}}, \quad n \geq 0$$

Finally, any periodic continued fraction is a *quadratic* irrational number, and conversely. [A quadratic irrational number is of the form $(p \pm \sqrt{d})/q$ and is a root of the quadratic equation $q^2x^2 - 2pqx + (p^2 - d)$, where d is a positive integer not a perfect square.]

Note that the Euclidean algorithm can be used only for the continued fraction expansion of rational numbers; from Theorem 2.2.14 we see that a more general procedure, which can be used for *both* rational and irrational numbers, is as follows. Let x be the given number (either rational or irrational). Calculate c_0, the greatest integer less than x, and express x in the form $x = c_0 + 1/x_1$, $0 < 1/x_1 < 1$, where the number $x_1 = 1/(x - c_0) > 1$ will be irrational if x is irrational. We next compute c_1, the largest integer less than x_1, and we have $x_1 = c_1 + 1/x_2$, $0 < 1/x_2 < 1$, $c_1 \geq 1$, where again the number $x_2 = 1/(x_1 - c_1) > 1$ could be irrational. Continuing this process, we obtain $x = (c_0; c_1, \ldots)$, which will be finite or not, depending on whether x is rational or irrational. Examples are presented below; we first prove the following.

Theorem 2.2.15. Consider the infinite continued fraction $\xi = (c_0; c_1, \ldots)$, and let $r_n = p_n/q_n$ be the nth convergent to ξ. Then the following hold:

a. $p_n q_{n-1} - p_{n-1} q_n = (-1)^{n-1}$, $\quad n \geq 1$.
b. $r_n - r_{n-1} = (-1)^{n-1}/q_n q_{n-1}$, $\quad n \geq 1$.
c. $r_n - r_{n-2} = (-1)^{n-2} c_n/q_n q_{n-2}$, $\quad n \geq 2$.
d. For the even values of n the convergents form a monotonically increasing sequence with ξ as a limit, whereas for the odd values of n the convergents form a monotonically decreasing sequence tending to ξ; moreover, every r_{2n} is less than every r_{2n-1} and each convergent r_n, $n \geq 2$, lies in between the two preceding convergents.

Proof

a. We will prove it by induction. For $n = 1$, and using Theorem 2.2.14, we have $p_1 q_0 - p_0 q_1 = (c_1 p_0 + p_{-1}) \cdot 1 - c_0 c_1 = (c_1 c_0 + 1) - c_1 c_0 = 1$, and so the equality is valid; note that $p_0 = c_0$, $q_0 = 1$, and $q_1 = c_1 q_0 \geq q_0$. Assume now that the equality is true for $n = k$; that is, $p_k q_{k-1} - p_{k-1} q_j = (-1)^{k-1}$. We will show that it is true for $k + 1$. Again, using Theorem 2.2.14, we obtain $p_{k+1} q_k - p_k q_{k+1} = (c_{k+1} p_k + p_{k-1}) q_k - p_k (c_{k+1} q_k - q_{k-1}) = -(p_k q_{k-1} - p_{k-1} q_k) = -(-1)^{k-1} = (-1)^k$.
b. To prove this part, divide both sides of $p_n q_{n-1} - p_{n-1} q_n = (-1)^{n-1}$ by $q_n q_{n-1}$, which yields $p_n/q_n - p_{n-1}/q_{n-1} = (-1)^{n-1}/q_n q_{n-1}$. Since $r_n = p_n/q_n$, the proof is completed.
c. Here we have $r_n - r_{n-2} = p_n/q_n - p_{n-2}/q_{n-2} = (p_n q_{n-2} - p_{n-2} q_n)/q_n q_{n-2}$. Replacing in the numerator p_n and q_n by their corresponding expressions (from Theorem 2.2.14) $c_n p_{n-1} + p_{n-2}$ and $c_n q_{n-1} + q_{n-2}$, respectively, we obtain $p_n q_{n-2} - p_{n-2} q_n = (c_n p_{n-1} + p_{n-2}) q_{n-2} - p_{n-2}(c_n q_{n-1} + q_{n-2}) = c_n(p_{n-1} q_{n-2} - p_{n-2} q_{n-1}) = (-1)^{n-2} c_n = (-1)^n c_n$.
d. This is derived from the previous parts of the theorem. Namely, from parts b and c we can infer that $r_{2n} < r_{2n+2}$, $r_{2n+1} < r_{2n-1}$, and $r_{2n} < r_{2n-1}$ because of q_n are positive for $n \geq 1$. Therefore, we have $r_0 < r_2 < r_4 < \cdots$ and $r_1 > r_3 > r_5 > \cdots$. To prove the last part, observe that the sequence with even subscripts is monotonically increasing and is bounded above by r_1; likewise, the sequence with odd subscripts is monotonically decreasing and bounded below by r_0. These two limits must be identical since the difference $r_n - r_{n-1}$ tends to zero as n tends to infinity, and the integers q_n are increasing with n. □

GREATEST COMMON DIVISORS OF INTEGERS

It should be noted that the results of Theorem 2.2.15 would be different if, in Theorem 2.2.14, we had defined p_n and q_n starting from p_{-1} and q_{-1}, instead of p_{-2} and q_{-2}. Namely, if we set

$$p_{-1} = 0, \quad p_0 = 1, \quad \text{and} \quad p_i = c_i p_{i-1} + p_{i-2}, \quad i \geq 1$$
$$q_{-1} = 1, \quad q_0 = 0, \quad \text{and} \quad q_i = c_i q_{i-1} + q_{i-2}, \quad i \geq 1$$

then the continued fraction expansion starts from c_1 instead of c_0, and what was previously a sequence of convergents with odd/even subscripts now becomes a sequence with even/odd subscripts. Moreover, the equation of part a becomes

$$p_n q_{n-1} - p_{n-1} q_n = (-1)^n, \quad n \geq 0$$

and parts b and c are valid for $n \geq 2$ and $n \geq 3$, respectively. In Chapter 7 we will use this last form of the convergents.

Example. Let us determine the continued fraction expansion of the rational number 144/89 using the algorithm described in Theorem 2.2.14; we will also observe the behavior of the convergents. At first we set $\xi_0 = 144/89$, $c_0 = [144/89] = 1$, $p_{-2} = 0$, $p_{-1} = 1$, $q_{-2} = 1$, and $q_{-1} = 0$, and using the relations $p_i = c_i p_{i-1} + p_{i-2}$ and $q_i = c_i q_{i-1} + q_{i-2}$, we obtain $p_0 = c_0$ and $q_0 = 1$; $p_0/q_0 = 1$ is the first even convergent to 144/89 approximating it from below. We then compute $\xi_1 = 1/(\xi_0 - c_0) = 89/55$, $c_1 = [89/55] = 1$, and using the same relations, $p_1 = 2$, and $q_1 = 1$; $p_1/q_1 = 2$ is the first odd convergent to 144/89 approximating it from above. Next we have $\xi_2 = 1/(\xi_1 - c_1) = 55/34$, $c_2 = 1$, $p_2 = 3$, $q_2 = 2$, and $p_2/q_2 = 3/2$ is the second even convergent to 144/89 approximating it again from below; note that $p_0/q_0 < p_2/q_2$ and also $p_0/q_0 < p_2/q_2 < p_1/q_1$. Continuing, we obtain $\xi_3 = 1/(\xi_2 - c_2) = 34/21$, $c_3 = 1$, $p_3 = 5$, and $q_3 = 3$, and $p_3/q_3 = 5/3$ is the second odd convergent to 144/89 approximating it again from above; we now have $p_3/q_3 < p_1/q_1$ and also $p_0/q_0 < p_2/q_2 < p_3/q_3 < p_1/q_1$. We leave it as an exercise for the reader to complete the example.

Theorem 2.2.16. Let ξ be an irrational number, and let $\xi = (c_0; c_1, \ldots, c_{n-1}, \xi_n)$, be its continued fraction expansion, where $\xi_n = (c_n; c_{n+1}, \ldots)$, $n \geq 0$. Then the following hold:

a. Each convergent is nearer to ξ then the preceding convergent.
b. $1/2q_{n+1} q_n < |\xi - p_n/q_n| < 1/q_{n+1} q_n < 1/q_n^2$, $n \geq 0$.
c. There exists an infinite number of rational fractions p/q, $(p, q) = 1$, such that $|\xi - p/q| < 1/q^2$.

Proof

a. From Theorem 2.2.14 we have $\xi = (p_{n-1}\xi_n + p_{n-2})/(q_{n-1}\xi_n + q_{n-2})$, from which we obtain $\xi(q_{n-1}\xi_n + q_{n-2}) = (p_{n-1}\xi_n + p_{n-2})$, or, rearranging, $\xi_n(\xi q_{n-1} - p_{n-1}) = -(\xi q_{n-2} - p_{n-2}) = -q_{n-2}(\xi - p_{n-2}/q_{n-2})$.

Dividing through by $\xi_n q_{n-1}$, and taking absolute values, we obtain $|\xi - p_{n-1}/q_{n-1}| = |q_{n-2}/\xi_n q_{n-1}| \cdot |\xi - p_{n-2}/q_{n-2}|$. However, we know that for $n \geq 1$, $\xi_n > 1$, and since the denominators q_n of the convergents form an increasing sequence of positive integers, we have $q_{n-1} > q_{n-2}$; therefore, $0 < |q_{n-2}/\xi_n q_{n-1}| < 1$, which implies that $|\xi - p_{n-1}/q_{n-1}| < |\xi - p_{n-2}/q_{n-2}|$, or, $|\xi - r_{n-1}| < |\xi - r_{n-2}|$, for $n \geq 2$.

b. From part b of Theorem 2.2.15 we have $|r_{n+1} - r_n| = 1/q_{n+1}q_n$, $n \geq 1$; moreover, we just proved that ξ is closer to r_{n+1} than it is to r_n, and so, $1/2q_{n+1}q_n < |\xi - p_n/q_n| < 1/q_{n+1}q_n < 1/q_n^2$, $n \geq 0$, where $1/q_{n+1}q_n < 1/q_n^2$, because $q_{n+1} > q_n$. (See also Figure 2.2.1.)

c. The proof of this part is immediate, because ξ is irrational, and there exists an infinite number of convergents p_n/q_n satisfying part b above. □

Note that the inequality $|x - p_n/q_n| < 1/q_n^2$ holds for rational numbers as well.

Example. Let us compute the first few terms of the continued fraction expansion of π, for which it can be shown that $\pi = (3; 7, 15, 1, 292, 1, 1, 1, 2, 1, 3, 1, 14, 2, \ldots)$. As we have stated above, in general, for any irrational number x, it may not be possible to give its complete continued fraction expansion, since the Euclidean algorithm cannot be applied; however, if a decimal approximation of x is known, then a corresponding part of the continued fraction expansion of x can be computed. In this case assume that we are satisfied with the approximation $\pi = \xi_0 = 3.14159$. Using the algorithm described in Theorem 2.2.14, we have

$$3.14159 = 3 + 0.14159, \quad c_0 = 3, \quad \xi_1 = 0.14159^{-1}$$

$$\frac{1}{0.14159} = 7 + 0.06264, \quad c_1 = 7, \quad \xi_2 = 0.06264^{-1}$$

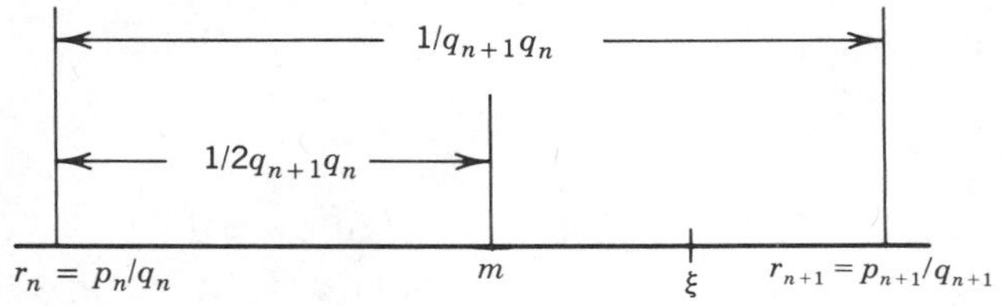

Figure 2.2.1. A geometric proof of part b of Theorem 2.2.16.

$$\frac{1}{0.06264} = 15 + 0.96424, \qquad c_2 = 15, \qquad \xi_3 = 0.96424^{-1}$$

$$\frac{1}{0.96424} = 1 + 0.03708, \qquad c_3 = 1, \qquad \xi_4 = 0.03708^{-1}$$

...

In this fashion we obtain $\pi = (3; 7, 15, 1, \ldots)$. So we see that the first four partial quotients of π are obtained from the first four partial quotients of the continued fraction expansion of the rational number 3.14159. Using the convergent equations we have the first few approximations to π as $r_0 = 3/1$, $r_1 = 22/7$, $r_2 = 333/106$, and $r_3 = 355/113$. Let us now check the validity of the formula $|\xi - p/q| < 1/q^2$. For r_2 we have $|3.14159 - 333/106| < 1/106^2$, which turns out to be $0.00007 < 0.000089$.

In passing we mention Euler's discovery that the continued fraction expansion of e, unlike that of π, has a noteworthy regularity: $e = (2; 1, 2, 1, 1, 4, 1, 1, 6, 1, 1, 8, 1, 1, \ldots)$; the proof is quite difficult and beyond the scope of this book.

Note that the partial quotients of π and e are not periodic, and that 1 appears more frequently than any other number. An interesting result by Lang and Trotter (1972) states that in the continued fraction expansion of almost all numbers, the probability that c_n, the nth partial quotient, is equal to a positive integer j is given by

$$\log_2 \frac{(j+1)^2}{j(j+2)}$$

For almost all numbers, this means that the probability for $c_n = 1$ is approximately .41.

Finally, let us now consider periodic, infinite continued fraction expansions of irrational numbers (see also Theorem 2.2.14); in other words, suppose that we have $\xi = (2; 3, 2, 3, 2, 3, \ldots)$. What does ξ look like? Observe that in this case we can write $\xi = (2; 3, \xi_2)$, according to Theorem 2.2.14; however, since $\xi = \xi_2 = \xi_4, \ldots$, we have $\xi = (2; 3, \xi)$. This implies that $\xi = 2 + 1/(3 + 1/\xi)$, or $3\xi^2 - 6\xi - 2 = 0$. This is a quadratic equation in ξ, and, discarding the negative root, we obtain $\xi = (3 + \sqrt{15})/3$, a *quadratic irrational* number.

Example. Let $\xi = \sqrt{7}$, and let us find its continued fraction expansion. Using the algorithm of Theorem 2.2.14, we have

$$\sqrt{7} = 2 + (\sqrt{7} - 2), \qquad\qquad c_0 = 2, \qquad \xi_1 = (\sqrt{7} - 2)^{-1}$$

$$\frac{1}{\sqrt{7}-2} = \frac{\sqrt{7}+2}{3} = 1 + \frac{\sqrt{7}-1}{3}, \qquad c_1 = 1, \qquad \xi_2 = [(\sqrt{7}-1)/3]^{-1}$$

$$\frac{3}{\sqrt{7}-1} = \frac{\sqrt{7}+1}{2} = 1 + \frac{\sqrt{7}-1}{2}, \qquad c_2 = 1, \qquad \xi_3 = [(\sqrt{7}-1)/2]^{-1}$$

$$\frac{2}{\sqrt{7}-1} = \frac{\sqrt{7}+1}{3} = 1 + \frac{\sqrt{7}-2}{3}, \qquad c_3 = 1, \qquad \xi_4 = [(\sqrt{7}-2)/3]^{-1}$$

$$\frac{3}{\sqrt{7}-2} = \sqrt{7}+2 = 4 + (\sqrt{7}-2), \qquad c_4 = 4, \qquad \xi_5 = (\sqrt{7}-2)^{-1}$$

Therefore, we have $\xi_5 = \xi_1$ and $\sqrt{7} = (2; 1, 1, 1, 4, 1, 1, 1, 4, \ldots)$.

2.3
Integer Factorization

Of all the problems in number theory to which computers have been applied, probably none has been so influenced as that of factoring. Because the problem is basic to the theory of numbers and so easily understood, it has fascinated people ever since antiquity (Dickson, 1952). Simply stated, given a number $n > 1$, find, if they exist, two integers a and b, such that $ab = n$. There are really two problems here: the first, called *primality testing*, is the problem of determining whether such a and b exist, and the second, called *factoring*, is the problem of finding them. In this section we will examine these two problems, and, in so doing, we will introduce additional mathematical concepts, which are necessary for a thorough understanding of the rest of the book.

2.3.1 Prime Numbers and the Sieve of Eratosthenes

We begin with certain definitions. Nonzero $a \neq \pm 1$ is *irreducible* if its only divisors are trivial; that is, the only divisors are ± 1 and $\pm a$. For example 13 and -7 are irreducible. Nonzero $a \neq \pm 1$ is *reducible* or *composite* if it has nontrivial divisors or, equivalently, if it can be written in the form $a = bc$ with b and c not equal ± 1 or $\pm a$. For instance, $276 = 12 \cdot 23$ is composite. Divisors are also called *factors*. A composite number can be written as a product of nontrivial factors. If, in turn, the factors are reducible, then they themselves can be written as products of nontrivial factors, and so on. The next theorem says the process eventually stops.

Theorem 2.3.1 (Existence of Irreducible Factorizations). Every nonzero $a \neq \pm 1$ can be written as $\pm$ the product of a finite number of positive irreducible integers, or $a = \pm u_1 \cdots u_r$, where $u_i > 1$ for $i = 1, \ldots, r$ and irreducible.

Proof. We can assume $a > 0$. If a is not already irreducible, then $a = bc$ with b and c not equal to 1 and $1 < b, c < a$. Continuing in this manner, if b and c

are composite, then their positive nontrivial factors will be strictly less than b and c, respectively. By the well-ordering principle, the nontrivial factorizations stop, since the factors form a strictly decreasing sequence of positive integers. □

For example, $1008 = 2 \cdot 2 \cdot 2 \cdot 2 \cdot 3 \cdot 3 \cdot 7$. Is this factorization unique—that is, can 1008 be written as a product of different irreducibles? Before we can answer this, we need another definition. An integer $p > 1$ is *prime* if, for every a, b, $p|ab$ implies $p|a$ or $p|b$. The next proposition characterizes primes as the positive irreducible elements. The proof of this proposition illustrates the type of reasoning required for some of the exercises, for this section.

Proposition 2.3.2. $p > 1$ is prime iff p is irreducible.

Proof. Suppose that $p > 1$ is prime and $a|p$; we must show that a is a trivial divisor. Write $p = ab$ for some b. Since p is a prime $p|a$ or $p|b$. If $p|a$, then $a = \pm p$ (see also Exercise 2 for Section 2.2.1). If $p|b$, then $b = pc$ for some c and $p = ab = apc$, which implies $1 = ac$, so $a = \pm 1$. Therefore, we have shown that any divisor of p is trivial, and so p is irreducible. Conversely, suppose that $p > 1$ is irreducible and that $p|ab$. If p does not divide a, then $(a, p) = 1$ since p, being irreducible, has only the positive divisors 1 and p. By Theorem 2.2.3 there exist x and y such that $ax + py = 1$. Multiplying through by b, we obtain $abx + pby = b$, and since p divides the left side, we have $p|b$. This proves the theorem, because we have shown that if $p|ab$ and p does not divide a, then $p|b$, and thus p is a prime. □

Theorem 2.3.3 (Uniqueness of Irreducible Factorizations). Every nonzero $a \neq \pm 1$ can be written as $\pm$ the product of primes in at most one way, unique up to the order of the factors.

Proof. Suppose positive $a = u_1 \cdots u_r = v_1 \cdots v_s$, where the u_i and v_j values are positive irreducibles and hence primes. We have $u_1 | v_1 \cdots v_s$ and since u_1 is prime, $u_1 | v_{j_1}$ for some j_1. However, v_{j_1} being irreducible has no nontrivial divisors so $u_1 = v_{j_1}$. Continuing, we have u_2 divides the product of the remaining v_j values and hence $u_2 = v_{j_2}$ for some j_2, and so on. Finally, we have $r = s$, and, after some reordering, $u_i = v_i$ for all i. □

Theorem 2.3.3 is also known as the *fundamental theorem of arithmetic*. Historically, mathematicians took Theorem 2.3.3 for granted; however, there are mathematical systems where unique factorization fails to hold. As an example of such a system, consider the set E, whose elements are the positive *even* numbers $2, 4, 6, 8, \ldots$ Note that E is closed under multiplication. Now assuming that the only numbers we know are members of E, we see that $8 = 2 \cdot 4$ is "composite," whereas 14 is "prime," since it is not the product of two or more "numbers"; moreover, the number 840 has two factorings into "primes," namely, $840 = 2 \cdot 14 \cdot 30 = 6 \cdot 10 \cdot 14$, and the unique factorization theorem fails.

Since Theorem 2.3.3 holds for integers, **Z** is called a *unique factorization domain*, a special kind of ring of which we will see more examples later.

Put together, Theorems 2.3.1 and 2.3.3 say that every nonzero integer $a \neq \pm 1$ can be uniquely written as a product of a finite number of prime factors (unique up to the order in which the factors occur). This is also known as the *prime-power decomposition* of a; the latter can be written as

$$a = \pm p_1^{e_1} p_2^{e_2} \cdots p_k^{e_k}$$

where the p_i are now *distinct* primes and the e_i are positive integers. However, as we will see immediately below, it is sometimes convenient to permit some exponents to be zero.

As a first application of the above, let us express d, the greatest common divisor of a and b, in terms of the prime factors of a and b.

Let

$$a = p_1^{e_1} p_2^{e_2} \cdots p_k^{e_k}, \qquad b = p_1^{f_1} p_2^{f_2} \cdots p_k^{f_k}$$

where $e_i \geq 0$ and $f_i \geq 0$. Then, since $p^m | p^n$ if and only if $m \leq n$, we see that the greatest common divisor of a and b is

$$d = (a, b) = p_1^{\min(e_1, f_1)} p_2^{\min(e_2, f_2)} \cdots p_k^{\min(e_k, f_k)}$$

where as usual, $\min(i, j)$ denotes the smallest of the numbers listed in the parentheses. In a similar fashion we see that the least common multiple of a and b is

$$[a, b] = p_1^{\max(e_1, f_1)} p_2^{\max(e_2, f_2)} \cdots p_k^{\max(e_k, f_k)}$$

where $\max(i, j)$ denotes the largest of the numbers listed in the parentheses. Now using the fact that

$$|ab| = p_1^{e_1 + f_1} p_2^{e_2 + f_2} \cdots p_k^{e_k + f_k}$$

together with the observation that $\min(a, b) + \max(a, b) = a + b$, we can easily obtain the result of Theorem 2.2.8, namely, that $[a, b] = |ab|/(a, b)$.

Applying the fact that any natural number factors uniquely into a product of primes, we have Theorem 2.3.4.

Theorem 2.3.4. $\sqrt{2}$ *is irrational.*

Proof. Suppose that there exist two natural numbers a and b such that $\sqrt{2} = a/b$, $(a, b) = 1$. Then, using the prime-power decomposition, we see that, in the equation $2b^2 = a^2$, 2 occurs to an odd power on the left side, and to an even power on the right side, which is impossible. □

Theorem 2.3.5 (Euclid). There is an infinite number of primes.

Proof. The proof is indirect. Suppose that there is a finite number of primes $p_1, p_2, \ldots, p_m$. Consider the number $n = p_1 \cdot p_2 \cdots p_m + 1$, which either will be a prime number, in which case it will be a new prime, or will have a prime factor p. (Note that the first few numbers n are prime; for example, $2 + 1 = 3$, $2 \cdot 3 + 1 = 7$, $2 \cdot 3 \cdot 5 + 1 = 31$, $2 \cdot 3 \cdot 5 \cdot 7 + 1 = 211$, $2 \cdot 3 \cdot 5 \cdot 7 \cdot 11 + 1 = 2311$. However, $2 \cdot 3 \cdot 5 \cdot 7 \cdot 11 \cdot 13 + 1 = 30031 = 59 \cdot 509$.) If p were one of the primes p_i, then p would divide $p_1 \cdot p_2 \cdots p_m$, and since p divides $n = p_1 \cdot p_2 \cdots p_m + 1$, it would divide the difference of these numbers, namely, 1, which is impossible. Therefore, p must be a new prime. □

Theorem 2.3.6. Given any positive integer n, there exist n consecutive composite integers. In other words, there are arbitrarily large gaps in the series of primes.

Proof. Consider the integers $(n + 1)! + 2$, $(n + 1)! + 3, \ldots, (n + 1)! + (n + 1)$, where $k! = 1 \cdot 2 \cdots k$. Clearly, every one of these numbers is composite because $(n + 1)! + i$ is divisible by i if $2 \le i \le n + 1$. □

As we recall, two numbers m and n are called relatively prime if $(m, n) = 1$; observe that neither m nor n have to be prime as, for example, is the case with 9 and 14. Given $n > 0$, $\phi(n)$ denotes the number of positive integers m such that $m \le n$, and $(m, n) = 1$; ϕ is called the *Euler ϕ-function*, or, the Euler *totient* function. Note that for any prime number p, we have $\phi(p) = p - 1$. For example, $\phi(5) = 4$, because all the integers 1, 2, 3, and 4 are relatively prime to 5.

Let us now find a formula that expresses $\phi(n)$ in terms of the prime-power decomposition of $n = p_1^{e_1} p_2^{e_2} \cdots p_k^{e_k}$. Basically, we want to count the positive integers m, $m \le n$, where m is not divisible by any of the primes p_i that appear in the decomposition of n. Obviously, there are n possibilities for m; however, we must throw away n/p_j, $j = 1, 2, \ldots, k$, of them because they are divisible by p_j. That is, we are left with $n - n/p_1 - n/p_2 - \cdots - n/p_k$ possible m values. At this point, though, we have thrown away too much, because an integer that is a multiple of, say, both p_1 and p_2 has been discarded at least twice. We correct the error by adding to the last expression the quantity $n/(p_1 p_2) + n/(p_1 p_3) + \cdots + n/(p_{k-1} p_k)$. Now, however, we have added too much, because an integer that is divisible by, say, $p_1 p_2 p_3$ has been reentered at least twice, and the last expression has to be corrected. Continuing in this way, we obtain the formula

$$\phi(n) = n - \frac{n}{p_1} - \frac{n}{p_2} - \cdots - \frac{n}{p_k} + \frac{n}{p_1 p_2} + \frac{n}{p_1 p_3} + \cdots + \frac{n}{p_{k-1} p_k}$$
$$+ \cdots + (-1)^k \frac{n}{p_1 p_2 \cdots p_k}$$

As the reader can easily verify, this formula can be simplified to

$$\phi(n) = n\left(1 - \frac{1}{p_1}\right)\left(1 - \frac{1}{p_2}\right) \cdots \left(1 - \frac{1}{p_k}\right)$$

Example. To compute $\phi(60)$, we first find the prime-power decomposition $60 = 2^2 \cdot 3 \cdot 5$. Then, applying our formula, we obtain

$$\phi(60) = 60(1 - 1/2)(1 - 1/3)(1 - 1/5) = 16$$

We conclude this section with a discussion on how to find prime numbers. Like gold nuggets, primes are found mostly by sieves. The first one was designed by the Greek mathematician Eratosthenes of Kyrenia, who lived in the third century B.C. Eratosthenes' sieve idea is charmingly simple.

To find all primes $\leq n$, write down the integers from 2 to n in order. Then, cross out all the even numbers, because they are divisible by 2 and therefore not prime (except 2 itself). Next, all multiples of 3 are crossed out, and so on. After the ith pass, all multiples of the first i primes $p_1, \ldots, p_i$ have been crossed out. The first $x > p_i$ that has not been crossed out is the $(i + 1)$st prime. (Why?) All multiples of p_{i+1} are then crossed out and the process terminates when, on the list, there are no more integers left uncrossed that are greater than the last prime found. The integers that remain uncrossed have passed the sieve and are the primes $\leq n$.

In the following example we find the primes ≤ 60 but have listed only the odd numbers ≥ 2:

2   3   5   7   ~~9~~   11   13   ~~15~~   17   19   ~~21~~   23   ~~25~~   ~~27~~   29
31   ~~33~~   ~~35~~   37   ~~39~~   41   43   ~~45~~   47   ~~49~~   ~~51~~   53   ~~55~~   ~~57~~   59

Multiples of $3 \geq 3^2$ are crossed out with a "/," multiples of $5 \geq 5^2$ with a "\," and multiples of $7 \geq 7^2$ with a "—." The next prime after 7 is 11 but $11^2 > 60$, so, for reasons explained immediately below, the process stops and the uncrossed integers are the primes ≤ 60. (In this context, see also Dudley, 1983 and Mills, 1947.).

We can actually use this method to determine whether a given number n is prime, that is, whether it will be crossed out. The procedure can be made more efficient by incorporating the fact that, if n is composite, n has a prime divisor $\leq \sqrt{n}$. (The reason is that factors are always found in pairs; if a number has a factor larger than the square root, it must also have one smaller.) Therefore, in crossing out multiples of p_i on the list, we need only consider primes $\leq \sqrt{n}$, and the process stops when $p_i^2 \geq n$, for some i. However, despite this acceleration, the above procedure is inadequate for testing the largest primes known. For example, to test the 13,395-digit number $2^{44,497} - 1$, which was proved to be prime in 1979, a computer performing a million multiplications per second would need $10^{6.684}$ years to finish, if it were to stop when the square root of the number was reached.

For some of our subsequent applications, we will need to find the n largest primes $\leq M$, where M is the maximum integer that can be represented by the computer hardware at our disposal. For example, 16- and 32-bit words are common word sizes and, ignoring the sign bit, the largest integer these machines can represent is $2^{15} - 1 = 32,767$ and $2^{31} - 1 = 2,147,483,647$, respectively. To find these large primes, we first construct a table of primes $\leq \sqrt{M}$, using the sieve method. The multiples of each of these primes are then crossed out form the list $L, L+1, \ldots, M$, where the interval (L, M) is large enough to contain at least n primes. The integers left uncrossed are prime.

For the method to be practical, the list of primes $\leq \sqrt{M}$ cannot be too large, and we must find an interval (L, M) that contains at least n primes and is not too big for small n. The celebrated prime number theorem can help with both these problems (Erdoes, 1949; Goldstein, 1973; Levinson, 1969 and see also historical note 3). Let $\pi(x)$ be the number of primes $\leq x$; from the sieve example we constructed above, we see that $\pi(60) = 17$. The theorem says $\lim_{x \to \infty} \pi(x)/(x/\ln x) = 1$. For large x, $x/\ln x$ approximates $\pi(x)$; in fact, it is an underestimation. Roughly speaking, the theorem says that one in every $\ln M$ integers is a prime. Thus, an interval well in excess of length $n \cdot \ln M$ should contain n primes. Furthermore, the size of the prime table is approximately $\sqrt{M}/\ln \sqrt{M}$. If $n = 10$, then for the 16-bit machine $10 \cdot \ln 2^{15} \approx 104$, so an interval of length 200 should do. The prime table should have about $\sqrt{2^{15}}/\ln \sqrt{2^{15}} \approx 35$ primes. In fact, $\pi(\sqrt{2^{15}}) = 42$. For the 32-bit machine, $10 \cdot \ln 2^{31} \approx 214$, so take an interval of length 500. We see that $\sqrt{2^{31}}/\ln \sqrt{2^{31}} \approx 4314$ is roughly the size of the table. The actual size is $\pi(\sqrt{2^{31}}) = 4691$. The following algorithm incorporates the above ideas.

GENPR (Generate Primes)

Input: Two single-precision integers k, and m, and a one-dimensional array A of length k; m is an *odd* integer ≥ 3.

Output: The single-precision prime numbers $p_1 < p_2 < \cdots < p_r$ enclosed in the closed interval $[m, m + 2k - 2]$.

1. [Initialize.] $n := m + 2k - 2$; for $i := 1, 2, \ldots, k$, do $A(i) := 1$; $d := 3$.
2. [If $d^2 > n$, obtain the primes and terminate.] If $d > [n/d]$, go to 6.
3. [Compute the least positive integer j such that $d | (m + 2j - 2)$ and $m + 2j - 2 \geq 3$.] $r := \mathbf{MOD}(m, d)$; $j := 1$; if $r > 0$ and r is even, then $j := j + d - r/2$; if $r > 0$ and r is odd, then $j := j + (d - r)/2$; if $m \leq d$, then $j := j + d$.
4. [Cross our composites.] For $i := j, j + d, j + 2d, \ldots$ until $i > k$, do $A(i) := 0$.
5. [Update d.] If $\mathbf{MOD}(d, 6) = 1$, then $d := d + 4$ else $d := d + 2$; go to 2.
6. [Obtain the primes.] For $i := k, k - 1, \ldots, 1$, do {if $A(i) = 1$, then output the prime $m + 2i - 2$}; exit.

Computing-Time Analysis of **GENPR**. All arithmetic in this algorithm is single precision, which implies that each arithmetic operation is performed in time $0(1)$.

Steps 1 and 6 are executed once, and there are k operations involved in each one of them; therefore, the execution time of both steps is $0(k)$.

Steps 2 through 5 form a loop for which, as can be seen from step 2, $\sqrt{n}$ where $n = m + 2k - 2$, is an upper bound on the number of its executions. Moreover, each time through the loop, in step 4 there are at most k operations involved, whereas in steps 2, 3, and 5 there are either one or two operations involved. Therefore, the whole loop is executed in time $0(k\sqrt{n})$.

Putting the above results together, and keeping in mind that $0(k + k\sqrt{n}) = 0(k\sqrt{n})$, we see that

$$t_{\mathbf{GENPR}}(A, k, m) = 0(k\sqrt{n})$$

where $n = m + 2k - 2$.

Example. Let us apply the preceding algorithm in order to find the primes between 3 and 21; in other words, in this case $m = 3$ and $k = 10$.

After step 1, $d = 3$ and $A(i) = 1$, $i = 1, \ldots, 10$. In step 2 the test for termination fails, so we execute step 3, where we obtain $r = 0$ and $j = 4$ (since $m = d$). In step 4 we update the A array by setting $A(4) := 0$, $A(7) := 0$, and $A(10) := 0$, and in step 5 we update d by setting $d := 5$. This completes the first execution of the loop formed by steps 2 through 5.

At the beginning of the second execution of the loop, the termination test succeeds and we go to step 6. In step 6 for $i = 10$ we output nothing since $A(10) = 0$; however, for $i = 9$, $A(9) = 1$, and we output the prime number 19, and so on. In this way we obtain the prime numbers 19, 17, 13, 11, 7, 5, and 3.

2.3.2 Integers Modulo m and the Greek–Chinese Remainder Algorithm

Recall that by fixing $m > 1$, we defined $b \equiv_m a$, if $b - a = mq$ for some q (which is the same as saying that b is in the arithmetic progression $\{\ldots, -3m + a, -2m + a, -m + a, a, m + a, 2m + a, 3m + a, \ldots\}$) or, in the language of this section, if m divides $b - a$. Instead of $b \equiv_m a$ we can also write $b \equiv a \pmod{m}$ and read "b is congruent to a modulo m." The notation is due to Gauss and m is called the *modulus*. The set of equivalence classes $\mathbf{Z}/\equiv_m$ is also written $\mathbf{Z}_m$ and is called the *integers modulo m*. The next theorem presents basic properties of congruences and enables us to conclude that important algebraic properties of the integers carry over to the integers modulo m.

Theorem 2.3.7

a. If $a \equiv b \pmod{m}$ and d divides m, then $a \equiv b \pmod{d}$.
b. If $a \equiv b \pmod{m}$ and $a \equiv b \pmod{n}$, then $a \equiv b \pmod{[m, n]}$.
c. If $a \equiv c \pmod{m}$ and $b \equiv d \pmod{m}$, then $a + b \equiv c + d \pmod{m}$, $a - b \equiv c - d \pmod{m}$, and $ab \equiv cd \pmod{m}$.
d. (Cancellation property.) If $ab \equiv ac \pmod{m}$, then $b \equiv c \pmod{m/d}$, where $d = (a, m)$. In particular, if $(a, m) = 1$, then $ab \equiv ac \pmod{m}$ implies $b \equiv c \pmod{m}$.

Proof. We prove part a and leave the rest as an exercise for the reader. From $a \equiv b \pmod{m}$ we see that $a - b = km$, for some integer k. Now if $d | m$, then d also divides km, which implies that $d | (a - b)$. Hence, $a - b = k'd$ for some integer k', and $a \equiv b \pmod{d}$. □

Example. From $3 \cdot 4 \equiv 3 \cdot 6 \pmod{6}$ we obtain $4 \equiv 6 \pmod{2}$, whereas from $3 \cdot 4 \equiv 3 \cdot 9 \pmod{5}$ we obtain $4 \equiv 9 \pmod{5}$.

A useful alternative characterization of integral domains is provided by the following cancellation law.

Theorem 2.3.8. A nontrivial commutative ring D is an integral domain if and only if $ab = ac$, $a \neq 0$ implies $b = c$.

Proof. We prove the theorem in one direction only, leaving the rest as an exercise for the reader. Let D be an integral domain with a, b in D; then, $ab = 0$ implies either $a = 0$ or $b = 0$. From $ab = ac$, $a \neq 0$, we have $a(b - c) = 0$, $a \neq 0$ and, since D has no zerodivisors, $(b - c) = 0$, yielding $b = c$. □

Consider the equivalence class **a** and **b** in Z_m; we define their sum and product, $\mathbf{a} + \mathbf{b}$ and $\mathbf{a} \cdot \mathbf{b}$, to be the sum and product of their representatives. That is, given the surjective map $s : Z \to Z_m$, $\mathbf{a} + \mathbf{b} = s(a + b)$ and $\mathbf{a} \cdot \mathbf{b} = s(a \cdot b)$. Theorem 2.3.7 assures us that addition and multiplication is well defined; in other words, if $c \in \mathbf{a}$ and $d \in \mathbf{b}$ are different representatives, we still obtain the same equivalence classes for the sum and the product, since $s(c + d) = s(a + b)$ and $s(c \cdot d) = s(a \cdot b)$ by the theorem. An equivalence relation that preserves algebraic properties (such as $\equiv_m$ on Z, which preserves addition and multiplication) is called a *congruence* relation.

Using the Euclidean property we can view arithmetic on the integers modulo m as *remainder* or *modular arithmetic*. A *complete residue system* modulo m consists of m integers, one representative from each residue class. Two are the most common residue systems: the *nonnegative* residue system modulo m, consisting of the integers $0, 1, 2, \ldots, m-1$, and the *least absolute value or symmetric* residue system, consisting of the integers $0, \pm 1, \pm 2, \ldots, \pm(m-1)/2$, for odd m.

Write an integer a as $a = mq + r$ with $0 \le r < m$. The remainder r, sometimes written $r_m(a)$ or $r(a)$, is called the *remainder* or *residue* modulo m. The next proposition says that two integers are congruent, exactly when they have the same residues modulo m.

Proposition 2.3.9. $b \equiv a \pmod{m}$ iff $r_m(b) = r_m(a)$.

Proof. We observe that for any a, $r(a) \equiv a \pmod{m}$. By transitivity it follows that $b \equiv a \pmod{m}$ iff $r(b) \equiv r(a) \pmod{m}$, that is, iff $m | [r(b) - r(a)]$. But, since $0 \le r(b), r(a) < m$, m divides $r(b) - r(a)$ exactly when $r(b) = r(a)$. □

The proposition says that the equivalence class of an integer a, $\mathbf{a} = a + mZ$, is precisely the set of elements with the same remainders or residues $r(a)$. The remainders $0, 1, \ldots, m-1$ are representatives of the equivalence classes. By abuse of notation we sometimes identify an equivalence class with its representative remainder and say that Z_m is simply the set $\{0, 1, \ldots, m-1\}$.

When we do residue or remainder arithmetic we simply work with the remainders modulo m. The additive inverse for any number a in Z_m is $m - a$, but the multiplicative inverse of a, defined as the solution to the congruence equation $ax \equiv 1 \pmod{m}$, does not always exist.

Theorem 2.3.10. Let a be a member of Z_m. Then a has a multiplicative inverse modulo m if and only if $(a, m) = 1$.

Proof. By the extended Euclidean algorithm we can find integers x and y such that $(a, m) = ax + my$, which implies that $ax \equiv (a, m) \pmod{m}$. If $(a, m) = 1$, it is obvious that x is the multiplicative inverse of a modulo m, whereas if $(a, m) > 1$, then no x exists such that $ax \equiv 1 \pmod{m}$ since $ax = 1 + km$ implies that $(a, m) = 1$. □

Theorem 2.3.11. The set $\mathbf{Z}_m = \{0, 1, 2, \ldots, m - 1\}$, with addition and multiplication defined modulo m, is a commutative ring with unity for any integer $m > 1$. Such a ring is a finite field if and only if m is a prime.

Proof. Having previously defined the surjective map $s : \mathbf{Z} \to \mathbf{Z}_m$, and with it addition, subtraction, and multiplication in $\mathbf{Z}_m$, we can now easily prove that $\mathbf{Z}_m$ is a commutative ring with identity, if we use the fact that $\mathbf{Z}$ is such a ring. For instance, to verify that the associative law for multiplication holds observe that for any a, b, and c in $\mathbf{Z}$ we have $s(a) \cdot [s(b) \cdot s(c)] = s(a) \cdot s(b \cdot c) = s[a \cdot (b \cdot c)] = s[(a \cdot b) \cdot c] = s(a \cdot b) \cdot s(c) = [s(a) \cdot s(b)] \cdot s(c)$.

Now, if m is a prime number p, then all the nonzero elements of $\mathbf{Z}_p$ are units, since they have multiplicative inverses, and hence, $\mathbf{Z}_p$ is a field. On the other hand, if m is not a prime, then $\mathbf{Z}_m$ is not a field. To see this, write $m = a \cdot b$, $a, b < m$; then $s(a) \cdot s(b) = s(m) = s(0)$, but $s(a) \neq s(0)$ and $s(b) \neq s(0)$, which implies that there are zerodivisors. □

Example. Consider $p = 5$. Then $\mathbf{Z}_5 = \{0, 1, 2, 3, 4\}$ is a field because all the nonzero elements 1, 2, 3, and 4 are units; that is, they all have inverses, which are 1, 3, 2, and 4 respectively. On the other hand, for $m = 8$, $\mathbf{Z}_8 = \{0, 1, 2, 3, 4, 5, 6, 7\}$ has zerodivisors, for instance, $2 \cdot 4 = 0 \pmod 8$, and hence it is not a field.

We mention in passing that the multiplicative group $\mathbf{Z}_m$ has $\phi(m)$ elements, where ϕ is the Euler function; in other words, $\mathbf{Z}_m$ is of *order* $\phi(m)$.

Theorem 2.3.12. The characteristic λ of a finite field is a prime number.

Proof. Suppose that the characteristic λ of a finite field is composite; that is, $\lambda = ab$, $1 < a, b < \lambda$. Then $a \cdot 1$ and $b \cdot 1$ are both nonzero, but $(a \cdot 1)(b \cdot 1) = (ab) \cdot 1 = \lambda \cdot 1 = 0$, contradicting the fact that in the finite field there are no zerodivisors. □

Applications of Linear Congruences One of the many applications of linear congruences is the ancient error-detecting algorithm that we are taught in elementary school, known as "casting out 9's." If we add decimal numbers column by column, then if in any one column the sum exceeds 9, we reduce modulo 10 and add 1 or 2 or 3, and so on to the next column. Thus, in terms of the sum of the decimal digits, we have added 1 or 2 or 3, and so forth, and subtracted 10 or 20 or 30, and so on. Therefore, the sum of the digits has not changed modulo 9. [Also think of $abc = 100a + 10b + c = 99a + 9b + (a + b + c)$.]

Example. We consider the addition of just two numbers:

```
   89   sum of digits = 17 ,    sum of sum of digits =  8
 + 89   sum of digits = 17 ,    sum of sum of digits =  8
  ───                                                 ──
  178                  34                             16
```

We see that the sum of digits of 178 is $16 \equiv 7 \pmod{9}$, and so is the sum of digits of 34, and the sum of digits of 16.

The same error-detection algorithm applies for multiplication as well.

```
    89   sum of digits = 17
  × 87   sum of digits = 15
  ────
  7743
```

The product of the sums of digits is $17 \cdot 15 = 255$, whose sum of digits is $12 \equiv 3 \pmod{9}$; likewise, the sum of the digits of the product 7743 is $21 \equiv 3 \pmod{9}$. The reason why the sums of digits when multiplied are congruent to the product of the numbers modulo 9 is that $10^k \equiv 1 \pmod{9}$, $k \geq 0$. The only problem with this ancient error-detecting algorithm is that it can fail to detect an error; for random errors, 10% go undetected.

Computation of Multiplicative Inverses Multiplicative inverses modulo m play an important role throughout this book, and we next study two ways to compute them.

As we saw in the proof of Theorem 2.3.10, the extended Euclidean algorithm can be used to compute the multiplicative inverse of nonzero a modulo m, denoted $a^{-1} \pmod{m}$, or simply a^{-1}. From $(a, m) = 1$ we have

$$ax + my = 1 ,$$

or, more appropriately

$$ax = 1 - my$$

from which we conclude that

$$ax \equiv 1 \pmod{m}$$

and hence x is the multiplicative inverse of a modulo m. [Note that in this case we are dealing with single-precision integers, and setting $L(b) = 1$ in the computing time of the Euclidean algorithm, we see that the multiplicative inverse is computed in time ~ 1.]

Example. In $GF(11)$ let us compute $4^{-1} \pmod{11}$, the multiplicative inverse of 4 modulo 11. Applying the extended Euclidean algorithm to 11 and 4, we obtain the following table:

Iteration	q	a_0	a_1	x_0	x_1	y_0	y_1
0	—	11	4	1	0	0	1
1	2	4	3	0	1	1	−2
2	1	3	1	1	−1	−2	3
3	3	1	0	−1	4	3	−11

from which we see that $1 = 11 \cdot (-1) + 4 \cdot 3$, where -1 and 3 appear in the columns under x_0 and y_0, respectively; hence, $4^{-1} \pmod{11} = 3$. We can speed up the computation of multiplicative inverses if we note that, in this example, we did not have to compute the x sequence. (In this case, the inverse appeared under the y_0 column; however, depending on how we set up the table, the inverse can also appear under the x_0 column.)

Another way to compute the multiplicative inverse of a nonzero element modulo a prime m is to use the following celebrated theorem (see historical note 4).

Fermat's "Little" Theorem (1640). If m is a prime and a is an integer not divisible by m, then $a^{m-1} \equiv 1 \pmod{m}$.

Proof. Consider the numbers $a, 2a, 3a, \ldots, ma$. When divided by m, no two of these numbers, say, ia and ja, give the same remainder. [If they did, then $(i-j)a$ would be a multiple of m, because subtraction causes the remainders to cancel; since m does not divide a, it would have to divide $i-j$, a contradiction, since both i and j are from the sequence $1, 2, 3, \ldots, m$.] Therefore, when divided by m, the numbers $a, 2a, 3a, \ldots, ma$, will give the remainders $0, 1, 2, \ldots, m-1$ in some order, the 0, of course, being obtained on dividing ma by m. In other words, omitting the 0, we have $a^{m-1}(m-1)! \equiv (m-1)! \pmod{m}$. Subtracting, we obtain $a^{m-1}(m-1)! - (m-1)! \equiv 0 \pmod{m}$, which implies that $(a^{m-1}-1)(m-1)!$ is a multiple of m. Since m does not divide $(m-1)!$, and it is prime, it must divide $(a^{m-1}-1)$, and this completes the proof. □

Corollary 2.3.13. If in $\mathbb{Z}_m$ m is prime, then $a^{-1} = a^{m-2}$.

The following is Euler's generalization of Fermat's theorem (m need not be a prime).

Theorem 2.3.14 (Euler). If $(a, m) = 1$, then $a^{\phi(m)} \equiv 1 \pmod{m}$.

Proof. The proof parallels that of Fermat's theorem. Consider the *prime residue system modulo* m $r_1, r_2, \ldots, r_{\phi(m)}$, and multiply each r_k with a, where $(a, m) = 1$. (This residue system is called "prime" because it consists only of those residue classes that are relatively prime to the modulus m and constitute a multiplicative group.) This multiplication changes the sequence of the above residues, but their total product is not affected, because they form a multiplicative group. Hence, $a^{\phi(m)} r_1 r_2 \cdots r_{\phi(m)} \equiv r_1 r_2 \cdots r_{\phi(m)}$ (mod m). Since by definition the residues are relatively prime to m, we can apply the cancellation property of Theorem 2.3.7 to obtain $a^{\phi(m)} \equiv 1$ (mod m). □

Corollary 2.3.15. If in $\mathbf{Z}_m$ $(a, m) = 1$, then $a^{-1} = a^{\phi(m)-1}$.

From the above we see that, computing the multiplicative inverse of a, means raising it to some power k, where k could be either $m - 2$ or $\phi(m) - 1$. There are two ways for computing these powers.

The first, called the "brute force" approach, requires that we do k multiplications; since we deal with single-precision integers, each of these multiplications takes time ~ 1, and hence this approach takes time $0(m)$. [Recall that $\phi(m) < m$.]

The second way, called the "binary method," is a more efficient way to raise a to the power k; this method, which was known in India 2000 years ago, works as follows. Write k in the binary system, omitting the leading zeros; that is, $k = \Sigma_{0 \leq i \leq n-1} k_i 2^i$. Then replace each "1" by the pair of letters "SM$_a$" and each "0" by the letter "S"; moreover, cross off the "SM$_a$" that appears at the left. The resulting sequence of letters is a rule for computing a^k, if we interpret "S" as *square and reduce mod m*, and "M$_a$" as *multiply times a and reduce mod m*.

Example. In $GF(11)$, to compute 4^{-1} (mod 11), we have to compute 4^9, where the binary representation of 9 is 1001. So we form the sequence SM$_4$ S S SM$_4$, and crossing off the leading SM$_4$, we obtain SSSM$_4$, which states that we should "square, square, square, and multiply times 4," reducing, of course, (mod 11) each time; that is, $4^9 = [(4^2)^2]^2 4$. In this way we successively compute in $GF(11)$ $4^2 = 5$, and $4^4 = 5^2 = 3$; to see the last step, that is, "square and multiply," observe that $4^9 = (4^4)^{2+1} = (3^2)4 = 9 \cdot 4 = 3$, from which we obtain the multiplicative inverse of 4 modulo 11.

The "binary" procedure described above works from left to right on the bit representation of k and, since it requires no temporary storage, is well suited for hardware implementation. However, it is easier to work from right to left, since we can get to the bits by simply shifting them out one at a time. The following algorithm incorporates the above ideas and works from right to left.

E (Exponentiate)

    Input: Nonzero a, k, and m; a is in $\mathbf{Z}_m$, and $k = \Sigma_{0 \leq i \leq n-1} k_i 2^i$ is either $m - 2$ or $\phi(m) - 1$.

Output: a^{-1}, the multiplicative inverse of a modulo m, where $a^{-1} = a^k$ in $\mathbf{Z}_m$.

1. [Initialize.] $K := k$; $B := 1$; $A := a$.
2. [Compute next bit.] $q := K/2$; $r := K - 2 \cdot q$; $K := q$; if $r = 0$, go to step 5.
3. [Multiply and reduce mod m.] $B := A \cdot B \pmod{m}$.
4. [Done?] If $K = 0$, return $a^{-1} \pmod{m} := B$.
5. [Square and reduce mod m.] $A := A^2 \pmod{m}$; go to step 2.

Computing-Time Analysis of E. We know that there are n bits in the binary representation of k, and let us assume that j of them are "1"-bits. Then we have to perform $n + j$ multiplications, each of which takes time ~ 1. Since j can be at most n, we have that $0(2 \cdot n) = 0(2 \cdot \log_2 m)$ (justify this) $= 0(\log_2 m)$, and

$$t_\mathbf{E}(a, k, m) = 0[L(m)]$$

an improvement over the "brute force" approach.

Suppose now that instead of over $\mathbf{Z}_m$, we apply algorithm **E** over the integers with inputs a, b. Then, if we change step 3 to "$B := A + B$" and step 5 to "$A := A + A$" and if in step 1 we initialize $B := 0$ instead of $B := 1$, and $K := b$ instead of $K := k$, then the algorithm returns $B = a \cdot b$. This is a practical way for doing multiplication and is often called the "Russian peasant" method; Russian peasants used it, since they "could" only multiply and divide by 2, and add.

Example. To multiply 38 by 19, we put them at the top of two columns, called a and b, and fill in the columns below by multiplying the a-number by 2 and dividing the b-number by 2. Whenever the b-number is odd, we subtract one from that b-number before dividing it by 2, and put the corresponding a-number in a third column, called the *sum*-column. When we reach the number 1 in the b-column, we add the numbers in the *sum*-column to obtain the answer. We obtain the following table:

a	b	Sum
38	19	38
76	9	76
152	4	
304	2	
608	1	608
		$722 = 38 \cdot 19$

The Russian peasant method is based on the fact that $a \cdot b = (2a) \cdot (b/2)$ if b is even and $a \cdot b = (2a) \cdot (b - 1)/2 + a$ if b is odd; of course, $a \cdot 1 = a$.

A Closer Look at Fermat's Theorem Fermat's theorem plays an important role in digital encryption, and we will now examine it somewhat closer. If for some number a the number $a^m - a$ [obtained from $a^{m-1} \equiv 1 \pmod{m}$] gives a nonzero remainder when it is divided by m, then m is certainly composite. Suppose, however, that $a^m - a$ is a multiple of m. Does it follow that m must be prime? Several examples suggest that the answer is "yes": $2^2 - 2$ is a multiple of 2, $2^3 - 2$ is a multiple of 3, $2^5 - 2$ is a multiple of 5, $2^7 - 2$ is a multiple of 7, and the numbers 2, 3, 5, and 7 are all primes.

The ancient Chinese had a test for primality that stated that a number m is prime *if and only if* m divides $2^m - 2$, or $2^{m-1} \equiv 1 \bmod(m)$. This theorem stood undisputed for many centuries, and Leibniz believed the result as well. Here the first "if" is all right, but the "and only if" is wrong, as was demonstrated in 1819 by the French mathematician Pierre Frédéric Sarrus. Sarrus pointed out that $2^{341} - 2$ is a multiple of 341, even though 341 is composite, the product of 11 and 31. This can be easily seen if we take advantage of the fact that

$$2^{10} \equiv 1 \pmod{341}$$

because then

$$2^{340} \equiv (2^{10})^{34} \equiv 1^{34} \equiv 1 \pmod{341}, \quad \text{or} \quad 341 | (2^{341} - 2)$$

Since Sarrus' discovery many other counterexamples involving many different values of a have been found; for instance, $3^{91} - 3$ is a multiple of the composite number 91, and so on. Of course, we will not always be as lucky as we were in calculating $2^{340} \pmod{341}$, where we used the fact that $2^{10} \equiv 1 \pmod{341}$. So, in general, we will have to raise a number to a certain power, but we already know how to do this very efficiently.

A composite number that masquarades as prime vis-à-vis Fermat's theorem with a given value of a is called *pseudoprime* to the *base a*. The number 341 is the smallest pseudoprime to the base 2, whereas 91 is a pseudoprime to the base 3. It turns out that for every base a there are infinitely many pseudoprimes (see Theorem 2.3.28). There are even composite numbers, such as $561 = 3 \cdot 11 \cdot 17$ and $1729 = 7 \cdot 13 \cdot 19$ that are pseudoprimes to *every* base a. Numbers of this kind, called *Carmichael numbers*, will be examined later (see historical note 5).

By contrast, the following theorem is true if and only if m is prime.

Theorem 2.3.16 (Wilson). $(m - 1)! \equiv -1 \pmod{m}$ if and only if m is a prime.

Proof. For a proof, not relying on group theory when m is *prime*, consider the product $2 \cdot 3 \cdot 4 \cdots (m-3) \cdot (m-2)$. Here each factor has its own multiplicative inverse modulo m (see also Exercise 7 of this section); for instance, for $m = 7$, we consider $2 \cdot 3 \cdot 4 \cdot 5$, where 2 and 4 form a pair of inverses modulo 7, and so do 3 and 5. By definition, the product of such pairs is $\equiv 1 \pmod{m}$, and so is the product of the all the $(m-3)/2$ pairs of multiplicative inverses. In other words, we have

$$2 \cdot 3 \cdot 4 \cdots (m-3) \cdot (m-2) \equiv 1 \pmod{m}$$

and multiplying by $m - 1$, we obtain

$$2 \cdot 3 \cdot 4 \cdots (m-3) \cdot (m-2)(m-1) = (m-1)! \equiv m - 1 \equiv -1 \pmod{m} \quad \square$$

Using Wilson's theorem, we see that if we construct the function $f(m) = \sin\{\pi \cdot [(m-1)! + 1]/m\}$, then $f(m)$ is zero if and only if m is prime. Unfortunately, to use Wilson's theorem as a primality test is of *no* practical value, becasue we have to calculate $(m-1)!$ first—and this can be an extremely large number even for small values of m.

Some Results on Group Theory We continue now with some interesting results on groups that will be used in the presentation of the Greek–Chinese remainder theorem. As was indicated in Theorem 2.3.11, while $\mathbf{Z}_m$ is a ring, it certainly need not be a field, because there may be some noninvertible elements. We have already seen the ring $\mathbf{Z}_8$, where the elements 2, 4, and 6 have no multiplicative inverse in $\mathbf{Z}_8$, whereas 1, 3, 5, and 7 *do* have such inverses. The difference, of course, stems from the fact that 1, 3, 5, and 7 are relatively prime to the modulus 8, whereas, 2, 4, and 6 are not.

The invertible elements of $\mathbf{Z}_m$ form a multiplicative group, called the *group of units* of $\mathbf{Z}_m$. This group is denoted by $U_m = \{\mathbf{a}: (a, m) = 1\}$ and has $\phi(m)$ elements, or order $\phi(m)$. Note that U_8 contains the four elements $\{1, 3, 5, 7\}$, each of which has a multiplicative inverse; moreover, observe that, in their multiplication table shown in Figure 2.3.1, each row contains a permutation of all the group elements.

Let G be an *abelian* group with n elements; that is, for any pair of a, b in G, commutativity holds. For a in G, a^k denotes $a \cdot a \cdots a$ (k times); $a^0 = e$, the identity element of G. The usual laws of exponents apply. We have the following absraction of Fermat's theorem.

Theorem 2.3.17. If G is an abelian group consisting of n elements, then for any a in G, $a^n = e$.

*	1	3	5	7
1	1	3	5	7
3	3	1	7	5
5	5	7	1	3
7	7	5	3	1

Figure 2.3.1. Multiplication table for the multiplicative group U_8.

Proof. We are going to simply sketch the proof. Let $a_1, a_2, \ldots, a_n$ be the elements of G. Then $a \cdot a_1, a \cdot a_2, \ldots, a \cdot a_n$ are all distinct and the set $\{a \cdot a_1, a \cdot a_2, \ldots, a \cdot a_n\}$ is the same as the set $\{a_1, a_2, \ldots, a_n\}$. To complete the proof along the lines of Fermat's theorem, we need to prove (by induction) that all possible ways of associating the product of any $n \geq 3$ elements of G are equal (generalized associativity), and that all possible ways of multiplying $n \geq 2$ elements of G, regardless of order, give the same result (generalized commutativity). □

This abstract version of Fermat's theorem generalizes to nonabelian groups, and a proof of this can be found elsewhere (e.g., Childs, 1979).

Let G be a group consisting of n elements $a \in G$ and $S = \{k \geq 1 : a^k = e\}$. Since $a^n = e$, it follows that S is nonempty and, by well-ordering, it has a *least* element k_0, called the *order* of the group element a. A group is *cyclic* if there is an element a whose powers $1, a, a^2, \ldots$ run through all the elements of the group. Such a group element is called *generator*, or *primitive root* modulo m, if $G = U_m$ for some m.

Primitive roots can be used as a good way to generate random numbers on a computer. To see this, choose a primitive root modulo the word size, and then, each time the user asks for a random number, output the next higher power of the primitive root. Having started with a primitive root ensures that the number of "random numbers" generated will be as large as possible. We have the following.

Theorem 2.3.18. Let G be a group consisting of n elements, and let k_0 be the order of a in G. Then $k_0 | n$.

Proof. It is easily seen that $e, a, a^2, \ldots, a^{k_0-1}$ are k_0 distinct elements of G. If these k_0 elements do not exhaust the group, there must be some other element in G, say, a_2. Then we can easily see that $a_2, a_2 \cdot a, a_2 \cdot a^2, \ldots, a_2 \cdot a^{k_0-1}$ are k_0 distinct elements, all different from the previous k_0 elements of G. If G is not exhausted, there must be another element a_3, and so on. This process of obtaining new elements a_i must eventually terminate since G has n elements. Clearly, then, we see that the order n of G is $k_0 \cdot j$, where a_j was the last element obtained. □

Theorem 2.3.19. If k_0 is the order of a in G and $a^k = e$, then $k_0 | k$.

Proof. Let $k = k_0 q + r$ with $0 \leq r < k_0$. If $r = 0$, we have nothing to prove, so suppose $r > 0$. Then $a^k = a^{k_0 q + r} = (a^{k_0})^q a^r = e$, which implies $a^r = e$. This, however, is a contradiction since k_0 is the least number such that $a^{k_0} = e$. □

Letting $G = U_m$, it follows from the above the every element of U_m has an order that divides $\phi(m)$. The primitive roots, if they exist, are exactly the elements of maximum possible order $\phi(m)$. It now becomes obvious that Theorem 2.3.14 (Euler's) is a corollary of these remarks.

Let us now see if the group U_8 is cyclic—that is, if there exists an element a of U_8 whose powers $1\ a, a^2, a^3, \ldots$ run through all the elements of the group.

Let us look at the order of the various elements of U_8 modulo 8:

$n =$	1	2	3	4	5	6	7	etc.
$3^n \equiv$	3	1	3	1	3	1	3	etc.
$5^n \equiv$	5	1	5	1	5	1	5	etc.
$7^n \equiv$	7	1	7	1	7	1	7	etc.

Obviously, the period after which the sequences repeat for the first time is 2, and therefore, the order of every element of U_8 is 2, which divides $\phi(8) = 4$ (Theorem 2.3.18). Therefore, U_8 is not cyclic, and 8 does not have primitive roots. The following theorem tells us why.

Theorem 2.3.20. The group U_m is cyclic if and only if m is 1, 2, 4, p^a, or $2p^a$, where p is an odd prime, and $a > 0$. Therefore, the integer m has a primitive root for precisely such values of m.

Proof. The proof can be found in LeVeque (1977). □

It turns out that 8 is the smallest integer not having a primitive root. On the other hand, according to Theorem 2.3.20 U_{18} is a cyclic group because $18 = 2 \cdot 3^2$. There are $\phi(18) = 6$ elements in U_{18}, namely, $U_{18} = \{1, 5, 7, 11, 13, 17\}$, and 5 is a primitive root because

$n =$	0	1	2	3	4	5	6	etc.
$5^n \equiv$	1	5	7	17	13	11	1	etc.

The reader is encouraged to find all the primitive roots modulo 18. An immediate consequence of Theorem 2.3.20 is the following.

Corollary 2.3.21. If m is an odd prime number, then U_m is cyclic, and the equation $x^2 = 1$, in U_m, has only the solutions $x = \pm 1$.

Once we have found a primitive root a modulo m in U_m, we can immediately find another one, a^{-1}, the multiplicative inverse of a modulo m. In the case U_{18}, $5^{-1} = 11$ modulo 18, and 11 is a primitive root as well. (This fact is justified below.)

How many primitive roots are there in U_m, for instance, in U_{18}? If we raise a given primitive root r to the power $k > 1$, where $\gcd[k, \phi(m)] = 1$, then $r' = r^k$ must be another primitive root because the order of r' is $\phi(m)$

(this is justified below); indeed, in U_{18}, $\phi(18) = 6$, and since $(5, 6) = 1$, we compute 5^5 in $\mathbf{Z}_{18}$ to obtain 11, another primitive root. Thus, in general there are $\phi[\phi(m)]$ primitive roots; for instance, in U_{18} we have $\phi[\phi(18)] = \phi(6) = 2$ primitive roots, 5 and 11.

If, however, $gcd[k, \phi(m)] = d > 1$, then the order of $r' = r^k$ is $\phi(m)/d$. To see this, first observe that $\phi(m)/d$ is a period of r'; namely, $(r')^{\phi(m)/d} = r^{\phi(m)k/d} \equiv 1^{k/d} = 1 \pmod{m}$. Second, note that $\phi(m)/d$ is the shortest period of r', because by introducing the least common multiple $[k, \phi(m)]$ we have: $(r')^{\phi(m)/d} = r^{\phi(m)k/d} = r^{[k,\phi(m)]} = r^{\phi(m)j} \equiv 1^j = 1 \pmod{m}$.

Congruence Equations Finally, we turn our attention to congruence equations and the Greek–Chinese algorithm. Observe that the equation $6x \equiv 4 \pmod 8$ has the solutions $x \equiv 2, x \equiv 6, \ldots \pmod 8$, while the equation $6x \equiv 5 \pmod 8$ has no solutions [try $x \equiv 0, 1, \ldots, 7 \pmod 8$]. The next theorem tells us exactly when we can solve such an equation.

Theorem 2.3.22. *The congruence equation $ax \equiv b \pmod m$ has a solution iff $(a, m) | b$. If the solution exists and $d = (a, m)$, then it is unique modulo m/d.*

Proof. The integer x satisfies $ax \equiv b \pmod m$ if and only if there exists an integer y such that $ax + my = b$. By Theorem 2.2.4, the equation $ax + my = b$ has solutions if and only if $(a, m) | b$. To prove the second part, suppose that x satisfies the congruence $ax \equiv b \pmod m$, and let z be congruent to x modulo m/d, where $d = (a, m)$. Then $z = x + w(m/d)$, for some w in $\mathbf{Z}$, and $az = ax + aw(m/d) = ax + mw(a/d) \equiv ax \equiv b \pmod m$; that is, $az \equiv b \pmod m$. Conversely, suppose that $ax \equiv az \equiv b \pmod m$. Then $ax - az \equiv b - b = 0 \pmod m$, and so $m | a(x - z)$. By Theorem 2.2.6 m/d divides $x - z$, and, hence, $x \equiv z \pmod{m/d}$.

Example. Let us find the solutions to the congruence equation $270x \equiv 36 \pmod{342}$. Applying the extended Euclidean algorithm, we find that $(-5) \cdot 270 + 4 \cdot 342 = 18$, and $18 | 36$. By Theorem 2.3.22, the congruence has a solution that is unique modulo $19 = 342/18$. To find that solution, multiply $(-5) \cdot 270 + 4 \cdot 342 = 18$ by $2 = 36/18$ to obtain $(-10) \cdot 270 + 8 \cdot 342 = 36$, from which we see that -10 is one solution modulo 342. Modulo 19, the unique solution, becomes 9 since $9 \equiv -10 \pmod{19}$. Modulo 342, the rest of the solutions, are 9, 28, 47, 66, 85, 104, 123, 142, and so on.

As a special case of Theorem 2.3.22 we have the following corollary.

Corollary 2.3.23. *The congruence equation $ax \equiv 1 \pmod m$ has solution iff $(a, m) = 1$. The solution, $a^{-1} \pmod m$, is unique modulo m and is the multiplicative inverse of a modulo m.*

Example. $2x \equiv 1 \pmod{26}$ does not have a solution because $(2, 26) = 2$. On a more intuitive level, the solution does not exist because we are looking for x such that $2x - 1 = k \cdot 26$; however, the left side is always odd whereas the right side is always even, and therefore, a solution is impossible.

We next consider the problem of solving a system of linear congruences; with this problem there is associated the Greek–Chinese remainder theorem and algorithm (see historical note 6). However, in order to present the celebrated Greek–Chinese remainder theorem in a general form, we will first discuss the fact that, if the integer m can be factored as $m = p_1^{e_1} p_2^{e_2} \cdots p_k^{e_k}$ (prime-power decomposition), then the ring $\mathbf{Z}_m$ can be also "factored" as a product of $\mathbf{Z}_{p_i^{e_i}}$. (This is fully justified in Theorem 2.3.25.)

For instance, we expect to see $\mathbf{Z}_6 = \mathbf{Z}_2 \times \mathbf{Z}_3$, since $6 = 2 \cdot 3$, which means that we have to consider ordered pairs (x_1, x_2), where x_1 is in $\mathbf{Z}_2$ and x_2 is in $\mathbf{Z}_3$; for example, the six elements of $\mathbf{Z}_6$ are $(0, 0)$, $(0, 1)$, $(0, 2)$, $(1, 0)$, $(1, 1)$, and $(1, 2)$. We do componentwise arithmetic with ordered pairs as follows. If we denote by $\otimes$ either one of the binary operators $+$ or $\cdot$, then $(x_1, x_2) \otimes (y_1, y_2) = (x_1 \otimes y_1, x_2 \otimes y_2)$, where "$x_1 \otimes y_1$" is done in $\mathbf{Z}_2$ (arithmetic modulo 2), while "$x_2 \otimes y_2$" is done in $\mathbf{Z}_3$ (arithmetic modulo 3); for example, $(0, 2) \cdot (1, 2) = (0, 1)$, where the multiplication $2 \cdot 2$ was done modulo 3. We now have the following.

Theorem 2.3.24 (Greek–Chinese Remainder Theorem). Let $m_1, m_2, \ldots, m_k$ be pairwise relatively prime integers >1, and let $M = m_1 m_2 \cdots m_k$. Then there is a unique nonnegative solution modulo M of the simultaneous congruences

$$x \equiv a_1 \pmod{m_1}$$
$$x \equiv a_2 \pmod{m_2}$$
$$\cdots$$
$$x \equiv a_k \pmod{m_k}$$

Stated in other words, the mapping that associates with each integer x, $0 \leq x \leq M - 1$, the k-tuple $(a_1, a_2, \ldots, a_k)$, where $x \equiv a_i \pmod{m_i}$, $i = 1, 2, \ldots, k$, is a bijection between $\mathbf{Z}_M$ and $\mathbf{Z}_{m_1} \times \mathbf{Z}_{m_2} \times \cdots \times \mathbf{Z}_{m_k}$.

Proof. We give a constructive proof of the theorem. To find the number x, $0 \leq x \leq M - 1$, that satisfies the simultaneous congruences $x \equiv a_i \pmod{m_i}$, $i = 1, 2, \ldots, k$, we solve the congruences two at a time.

We begin with the first two. The first congruences $x \equiv a_1 \pmod{m_1}$ is true for any x of the form $x = a_1 + m_1 q$, for some q. To compute q, we substitute the value of x in the second congruence, $x \equiv a_2 \pmod{m_2}$ and obtain $x = a_1 + m_1 q \equiv a_2 \pmod{m_2}$, which yields $q \equiv (m_1)^{-1}(a_2 - a_1) \pmod{m_2}$. (Of course, we have first to compute m_1^{-1}, the multiplicative inverse of m_1 modulo m_2; see also Programming Exercise 2 for this section for a description of the procedure **MODINV**.) Therefore, $q = m_1^{-1}(a_2 - a_1) + rm_2$, for some r, and the solution x to the first *two* congruences is $x = a_{12} + r(m_1m_2)$, for some r, obtained by replacing q in $x = a_1 + m_1q$.

Therefore, the first two congruence equations can be replaced by the single congruence $x \equiv a_{12}$ (mod m_1m_2), where now the modulo is the product m_1m_2. We then apply the above procedure to $x \equiv a_{12}$ (mod m_1m_2) and what was originally the third congruence and repeat the process until we have found x satisfying all the congruences.

To show uniqueness, suppose that there is x', $0 \leq x' \leq M - 1$, such that $x' \equiv a_i$ (mod m_i), for each i. Then $x - x' \equiv 0$ (mod m_i) for each i, which implies that $m_i | (x - x')$ for each i. But then $M | (x - x')$, and since $|x - x'| < M$, we have $x = x'$. $\square$

It is interesting to note that, if the moduli m_i, in Theorem 2.3.24, are not relatively prime, then a solution exists if and only if $(m_i, m_j) | (a_i - a_j)$, for all i and j. If a solution exists, it is unique modulo $[m_1, m_2, \ldots, m_k]$, the least common multiple of the m_i values. See also Exercise 10 for this section.

Example. Let us solve the system of congruences

$$x \equiv 1 \pmod{2}, \quad x \equiv 2 \pmod{5}, \quad x \equiv 5 \pmod{7}$$

Following the procedure outlined in the proof of Theorem 2.3.24, we see that the first congruence is satisfied by $x = 1 + 2q$. To compute q, we replace x in the second congruence yielding $1 + 2q \equiv 2$ (mod 5), or $2q \equiv (2 - 1)$ (mod 5). We then compute the multiplicative inverse of 2 (mod 5), which is 3, and, therefore, $q \equiv 3$ (mod 5), or $q = 3 + 5r$, for some r. Hence, the solution x to the first two congruences is $x = 1 + 2(3 + 5r) = 7 + 2 \cdot 5r$, or $x \equiv 7$ (mod $2 \cdot 5$).

We now have to solve the system of the two congruences $x \equiv 7$ (mod $2 \cdot 5$), and $x \equiv 5$ (mod 7). As before, we have $x = 7 + 2 \cdot 5q \equiv 5$ (mod 7), or $2 \cdot 5q \equiv (5 - 7) = -2 \equiv 5$ (mod 7). The multiplicative inverse of 10 modulo 7 is the same as that of 3 modulo 7, that is, 5. We then have $q \equiv 5 \cdot 5$ (mod 7) $\equiv 4$ (mod 7), or $q = 4 + 7r$, for some r. Hence, the solution x to the three congruences is $x = 7 + 2 \cdot 5(4 + 7r)$, or $x \equiv 47$ (mod $2 \cdot 5 \cdot 7$).

Note that $47 = 1 + 3 \cdot (2) + 4 \cdot (2 \cdot 5)$, where the coefficients 3 and 4 are the values of q.

Before we describe the Greek–Chinese algorithm, which is based on Theorem 2.3.14, we present the following two related propositions.

Theorem 2.3.25. Let $m = p_1^{e_1} p_2^{e_2} \cdots p_k^{e_k}$. Then the function that associates with each x in $\mathbf{Z}_m$ the k-tuple $(x_1, x_2, \ldots, x_k)$, where $x \equiv x_i$ (mod $p_i^{e_i}$), $i = 1, 2, \ldots, k$, is a ring isomorphism (i.e., a homomorphism that is one-to-one and onto) of Z_m with the ring of k-tuples $(x_1, x_2, \ldots, x_k)$, and $x_i \in Z_{p_i^{e_i}}$ for $i = 1, 2, \ldots, k$. Moreover, if we denote by $\otimes$ either one of the binary operators $+$ or $\cdot$, we have $(x_1, x_2, \ldots, x_k) \otimes (y_1, y_2, \ldots, y_k) = (x_1 \otimes y_1, x_2 \otimes y_2, \ldots, x_k \otimes y_k)$, where the ith sign "$\otimes$" on the right side of this equality is the operation sign of $Z_{p_i^{e_i}}$, $i = 1, 2, \ldots, k$.

Proof. The proof follows immediately from Theorem 2.3.24 and is left as an exercise for the reader. □

The factorization described in Theorem 2.3.25 is written as $Z_m \cong \times_{1 \leq i \leq k} Z_{p_i^{e_i}}$. This factorization of the ring Z_m induces a factorization of the group of units $U_m \cong \times_{1 \leq i \leq k} U_{p_i^{e_i}}$.

One application of Theorem 2.3.25 (and of the Greek–Chinese remainder theorem in general) that will be examined in detail in Section 2.4 is the *Greek–Chinese representation* of a number x. That is, any positive integer x, $0 < x < M$, where $M = m_1 m_2 \cdots m_k$, with $(m_i, m_j) = 1$ for $i \neq j$, is uniquely represented by its least nonnegative residues modulo the m_i, and additions and multiplications are performed column by column.

Example. For $m_1 = 3$ and $m_2 = 5$, we have $6 = (0, 1)$ and $7 = (1, 2)$; then their sum is $[0 + 1 \pmod{3}, 1 + 2 \pmod{5}] = (1, 3)$, which turns out to be 13. What can you say about their product?

Note that given the pair $(1, 3)$ we can find the corresponding integer by using either the Greek–Chinese remainder algorithm described below or look-up tables.

The next theorem is a generalization of Fermat's theorem. Note that a natural number is *squarefree* if it is the product of distinct primes; that is, it is not divisible by any square >1; for example, the numbers $1, 2, 3, 5, 6, \ldots$ are all squarefree.

Theorem 2.3.26. A natural number m is squarefree if and only if there is some $q > 1$ (defined below) such that for all integers b, $b^q \equiv b \pmod{m}$. That is, if $m = p_1 p_2 \cdots p_k$, for distinct primes, and $q = \lambda[p_1 - 1, p_2 - 1, \ldots, p_k - 1] + 1$, for any natural number λ, then $b^q \equiv b \pmod{m}$ for all b.

Proof. Suppose that m is squarefree, and $m = p_1 p_2 \cdots p_k$, for distinct primes. Then, by the uniqueness part of the Greek–Chinese remainder theorem, we have that for any b in Z and any $q > 1$, $b^q \equiv b \pmod{m}$ iff $b^q \equiv b \pmod{p_i}$, for each $i = 1, 2, \ldots, k$. Let $q = q_m + 1$, where $q_m = [p_1 - 1, p_2 - 1, \ldots, p_k - 1]$, or any other common multiple of $p_1 - 1, p_2 - 1, \ldots, p_k - 1$. This means that $q_m = (p_i - 1)q_i$, for each i and for some q_i, and so $b^q = b^{q_m+1} = b^{(p_i-1)q_i+1} = b \cdot b^{(p_i-1)q_i}$. If $(b, p_i) = 1$, then by Fermat's theorem $b^{p_i-1} \equiv 1 \pmod{p_i}$, whereas otherwise $b \equiv 0 \pmod{p_i}$. In either case we obtain $b^q \equiv b$ (mod p_i), for each $i = 1, 2, \ldots, k$, and hence, $b^q \equiv b$ (mod m).

Next suppose that m is not squarefree; that is, $m = p_1^{e_1} p_2^{e_2} \cdots p_k^{e_k}$, where now $e_j > 1$, $1 \le j \le k$. Then, using the Greek–Chinese remainder theorem, we find $x = b$, a solution to the system $\{x \equiv 0 \pmod{p_i}, i = 1, 2, \ldots, k, i \ne j, x \equiv p_j^{e_j-1} \pmod{p_j}\}$. Then, m does not divide b because $p_j^{e_j}$ does not divide b, but m does divide b^q for any $q > 1$. Therefore, $b^q \equiv b$ (mod m) does *not* hold for any $q > 1$. □

We finally present the Greek–Chinese remainder algorithm for solving a system of congruence equations, for the case when the moduli are pairwise relatively prime (then the solution always exists by Theorem 2.3.24). A more general algorithm for the case when the moduli are not pairwise relatively prime is intrinsically no more difficult, but for the applications we have in mind the moduli will be distinct odd primes.

The first case to consider is a system of two equations:

$$x \equiv a \pmod{m_1}$$
$$x \equiv b \pmod{m_2}, \qquad (m_1, m_2) = 1$$

This system is solved by the method outlined in the proof of Theorem 2.3.24. We also make use of the procedure **MODINV**, which computes the (least nonnegative) multiplicative invese of m_1 modulo m_2 (see Programming Exercise 2 for this section). Formulating our previous discussion into an algorithm, we have

GCRA2 (**G**reek–**C**hinese **R**emainder **A**lgorithm-2 equations)

Input: a, m_1, b, m_2 such that $x \equiv a \pmod{m_1}$ and $x \equiv b \pmod{m_2}$, where m_1, m_2 are single-precision integers such that $(m_1, m_2) = 1$, and m_1 and m_2 both >1.

Output: x, the unique, least nonnegative solution modulo $m_1 m_2$, of the system of congruences.

1. [No changes if $a > 0$.] $x := \mathbf{MOD}(a, m_1)$.
2. [Compute m_1^{-1}.] $m_1^{-1} := \mathbf{MODINV}(m_1, m_2)$.
3. [Compute q.] $q := \mathbf{MOD}[m_1^{-1} \cdot (b - x), m_2]$.
4. [Exit.] Return $x := x + m_1 q$ (since $0 \le q < m_2$, the value of x returned satisfies $0 \le x < m_1 m_2$).

Computing-Time Analysis of GCRA2. We observe that steps 1 and 2 are executed in time ~ 1 since we are dealing with single-precision integers (use the extended Euclidean algorithm to compute the multiplicative inverse). In step 3 a multiplication and a division take place, whereas in step 4 one multiplication is performed. The time to execute each one of these operations is dominated by the time it takes to compute the product $m_1 m_2$ (justify this) and, hence

$$t_{\text{GCRA2}}(a, m_1, b, m_2) = 0[L(m_1)L(m_2)]$$

We next consider the general case, where we have to solve the following system of k congruence equations with pairwise relatively prime moduli:

$$x \equiv a_1 \pmod{m_1}$$
$$x \equiv a_2 \pmod{m_2}$$
$$\cdots$$
$$x \equiv a_k \pmod{m_k}, \quad (m_i, m_j) = 1 \text{ for } i \neq j$$

The idea is to use **GCRA2** to solve successively pairs of congruence equations. The first pass returns the solution x_0 to the first two congruences, where x_0 is the least nonnegative solution modulo $m_1 m_2$. The next pass returns the solution, modulo $m_1 m_2 m_3$, to the pair

$$x \equiv x_0 \pmod{m_1 m_2}, \quad x \equiv a_3 \pmod{m_3}$$

and the process continues in this manner. We have the following algorithm.

GCRAk (Greek–Chinese Remainder Algorithm-k congruences)

Input: The pairs a_i, and m_i, such that $x \equiv a_i \pmod{m_i}$, $i = 1, 2, \ldots, k$; each m_i is a single-precision integer, $m_i > 1$ and $(m_i, m_j) = 1$ for $i \neq j$.

Output: x, the unique, least nonnegative solution modulo $m_1 m_2 \cdots m_k$ of the system of k congruences.

1. [Initialize.] $m := 1; x := \text{MOD}(a_1, m_1)$.
2. [Main loop applying **GCRA2**.] For $i = 1$ until $k - 1$, do $\{m := m \cdot m_i;$ $m^{-1} := \text{MODINV}(m, m_{i+1}); q := \text{MOD}[m^{-1}(a_{i+1} - x), m_{i+1}]; x := x + mq\}$.
3. [Exit.] Return x.

Computing-Time Analysis of GCRAk. The execution time of **GCRAk** is clearly dominated by the execution time of step 2, which is basically **GCRA2**. If $M = m_1 \cdots m_k$, then the ith execution of step 2 takes time $\sim L(m_1 \cdot m_2 \cdots m_i)L(m_{i+1}) \leq L(M)L(m_{i+1})$. Therefore, the $(k-1)$ executions of step 2 are performed in time $\leq L(M)\{\Sigma_{2 \leq i \leq k} L(m_i)\} \sim L(M) \cdot L(\Pi_{2 \leq i \leq k} m_i)$ (recall that L behaves just as the log function) $\leq L^2(M)$, and we have

$$t_{\text{GCRAk}}(a_i, m_i, i = 1, 2, \ldots, k) = 0[L^2(m_1 \cdot m_2 \cdots m_k)]$$

An example of this algorithm was presented immediately below Theorem 2.3.24. The reader should try this procedure on the system $x \equiv -2 \pmod{15}$, $x \equiv 6 \pmod 8$, and $x \equiv 11 \pmod 7$, where $x = 718$ is the unique solution modulo 840; again, it is worth noting that the answer is formed as follows: $718 = 13 + 7 \cdot (15) + 5 \cdot (15 \cdot 8)$. In general, if we have k congruence equations, the answer is formed as follows

$$x = q_1 + q_2 \cdot (m_1) + q_3 \cdot (m_1 \cdot m_2) + \cdots + q_k \cdot (m_1 \cdot m_2 \cdots m_{k-1})$$

where each q_i is in the range of the corresponding modulo m_i and $q_1 = \text{MOD}(a_1, m_1)$.

We close this section with an application of the Greek–Chinese remainder algorithm to *key safeguarding* (Asmuth, C., and J. Bloom: A modular approach to key safeguarding. Mathematics Department, Texas A&M University, College Station, TX, 77844).

Let K be the key to be safeguarded. We require L persons [out of a total of k, $(k > L)$ who have received "information" about the key] to have to come together in order to reconstruct the key; moreover, this task should be impossible for any group of $L - 1$ or fewer persons. The "information" passed to the various persons is described below.

To solve this problem, choose a set of integers $\{p, d_1, d_2, \ldots, d_k\}$ such that:

a. $p > K$.
b. $d_1 < d_2 < \cdots < d_k$.
c. $\gcd(p, d_i) = 1$, for all $i = 1, 2, \ldots, k$.
d. $\gcd(d_i, d_j) = 1$, for $i \neq j$.
e. $d_1 \cdot d_2 \cdots d_L > p \cdot d_{k-L+2} \cdot d_{k-L+3} \cdots d_k$.

Requirement e implies that the product of the L smallest d_i is larger than the product of p and the $L - 1$ largest d_i. Let $D = d_1 \cdot d_2 \cdots d_L$; then D/p is larger than the product of any $L - 1$ of the d_i. Now pick a random number r in the range $[0, (D/p) - 1]$ and compute $K' = K + r \cdot p$ to put K in the range $[0, D - 1]$. Then part of the information that is distributed to the various persons are the images

$$K_i \equiv K' \pmod{d_i}, i = 1, 2, \ldots, k$$

Example. $K = 5$, $L = 2$, $k = 3$, $p = 7$, $d_1 = 11$, $d_2 = 13$, $d_3 = 17$. Then, $D = d_1 \cdot d_2 = 11 \cdot 13 = 143 > 119 = 7 \cdot 17 = p \cdot d_3$ as required. Pick a random number r in the interval $[0, (143/7) - 1] = [0, 19]$, say, $r = 2$; we then have

$$K' = K + r \cdot p = 5 + 2 \cdot 7 = 19$$

Then the distributed images are

$$K_1 = 19 \pmod{11} = 8, \qquad K_2 = 19 \pmod{13} = 6, \qquad K_3 = 19 \pmod{17} = 2$$

Given any two images, we can compute K. For example, using K_1 and K_3, we have

$$K' \equiv 8 \pmod{11}, \qquad K' \equiv 2 \pmod{17}$$

and applying the Greek–Chinese remainder algorithm, we obtain $K' = 19$, from which we obtain $K = K' - r \cdot p = 5$.

2.3.3 Primality Testing

Basically the problem in this section is just this: *Given a positive integer m, is it prime?* Note that no factors of m are asked for. The most obvious approach to this problem is the following.

First Deterministic Test. Divide the number m by the integers in sequence: $2, 3, 4, \ldots, \lfloor \sqrt{m} \rfloor$. If any of the divisions comes out even (i.e., leaves no remainder), the test number m is composite and the divisor and the quotient are factors of m; otherwise, m is a prime. (Why do we need test only up to $\lfloor \sqrt{m} \rfloor$?)

What is the computing time of the above test? Obviously, we need to perform $\sqrt{m}$ divisions, and so the time of test m for primality is $0(\sqrt{m})$. However, this computing bound is not polynomial because, if we take into consideration the length of m, it becomes $0(2^{L(m)/2})$; that is, it is an exponential test and, hence, very slow. So this problem, which seemed very easy at first, turns out to be extremely difficult. Clearly, the problem with this first test (also known as "trial division" method) is that it does far more than is required; that is, it not only decides whether a number is prime or composite, but also gives factors of any composite number.

Remarkable progress has been made in recent years. There exist two groups of algorithms for primality testing: deterministic and *probabilistic*. The literature on the subject is quite extensive, (Adleman et al., 1983; Cohen et al., 1982; Dixon, 1984; Lucas, 1961; Pomerance, 1981; Pratt, 1975; Rabin, 1980; Solovay et al., 1977; Williams, 1978) and no attempt is made to present all the various algorithms. Instead, we present several primality tests that are representative of the two groups; invariably, almost all the tests trace their lineage to Fermat's theorem.

We continue with deterministic primality testing.

Second Deterministic Test. A number m is prime if and only if $m | \{(m-1)! + 1\}$.

This test is based on Theorem 2.3.16 (Wilson's). As we have already remarked, the factorial $(m-1)!$ destroys any attraction for this test; the slowest sieve method will be extremely fast compared to checking the divisibility of $(m-1)! + 1$ for large m. If m has 100 digits, then $(m-1)!$ has approximately 100^{102} *digits*—not to be confused with the number 100^{102}.

For the third test below we need the following: if m is prime, then there exists some natural number $b < m$ whose order modulo m is $m-1$; that is, $b^{m-1} \equiv 1 \pmod{m}$ and no smaller positive power of b is congruent to 1 mod m. The converse of the above result is as follows.

Theorem 2.3.27. Let m be an integer ≥ 2. If there is some $b < m$ such that the order of b mod m is exactly $m-1$, then m is a prime number.

Proof. If m is not prime, then $\phi(m) < m - 1$. For a given natural number $b < m$ either $(b, m) = 1$, in which case its order must divide $\phi(m) < m - 1$, or $(b, m) > 1$, which implies that no power of b is congruent to 1 mod m.
□

We therefore have the following.

Third Deterministic Test. A number m is prime if and only if there is an element b of order exactly $m-1$ modulo m. Or, equivalently, m is prime if and only if there exists b, $(b, m) = 1$, such that $b^{m-1} \equiv 1 \pmod{m}$ and $b^{(m-1)/p}$ is not congruent to 1 (mod m) for *each* prime factor p of $(m-1)$.

This test is also known as the "Lucas test" (1961) and to decide whether m is a prime number we do the following. If b^{m-1} is not congruent to 1 (mod m) for some $b < m$, then m is not prime (*negative* test by Fermat's theorem). On the other hand, if for some b the order of b is $m-1$, then m is prime (*positive* test by Theorem 2.3.27).

Unfortunately, it is very hard to prove that the order b is $m-1$ modulo m, because we have to show that for *each* prime p dividing $m-1$, $b^{(m-1)/p}$ is not congruent to 1 (mod m).

Here are two small examples. First let $m = 899$. After checking that 899 is not divisible by $2, 3, 5, 7$, and 11, we suspect that 899 might be prime. However, to be completely sure we apply the test; so, we take $b = 2$ and see that $2^{898} \equiv 683 \pmod{899}$. Therefore, 899 is not a prime; indeed, it is $29 \cdot 31$. Here we make use only of the negative test. Consider now the number 341 (old aquaintance), for which we know that $2^{340} \equiv 1 \pmod{341}$. To apply the positive test we must find the order of 2 modulo 341, which means that we must factor 340. It turns out that $340 = 2^2 \cdot 5 \cdot 17$, and $2^{340/17} \equiv 1 \pmod{341}$. Therefore, 341 is composite.

If we took, instead of 341, some 40-digit number m, then to use the positive test we would have to factor $m-1$, and chances are good that after factoring out some small primes we would be left with a factor of $m-1$ containing 35 or more digits, which we would have to factor—almost as difficult a problem as factoring m itself. Therefore, the Lucas test is not generally feasible unless $m-1$ factors easily.

Fourth Deterministic Test. This test was originally developed in 1980 by Adleman, Pomerance and Rumely (Adleman et al., 1983), and was later improved by Cohen and Lenstra (1982). Its details require a technical understanding of algebraic number theory, but in its essence it is quite similar to Fermat's test.

It has been shown that the computing time of this test is

$$O(L(m)^{L\{L[L(m)]\}})$$

Despite the fact that we are faced with an exponential test, the expression $L\{L[L(m)]\}$ drifts toward infinity at an extremely slow rate; for instance, the first number for which $L\{L[L(m)]\}=2$ is the number $10^{999,999,999}$. Therefore, all numbers less than $10^{999,999,999}$ can be tested for primality in polynomial time; hence, the computing time of the above test is tantalizingly close to polynomial time, and is the best that can be done today. It is not known if a polynomial-time deterministic algorithm for primality testing exists.

The above tests are all deterministic; that is, given a number m, we can find with certainty whether it is prime or composite. If we now replace "certainly" by "probably," then it is possible to construct probabilistic primality tests that work in polynomial time; these tests are also called *pseudoprimality* tests. Below we examine several of the pseudoprimality tests.

More precisely, by a pseudoprimality test we mean a test that is applied to a pair of integers (b, m) and that has the folloing properties:

1. The possible outcomes of the test are: "m is composite" or "inconclusive."
2. If the outcome is "m is composite," then m is composite.
3. The test runs in a time that is polynomial in $L(m)$.

In a *good* pseudoprimality test there is a fixed positive number k such that every composite integer m is declared to be composite for at least km choices of the base b, where $1 \leq b \leq m$. Moreover, we will say that an integer m is *very probably prime* if we have subjected it to a good pseudoprimality test and have found it to be pseudoprime to all the bases.

First Probabilistic Test. Given m, randomly choose b, $1 < b < m$. If $b | m$, then the outcome is "m is composite"; otherwise, it is "inconclusive."

The probability that the outcome is "m is composite" equals the probability that we happen to have found a number b, $b \neq 1, m$, such that $b | m$. If

$d(m)$ is the number of divisors of m, and if b is randomly chosen from $1 < b < m$, the probability of this event is $p = [d(m) - 2]/m$. Clearly, this is a poor test.

Second Probabilistic Test. Given m, randomly choose b, $1 < b < m$. If $(b, m) \neq 1$, then the outcome is "m is composite"; otherwise, it is "inconclusive."

Obviously, if m is composite, the number of such integers $b < m$ for which the outcome of this test will be "m is composite" is $m - \phi(m)$. This number will be large if m has some small prime factors. However, if $m = pq$, for large prime numbers p and q, then the proportion of useful bases is very small and, therefore, this test is as poor as the previous one.

Third Probabilistic Test. If, for given b and m, b^{m-1} is not congruent to 1 (mod m), then the outcome is "m is composite"; otherwise, it is "inconclusive."

This test is much better than the previous two, but it still leaves a lot to be desired because all pseudoprimes to the base b will produce the outcome "inconclusive." As you recall, 341 is a pseudoprime to the base 2. The following theorem tells us that there are infinitely many pseudoprimes to the base 2.

Theorem 2.3.28. If m is a pseudoprime to the base 2, then so is $n = 2^m - 1$.

Proof. Let $m = ab$, $a > 1$, $b > 1$, be a pseudoprime to the base 2; that is, $2^{m-1} \equiv 1 \pmod{m}$. We prove that $n = 2^m - 1$ is pseudoprime to the base 2 using the identity

$$2^{ab} - 1 = (2^a - 1)(1 + 2^a + \cdots + 2^{a(b-1)})$$

More precisely, to prove $2^{n-1} \equiv 1 \pmod{n}$, we show that $n = 2^m - 1$ divides $2^{n-1} - 1$. According to this identity, we need only prove that $m|(n-1)$. Now since m is a pseudoprime to the base 2, it follows that $m|(2^{m-1} - 1)$; moreover, $n - 1 = 2^m - 1 - 1 = 2(2^{m-1} - 1)$, which yields $m|(n-1)$. $\square$

As we have already seen, besides pseudoprimes there also exist some composites numbers m called *absolute pseudoprimes* or *Carmichael numbers*, defined by $m|(b^{m-1} - 1)$ for all b such that $(b, m) = 1$.

How can it be possible for $b^{m-1} - 1$ to be congruent to 1 modulo m for all b, $(b, m) = 1$? After all, for $m = 561$ there are $\phi(561) = 320$ such values of b. The answer turns out to be quite simple. All we need are three or more odd primes p_i such that $(p_i - 1)|(m - 1) = (\Pi p_i - 1)$ for each prime p_i; verify this for $561 = 3 \cdot 11 \cdot 17$. This is sufficient because, at first, according to Fermat's theorem $b^{p_i - 1} \equiv 1 \pmod{p_i}$, for all b such that $\gcd(b, p_i) = 1$. These form a system of congruence equations and, writing them as $x \equiv$

$b^{p_i-1} - 1$ (mod p_i), we see that they have a unique solution mod $m = \Pi p_i$ (Greek–Chinese remainder theorem). Of course, this unique solution is $x = 0$, and we have $0 \equiv b^{p_i-1} - 1$ (mod m), or $b^{p_i-1} \equiv$ (mod m). Since $p_i - 1$ divides $m - 1$ we have $m - 1 = k_i(p_i - 1)$ for some k_i, and $b^{m-1} \equiv 1$ (mod m). This last congruence holds for all b such that $(b, p_i) = 1$, which implies for all b such that $(b, m) = 1$. That is, these numbers are pseudoprimes to every base b, and they will also produce the outcome "inconclusive" in the third probabilistic test. The smallest such number is $561 = 3 \cdot 11 \cdot 17$; another Carmichael number is $1729 = 7 \cdot 13 \cdot 19$. The following test takes care of such numbers.

Fourth Probabilistic Test. This is the strong pseudoprimality test. Given b and m, let $m - 1 = t2^s$, t odd integer, and consider $x_r \equiv b^{t2^r}$ (mod m), for $0 \le r < s$ (x_r being the least absolute value remainder modulo m). If either $x_0 = 1$ or there is an integer i, $0 \le i < s$, such that $x_i = -1$, then m is called a *strong pseudoprime* to the base b, and the outcome is "inconclusive"; otherwise, it is "m is composite."

This test can unmask pseudoprimes even such as $m = 561$. For example, $560 = 35 \cdot 2^4$; then for $r = 4$, and 3 we have $2^{35 \cdot 2^4} = 2^{560} \equiv 1$ (mod 561), and $2^{35 \cdot 2^3} = 2^{280} \equiv 1$ (mod 561), respectively, whereas, for $r = 2$, we have $2^{35 \cdot 2^2} = 2^{140} \equiv 67$ (mod 561), and, hence, 561 is composite. We now have the following.

Theorem 2.3.29. If the outcome of the strong pseudoprimality test is "m is composite," then m is composite.

Proof. The proof is obtained by contradiction. Suppose that m is an odd prime. Then we will prove by induction that $b^{t2^r} \equiv 1$ (mod m) for all r, $0 \le r \le s$, and this will contradict the outcome of the test. Obviously, our claim is true for $r = s$, by Fermat's theorem. Assuming that it is true for i, then it is easily seen that it is true for $i - 1$, because $(b^{t2^{i-1}})^2 = b^{t2^i} \equiv 1$ (mod m) implies that the quantity being squared is ± 1. By assumption m is an odd prime and hence, by Corollary 2.3.21, U_m is cyclic and the equation $x^2 = 1$ in U_m has only the solutions $x = \pm 1$. But -1 is ruled out by the outcome of the test, and the proof is completed. □

Computing Time Analysis of the Fourth Probabilistic Test. There are at most $\log m \sim L(m)$ possible values of r to check, and for each one of them we do one multiplication of two integers and raise b to a certain power modulo m. The length of each of the two integers to be multiplied is $\sim L(m)$, and therefore, their product is computed in time $0[L^2(m)]$; moreover, this time dominates the time it takes to raise b to the corresponding power. Therefore, the overall time of the test is $0[L^3(m)]$.

It has been proved that the fourth probabilistic test has the following property: if m is composite, the chance that the outcome of the test will be "m is composite" is at least $\frac{1}{2}$. The basic idea of the proof is that a subgroup of a group that is not the entire group can consist of at most half of the elements of that group; for details see Wilf (1986). Moreover, it has been shown by Rabin that no odd composite integer m is a strong pseudoprime to more that $\frac{1}{4}$ of the bases that are less than m. In practice, for a given m we apply the test 100 times using 100 different bases b_i independently chosen at random, $0 \leq b_i \leq m$. If m is composite, the probability that it will be so declared is at least $1 - 2^{-100}$, and each test is done in polynomial time.

2.3.4 Factorization of Large Integers

The problem of finding the factors of large integers is in a much more primitive condition than is primality testing. We do not even know of a polynomial-time probabilistic algorithm that will return a factor of a large composite integer, with probability $> \frac{1}{2}$. Again the literature on the subject is quite extensive (Dixon, 1981 and 1984; Guy, 1976; Knuth, 1969; Lehman, 1974; Morrison et al., 1975; Williams, 1982 and 1984; Wunderlich, 1979); we present the following (simple variant of the) general purpose factoring method that is still the most powerful.

The method is based on Legendre's idea (1798) that if $u^2 \equiv v^2 \pmod{m}$, $0 < u, v < m, u \neq v$, and $u + v \neq m$, then m is a divisor of $(u - v)(u + v)$, but m divides neither $(u - v)$ nor $(u + v)$; therefore, $(u - v, m)$, the greatest common divisor of $u - v$ and m, is a nontrivial factor of m and can be easily computed using Euclid's algorithm. The search for u and v is carried out in two steps, as described below.

Let m be the number whose factorization is desired. Set $n = [\sqrt{m}]$, the largest integer less than $\sqrt{m}$, and compute $a_k = (n + k)^2 - m$, for small k. [Instead of using a_k, the quantity Q_k was used by Morrison and Brillhart (1975); if $\sqrt{(cm)} = (b_0; b_1, b_2, \ldots, b_{k-1}, \xi_k)$ is the continued fraction expansion of $\sqrt{(cm)}$, c a small multiplier, than Q_k is defined by $\xi_k = [P_k + \sqrt{(cm)}]/Q_k$.]

Let $\{q_i, i = 1, 2, \ldots, j\}$ be a set of small primes such that each q_i can divide an expression of the form $x^2 - m$; this set is also known as a *factor base B*. Keep all those a_k that can be written as

$$a_k = (-1)\omega_{k_0} \Pi_{1 \leq i \leq j} q_i^{\omega_{k_i}}$$

these a_k are also called the *B-numbers*. With each a_k there is associated an exponent vector

$$\mathbf{e_k} = (w_{k_0}, w_{k_1}, w_{k_2}, \ldots, w_{k_j}), \quad w_{k_i} \equiv \omega_{k_i} \pmod{2}, \quad i = 0, 1, 2, \ldots, j$$

If we find enough B-numbers so that the resulting set of exponent vectors is a linearly dependent set modulo 2 (any set of $j + 2$ B-numbers would certainly have that property), then we can represent the zero vector as a sum of a certain set S of exponent vectors, say, $\Sigma_{s \in S} \mathbf{e}_s \equiv (0, 0, \ldots, 0)$ (mod 2). Now, define the integers

$$e'_i = (\tfrac{1}{2}) \Sigma_{k\,:\,a_k \in S} w_{k_i}, \qquad i = 0, 1, 2, \ldots, j$$

$$u = \Pi_{a \in S} a \pmod{m}, \qquad v = \Pi_{1 \leq i \leq j} q_i^{e'_i}$$

It then follows that $u^2 \equiv v^2 \pmod{m}$, and $(u - v, m)$ may be a nontrivial factor of m.

Example. Let us factor 1729, the second Carmichael number. In this case $m = 1729$, $n = 41$, and we then compute $a_k = (n + k)^2 - m$, for small k. We have $a_1 = 35$, $a_2 = 120$, $a_3 = 207$, $a_4 = 296$, $a_5 = 387$, $a_6 = 480$, $a_7 = 575$, $a_8 = 672$, $a_9 = 771$, and so on. We next pick the set of small primes $\{2, 3, 5, 7\}$ and we see that only 35, 120, 480, and 672 are the B-numbers, namely, $35 = (-1)^0 \cdot 2^0 \cdot 3^0 \cdot 5^1 \cdot 7^1$, $120 = (-1)^0 \cdot 2^3 \cdot 3^1 \cdot 5^1 \cdot 7^0$, $480 = (-1)^0 \cdot 2^5 \cdot 3^1 \cdot 5^1 \cdot 7^0$, and $672 = (-1)^0 \cdot 2^5 \cdot 3^1 \cdot 5^0 \cdot 7^1$. Except for 480, we include these B-numbers in the set S; observe that, except for 480, the modulo 2 sum of the exponent vectors for the numbers in the set S are $(0, 0, 0, 0, 0)$. We next compute $e'_0 = 0$, $e'_1 = 8/2 = 4$, $e'_2 = 2/2 = 1$, $e'_3 = 2/2 = 1$, $e'_4 = 2/2 = 1$, and the quantities $u = 35 \cdot 120 \cdot 672 = 2822400 \equiv 672 \pmod{1729}$ and $v = (-1)^0 \cdot 2^4 \cdot 3^1 \cdot 5^1 \cdot 7^1 = 1680$. Then $(1680 - 672, 1729) = 7$, and 7 is a factor of 1729.

It has been proved by Dixon (1981) that we can expect to factor m using the above method in time

$$O[\exp(2L\{L[L(m)]\})].$$

We mention in passing that numbers of the form $F_m = 2^{2^m} + 1$, and $M_m = 2^m - 1$, called *Fermat* and *Mersenne numbers*, respectively, have had a profound effect on the development of factoring and primality testing techniques. As early as 1640 Fermat conjectured that every F_m, $m \geq 1$, is a prime. However, this conjecture was shown to be false in 1729 when Euler found that 641 is a divisor of F_5.

2.4
Exact Computations using Modular Arithmetic

Based on the principles of number theory that we have seen in the previous sections, we are now going to discuss an interesting way for doing exact arithmetic on (large) integers. The idea is to use a single modulus, or several moduli (Theorem 2.3.25) that contain no common factors, and to work indirectly with residues instead of directly with the integers themselves (Gregory, 1980; Knuth, 1969; Scott, 1985).

We first examine the case of a single modulus. Suppose that we are given an expression over $\mathbf{Z}$, $e(i_1, i_2, \ldots, i_h)$ for integer arguments $i_1, i_2, \ldots, i_h$, and that we are asked to evaluate it. The straightforward approach is to simply evaluate e over $\mathbf{Z}$; this, however, could result in intermediate results not being integers (having infinite expansions as for example $\frac{1}{3} = 0.333\ldots$), in which case truncating or rounding them could lead to inaccuracies in the final result. In order to avoid these problems, we have to use exact arithmetic operations, and to do so we must take the roundabout route in evaluating e. This roundabout approach is illustrated in Figure 2.4.1.

That is, instead of evaluating e over $\mathbf{Z}$, we first obtain from $e(i_1, i_2, \ldots, i_h)$ the equivalent expression $e_m(i'_1, i'_2, \ldots, i'_h)$ over $\mathbf{Z}_m$, for some m, where $i'_j \equiv i_j \pmod{m}$, or equivalently, $i'_j = r_m(i_j)$, $j = 1, 2, \ldots, h$. Provided e_m is defined over $\mathbf{Z}_m$, we then evaluate it over $\mathbf{Z}_m$ to obtain the equivalent result res_m, where $\text{res}_m \equiv \text{res} \pmod{m} = r_m(\text{res})$; finally, we map res_m back to the integers.

It should be noted that the congruence equation $\text{res} \equiv \text{res}_m \pmod{m}$ does not *uniquely* determine the final result res. [This is analogous to the problem "what is the value of $x \equiv 7 \pmod{13}$". Obviously, x is not uniquely determined since it could be 7, or 20, and so on. However, if we somehow knew that $x < 13$, then x could only be 7.] Likewise, in order to uniquely determine res, we need to have an a priori bound on it; this bound is then chosen as the modulus m, and all operations are performed in $\mathbf{Z}_m$. If the available bound is on res, then we solve $\text{res} \equiv \text{res}_m \pmod{m}$ for the least nonnegative solution, whereas if the bound is on $|\text{res}|$, then we solve for the least absolute value solution.

The problem with the roundabout method mentioned above is when division is involved. However, as we recall, when p is prime, $(\mathbf{Z}_p, +, \cdot)$ is a finite field [also denoted as $GF(p)$] where we can perform all the arithmetic operations; since the inverse of every nonzero element exists, we define division modulo p as follows:

$$\frac{a}{b} \pmod{p} = a \cdot [b^{-1} \pmod{p}] \pmod{p}$$

where, as we mentioned before, $b^{-1} \pmod{p}$ is the multiplicative inverse of b modulo p and will also be denoted as simply b^{-1}. The quotient of two integers in $GF(p)$, when it exists, is always an integer, even in those cases where b does not divide a in $\mathbf{Z}$.

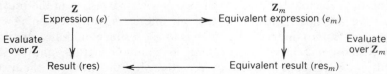

Figure 2.4.1. The roundabout route taken to evaluate the expression (e); a single modulus is used.

Example

$$3/4 \pmod{11} = 3 \cdot 4^{-1} \pmod{11}$$
$$= 3 \cdot 3 \pmod{11}$$
$$= 9$$

Seen as an intermediate result, the integer we obtained in this example makes sense because

$$(3/4) \cdot 4 \pmod{11} = [3/4 \pmod{11}] \cdot 4 \pmod{11}$$
$$= 9 \cdot 4 \pmod{11} = 3$$

Assume now that the modulus p bounds the final result res. If res is not an integer in $GF(p)$, then $res_p \neq res$, and to obtain the latter, additional a priori information is needed; this a priori information differs from case to case and is illustrated in the two examples below. If, however, res is in $GF(p)$, then $res_p = res$. [Recall that res is the result obtained on evaluating the expression e over the integers, whereas res_p is the result obtained on evaluating the expression e_p in $GF(p)$.] Therefore, single-modulus arithmetic can be used in carrying out a sequence of exact arithmetic operations on integers in $GF(p)$, even though the sequence involves one or more divisions; the major difficulty could be in interpreting the results.

When we want to evaluate an expression where the final result could be either positive or negative, we have to be able to handle negative numbers. This can be done by using the symmetric set $\mathbf{Z}_p = \{-(p-1)/2, \ldots, -2, -1, 0, 1, 2, \ldots, (p-1)/2\}$, which, as can be easily shown, is isomorphic to $\{0, 1, 2, \ldots, p-1\}$; the mapping between the two sets for the case $p = 11$ is shown in Figure 2.4.2.

For example, $x \equiv 17 \pmod{11}$ has the least nonnegative solution $x = 6$ and the least absolute value solution $x = -5$.

Of course, exact arithmetic operations can be performed with either set, but it is easier to map our data (especially if they contain negative integers) on the nonnegative set, do all our operations there, and then map the results back on the symmetric set.

Example. Let us use exact arithmetic operations in $GF(11)$ to evaluate $x = 1/3 - 4/3$; here the a priori informaiton is the fact that we are looking for the final result res in the symmetric set.

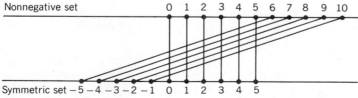

Figure 2.4.2. Mapping between the symmetric and the nonnegative sets for $p = 11$.

$$x \pmod{11} = (1/3 + (-4)/3) \pmod{11}$$
$$= (1/3 + 7/3) \pmod{11} \quad \text{(in the nonnegative set)}$$
$$= (1 \cdot 3^{-1} + 7 \cdot 3^{-1}) \pmod{11} \quad (3^{-1} \text{ is the multiplicative inverse of 3 modulo 11})$$
$$= (1 \cdot 4 + 7 \cdot 4) \pmod{11}$$
$$= 32 \pmod{11}$$

and mapping this result back into the symmetric set we obtain the correct answer -1.

In this above example $\text{res}_p = \text{res}$ because we knew from the beginning that res is in the symmetric set of $GF(11)$. However, this is not the case in our next example.

Example. Let us evaluate $x = 1/2 - 2/3$ using the same a priori information as in the above example. Proceeding as in the previous example, we have

$$x \pmod{11} = (1/2 + (-2)/3) \pmod{11}$$
$$= (1/2 + 9/3) \pmod{11}$$
$$= (1 \cdot 2^{-1} + 9 \cdot 3^{-1}) \pmod{11}$$
$$= (1 \cdot 6 + 9 \cdot 4) \pmod{11}$$
$$= 42 \pmod{11}$$
$$= 9$$

If at this point we map the result back into the symmetric set, we obtain the wrong answer $x = -2$. So we have to have additional information; namely, we have to know that we are looking for a rational number $x \pmod{11} = (a/b) \pmod{11}$, where $b = 6$, the least common multiple of the denominators of the two fractions. Then, $a = [x \pmod{11}] \cdot b \pmod{11} = 9 \cdot 6 \pmod{11} = 10 \pmod{11} = -1$ (mapped on the symmetric set). Therefore, we have $x = (-1)/6$, which is the correct answer.

We next turn to multiple-moduli arithmetic. Using multiple-moduli arithmetic is the "natural" solution to a problem encountered in single-modulus arithmetic; namely, we saw that a vital requirement for the single-modulus arithmetic to work is to pick the modulus m large enough so that we can uniquely determine res from its residue value res_m ($m > \text{res}$). However, if we are forced to choose m larger than the wordsize of our computer, then our scheme looses all its attractiveness. Therefore, we use multiple moduli, and take a roundabout route; this scheme is illustrated in Figure 2.4.3.

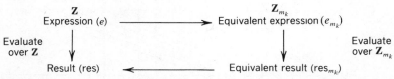

Figure 2.4.3. The roundabout route taken to evaluate the expression (e); n moduli are used, $(k = 1, 2, \ldots, n)$.

That is, given $e(i_1, i_2, \ldots, i_h)$ for integer arguments $i_1, i_2, \ldots, i_h$, we first compute $e_{m_k}(i_{1k}, i_{2k}, \ldots, i_{hk})$, where $i_{jk} = r_{m_k}(i_j)$, $j = 1, 2, \ldots, h$, and $k = 1, 2, \ldots, n$, for single-precision modui m_k. Provided e_{m_k} is defined over $\mathbf{Z}_{m_k}$, we then evaluate it over $\mathbf{Z}_{m_k}$ to obtain the equivalent result res_{m_k}, $k = 1, 2, \ldots, n$, and finally, using the Greek–Chinese remainder algorithm, or tables, we obtain the final result res. Here again the moduli m_k have to be appropriately chosen so that $m_1 \cdot m_2 \cdots m_n > \text{res}$. Moreover, if the bound is on res, then we solve the Greek–Chinese remainder problem for the least nonnegative solution, whereas if the bound is on $|\text{res}|$, we solve for the least absolute value solution.

Let us now closer examine multiple-moduli arithmetic. The number systems that we are all familiar with are linear, positional and weighted, meaning that all positions derive their weight from the same radix or base. In the decimal system for example, the weights are 10^0, 10^1, 10^2, 10^3, and so forth. Instead, the multiple-residue number system uses positional bases that are relatively prime to each other, for example, 3, 5, 7. Table 2.4.1 lists the numbers 0–29 and their residues to bases 3, 5, and 7.

The residues in Table 2.4.1 uniquely identify a number; there are $3 \cdot 5 \cdot 7 = 105$ different numbers that can be represented in this system. For example, the vector [2, 3, 1] uniquely identifies the decimal number 8 and is called the *standard residue digits of 8 with respect to the given base vector*. In our case the vector $\beta = [3, 5, 7]$ is the *base vector*. We will use the notation

$$8 \,(\text{mod } \beta) = [8 \,(\text{mod } 3), 8 \,(\text{mod } 5), 8 \,(\text{mod } 7)] = [2, 3, 1]$$

As in the case of the single modulus, here, too, we can define *the least nonnegative number system*, and provided all the moduli are odd, the *least absolute value* (or, *symmetric residue*) *number system* with respect to the base vector β.

Consider now the general base vector $\beta = [m_1, m_2, \ldots, m_n]$, $(m_i, m_j) = 1$ for $i \neq j$, and let $M = m_1 \cdot m_2 \cdots m_n$. (Note that since the moduli are

TABLE 2.4.1 The Residues, to the Bases 3, 5, 7, of the Numbers 0–29.

	Residue to Base				Residue to Base				Residue to Base		
N	3	5	7	N	3	5	7	N	3	5	7
0	0	0	0	10	1	0	3	20	2	0	6
1	1	1	1	11	2	1	4	21	0	1	0
2	2	2	2	12	0	2	5	22	1	2	1
3	0	3	3	13	1	3	6	23	2	3	2
4	1	4	4	14	2	4	0	24	0	4	3
5	2	0	5	15	0	0	1	25	1	0	4
6	0	1	6	16	1	1	2	26	2	1	5
7	1	2	0	17	2	2	3	27	0	2	6
8	2	3	1	18	0	3	4	28	1	3	0
9	0	4	2	19	1	4	5	29	2	4	1

pairwise relatively prime, M is their least common multiple.) We have the following important result.

Theorem 2.4.1. Two integers n_1 and n_2 have the same standard residue representation with respect to the base vector $\beta = [m_1, m_2, \ldots, m_n]$ iff $n_1 \equiv n_2 \pmod{m_1 \cdot m_2 \cdots m_n}$.

Proof. The proof is left as an exercise for the reader. □

Example. Consider $\beta = [3, 5, 7]$, as in Table 2.4.1, where $M = 105$. We have $9 \equiv 114 \pmod{105}$ and, hence, $9 \pmod{\beta} = [0, 4, 2] = 114 \pmod{\beta}$.

From Theorem 2.4.1 we conclude that the set $\mathbf{Z}_\beta = \{n \pmod{\beta} : n \in \mathbf{Z}\}$ contains M elements that are in one-to-one correspondence with the elements of $\mathbf{Z}_M$. It is easily shown that the two sets $\mathbf{Z}_\beta$ and $\mathbf{Z}_M$, along with the corresponding operations of addition and multiplication, constitute *isomorphic* finite commutative rings and, hence, *multiple-residue arithmetic is equivalent to arithmetic modulo M.*

The main advantage of the multiple-residue number system is the absence of carries in the operations of addition and multiplication. Arithmetic is closed (done completely) within each residue position. Therefore, it is possible to perform addition and multiplication on long integers as fast as on single-precision integers.

On binary computers it is advantageous to choose moduli of the form $m = 2^e - 1$; that is, each modulus is one less than a power of 2. We then have

$$u + v \pmod{2^e - 1} = \begin{cases} u + v, & \text{if } u + v < 2^e_1 \\ \{u + v \pmod{2^e - 1}\} + 1, & \text{if } u + v \geq 2^e \end{cases}$$

$$u \cdot v \pmod{2^e - 1} = (uv \bmod 2^e) + \lfloor uv/2^e \rfloor$$

In the last expression we have the sum of the upper half and the lower half of the product. We see, therefore, that it is relatively easy to work modulo $2^e - 1$, as in ones' complement arithmetic. (If the modulo m is not chosen as above, then for the addition we would have $u + v - m$, whenever $u + v \geq m$.)

There are cases when we need to know whether the moduli are relatively prime. If they are of the form $2^e - 1$, then the simple rule $(2^e - 1, 2^f - 1) = 2^{(e,f)} - 1$ can be used to determine whether this is the case; that is, the moduli are relatively prime iff e and f are relatively prime. This follows from Euclid's algorithm and the identity $2^e - 1 \pmod{2^f - 1} = 2^{e(\bmod f)} - 1$.

Example. From Table 2.4.1 consider $4 = [1, 4, 4]$ and $5 = [2, 0, 5]$; their sum is $[1 + 2 \pmod 3, 4 + 0 \pmod 5, 4 + 5 \pmod 7] = [0, 4, 2] = 9$. Likewise, their product is $[1 \cdot 2 \pmod 3, 4 \cdot 0 \pmod 5, 4 \cdot 5 \pmod 7] = [2, 0, 6] = 20$. Note that in the absence of a table, the answer is recovered with the help of the Greek–Chinese remainder algorithm. Moreover, if the result of the operation is greater than M, we will not obtain the correct answer; so the choice of the base vector is critical.

We see, therefore, that the amount of time required to add, subtract, and multiply two n-digit numbers using multiple-residue arithmetic is $0[L(n)]$, not counting the time to convert in and out of the modular representation. For addition and subtraction this is no real advantage, but it can be a considerable advantage with respect to multiplication since the conventional method requires time $0[L^2(n)]$. Multiple-residue arithmetic is ideally suited for highly parallel computers.

In order to be able to perform division, we define b^{-1} (mod β), the multiplicative inverse of $b = [b_1, b_2, \ldots, b_n]$ modulo the base vector $\beta = [m_1, m_2, \ldots, m_n]$, as follows:

$$b^{-1} \text{ (mod } \beta) = [b_1^{-1} \text{ (mod } m_1), b_2^{-1} \text{ (mod } m_2), \ldots, b_n^{-1} \text{ (mod } m_n)]$$

Then, if $a = [a_1, a_2, \ldots, a_n]$ and we represent $(b_j)^{-1}$ (mod m_j) by $(b_j)^{-1}(m_j)$, we have

$$\frac{a}{b} \text{ (mod } \beta) = [a_1 \cdot (b_1)^{-1}(m_1) \text{ (mod } m_1), a_2 \cdot (b_2)^{-1}(m_2) \text{(mod } m_2),$$
$$\ldots, a_n \cdot (b_n)^{-1}(m_n) \text{ (mod } m_n)]$$

Of course, as in the case of the single-modulus arithmetic, if b does not divide a, the result cannot be interpreted without further information; however, it is valid as an intermediate result.

The major difficulty with the multiple-residue number system is magnitude comparison of two integers. Of course, with the introduction of the symmetric residue system we can perform a subtraction and then check for a positive or negative result. However, we still have a problem since the numbers in this system do not carry with them indication of their sign; so, one way to determine the sign of number x is to convert it back to its conventional form, a self-defeating idea. The problem of sign detection can be solved in a much better way by converting the number x in the *mixed-radix* representation, using *only* multiple-moduli arithmetic operations. We have already seen an example of mixed-radix form when we expressed the solution x of the Greek–Chinese remainder algorithm as

$$x = q_1 + q_2 \cdot m_1 + q_3 \cdot m_1 \cdot m_2 + \cdots + q_n \cdot m_1 \cdot m_2 \cdots m_{n-1} \quad (*)$$

where each q_i is in the range of the corresponding modulo m_i; q_n is called the *leading term* of x. In this form the sign of x is that of its leading term. (Note that in the mixed-radix form we have $x = \Sigma_{1 \le i \le n} q_i \cdot M_i$, where $M_i = m_1 \cdot m_2 \cdots m_{i-1}$ and $M_1 = 1$, the ratios M_i/M_{i-1} being different for the various positions i. If $m_1 = m_2 = \cdots = m_n$, we have the *fixed-radix* representation and, if, $m_1 = m_2 = \cdots = m_n = 10$, we have the well-known decimal representation.) For sign detection it is helpful to let the last modulus in the base vector be 2, since we want to know in which half of our number range lies the result. For example, in Figure 2.4.2, the numbers $0, 1, \ldots, 5$ constitute the lower half of the number range, corresponding to $q_n = 0$, whereas the numbers $6, 7, \ldots, 10$ constitute the upper half of the number range, corresponding to $q_n = 1$.

Suppose now that we are given $x = [a_1, a_2, \ldots, a_n]$ with respect to the base vector $\beta = [m_1, m_2, \ldots, 2]$ and we want to determine its sign; then we have to convert x to the mixed-radix representation and determine the sign of its leading term. Basically we have to find the digits $q_1, q_2, \ldots, q_n$, mentioned above. Clearly, from (∗) we have

$$x \equiv q_1 \pmod{m_1}$$

and hence $q_1 = a_1$, the first residue digit. We then form the difference $x - q_1$ (by substracting q_1 from every residue digit of x) and have

$$x - q_1 = q_2 \cdot m_1 + q_3 \cdot m_1 \cdot m_2 + \cdots + q_n \cdot m_1 \cdot m_2 \cdots m_{n-1}$$

Note that, from now on, the first residue digit of $x - q_1$ is zero and, hence, the first residue digits of all other numbers need not be considered; therefore, we consider $n - 1$ to be the dimension of the vector $x - q_1$. We next compute $(m_1)^{-1} \pmod{\beta_r}$, the (multiple-moduli) multiplicative inverse of m_1 modulo β_r, where $\beta_r = [m_2, \ldots, 2]$ (also of dimension $n - 1$) and (multiple-moduli) calculate the product $(x - q_1) \cdot (m_1)^{-1}$ to obtain q_2 in the position of the second residue. Repeat this process until we obtain q_n; if its value is 0/1, the number is positive/negative since it lies in the lower/upper half of the number range.

Example. Let us determine the sign of $x = [4, 2, 0, 1]$, using $\beta = [7, 5, 3, 2]$. Obviously, $q_1 = 4$, and $x' = x - 4 = [0, 3, 2, 1]$, or, for the reasons explained above, $x' = [3, 2, 1]$; at this point, the base vector has been reduced to $\beta_r = [5, 3, 2]$. Next, to obtain q_2 in the position of the second residue, we compute $(m_1)^{-1} \pmod{\beta_r} = 7^{-1} \pmod{\beta_r} = [3, 1, 1]$, and multiplying x' by it we have $[4, 2, 1]$; hence, $q_2 = 4$, and subtracting it from the last modular expression, we obtain $x'' = [0, 1, 1]$ or $x'' = [1, 1]$. (At this point, for the same reasons as above, $\beta_r = [3, 2]$.) We next compute $(m_2)^{-1} \pmod{\beta_r} = 5^{-1} \pmod{\beta_r} = [2, 1]$, and multiplying x'' by it we obtain $[2, 1]$; therefore, $q_3 = 2$, and subtracting it from the last modular expression we obtain $x''' = [0, 1]$ or $x''' = [1]$. (Now $\beta_r = [2]$.) Finally, we compute $(m_3)^{-1} \pmod{\beta_r} = 3^{-1} \pmod{\beta_r} = [1]$ and multiplying x''' by it, we obtain $[1]$, from which we see that $q_4 = 1$. Therefore, x is a negative number; indeed, we have $x = -3$.

Exercises

Section 2.1.1

1. Complete the proof of Proposition 2.1.2.
2. Let A, B, and C denotes sets.
   (a) Show that $A \cup (B \cap C) = (A \cup B) \cap (A \cup C)$ (union distributes over intersection).
   (b) Show that $B \subset A$ iff $B \cap A = B$.
   (c) Show that the sets $A - B$, $A \cap B$, and $B - A$ are pairwise disjoint and that $A \cup B = (A - B) \cup (A \cap B) \cup (B - A)$.
3. (Induction really works.) Let $p(n)$ be a statement about a natural number n. Assume that every nonempty set of natural numbers has a smallest element (this is the well-ordering principle whose proof can be found in Section 2.2.1). Show that if (i) $p(0)$ is true, and (ii) $p(n)$ true implies $p(n + 1)$ true, then for every n, $p(n)$ is true. [*Hint*: If not, then $S = \{m : p(m) \text{ false}\}$ has a smallest element m_0.]

Section 2.1.2

1. Complete the proof of Proposition 2.1.4.
2. (a) Show that $\equiv_m$ is an equivalence relation on $\mathbf{Z}$ and that the equivalence class of an integer a is the set $\mathbf{a} = a + m\mathbf{Z}$.
   (b) Show that the set of equivalence classes is $\mathbf{Z}/\equiv_m = \{\mathbf{0}, \mathbf{1}, \ldots, \mathbf{M - 1}\}$.

Section 2.1.3

Notation for Exercises 2 and 3: for a natural number n, $\tilde{n} = \{1, \ldots, n\}$ the set with n elements: $\tilde{0} = \emptyset$.

1. Complete the proof of Proposition 2.1.9.
2. Show by induction that the set of bijections on n elements, $Bij(\tilde{n})$, has $n!$ elements.
3. Show by induction that the power set $\wp(\tilde{n})$ has 2^n elements.
4. Show that the logarithm to any base, $\log : R^+ \to R$ is a bijection. What is $\log^{-1}$?

Section 2.2.1

1. Prove that every nonempty set of real numbers has a least and a maximal element.

2. If $a|b$ and $b|a$, show that $a = \pm b$.
3. If $d|a, b$, show that $d|ax + by$ for any x and y.
4. If $a|bc$ and $(a, b) = 1$, show that $a|c$. (*Hint*: Use Theorem 2.2.3.)
5. If a and b are both nonzero, show that $m = ab/(a, b)$ is the least common multiple of a and b.
6. Prove part b of Theorem 2.2.4. (*Hint*: First show that any solution to the homogeneous equation $ax + by = 0$ is of the form $x = -nb/d$, $y = na/d$ for some n.)
7. Show that $gcd(a, b) = gcd(a, b + ax)$ for any integer x.
8. Prove that if $d = gcd(a, b)$ and $d = ax + by$, then $gcd(x, y) = 1$.
9. Let $a_1, \ldots, a_n$, b be integers. Show that the equation $a_1 x_1 + \cdots + a_n x_n = b$ has an integer solution if and only if $gcd(a_1, \ldots, a_n)|b$, where $gcd(a_1, \ldots, a_n) = gcd[a_1, gcd(a_2, \ldots, a_n)]$, $n \geq 2$.

Section 2.2.2

1. The Fibonacci sequence is defined by $f_0 = 0$, $f_1 = 1$, and $f_{n+2} = f_{n+1} + f_n$, $n \geq 0$. How many operations are needed to compute $gcd(f_{n+1}, f_n)$? What is $gcd(f_{n+1}, f_n)$?
2. (Lucas' theorem.) Show that $gcd(f_m, f_n) = f_{gcd(m,n)}$. [*Hint*: Apply the Euclidean algorithm on m and n and use the following relations (that can be proved by induction): $f_{m-1} f_n + f_m f_{n+1} = f_{m+n}$ and $f_n | f_{kn}$.]

Section 2.2.3

1. Apply **XEA** to the pair 217, 413.
2. Find integers x, y such that
   (a) $1 = 3x + 5y$.
   (b) $1 = 12x + 21y$.
3. If $d = ax + by$, where a, b, x, y, d are all integers, must $d = gcd(a, b)$? What, if anything, can be said about the relationship of d to $gcd(a, b)$?
4. Find *all* solutions to Exercises 2a and 2b above.
5. For each of the following equations, find one integer solution or show that no integer solution exists: (a) $3x + 2y = 5$, (b) $2x + 6y = 7$.

Section 2.2.4

1. Complete the proofs of Theorems 2.2.12 and 2.2.13.
2. Complete the example following Theorem 2.2.15.

3. Show that the continued fraction expansion of $\sqrt{5} = (2; 4, 4, \ldots)$. [*Hint*: Note that in this case it is $\xi = 2 + 1/[4 + (\xi - 2)]$.]

Section 2.3.1

1. Show that for any n there is a prime p with $n < p \leq n! + 1$.
2. To check that n is a prime, justify that it is enough to show that p does not divide n for any $p \leq \sqrt{n}$.
3. If $p_1 = 2$, $p_2 = 3$, $p_3, \ldots$ are the primes in increasing order, then show that:
   (a) $p_{n+1} \leq 2p_n + 1$. [*Hint*: Use **Bertrand's postulate**, which states that for any $n > 1$ there is a prime p with $n \leq p < 2n$; a proof of this can be found in Niven and Zuckerman (1980).]
   (b) $p_{n+1} \leq p_1 + p_2 + \cdots + p_n$ for all $n > 1$.
4. Show that the square root of any prime p is not rational; in other words, $\sqrt{p} \neq r/s$.
5. For what values of n is $\phi(n)$ odd?
6. If $\phi(n) = 2$, what can you infer about n?

Section 2.3.2

1. Complete the proof of Theorem 2.3.7.
2. Complete the proof of Theorem 2.3.8.
3. Show that if 5 does not divide n, then 5 divides $n^8 - 1$.
4. Find $2^{47} \pmod{23}$; use Fermat's theorem as well.
5. Show that $n^7 - n$ is divisible by 2, 3, 5, and 7 for all n.
6. Using the two ways discussed in the text, compute the multiplicative inverse of 7, 8, and 9 modulo 19. [*Hint*: Use algorithm **E**.]
7. Prove that for a given m, the multiplicative inverse of $(m - 1)$ in $\mathbf{Z}_m$ is $(m - 1)$.
8. If G is a group, a in G, and the order of a is s, then show that the order of a^r is $s/(r, s)$.
9. Find all the primitive roots modulo 27.
10. Use Theorem 2.2.4 to prove that the system of congruences $x \equiv a \pmod{m_1}$ and $x \equiv b \pmod{m_2}$ has a solution iff $(m_1, m_2) | (b - a)$. If the solution exists, it is unique modulo $[m_1, m_2]$, the least common multiple of m_1, and m_2. (Generalize to the case of more than two congruences. Prove uniqueness by showing directly that $x \equiv x_0 \pmod{m_1}$ and $x \equiv x_0 \pmod{m_2}$ iff $x \equiv x_0 \pmod{[m_1, m_2]}$.)

11. Prove Theorem 2.3.25.

12. (Nikomachos; see historical note 6.) Solve the following system of congruence equations:

$$x \equiv 2 \pmod{3}, \qquad x \equiv 3 \pmod{5}, \qquad x \equiv 2 \pmod{7}$$

13. (a) Show that there are infinitely many squarefree numbers.

  (b) Show that every positive integer is uniquely represented as the product of a square and a squarefree number.

Section 2.4

1. Prove Theorem 2.4.1.

2. If the integer d is the greatest common divisor of the integers m and n, then show that the polynomial $x^d - 1$ is the greatest common divisor of the polynomials $x^m - 1$ and $x^n - 1$, over any field.

3. Use multiple-moduli arithmetic and let $\beta = [3, 5, 7]$ be the base vector. Compute $a + b$, $a - b$, $a \cdot b$, and $23^{-1} \pmod{\beta}$, where $a = 19$ and $b = 23$.

4. Using the same base vector as above, $\beta = [3, 5, 7]$, give the multiple-moduli representation of the following integers, employing both the symmetric and the least nonnegative number systems: **(a)** 127, **(b)** -127, **(c)** 537, **(d)** -537.

5. Let $\beta = [7, 5, 3, 2]$ be the base vector. What is the sign of $x = [6, 3, 1, 1]$?

Programming Exercises

Section 2.2.1

1. Given a and nonzero b, the Euclidean property says there exists a unique quotient q and remainder r such that $a = bq + r$, $0 \le r < |b|$.

  (a) Write a procedure **QUO** that returns the quotient q.

  (b) Write a procedure **MOD** that returns the least nonnegative remainder r.

  (c) Write a procedure **LAVMOD** that returns the least absolute value remainder r, such that $-|b|/2 < r \le |b|/2$.

Section 2.2.2

1. Using **EA**, write a procedure **GCDLCM** that, given a and b, returns $d = gcd(a, b)$ and $m = lcm(a, b)$. The procedure should return $(0, a) =$

$(a, 0) = |a|$ and $[0, a] = [a, 0] = 0$ for any a. Run this procedure with the following inputs:

a	b
18755	6727
4199	407
69	9453
1463	14098

If there is a problem of overflow (e.g., if the maximum representable integer is $2^{15} - 1 = 32{,}767$), then output the least common multiple as the expression $[a/(a, b)]b$ (e.g., $[21, 35] = 3 \cdot 35$).

2. Choose 500 pairs of integers a, b at random between 1 and 500. For each pair, compute the $\gcd(a, b)$ using the procedure developed in the exercise above. What percentage of the 500 pairs that you chose were relatively prime? Compare your observed percentage with $100 \cdot (6/\pi^2)$; that is, how close is your answer to .60793 (Dirichlet)?

Section 2.2.3

1. Write a subrountine **GLE** (general linear equation) that, given the input a, b, c with a and b not both zero, returns x, y, and t where t is a tag,

$$t = \begin{cases} 0, & \text{if } ax + by = c \text{ has a solution} \\ 1, & \text{if the equation has no solution} \end{cases}$$

and if $t = 0$, then x, y is the solution with the least positive x. The subroutine should use **XEA** to first find a solution to $ax + by = (a, b)$ and then use Theorem 2.2.4. Run the program with the following inputs:

$168x - 66y = 42$, $343x + 407y = 7$,
$426x - 156y = 128$, $1463x + 4235y = 11$

Section 2.2.4

1. (Lagrange, 1798.) Let $p(x) = c_n x^n + \cdots + c_0$, $c_n > 0$, be a polynomial with integer coefficients, having no rational roots, and having exactly one real root $\xi > 1$. Write a procedure to find (along with any other information you wish to obtain) the first hundred or so *partial quotients* (pq) of ξ, using the following exponential-time algorithm (which basically involves only addition and will be fully understood after we have covered the material in Chapter 7):

  1. [Initialize.] Set $pq := 1$.

2. $[x := 1 + x.]$ For $i = 0, 1, \ldots, n - 1$ (in this order) and for $j = n - 1, \ldots, i$ (in this order), set $c_j := c_{j+1} + c_j$. [This step is basically the Ruffini–Horner method discussed in Section 3.1.2; the roots of $p(1 + x)$ are one less than those of $p(x)$.]
3. [Continue?] If $c_n + c_{n-1} + \cdots + c_0 < 0$, set $pq := pq + 1$ and go to 2.
4. $[x := 1/(1 + x).]$ Output pq, the value of the next partial quotient, replace the coefficients $(c_n, c_{n-1}, \ldots, c_0)$ by $(-c_0, -c_1, \ldots, -c_n)$, and go to 1. [The roots of $p(1/(1 + x))$ are reciprocals of those of $p(x)$.]

Try the above algorithm on $p(x) = x^3 - 2$, and similar polynomials.

Section 2.3.1

1. Using the sieve procedure outlined in the text, construct a vector of primes not exceeding 100.
2. Find the 20 largest primes $< 2^{15}$, using the method outlined in the text.

Section 2.3.2

1. Write a procedure **LCE** (linear congruence equation) that, given the input nonzero a, b, and $m > 1$, returns x, n, and t where

$$t = \begin{cases} 0, & \text{if } ax \equiv b \pmod{m} \text{ has a solution} \\ 1, & \text{if the congruence has no solution} \end{cases}$$

and if $t = 0$, then x is the least nonnegative solution with modulus $n = m/(a, m)$; see Theorem 2.3.22. The procedure should use **GLE** (Section 2.2.3). Run the program with the following inputs:

$$36x \equiv -20 \pmod{16}, \qquad 22x \equiv 253 \pmod{143}$$
$$57x \equiv 148 \pmod{38}, \qquad 35x \equiv 9973 \pmod{12}$$

2. Write a procedure **MODINV** using **LCE** that, given the input nonzero a and $m > 1$, returns x and t where

$$t = \begin{cases} 0, & \text{if } ax \equiv 1 \pmod{m} \text{ has a solution} \\ 1, & \text{if the congruence has no solution} \end{cases}$$

and if $t = 0$, then x is the least nonnegative solution modulo m. Run this program with the following inputs:

$$32x \equiv 1 \pmod{35}, \qquad 74x \equiv 1 \pmod{128}$$
$$119x \equiv 1 \pmod{12}, \qquad 486x \equiv 1 \pmod{1033}$$

3. (a) Implement **GCRAk** and use it to solve the following systems:

  (i)
  $x \equiv -28 \pmod{15}$
  $x \equiv 18 \pmod{22}$

  (ii)
  $x \equiv 24 \pmod{13}$
  $x \equiv 52 \pmod{17}$
  $x \equiv -1 \pmod{19}$

  (iii)
  $x \equiv 6 \pmod{23}$
  $x \equiv -50 \pmod{12}$
  $x \equiv 27 \pmod{31}$

  (b) Modify the algorithm so that it returns the least absolute value remainder, **LAVGCRA**, and run it on the above input.

Section 2.3.4

1. We outline a procedure for determining the prime factorization of an integer $n > 1$. Using the sieve method, construct a table of primes $\leq \sqrt{n}$ and test for successive prime divisors $2, 3, \ldots, \leq \sqrt{n}$. If none is found, then n is prime (see also Exercise 2 for Section 2.3.1). If some prime $p \mid n$, then find the highest power e of p that divides n. Store p, e and for the new quotient $q = n/p^e$ check the next successive primes that are $\leq \sqrt{q}$. When no more divisors are found, then either $q = 1$ and all prime factors have been found or $\sqrt{n} < q < n$ and q is an additional prime (n has at most one prime factor $> \sqrt{n}$). Write a procedure **FACT** that determines the prime factorization of an integer $n \leq 2^{15} - 1 = 32{,}767$. You will need a table of primes $< \lfloor \sqrt{2^{15}} \rfloor = 181$. Use this procedure to factor 9583, 9973, 16384, 17017, and 29957.

Historical Notes and References

1. Our presentation of set theory is "naive" because we assume that any plausible description defines a set. Early in this century Bertrand Russell surprised logicians and mathematicians by deducing some logical contradictions in naive set theory. Consider the set $y = \{x : x \notin x\}$, that is, the set that contains all sets that do not contain themselves as elements. It is not important for our purposes to know whether there do exist sets that contain themselves. For all we care, y could be the set of all sets. The question is, does $y \in y$? If $y \in y$, then $y \notin y$ by the definition of y. On the other hand, if $y \notin y$, then $y \in y$. Either $y \in y$ or $y \notin y$, but both cases lead to contradiction. This is one of Russell's paradoxes, and it is analogous to the barber's paradox: there is a barber in a village who shaves only those men who do not shave themselves. Who shaves the barber? Either the barber does or does not shave himself, and again both cases lead to contradiction. In an effort to develop consistent set theories mathematicians have constructed axiomatic set theories that specify precise ways in which sets can be constructed. However, if one does not push too far and uses set theory primarily as a tool, naive set theory is fine.

2. Leonardo da Pisa, widely known as Fibonacci, was a lone star of the first magnitude in the dark mathematical sky of the Middle Ages; he lived around 1200. He traveled widely in Arabia and, through his book *Liber Abaci*, brought to Europe the Hindu–Arabic number system and other superior methods of the East.

3. It is interesting to note that it was in 1896, almost a hundred years after C. F. Gauss' and Legendre's conjectures, that Hadamard and de la Vallée Poussin proved the "Prime Number Theorem" using "analytic" methods, that is, mathematical tools from outside the domain of integers. The first proof not using such tools came in 1949 (Erdoes).

4. Pierre de Fermat stated his "little theorem" in a letter to his friend Bernard Frénicle de Bessy on October 18, 1640. Modular arithmetic was invented and formulated by Carl Friedrich Gauss.

5. The Carmichael numbers are named after the American mathematician R. D. Carmichael, who discovered their properties in 1909.

6. In the literature the Greek–Chinese remainder theorem is referred to as simply the "Chinese remainder theorem" and is attributed to Sun-Tsu, who lived in the first century of our era. In his book *Suan-ching* (=Arithmetic), Sun-Tsu gave a rule, t'ai-ten (=great generalization), for the determination of a natural number that yields the remainders 2, 3, 2 when divided by 3, 5, 7, respectively. However, the Greek mathematician Nikomachos from Gerasa, who also lived in the first century of our era, mentioned a similar problem in his book *Arithmetiki Isagogi* (=Arithmetical Introduction); namely, he introduced as a game a method for the determination of a natural number from the remainders obtained on dividing this number by other natural numbers (*Nicomachi Geraseni Pythagorei Introductionis Arithmeticae*, Libri II, Lipsiae, 1866). The two methods are similar and we give credit to both. (It should be also noted that Nikomachos was a Phythagorean philosopher and that the sieve of Eratosthenes is described in his book.)

Adleman, L. M., C. Pomerance, and R. S. Rumely: On distinguishing prime numbers from composite numbers. *Annals of Mathematics* **117**, 173–206, 1983.

Bradley, G. H.: Algorithm and bound for the greatest common divisor of n integers, *Communications of the ACM* **13**, 433–436, 1970.

Childs, L.: *A concrete introduction to higher algebra*. Springer Verlag, New York, 1979.

Cohen, H., and H. W. Lenstra Jr.: *Primality testing and Jacobi sums*. Report 82-18, Mathematical Institut of the University of Amsterdam, Amsterdam, 1982.

Collins, G. E.: The computing time of the Euclidean algorithm. *SIAM Journal of Computing* **3**, 1–10, 1974.

Dickson, L. E.: *History of the theory of numbers*. Chelsea, New York, 1952.

Dirichlet, P. G. L.: *Abhandlungen der koeniglichen preussischen Akademie der Wissenschaften*, 1849, 69–83.

Dixon, J. D.: Asymptotically fast factorization of integers. *Mathematics of Computation* **36**, 255–260, 1981.

Dixon, J. D.: Factorization and primality tests. *American Mathematical Monthly* **91**, 333–352, 1984.

Dudley, U.: Formulas for primes. *Mathematics Magazine* **56**, 17–22, 1983.

Erdoes, P.: On a new method in elementary number theory which leads to an elementary proof of the prime number theorem. *Proceedings of the National Academy of Sciences* (USA) **35**, 374–384, 1949.

Goldstein, L. J.: A history of the prime number theorem. *American Mathematical Monthly* **80**, 599–614, 1973.

Gregory, R. T.: *Error-free computation: Why it is needed and methods for doing it.* Krieger, Huntington, NY, 1980.

Guy, R. K.: How to factor a number. Congressus Numerantium XVI, *Proceedings of the Fifth Manitoba Conference on Numerical Mathematics* (Winnipeg, 1976) pp. 49–89.

Knuth, D.: *The art of computer programming.* Vol. 2. *Seminumerical algorithms.* Addison-Wesley, Reading, MA, 1969.

Lagrange, J. L.: *Traité de la resolution des equations numeriques.* Paris (n.p.), 1798.

Lamé, Gabriel: Note sur la limite du nombre des divisions dans la recherche du plus grand commun diviseur entre deux nombres entiers. *Comptes Rendus de l'Académie des Sciences* (Paris) **19**, 867–870, 1844.

Lang, S., and H. Trotter: Continued fractions for some algebraic numbers. *Journal fuer die reine und angewandte Mathematik* **255**, 122–134, 1972.

Legendre, A. M.: Théorie des nombres. Paris (n.p.), 1798.

Lehman, R. S.: Factoring large integers. *Mathematics of Computation* **28**, 637–646, 1974.

LeVeque, W. J.: *Fundamentals of number theory.* Addison-Wesley, Reading, MA, 1977.

Levinson, N.: A motivated account of an elementary proof of the prime number theorem. *American Mathematical Monthly* **76**, 225–245, 1969.

Lipson J. D.: *Elements of algebra and algebraic computing.* Addison-Wesley, Reading, MA, 1981.

Lucas, E.: *Théorie des nombres.* Blanchard, Paris, 1961.

Mills, W. H.: A prime representing function. *Bulletin of the American Mathematical Society* **53**, 604, 1947.

Morrison, M. A., and J. Brillhart: A method of factoring and the factorization of F_7. *Mathematics of Computation* **29**, 183–205, 1975.

Motzkin, T. S.: The Euclidean algorithm. *Bulletin of the American Mathematical Society* **55**, 1142–1146, 1949.

Niven, I., and H. S. Zuckerman.: *An introduction to the theory of numbers*, 4th ed. Wiley, New York, 1980.

Olds, C. D.: *Continued fractions*, Random House, New York, 1963.

Pomerance, C.: Recent developments in primality testing. *The Mathematics Intelligencer* **3**, 97–105, 1981.

Pratt, V. R.: Every prime has a succinct certificate. *SIAM Journal of Computing* **4**, 214–220, 1975.

Rabin, M. O.: Probabilistic algorithm for testing primality. *Journal of Number Theory* **12**, 128–138, 1980.

Richards, I.: Continued fractions without tears. *Mathematics Magazine* **54**, 163–171, 1981.

Schroeder, M. R.: *Number theory in science and communication.* Springer Verlag, New York, 1984 and 1986 (2nd ed.).

Scott, N. R.: *Computer number systems and arithmetic.* Prentice-Hall, Englewood Cliffs, N.J., 1985.

Sims, C. C.: *Abstract algebra, a computational approach.* Wiley, New York, 1984.

Solovay, R., and V. Strassen: A fast Monte Carlo test for primality. *SIAM Journal of Computing* **6**, 1977, 84–85; erratum ibid., **7**, 118, 1978.

Wilf, H. S.: *Algorithms and complexity.* Prentice-Hall, Englewood Cliffs, NJ, 1986.

Williams, H. C.: Primality testing on a computer. Ars Combinatoria **5**, 127–185, 1978.

Williams, H. C.: The influence of computers in the development of number theory. *Computers and Mathematics with Applications* **8**, 75–93, 1982.

Williams, H. C.: Factoring on a computer. *The Mathematics Intelligencer* **6**, 29–36, 1984.

Wunderlich, M.: A running time analysis of Brillhart's continued fraction factoring method. Number theory Carbondale 1979. *Lecture notes in Mathematics no.* 751. Springer Verlag, Berlin, 1979, pp. 328–342.

3
Polynomials

3.1 Fundamental Concepts

3.2 Greatest Common Divisors of Polynomials over a Field

3.3 Galois Fields $GF(p^r)$

Exercises

References

In this chapter we become acquainted with polynomials; we will discover that in many ways, polynomial arithmetic is akin to that of integers: there is a version of Euclid's algorithm and even a Greek–Chinese remainder theorem for polynomials. Moreover, many of the things we have learned about the integers (residue systems, etc.) can be generalized to polynomials, with virtually an infinite number of applications. The most important application of polynomials with integer coefficients is in the construction of Galois fields, which play a dominant role in the digital world.

3.1
Fundamental Concepts

In this section we formally introduce polynomials and study various algorithms for doing operations on them. We will be interested mainly in polynomials over the ring of integers and over a (finite) field. The reader should keep this in mind as we proceed, because polynomials "behave" differently in these two different algebraic structures.

3.1.1 Basic Facts of Polynomials

Let J be an integral domain and x an indeterminate; that is, x is not an independent element of J, but simply a formal symbol. An expression of the form

$$p(x) = c_n x^n + c_{n-1} x^{n-1} + \cdots + c_1 x + c_0 \quad \text{or}$$
$$p(x) = c_0 x^n + c_1 x^{n-1} + \cdots + c_{n-1} x + c_n$$

where $c_i \in J$, $i = 0, 1, \ldots, n$ is called a *polynomial in x* with *coefficients in J*, or a *polynomial in x over J*; in this book we use both forms of polynomials. Each expression $c_i x^i$ is called a *term* (of degree i) of $p(x)$. If the coefficient of x^n is not equal to 0, then n is the *degree* of the polynomial and is denoted by $\deg[p(x)]$; the coefficient of x^n is called the *leading* coefficient of $p(x)$ and is denoted by $lc[p(x)]$; if $lc[p(x)] = 1$ (the unity element of J), the polynomial is called *monic*. We shall ordinarily use an indeterminant in describing polynomials, but there are cases, especially when we multiply or divide, when the polynomials will be represented only by their coefficients; that is, polynomials will be considered as a tuple of coefficients.

Two polynomials, $p_1(x) = \Sigma_{0 \leq i \leq m} c_i x^i \neq 0$ and $p_2(x) = \Sigma_{0 \leq i \leq n} d_i x^i \neq 0$, with coefficients in J, are equal precisely when $c_0 = d_0, c_1 = d_1, \ldots, c_i = d_i, \ldots$ for all i. For these polynomials $p_1(x)$ and $p_2(x)$, we define their sum and product as follows:

$$p_1(x) + p_2(x) = \Sigma_{0 \leq i \leq m} c_i x^i + \Sigma_{0 \leq i \leq n} d_i x^i = \Sigma_{0 \leq i \leq \max(m,n)} (c_i + d_i) x^i$$
$$p_1(x) \cdot p_2(x) = (\Sigma_{0 \leq i \leq m} c_i x^i)(\Sigma_{0 \leq j \leq n} d_j x^j) = \Sigma_{0 \leq h \leq m+n} (\Sigma_{i+j=h} c_i d_j) x^h$$

Clearly, both the sum and the product of $p_1(x)$, $p_2(x)$, constitute another polynomial in x with coefficients from the same integral domain J. We can now easily show that the set of polynomials in x with coefficients in the integral domain J, *is itself an integral domain, denoted by* $J[x]$. ($J[x]$ contains both J and the indeterminate x.) To see this, consider the polynomials $p_1(x)$ and $p_2(x)$ defined above; we will show that $p_1(x) \cdot p_2(x) \neq 0$, where 0 denotes the zero polynomial $0 + 0x + 0x^2 + \cdots$. Assume that $\deg[p_1(x)] = m$ and $\deg[p_2(x)] = n$, so that $c_m \neq 0$ and $d_n \neq 0$. It follows that $c_m d_n \neq 0$ since c_m and d_n are both elements of J, an integral domain. However, $c_m d_n$ is the coefficient of x^{m+n} in the product $p_1(x) \cdot p_2(x)$; therefore, $p_1(x) \cdot p_2(x) \neq 0$; which implies that $J[x]$ has no zerodivisors.

It should be noted that equality of polynomials, as was defined above, is not the same as equality of functions, despite the fact that any polynomial $p(x)$ in $J[x]$ defines a function on J by substitution; that is, the function defined by $p(x)$ maps j in J to $p(j)$. However, as you recall, two functions are equal precisely when they have equal values for every element of J, and it is possible for two polynomials to define equal functions on J without being equal as polynomials. For instance, in the field $\mathbf{Z}_2[x]$, the polynomials $p_1(x) = x^3 - 1$, and $p_2(x) = x^5 - 1$ are not equal as polynomials, but $p_1(0) = p_2(0) = 1$, $p_1(1) = p_2(1) = 0$, and so they are equal as functions on $\mathbf{Z}_2[x]$. In general, if J is an integral domain with n elements, $a_0, a_1, \ldots, a_{n-1}$, where a_0, a_1 are the zero and unit elements respectively, then the polynomial

$$p(x) = x(x - a_1)(x - a_2) \cdots (x - a_{n-1}) \neq 0$$

is equal to zero for all x in J and, therefore, defines a function equal to the one defined by the zero polynomial. We will later prove that if an integral domain J has infinitely many elements, like $J = \mathbf{R}$, then two polynomials that define the same function on J must be equal as polynomials.

The fact that we can assign to a polynomial an integer ≥ 0, its degree, is extremely useful, because it permits us to carry out induction proofs for polynomials of the same kind that we did for integers. An application follows below.

Consider a polynomial $p_2(x) \neq 0$. [In words, we say that $p_2(x)$ is not *identically* zero.] A polynomial $p_1(x)$ is *divisible* by the polynomial $p_2(x)$, if there exists a polynomial $q(x)$ such that $p_1(x) = p_2(x) \cdot q(x)$, and we write $p_2(x) | p_1(x)$. The degree of $p_2(x)$ is not greater than the degree of $p_1(x)$, unless, of course, $p_1(x) = 0$; $p_2(x)$ is called a *divisor* or *factor* of $p_1(x)$. Note that the divisibility defined above is not the same as the divisibility defined earlier for integers; for example, $5|7$ holds if $J = \mathbf{Q}$ and 5 and 7 are considered to be polynomials of degree zero, whereas it is not true that the integer 5 divides 7.

Theorem 3.1.1 (Euclidean Property). Let J be an integral domain; moreover, let $p_1(x) = c_m x^m + c_{m-1} x^{m-1} + \cdots + c_1 x + c_0$ and $p_2(x) = d_n x^n + d_{n-1} x^{n-1} + \cdots + d_1 x + d_0 \neq 0$ be two polynomials of degree m and n in $J[x]$, and let

d_n be a unit in J. Then there exist unique $q(x)$, $r(x)$ in $J[x]$ (the *quotient* and *remainder* polynomials, respectively) such that

$$p_1(x) = p_2(x)q(x) + r(x), \qquad \deg[r(x)] < \deg[p_2(x)]$$

Proof. We use induction on the degree of the divident $p_1(x)$. If $p_1(x) = 0$ or $\deg[p_1(x)] < \deg[p_2(x)]$, then define $q(x) = 0$ and $r(x) = p_1(x)$. Otherwise, let $\deg[p_1(x)] = \deg[p_2(x)] + k$, $k \geq 0$, and form the polynomial $p_1'(x) = p_1(x) - (c_m/d_n)x^k p_2(x)$. Then, $\deg[p_1'(x)] < \deg[p_1(x)]$ since we have cancelled off the highest power of x. If $p_1'(x) = 0$ or $\deg[p_1'(x)] < \deg[p_2(x)]$, then we are done; otherwise, by induction $p_1'(x) = p_2(x)q_0(x) + r(x)$, for some $q_0(x)$ and $r(x)$ with $\deg[r(x)] < \deg[p_2(x)]$. Therefore, $p_1(x) = p_2(x)[q_0(x) + (c_m/d_n)x^k] + r(x)$, proving the existence of $q(x)$, $r(x)$. Clearly, $q(x)$ and $r(x)$ are polynomials in $J[x]$ and either $r(x) = 0$ or $\deg[r(x)] < \deg[p_2(x)]$. The uniqueness is proved if we assume that there were another pair $q_1(x)$, $r_1(x)$ such that $p_1(x) = p_2(x)q_1(x) + r_1(x)$, $\deg[r_1(x)] < \deg[p_2(x)]$. Then, $r(x) - r_1(x) = p_2(x)\{q_1(x) - q(x)\}$ and $p_2(x)|[r(x) - r_1(x)]$, which can be true only if $r(x) - r_1(x) = 0$. Hence, $r(x) = r_1(x)$ and $q(x) = q_1(x)$. □

In Chapter 5 we discuss what happens when d_n is not a unit in J. Note that if J is a field, the Euclidean or division property holds provided *only* that the divisor polynomial $p_2(x)$ is nonzero.

We have the following procedure, which is the well-known synthetic division algorithm.

PDF (Polynomial Division over a Field)

Input: $p_1(x) = \Sigma_{0 \leq i \leq m} c_i x^i$ and $p_2(x) = \Sigma_{0 \leq i \leq n} d_i x^i$ over a field, $m \geq n \geq 0$ and $d_n \neq 0$. (This algorithm will work over an integral domain J provided d_n is a unit in J.)

Output: $q(x) = \Sigma_{0 \leq i \leq m-n} q_i x^i$ and $r(x) = \Sigma_{0 \leq i \leq n-1} r_i x^i$, satisfying the Euclidean property (Theorem 3.1.1).

1. [Main loop.] For $k = m - n$ down to 0, do $\{q_k := c_{n+k}/d_n$; for $j = n + k - 1$ down to k, do $c_j := c_j - q_k d_{j-k}\}$.
2. [Exit.] Return q_i, $i = 0, 1, 2, \ldots, m - n$, the coefficients of $q(x)$ computed in step 1, and r_i, $i = 0, 1, 2, \ldots, n - 1$, the coefficients of $r(x)$, where $r_i = c_i$ (c_i was also computed in step 1).

Computing-Time Analysis of PDF. Throughout this algorithm coefficient additions, multiplications, and divisions are performed in a field (single-precision arithmetic) and, therefore, each operation is executed in time ~ 1.

Step 1, the main loop, clearly dominates the computing time of the algorithm. This loop is, clearly, executed $(m - n + 1)$ times, and during each execution one division and $n-1$ coefficient updates take place. Therefore, step 1 is executed in time $0[(n-1)(m-n+1)] = 0[n(m-n+1)]$, and

$$t_{\text{PDF}}[p_1(x), p_2(x)] = 0[n(m-n+1)]$$

That is, $t_{\text{PDF}}[p_1(x), p_2(x)]$ equals the time it takes to compute the product $p_2(x)q(x)$ over a field (just as in the case of integer division).

Example. Consider the polynomials $p_1(x) = 7x^5 + 4x^3 + 2x + 1$ and $p_2(x) = x^3 + 2$ with integer coefficients. Since the leading coefficient of $p_2(x)$ is 1, we can apply **PDF** to obtain

$$
\begin{array}{r}
7x^2 +4 \\
x^3+2\overline{)7x^5 +4x^3 +2x+1}\\
\underline{7x^5 14x^2 }\\
4x^3 -14x^2 +2x+1\\
\underline{4x^3 -8}\\
-14x^2 +2x-7
\end{array}
$$

If we write only the coefficients, we have the following table:

$$
\begin{array}{r}
7\;04\\
1\;0\;0\;2\overline{)7\;0\;40\;21}\\
\underline{7\;0\;014}\\
0\;4\;-14\;21\\
\underline{0\;00\;0}\\
4\;-14\;21\\
\underline{40\;0\;-8}\\
-14\;2\;-7
\end{array}
$$

Let J be an integral domain and consider $p(x)$ in $J[x]$. If $\alpha \in J$, we can then divide $p(x)$ by $x - \alpha$ (this can be done since the coefficient of x is a unit) to obtain

$$p(x) = (x-\alpha)q(x) + r(x), \quad \deg[r(x)] < \deg(x-\alpha) = 1$$

that is, $r(x)$ is a constant in J. We say that $\alpha \in J$ is a *root* or *zero* of $p(x)$ if $p(\alpha) = 0$.

We have the following.

Theorem 3.1.2. Let J be an integral domain, $p(x) \in J[x]$ and $\alpha \in J$. Then α is a root of $p(x)$ iff $(x-\alpha) | p(x)$.

Proof. Obvious from the above. $\square$

Corollary 3.1.3. Let J be an integral domain, $p(x) \in J[x]$ and $\alpha \in J$. Then, if we divide $p(x)$ by $(x - \alpha)$, the remainder is $p(\alpha)$.

Example. Consider $x^2 - x - 2 \in \mathbf{Z}[x]$, which has the roots -1 and 2. Then clearly $(x + 1) | (x^2 - x - 2)$ and $(x - 2) | (x^2 - x - 2)$; moreover, if we divide $x^2 - x - 2$ by $x - 3$, the remainder is $4 = (3)^2 - (3) - 2$. In the next section we will see how to efficiently evaluate a polynomial at a given point.

If α is a root of $p(x)$ and $p(x) = (x - \alpha)^m q(x)$, $m \geq 1$, $q(\alpha) \neq 0$, then m is called the *multiplicity of the root* α; if $m = 1$, then α is called a *simple* root. We have the following.

Theorem 3.1.4. Let J be an integral domain, and $p(x) \neq 0$ in $J[x]$. If the degree of $p(x)$ is n, then $p(x)$ has at most n roots, counting multiplicities. These roots are in J, or in a bigger domain.

Proof. Let us assume that $\alpha_1, \alpha_2, \ldots, \alpha_m$, in J or in a bigger domain, are distinct roots of $p(x)$. (The case of multiple roots is left to the reader.) We will prove by induction that $p(x)$ is divisible by $(x - \alpha_1)(x - \alpha_2) \cdots (x - \alpha_m)$. For $m = 1$ the claim is true by Theorem 3.1.2. Assuming that our claim is true for $m - 1$, then $p(x)$ can be expressed as $p(x) = (x - \alpha_1)(x - \alpha_2) \cdots (x - \alpha_{m-1}) q(x)$, for some $q(x) \in J[x]$. However, for $x = \alpha_m$ we obtain $(\alpha_m - \alpha_1)(\alpha_m - \alpha_2) \cdots (\alpha_m - \alpha_{m-1}) q(\alpha_m) = 0$; since the roots are all distinct, this implies that $q(\alpha_m) = 0$. Using Theorem 3.1.2 again, we obtain $q(x) = (x - \alpha_m) r(x)$, and hence, $p(x) = (x - \alpha_1)(x - \alpha_2) \cdots (x - \alpha_{m-1})(x - \alpha_m) r(x)$; moreover, from the last expression we see that m cannot be greater than n, since for $m = n$ the polynomial $r(x)$ is a constant, and this completes the proof of the theorem. □

The following example demonstrates what happens in Theorem 3.1.4 if J is *not* an integral domain.

Example. Consider $x^2 - 1$ with coefficients from $\mathbf{Z}_8$, which contains zerodivisors. It has four roots, namely, 1, -1 (or 7), 3 and 5.

Corollary 3.1.5. Let J be an integral domain, and $p(x) \in J[x]$. If the degree of $p(x)$ is n, and $p(x)$ has more than n roots, then $p(x) = 0$.

We can now prove the following.

Theorem 3.1.6. Let J be an integral domain with unit element. If J has an infinite number of elements, then two different polynomials $p(x)$ and $q(x)$ in $J[x]$ always define different polynomial functions.

Proof. The difference of the given polynomials $p(x) - q(x) = d(x)$ defines a function $d_f(x)$ that is the difference of the polynomial functions $p_f(x)$ and

$q_f(x)$, defined by $p(x)$ and $q(x)$, respectively. If $p_f(x)$ and $q_f(x)$, were equal, then their difference would be 0 for every x in J. However, this implies that the polynomial $d(x)$ would have as root every element of J; that is, if $\deg[d(x)] = n_d$, then $d(x)$ has more roots than n_d, which implies that $d(x) = 0$. Therefore, $p(x) = q(x)$. □

3.1.2 The Ruffini–Horner Method

In this section we consider two efficient algorithms that help us to (1) evaluate a polynomial $p(x)$ at a given point $x = \alpha$ and (2) compute a new polynomial $p(y)$, where $x = \alpha + y$.

Consider an integral domain $J[x]$ and the polynomial

$$p(x) = c_0 + c_1 x + \cdots + c_n x^n$$

in it, which we want to *evaluate* at the point $x = \alpha$. Obviously, the most straightforward way to achieve this is to replace x by α in the polynomial expression above and to evaluate each term separately; however, this is a rather time-consuming process. There is another way to proceed, though, namely, by use of the Ruffini–Horner method. [This is widely known as just *Horner's method*, but Ruffini anticipated Horner by 15 years; see the article by Cajori (1911).]

The Ruffini–Horner method for efficiently computing $p(\alpha)$ works as follows. Set $p_0 := c_n$; then multiply it by α and add c_{n-1} to obtain $p_1 := \alpha p_0 + c_{n-1}$. Next multiply p_1 by α and add c_{n-2} to obtain $p_2 := \alpha p_1 + c_{n-2}$, and so on. That is, the recursive scheme for this process is

$$\begin{cases} p_0 := c_n \\ p_k := \alpha p_{k-1} + c_{n-k}, \quad k > 0 \end{cases} \quad \textbf{(RH)}$$

and we obtain the *nested* form

$$p(\alpha) = c_0 + \alpha \{ c_1 + \alpha [c_2 + \cdots + \alpha (c_{n-1} + \alpha c_n) \cdots] \}$$

*Computing-Time Analysis of the **Ruffini–Horner Method**.* We consider the following two cases:

Case a. The coefficients of $p(x)$ and the point α belong to a (finite) field. In this case we have n single-precision additions and multiplications, where $n = \deg[p(x)]$; therefore, $p(\alpha)$ is computed in time $O(n)$.

Case b. The coefficients of $p(x)$ and the point α belong to the ring of integers. In this case we have n long-integer additions and multiplications, where, as we recall from Section 1.2, the time for multiplications dominates the time for additions; therefore, we assume that we only multiply n various terms by α, where $n = \deg[p(x)]$. If μ is the value of the largest term obtained during the execution of the Ruffini–Horner method, then, clearly

$$t_{R-H}[p(x), \alpha] = 0[nL(\alpha)L(\mu)]$$

To obtain μ itself, let $d = |p(x)|_\infty$ [the maximum coefficient of $p(x)$ in absolute value] and consider the "worst" possible case, where all the coefficients of $p(x)$ are equal to d; then μ is obtained from the following scheme

$$\begin{array}{ccccc} d & d & d & \cdots & d \\ d & d(\alpha+1) & d(\alpha^2+\alpha+1) & \cdots & d(\alpha^n+\alpha^{n+1}+\cdots+\alpha+1) \end{array}$$

That is, the largest term obtained during the computations is the value of $p(x)$ at $x = \alpha$ and $\mu = d(\alpha^n + \alpha^{n-1} + \cdots + \alpha + 1)$. However, $\mu = d(\alpha^n + \alpha^{n-1} + \cdots + \alpha + 1) \leq d[(n+1)\alpha^n]$ and hence

$$t_{R-H}(p(x), \alpha) = 0\{nL(\alpha)L[d(n+1)\alpha^n]\}$$
$$= 0(nL(\alpha)\{L(d) + L(n+1) + nL(\alpha)\})$$

However, for all practical purposes $L(n+1) = 1$ and, since $L(d) + nL(\alpha) \leq nL(\alpha)L(d) + 1$, we have

$$t_{R-H}[p(x), \alpha] = 0\{n^2 L^2(\alpha) L[|p(x)|_\infty]\}$$

Example. Let us evaluate $p(x) = x^3 - 7x + 7$ at the point $\alpha = 3$, working over the integers. Using the Ruffini–Horner method, we obtain $p(3) = 13$.

$$\begin{array}{cccc} 1 & 0 & -7 & 7 \\ 1 & 3 & 2 & 13 \end{array}$$

We work as follows. In the first row we write all the coefficients of $p(x)$, including those equal to zero; the leading coefficient is at the left side. The second row is obtained as follows. The first (leftmost) term of the second row is just the leading coefficient of $p(x)$. Then multiply this first term by α, add to the product the second coefficient (in the first row), and write the sum as the second term of the second row; in our example $\alpha = 3$ and we have $1 \cdot 3 + 0 = 3$. In general, to compute the "next" term of the second row, multiply the most recently computed term of the second row by α and add to the product the "next" coefficient of the first row. Thus, continuing with out example, we have $3 \cdot 3 - 7 = 2$ and $2 \cdot 3 + 7 = 13$, which is $p(3)$.

Note that $p(3) = 13$, the last term of the second row in this example, is the remainder obtained on dividing $x^3 - 7x + 7$ by $x - 3$. The quotient $q(x)$ of this division is also computed in the above scheme and it is obtained from the other terms of the second row; namely, $q(x) = x^2 + 3x + 2$. This is also known as the *synthetic division* algorithm.

Consider now, in an integral domain $J[x]$, the following polynomial in x

$$p_n(x) = \Sigma_{0 \le i \le n} c_i x^{n-i} = c_0 x^n + c_1 x^{n-1} + \cdots + c_n \tag{RH1}$$

from which we wish to obtain another one in y, where $x = \alpha + y$. Obviously, the polynomial in y can be computed using Taylor's expansion theorem, according to which $p(\alpha + y) = \Sigma_{0 \le i \le n}[p^{(i)}(\alpha)/i!]y^i$. However, we can do even better; note that this substitution produces an equation in y, with coefficients b_i, as shown by relations

$$\Sigma_{0 \le i \le n} c_i x^{n-i} = \Sigma_{0 \le i \le n} c_i (y + \alpha)^{n-i} = \Sigma_{0 \le i \le n} b_i y^{n-i} = \Sigma_{0 \le i \le n} b_i (x - \alpha)^{n-i} \tag{RH2}$$

Using the first and last expressions in (RH2), we will show that the coefficients b_i of the transformed polynomial can be obtained with a sequence of applications of the synthetic division algorithm.

We can write

$$p_n(x) = \Sigma_{0 \le i \le n} c_i x^{n-i} = (x - \alpha) p_{n-1}(x) + r_n \tag{RH3}$$

where $p_{n-1}(x)$ is a plolynomial of degree $n - 1$ and r_n is the coefficient b_n in (RH2). If we express $p_{n-1}(x)$ as

$$p_{n-1}(x) = a_0 x^{n-1} + a_1 x^{n-2} + \cdots + a_{n-1} \tag{RH4}$$

substitute in (RH3), and equate the coefficients of equal powers of x in both sides, we obtain

$$a_0 = c_0, \quad a_j = c_j + \alpha a_{j-1}, \quad j = 1, 2, \ldots, n \tag{RH5}$$

which is the synthetic division algorithm examined above. Note that the last coefficient a_n is precisely the remainder r_n in (RH3), or the coefficient b_n in (RH2).

Further application of the same process on $p_{n-1}(x)$ gives

$$p_{n-1}(x) = (x - \alpha) p_{n-2}(x) + r_{n-1} \tag{RH6}$$

where $p_{n-2}(x)$ is of degree $n - 2$. Combining (RH3) and (RH6), we obtain

$$p_n(x) = (x - \alpha)^2 p_{n-2}(x) + r_{n-1} \cdot (x - \alpha) + r_n$$

If the process is repeated n times, we obtain

$$p_n(x) = r_0 \cdot (x - \alpha)^n + r_1 \cdot (x - \alpha)^{n-1} + \cdots + r_n$$

in which the coefficients r_i are the coefficients b_i themselves, and appear as remainders in each application of algorithm (RH5). In particular, for $\alpha = 1$, this algorithm does not require any multiplications, and so the transformation $x = 1 + y$ can be performed in a very efficient manner. We leave it as an exercise for the reader to show that the time needed to compute $p(y)$, where $x = \alpha + y$, is

$$0\{n^3 L^2(\alpha) + n^2 L(\alpha) L[|p(x)|_\infty]\}$$

Example. Consider the polynomial equation $p(x) = x^3 - 7x + 7 = 0$, from which we want to compute the polynomial $p(1 + y)$. By repeated application of the synthetic division algorithm, the transformed polynomial in y may be obtained in the following two ways:

```
    Ruffini (1804)              Horner (1819)
 1   0   -7    7             1   0   -7   7
     1    1   -6   1             1    1  -6   1
          1    2  -4                  2  -4
               1   3                  3
                   1             1 .
```

In both cases the transformed equation in y is $y^3 + 3y^2 - 4y + 1 = 0$. (In each case, read the coefficients of the last column upward.)

3.1.3 Interpolation over a Field

Let J be a field now, and consider a collection of $n + 1$ "sample" points, $(a_i, b_i) \in J \times J$, $i = 1, 2, \ldots, n + 1$, and a_i values distinct. Then, the interpolation problem over J is to find a polynomial $p(x) \in J[x]$ such that

$$p(a_i) = b_i, \quad i = 1, 2, \ldots, n + 1$$

The interpolation problem is of great importance in many areas of mathematics, and in this section we will consider two methods for solving it. (In Chapter 6 we will use interpolation to obtain the factors of a polynomial.)

We begin with Lagrange's interpolation.

Theorem 3.1.7. Let $(a_i, b_i) \in J \times J$, J a field, $i = 1, 2, \ldots, n + 1$ and a_i values distinct. Then there exists a unique polynomial $p(x) \in J[x]$ of degree $\leq n$ such that $p(a_i) = b_i$, $i = 1, 2, \ldots, n + 1$.

Proof. (Existence.) For the existence of $p(x)$ we use *Lagrange's interpolation formula*

$$p(x) = \Sigma_{1 \le i \le n+1} b_i L_i(x)$$

where

$$L_i(x) = \frac{(x-a_1)\cdots(x-a_{i-1})(x-a_{i+1})\cdots(x-a_{n+1})}{(a_i-a_1)\cdots(a_i-a_{i-1})(a_i-a_{i+1})\cdots(a_i-a_{n+1})}$$

By inspection we have $L_i(a_j) = 0$ for $i \ne j$, because of the factor $(a_j - a_j)$ in the nominator of $L_i(x)$, whereas $L_i(a_i) = 1$; hence we have $p(a_i) = b_i$, as required. Moreover, by construction we have $\deg[p(x)] \le n$.

(Uniqueness.) Suppose that there was another polynomial $p'(x) \in J[x]$ such that $p'(a_i) = p(a_i) = b_i$, $i = 1, 2, \ldots, n+1$. Then the polynomial $p'(x) - p(x)$ has obviously $n+1$ roots but is of degree $\le n$. Clearly, as a result of Theorem 3.1.4, this cannot be the case unless $p'(x) - p(x) = 0$. Hence $p'(x) = p(x)$. □

Example. Consider the "sample" points $a_1 = 0$, $b_1 = 0$, $a_2 = 1$, $b_2 = 1$, $a_3 = -1$, $b_3 = 1$, $a_4 = 3$, and $b_4 = 9$ and let $J = \mathbf{R}$; observe that the b_i values do not have to be all distinct, and they are not. Using Lagrange's formula and performing computations in $\mathbf{R}$, we obtain the following polynomial $p(x) \in \mathbf{R}[x]$:

$$p(x) = 0\left[\frac{(x-1)(x+1)(x-3)}{(0-1)(0+1)(0-3)}\right] + 1\left[\frac{(x-0)(x+1)(x-3)}{(1-0)(1+1)(1-3)}\right]$$
$$+ 1\left[\frac{(x-0)(x-1)(x-3)}{(-1-0)(-1-1)(-1-3)}\right] + 9\left[\frac{(x-0)(x-1)(x+1)}{(3-0)(3-1)(3+1)}\right]$$
$$= x^2.$$

Check: $p(0) = 0$, $p(1) = 1$, $p(-1) = 1$, $p(3) = 9$.

Example. Compute $p(x) \in \mathbf{Z}_{11}[x]$ such that $p(0) = 3$, $p(1) = 2$, $p(2) = 1$, and $p(3) = 2$. Working with the nonnegative set of integers, we have

$$p(x) = 3\left[\frac{(x-1)(x-2)(x-3)}{(0-1)(0-2)(0-3)}\right] + 2\left[\frac{(x-0)(x-2)(x-3)}{(1-0)(1-2)(1-3)}\right]$$
$$+ 1\left[\frac{(x-0)(x-1)(x-3)}{(2-0)(2-1)(2-3)}\right] + 2\left[\frac{(x-0)(x-1)(x-2)}{(3-0)(3-1)(3-2)}\right]$$
$$= 3 \cdot 5^{-1}(x^3 + 5x^2 + 5) + (x^3 + 6x^2 + 6x)$$
$$+ 9^{-1}(x^3 + 7x^2 + 3x) + 2 \cdot 6^{-1}(x^3 + 8x^2 + 2x)$$
$$= 4x^3 + 10x^2 + 7x + 3$$

Check: $p(0) = 3$, $p(1) = 2$, $p(2) = 1$, $p(3) = 2$ over $\mathbf{Z}_{11}$.

Let us now see how to solve the interpolation problem using the Greek–Chinese remainder algorithm.

From Corollary 3.1.3 we recall that, if we divide $p(x) \in J[x]$ by $(x - a)$, $a \in J$, the remainder is $p(a)$; that is

$$p(x) = (x - a)q(x) + p(a)$$

Therefore, $(x - a) | [p(x) - p(a)]$, and hence we have

$$p(a) = b \quad \text{iff} \quad p(x) \equiv b \ [\text{mod}(x - a)]$$

If we think of b as a constant polynomial, we see that we have introduced in $J[x]$ *congruence* modulo a polynomial. Congruence modulo a polynomial $m(x)$ has identical properties to congruence mod m of integers, and all the facts about congruences obtained there remain valid for congruences modulo $m(x)$.

Considering that since we want to find $p(x) \in J[x]$ such that

$$p(x) \equiv b_i \ [\text{mod}(x - a_i)], \quad i = 1, 2, \ldots, n + 1$$

we easily conclude that the interpolation problem is a special case of the Greek–Chinese remainder problem over $J[x]$.

Now, of course, the moduli are linear polynomials, but since the a_i values are all distinct, so are the $(x - a_i)$ values and, hence, they satisfy the restriction that they be pairwise relatively prime. (See Theorem 6.2.12 and surrounding text for the formal extension of the Greek–Chinese remainder theorem to polynomials. Also, just like the integers, two polynomials are relatively prime if any greatest common divisor is a constant; see also the next section.)

The following theorem will show us how to adapt to our needs the Greek–Chinese remainder algorithm that we presented in Chapter 2.

Theorem 3.1.8. Given a field J, $p(x) \in J[x]$ and $a \in J$, then:

a. $p(x) \ [\text{mod}(x - a)] = p(a)$.
b. If $p(a) \neq 0$, then $p^{-1}(x) \ [\text{mod}(x - a)] = p^{-1}(a)$.

Proof

a. The proof follows from Corollary 3.1.3, which states that in order to find the remainder on dividing $p(x)$ by $(x - a)$, we need only evaluate $p(x)$ at $x = a$.
b. By definition of the multiplicative inverse, what we want is a unique polynomial $v(x)$ of degree $<1[= \deg(x - a)]$ such that

$$p(x)v(x) \equiv 1 \ [\text{mod}(x - a)]$$

Since $\deg[v(x)] < 1$, $v(x)$ is a constant; this constant is $p^{-1}(a)$ because

$$p(x) \cdot p^{-1}(a) \equiv (x-a)\{p(x) \; [\mathrm{mod}(x-a)]\} p^{-1}(a)$$
$$= p(a) p^{-1}(a) = 1 \qquad \square$$

We now present the Greek–Chinese remainder algorithm adapted for the interpolation problem.

GCRAn-Interpolation

Input: $(a_i, b_i) J \times J$, $i = 1, 2, \ldots, n+1$, the a_i values are distinct, and J is a field.
Output: $p(x) \in J[x]$ such that $p(a_i) = b_i$, $i = 1, 2, \ldots, n+1$.
1. [Initialize.] $m(x) := 1$; $p(x) := b_1$.
2. [Main loop.] For $i = 1$ until n, do $\{m(x) := m(x) \cdot (x - a_i)$; $c := m^{-1} \cdot (a_{i+1})$; $q := [b_{i+1} - p(a_{i+1})] \cdot c$; $p(x) := p(x) + q \cdot m(x)\}$.
3. [Exit.] Return $p(x)$.

Computing-Time Analysis of **GCRAn-Interpolation**. Since the above algorithm is used exclusively over (finite) fields, coefficient additions and multiplications are executed in time ~ 1; moreover, in this case, the extended Euclidean algorithm for the integers computes the multiplicative inverse of a given number in time ~ 1.

We clearly see that step 2 dominates the execution time of our algorithm. Moreover, at the ith iteration of step 2 we have the following four operations:

1. $m(x) := m(x)(x - a_i)$. After this multiplication, $m(x)$ is of degree i; obviously, this multiplication is executed in time $0(i)$.
2. $c := m^{-1}(a_{i+1})$. As we have seen, over a field the evaluation of a polynomial of degree i at a given point takes time $0(i)$; since inversion is done in time ~ 1, clearly, $c := m^{-1}(a_{i+1})$ is executed in time $0(i)$.
3. $q := [b_{i+1} - p(a_{i+1})]c$. The polynomial $p(x)$ at this point is of degree $\leq i$ and, hence, $p(a_{i+1})$ is computed in time $0(i)$; this is an upper bound on the time needed to compute q, since the subtraction and the multiplication take time ~ 1.
4. $p(x) := p(x) + qm(x)$. This is done in time $0(i)$.

From the above we see that the ith iteration of step 2 is executed in time $0(i)$ and therefore the n iterations, of step 2, are executed in time

$$0(\Sigma_{i \leq i \leq n} i) = 0\left[\frac{n(n+1)}{2}\right] = 0(n^2)$$

which is the execution time of **GCRAn-Interpolation**.

Example. Compute $p(x) \in \mathbf{Z}_{11}[x]$ such that $p(0) = 3$, $p(1) = 2$, $p(2) = 1$, $p(3) = 0$. Applying the above algorithm, and working with the nonnegative set of integers, we obtain the following tableau:

i	a_i	b_i	$m(x)$	c	q	$p(x)$
1	0	3	1	—	—	3
2	1	2	x	1	10	$10x + 3$
3	2	1	$x^2 + 10x$	6	0	$10x + 3$
4	3	0	$x^3 + 8x^2 + 2x$	2	0	$10x + 3$

Thus $p(x) = 10x + 3$ is a solution to the given interpolation problem. Check: $p(0) = 3$, $p(1) = 2$, $p(2) = 1$, and $p(3) = 0$ in $\mathbf{Z}_{11}$.

However, if the last "sample" point was $a_4 = 3$, $b_4 = 2$, instead of $a_4 = 3$, $b_4 = 0$, then from step 2 we would obtain $q = (2 - 0)2 = 4$ and $p(x) = 10x + 3 + 4(x^3 + 8x^2 + 2x) = 4x^3 + 10x^2 + 7x + 3$, which would be the solution to our problem. Check: $p(x) = 3$, $p(1) = 2$, $p(2) = 1$, and $p(3) = 2$ in $\mathbf{Z}_{11}$.

The reader should note that, just as in the case of **GCRAk**, the solution to the $(n + 1)$ point interpolation problem is built as follows:

$$p(x) = q_1 + q_2(x - a_1) + q_3(x - a_1)(x - a_2)$$
$$+ q_4(x - a_1)(x - a_2)(x - a_3)$$
$$+ \cdots + q_{n+1}(x - a_1)(x - a_2) \cdots (x - a_n)$$

where $q_1 = b_1 \in J$, and q_i, $i > 1$, is computed in the $(i - 1)$th iteration of the above algorithm. On the other hand, Lagrange's interpolation formula gives us

$$p(x) = \Sigma_{1 \leq i \leq n+1} b_i \Pi_{\{1 \leq j \leq n+1, j \neq i\}} \frac{x - a_j}{a_i - a_j}$$

Comparing the two different forms of $p(x)$, we observe that only the first form is *extensible* in the sense that it can accommodate a new "sample" point by executing one more iteration of the **GCRAn-Interpolation** algorithm. On the other hand, the Lagrangian coefficients would all have to be recomputed with the addition of a new sample point.

3.1.4 Computations Using the Evaluation–Interpolation Scheme

For our discussion below we need the following. A polynomial $p(x)$ in $J[x]$, J an integral domain, is called *irreducible* or *prime* if, whenever $p(x) = p_1(x)p_2(x)$, then either $p_1(x)$ or $p_2(x)$ is a unit. [A polynomial $p(x)$ is a *unit* if there exists another polynomial $q(x)$ such that $p(x)q(x) = 1$.] Note that if J is a field, the units are all given by the zero-degree polynomials (constants); hence, in this case either $p_1(x)$ or $p_2(x)$ would have to be constant in order for $p(x)$ to be irreducible.

Example

1. $3x^2 + 3$ is irreducible over **R**.
2. $3x^2 + 3 = 3(x^2 + 1)$ is reducible over **Z**, since neither 3 nor $x^2 + 1$ is a unit in **Z**$[x]$.
3. $x^2 + 1$ is irreducible over **Z**, **Q**, or **R**, but it is reducible over **C**, $x^2 + 1 = (x + i)(x - i)$.
4. $x^2 - 2$ is irreducible over **Q**, since $\sqrt{2} \notin$ **Q**.

Just as we defined $\equiv_m$ in **Z**, so now in $J[x]$, J *a field*, we can define an equivalence relation.

Let $m(x) = x^n + \cdots + \mu_2 x^2 + \mu_1 x + \mu_0$ be a monic polynomial in $J[x]$, of degree $n > 0$. Then for any $p(x) \in J[x]$ we have

$$p(x) = m(x)q(x) + r(x), \quad \deg[r(x)] < \deg[m(x)]$$

The remainder $r(x)$ above is denoted by $r_{m(x)}[p(x)]$ or equivalently by $p(x)$ [mod $m(x)$]. If $p(x)$ [mod $m(x)$] = 0, then $m(x)$ exactly divides $p(x)$ and we write $m(x) | p(x)$.

We now define the equivalence relation mod $m(x)$ or, $\equiv_{m(x)}$, on $J[x]$ by

$$p_1(x) \equiv_{m(x)} p_2(x) \quad \text{iff} \quad m(x) | [p_1(x) - p_2(x)]$$

The reader should verify that this is an equivalence relation. We next define $J[x]/\equiv_{m(x)}$ or simply $J[x]_{m(x)} = \{p(x) \in J[x] : \deg[p(x)] < \deg[m(x)]\}$; each congruence class, denoted by $[p(x)]_{m(x)}$ or simply $[p(x)]$, has a unique representative in $J[x]_{m(x)}$, namely, $r_{m(x)}[p(x)]$, $\deg[r(x)] < \deg[m(x)]$. Define addition and multiplication of congruence classes by

$$[p_1(x)]_{m(x)} + [p_2(x)]_{m(x)} = [p_1(x) + p_2(x)]_{m(x)}$$
$$[p_1(x)]_{m(x)} \cdot [p_2(x)]_{m(x)} = [p_1(x) \cdot p_2(x)]_{m(x)}$$

It is easy to show that the operations defined above make $J[x]_{m(x)}$ a ring (analogous to Theorem 2.3.11).

We can think of J as being a subset of $J[x]_{m(x)}$ by identifying $a \in J$ with $[a]_{m(x)}$. [This can be done because, if $[a]_{m(x)} = [b]_{m(x)}$ for a, b in J, then $m(x)$ must divide $a - b$; however, $\deg[m(x)] \geq 1$, and so $a - b$ must be the zero polynomial, which implies $a = b$ in J. Therefore, the function that maps a in

116 POLYNOMIALS

J to $[a]_{m(x)}$ in $J[x]_{m(x)}$ is one-to-one, and we may identify a with $[a]_{m(x)}$.] Hence, every element of $J[x]_{m(x)}$ is an element of J plus a J-multiple of $[x]_{m(x)}$. If we let $\alpha = [x]_{m(x)}$, then every element of $J[x]_{m(x)}$ is of the form $m_{n-1}\alpha^{n-1} + \cdots + m_2\alpha^2 + m_1\alpha + m_0$, with m_i in J for all i; this is the same as saying that every element of $J[x]_{m(x)}$ is a unique J-linear combination of $1, \alpha, \alpha^2, \ldots, \alpha^{n-1}$. It is, therefore, convenient to think of elements of $J[x]_{m(x)}$ as polynomials in $J[x]$ evaluated at $\alpha = [x]_{m(x)}$, where $m(\alpha) = 0$—since $m(\alpha) = [m(x)]_{m(x)} = [0]_{m(x)}$.

If we think of $J[x]_{m(x)}$ (also denoted as $J[\alpha]$) as polynomials in α, then addition and multiplication in $J[\alpha]$ is the same as addition and multiplication of polynomials evaluated at α. Moreover, when multiplying in $J[\alpha]$, it is convenient to have a table expressing powers of α between n and $2n - n$ in terms of polynomials in α of degree $<n$. (This fact will be appreciated later.)

Example. If $J = \mathbf{R}$, then the monic polynomial $m(x) = x^2 + 1$ is irreducible in $\mathbf{R}[x]$. From the previous discussion it follows that every element of $\mathbf{R}[x]_{(x^2+1)}$ is of the form $a + b[x]_{(x^2+1)}$, $a, b \in \mathbf{R}$. If we let $i = [x]_{(x^2+1)}$, we observe that

$$i^2 = [x]^2_{(x^2+1)} = [x^2]_{(x^2+1)} = [x^2 + 1 - 1]_{(x^2+1)}$$
$$= [x^2 + 1]_{(x^2+1)} - [1]_{(x^2+1)} = -1$$

Then addition and multiplication is the same as for polynomials; for instance,

$$(a + ib)(c + id) = ac + i(ad + bc) + i^2bd$$

To express the above result as a polynomial in i of degree ≤ 1, we use the fact that $i^2 = -1$ to obtain

$$(a + ib)(c + id) = ac - bd + i(ad + bc)$$

Moreover, we note that $\mathbf{R}[x]_{(x^2+1)}$ is a field, because if a or $b \neq 0$, then

$$(a + ib)\left(\frac{a - ib}{a^2 + b^2}\right) = 1$$

Therefore, $\mathbf{R}[x]_{(x^2+1)}$ looks just like $\mathbf{C}$.

Example. Consider the monic polynomial $m(x) = x^2 + x + 1$, which is irreducible over $\mathbf{Z}_2$, and let $J[x] = \mathbf{Z}_2[x]$. Then the set $\mathbf{Z}_2[x]/\equiv_{m(x)}$ contains the elements $\{[0], [1], [x], [1 + x]\}$. All operations on these elements are carried out modulo $m(x)$; that is, they are all performed as defined by the rules of $\mathbf{Z}_2[x]$ and the results are replaced by their remainders on division by $m(x)$. So, we have

$$[x + 1] \cdot [x + 1] + [x] = (x + 1)(x + 1) + x = (x^2 + 1) + x = (x^2 + x + 1) = [0]$$

Also note that

$$[x] \cdot [x + 1] = (x^2 + x) = (x^2 + x + 1) + 1 = [1]$$

so that $[x]^{-1} = [x + 1]$ in $\mathbf{Z}_2[x]/\equiv_{m(x)}$. Thus, we have constructed a field with four elements.

Theorem 3.1.9. If J is a field, then $J[x]_{m(x)}$ is a commutative ring with unity. It is a field iff $m(x)$ is irreducible in $J[x]$.

Proof. The proof is analogous to the one of Theorem 2.3.11. For instance, the fact that $J[x]_{m(x)}$ satisfies the axioms for a commutative ring with unity follows from the fact that $J[x]$ satisfies those same axioms.

Next, to show that $J[x]_{m(x)}$ is a field if $m(x)$ is irreducible, we must verify that nonzero elements have inverses. Let $p(x)$ be any nonzero polynomial of degree $< n = \deg[m(x)]$; then $p(x)$ and $m(x)$ are relatively prime, and, as we will see in the next section, there exist polynomials $f(x)$ and $g(x)$ such that $p(x)g(x) + m(x)f(x) = 1$. Then, evaluating the last expression at α, where $\alpha = [x]_{m(x)}$ and $m(\alpha) = 0$, we obtain $p(\alpha)g(\alpha) = 1$ in $J[x]_{m(x)}$, which implies that $p(\alpha)$ has an inverse in $J[x]_{m(x)}$. Since this holds for every nonzero element of $J[x]_{m(x)}$, the latter is a field. We leave it up to the reader to complete the proof of the converse. □

From the above we see that we have a way for inventing new fields. We simply take an irreducible polynomial $m(x)$ in $J[x]$ and consider $J[x]_{m(x)}$—it will be a field.

A field of the form $J[x]_{m(x)}$ is called a *simple field extension* of J, and below we see some applications.

As we saw in the study of integers, when we want to compute the final result, res, of an expression with integer arguments, it is at times easier to follow a roundabout route; the same is true in evaluating expressions with polynomial arguments. Suppose, for example, that we want to compute the final result $\text{res}(x)$ of the expression $e[i_1(x), i_2(x), \ldots, i_h(x)]$ over $J[x]$, J a field, and $i_1(x), \ldots, i_h(x) \in J[x]$. If working over $J[x]$ is "hard," what we can do is work over the fields $J[x]_{m_k(x)}$, for various k, where $m_k(x) = x - a_k$, obviously an irreducible polynomial $a_k \in J$. In this case, however, we need to know a bound on $\deg[\text{res}(x)]$; for example, if we know that $\deg[\text{res}(x)] \leq n$, then the index k will take the values $1, 2, \ldots, n + 1$. Moreover, observe that, as a result of our choice of the polynomials $m_k(x)$, we have $J[x]_{m_k(x)}$ basically being J itself for all k, and thus computations are easy.

This roundabout approach works as follows. For each $k = 1, 2, \ldots, n + 1$, we first compute $i_j(a_k)$, $j = 1, 2, \ldots, h$ and then, provided we have chosen the points a_k so that the expression e_{m_k} is defined, we obtain $\text{res}_{m_k} = b_k = e_{m_k}[i_1(a_k), \ldots, i_h(a_k)]$. The final result, res($x$), is then obtained by interpolation over the sample points (a_k, b_k), $k = 1, 2, \ldots, n + 1$, and it satisfies $\text{res}(\alpha_k) = b_k$, $k = 1, 2, \ldots, n + 1$. (See Figure 3.1.1.)

Example. Working over $\mathbf{Z}_{11}[x]$, let us compute the product of $p_1(x) = 5x + 2$ and $p_2(x) = 8x^2 + 3$ using the scheme shown in Figure 3.1.1. We note that the degree of the polynomial $p(x) = p_1(x) \cdot p_2(x)$ is less than or equal to 3, and hence we need to evaluate at four points, that is, $k = 1, 2, 3, 4$. Taking $a_1 = 0$, $a_2 = 1$, $a_3 = 2$, and $a_4 = 3$ and computing $\{p_1(x) \ [\text{mod}(x - a_k)]\}\{p_2(x) \ [\text{mod}(x - a_k)]\}$, $k = 1, 2, 3, 4$ over $\mathbf{Z}_{11}$, we obtain the following points:

$$p_1(0) \cdot p_2(0) = 2 \cdot 3 = 6 = b_1$$
$$p_1(1) \cdot p_2(1) = 7 \cdot 11 = 0 = b_2$$
$$p_1(2) \cdot p_2(2) = 1 \cdot 2 = 2 = b_3$$
$$p_1(3) \cdot p_2(3) = 6 \cdot 9 = 10 = b_4$$

Having all four sample points, we now apply the **GCRAn-Interpolation** algorithm.

i	a_i	b_i	$m(x)$	c	q	$p(x)$
1	0	6	1	—	—	6
2	1	0	x	1	5	$5x + 6$
3	2	2	$x^2 + 10x$	6	4	$4x^2 + x + 6$
4	3	10	$x^3 + 8x^2 + 2x$	2	7	$7x^3 + 5x^2 + 4x + 6$

and we obtain $p(x) = 7x^3 + 5x^2 + 4x + 6 = (5x + 2)(8x^2 + 3)$ as our answer. Check: $p(0) = 6$, $p(1) = 0$, $p(2) = 2$, $p(3) = 10$ over $\mathbf{Z}_{11}$.

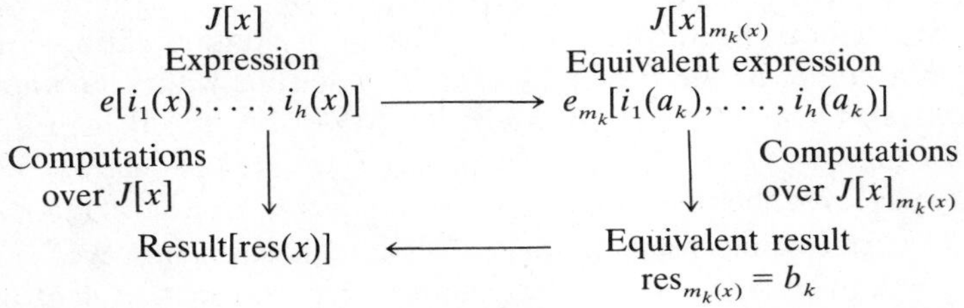

$$
\begin{array}{ccc}
J[x] & & J[x]_{m_k(x)} \\
\text{Expression} & & \text{Equivalent expression} \\
e[i_1(x), \ldots, i_h(x)] & \longrightarrow & e_{m_k}[i_1(a_k), \ldots, i_h(a_k)] \\
\text{Computations} & & \text{Computations} \\
\text{over } J[x] \downarrow & & \downarrow \text{over } J[x]_{m_k(x)} \\
\text{Result}[\text{res}(x)] & \longleftarrow & \text{Equivalent result} \\
& & \text{res}_{m_k(x)} = b_k
\end{array}
$$

Figure 3.1.1. The roundabout scheme for computing an expression using the evaluation–interpolation scheme.

It should be noted that we were able to use the **GCRAn-Interpolation** algorithm above because we worked in a field. Our approach has to be somewhat modified if J is just an integral domain, and this is the topic of the following discussion.

In this case we want to compute the final result $\mathrm{res}(x)$ of the expression $e[i_1(x), i_2(x), \ldots, i_h(x)]$ over $\mathbf{Z}[x]$, where $\mathrm{res}(x), i_1(x), \ldots, i_h(x) \in \mathbf{Z}[x]$. It should be observed that if $p(x) = \Sigma_{0 \le i \le n} c_i x^i \in \mathbf{Z}[x]$, then

$$r_m[p(x)] = \Sigma_{0 \le i \le n} r_m(c_i) x^i$$

where for an integer c, $r_m(c)$ indicates the remainder on dividing c by m. Suppose now that all the coefficients of $\mathrm{res}(x)$, res_i, satisfy $0 \le \mathrm{res}_i \le m_1 m_2 \cdots m_j$, where $m_1, \ldots, m_j$ are j moduli pairwise relatively prime. Then to compute $\mathrm{res}(x)$ following a roundabout route, we can do the following. For $k = 1, 2, \ldots, j$, and provided e_{m_k} is defined, obtain $\mathrm{res}_{m_k}(x) = e_{m_k}(r_{m_k}[i_1(x)], \ldots, r_{m_k}[i_h(x)])$ over $\mathbf{Z}_{m_k}[x]$, and then solve the polynomial Greek–Chinese remainder problem

$$\mathrm{res}(x) \equiv \mathrm{res}_{m_k}(x) \pmod{m_k}, \quad k = 1, 2, \ldots, j$$

for the least nonnegative solution. [If we had a bound on the absolute values of res_i, the coefficients of $\mathrm{res}(x)$, we would solve for the least absolute value solution.]

Example. We want to compute $p(x) = p_1(x) \cdot p_2(x)$ where $p_1(x) = 5x + 2$ and $p_2(x) = 8x^2 + 3$ over $\mathbf{Z}[x]$. We observe that all the coefficients of the product polynomial $p(x)$ are ≤ 40 and since $3 \cdot 5 \cdot 7 > 40$, we will use only these three moduli. For $m_1 = 3$ we obtain

$$r_3[p_1(x)] \cdot r_3(p_2(x)) = (2x + 2)(2x^2) = x^3 + x^2$$

For $m_2 = 5$ the product is

$$r_5[p_1(x)] \cdot r_5[p_2(x)] = (2)(3x^2 + 3) = x^2 + 1$$

Finally, for $m_3 = 7$ we have

$$r_7[p_1(x)] \cdot r_7[p_2(x)] = (5x + 2)(x^2 + 3) = 5x^3 + 2x^2 + x + 6$$

So we now know what $p(x) \pmod{m_k}$ looks like for the various moduli. To recover $p(x)$ itself, we have to solve the system

$$p(x) \equiv x^3 + x^2 \pmod{3}$$
$$p(x) \equiv x^2 + 1 \pmod{5}$$
$$p(x) \equiv 5x^3 + 2x^2 + x + 6 \pmod{7}$$

Clearly, $p(x) = ax^3 + bx^2 + cx + d$, where now the coefficients are obtained by solving the following four systems:

$a \equiv 1 \pmod 3$ $b \equiv 1 \pmod 3$ $c \equiv 0 \pmod 3$ $d \equiv 0 \pmod 3$
$a \equiv 0 \pmod 5$ $b \equiv 1 \pmod 5$ $c \equiv 0 \pmod 5$ $d \equiv 1 \pmod 5$
$a \equiv 5 \pmod 7$ $b \equiv 2 \pmod 7$ $c \equiv 1 \pmod 7$ $d \equiv 6 \pmod 7$

Applying the Greek–Chinese remainder algorithm 4 times, we obtain $a = 40$, $b = 16$, $c = 15$, and $d = 6$. In other words, $p(x) = 40x^3 + 16x^2 + 15x + 6$, the correct answer.

Generalizing what we have just seen, there are applications where the final result itself is a polynomial, given as the solution to a system of polynomial congruences; that is, we have

$$p(x) \equiv p_1(x) \pmod{m_1}$$
$$\cdots$$
$$p(x) \equiv p_k(x) \pmod{m_k}$$

where each p_i, $i = 1, 2, \ldots, k$ is given by $p_i = p_{0i} + p_{1i}x + p_{2i}x^2 + \cdots + p_{ni}x^n \in \mathbf{Z}[x]$.

In order to recover $p(x)$ we observe that, for $q(x) = \Sigma_{0 \le j \le n} a_j x^j$ and $w(x) = \Sigma_{0 \le j \le n} d_j x^j$ both in $\mathbf{Z}[x]$, and $m \in \mathbf{Z}^+$, we have

$$q(x) \equiv w(x) \pmod m \quad \text{iff} \quad a_j \equiv d_j \pmod m$$

since $m | [q(x) - w(x)]$ iff $m | (a_j - d_j)$, $j = 0, 1, 2, \ldots, n$.

From the above observation we conclude that, if the moduli m_i (of the system of congruences above) are pairwise relatively prime, and we set $M = m_1 m_2 \cdots m_k$, then the system of polynomial congruences has two unique solutions, one least-nonnegative coefficient solution, and another least absolute value coefficient solution. Either of these solutions is obtained by applying **GCRAk** (the Greek–Chinese remainder algorithm discussed in our study of the integers) $n + 1$ times. Clearly, the computing time of this process is $0[nL^2(m_1 m_2 \cdots m_k)]$.

3.2
Greatest Common Divisors of Polynomials over a Field

In this section we restrict our study to polynomials over a field. Chapter 5 is devoted entirely to the computation of greatest common divisors of polynomials over the ring of integers—a much more challenging topic (of great interest is the book by Netto, 1896).

3.2.1 Divisibility of Polynomials

We begin with the following definition.

Definition 3.2.1. A *Euclidean domain* is an integral domain J together with a "degree" (or "order") function $d: J - \{0\} \to N$ such that

1. $d(p_1 p_2) \geq d(p_1)$ ($p_1, p_2 \neq 0$).
2. For every p_1 and p_2 in J ($p_2 \neq 0$) there exist q and r in J that satisfy the Euclidean property $p_1 = p_2 q + r$, $d(r) < d(p_2)$ or $r = 0$.

Example (Euclidean Domains). $J = \mathbf{Z}$ with $d(p) = |p|$ is a Euclidean domain because for $p_1, p_2 \in \mathbf{Z}$, $p_2 \neq 0$ there exist q, r such that

$$p_1 = p_2 q + r, \qquad 0 \leq r < |p_2|$$

Note that if $r \neq 0$, $r - p_2$ also satisfies the Euclidean property; uniqueness here is obtained by taking r to be nonnegative. Also, if J is a *field*, $J[x]$ with $d[p(x)] = \deg[p(x)]$ is a Euclidean domain, because as we saw in the previous section for every $p_1(x), p_2(x) \in J[x]$, $p_2(x) \neq 0$, there exist *unique* $q(x), r(x)$ in $J[x]$ such that

$$p_1(x) = p_2(x) q(x) + r(x), \qquad \deg[r(x)] < \deg[p_2(x)]$$

Example (Non-Euclidean Domains). $J = \mathbf{Q}$, the rationals, with $d(p) = |p|$ is not a Euclidean domain because $d[5 \cdot (1/5)] = d(1) < d(5)$ and the first clause of Definition 3.2.1 does not hold. Also, $\mathbf{Z}[x]$ with $d[p(x)] = \deg[p(x)]$ is *not* a Euclidean domain because, for example, if we divide $7x^5 + 4x^3 + 2x + 1$ by $5x^3 + 2$, the quotient is not in $\mathbf{Z}[x]$.

Definition 3.2.2. Let $J[x]$ be an integral domain and $p_1(x), p_2(x)$ in $J[x]$, $p_2(x) \neq 0$. Then $p_h(x)$ in $J[x]$ is a *greatest common divisor* of $p_1(x)$ and $p_2(x)$, and we write $p_h(x) = \gcd[p_1(x), p_2(x)]$, iff:

a. $p_h(x) | p_1(x)$ and $p_h(x) | p_2(x)$.
b. Whenever $q(x) | p_1(x)$ and $q(x) | p_2(x)$, then $\deg[q(x)] \leq \deg[p_h(x)]$ and $q(x) | p_h(x)$.

Below we shall focus our attention on univariate polynomials (polynomials in one variable). The study of multivariate polynomials is omitted because, with the help of the evaluation and interpolation techniques that we discussed in the previous section, it is reduced to the study of univariate polynomials.

We can find a greatest common divisor of two polynomials $p_1(x), p_2(x)$ in $J[x]$, where $J[x]$ is an integral domain and $p_2(x) \neq 0$, by using the division theorem (Theorem 3.1.1) several times. The process is called *Euclid's algorithm for polynomials* and works as follows:

$$p_1(x) = p_2(x)q_1(x) + p_3(x) \quad \deg[p_3(x)] < \deg[p_2(x)]$$
$$p_2(x) = p_3(x)q_2(x) + p_4(x) \quad \deg[p_4(x)] < \deg[p_3(x)]$$
$$\cdots \qquad\qquad\qquad\qquad \cdots$$
$$p_{h-2}(x) = p_{h-1}(x)q_{h-2}(x) + p_h(x) \quad \deg[p_h(x)] < \deg[p_{h-1}(x)]$$
$$p_{h-1}(x) = p_h(x)q_{h-1}(x) + 0$$

Since $\deg[p_i(x)] < \deg[p_{i-1}(x)]$ for $i = 3, 4, \ldots, h$, it is guaranteed that eventually the sequence of divisions ends after at most $\deg[p_2(x)]$ steps.

Theorem 3.2.3. Let $J[x]$ be an integral domain and $p_1(x)$ and $p_2(x)$ in $J[x]$, $p_2(x) \neq 0$. In the Euclidean algorithm for polynomials, described above, the last nonzero remainder $p_h(x)$ is a greatest common divisor of $p_1(x)$ and $p_2(x)$.

Proof. From the above we easily see that any divisor of $p_2(x)$ and $p_3(x)$ is also a divisor of $p_1(x)$, and also that any divisor of $p_1(x)$ and $p_2(x)$ is a divisor of $p_3(x)$. Therefore, the common divisors of $\{p_1(x), p_2(x)\}$ are the same as the common divisors of $\{p_2(x), p_3(x)\}$, and hence, $gcd[p_1(x), p_2(x)] = gcd[p_2(x), p_3(x)]$. Proceeding in the same way, we have:

$$gcd[p_1(x), p_2(x)] = gcd[p_2(x), p_3(x)] = \cdots = gcd[p_{h-1}(x), p_h(x)] = p_h(x)$$
$\square$

The sequence of the remainder polynomials obtained during the execution of Euclid's algorithm is called a *polynomial remainder sequence* (PRS).

It should be noted, however, that it does not make sense (in general) to speak about "the" greatest common divisor of two polynomials since the algebraic system J may have many units; that is, if $p_h(x)$ is a greatest common divisor of $p_1(x)$ and $p_2(x)$, then so is $ap_h(x)$, when a is a unit, and conversely, if $p_h(x)$ and $p_m(x)$ are both greatest common divisors of the same polynomials $p_1(x)$, $p_2(x)$, then $p_h(x) = ap_m(x)$ for some unit a.

We will say that two polynomials $p(x)$ and $q(x)$ are *associates* if each is a nonzero scalar multiple of the other. Any polynomial is an associate of exactly one monic polynomial, and therefore, when $p_h(x)$ is monic, we can talk about *the* greatest common divisor. (If we lapse into sloppy terminology, this is what we have in mind.)

It is worth mentioning that, in **Z**, the greatest common divisor of two integers would not be unique either had we defined it as "greatest in absolute value"; for example, 6 and 9 would have two greatest common divisors in absolute sense: 3 and -3.

Two polynomials in $J[x]$ are *relatively prime* if any greatest common divisor is a constant in J. In that case we will say that the unit element of J is their greatest common divisor.

Below we examine the Euclidean algorithm for polynomials over a field—a relatively simple procedure. On the contrary, computing a greatest common divisor of $p_1(x)$ and $p_2(x)$ in $\mathbf{Z}[x]$ can be quite complicated mainly because $\mathbf{Z}[x]$ is *not* a Euclidean domain. Trying to compute polynomial gcd values in $\mathbf{Z}[x]$, the coefficients of the polynomials in the remainder sequence can become very large and thus slow down the computations. We will examine ways to avoid this problem in Chapter 5.

3.2.2 Euclid's Algorithm for Polynomials over a Field

Let J be a *field* now, and let $p_1(x)$ and $p_2(x) \neq 0$ be two polynomials in $J[x]$. As we saw above, by repeated application of the division algorithm **PDF**, described in Section 3.1.1, we can easily compute the greatest common divisor of $p_1(x)$ and $p_2(x)$.

If $p_h(x)$ is a greatest common divisor of $p_1(x)$ and $p_2(x)$, then clearly $p_h(x)$ is a divisor of every polynomial in the set

$$p_1(x)v(x) + p_2(x)u(x)$$

where $v(x)$ and $u(x)$ are arbitrary polynomials in $J[x]$. The question arises as to whether $p_h(x)$, itself, is in this set, that is, whether it is possible to find two polynomials $f(x)$ and $g(x)$ in $J[x]$ such that

$$p_1(x)g(x) + p_2(x)f(x) = p_h(x)$$

We have the following theorem.

Theorem 3.2.4. Let J be a field, and consider $p_1(x)$ and $p_2(x) \neq 0$ in $J[x]$. If $p_h(x) = gcd[p_1(x), p_2(x)]$, then there exist two polynomials $u(x)$, $v(x)$ in $J[x]$ such that

$$p_1(x)v(x) + p_2(x)u(x) = p_h(x) \tag{F}$$

Proof. From all the polynomials of the form (F) that are not identically zero, choose any one of least degree and denote it by $p_h(x)$. If $p_h(x)$ did not divide $p_1(x)$, then by Theorem 3.1.1 we would have $p_1(x) = p_h(x)q(x) + r(x)$, $r(x) \neq 0$ and $\deg[r(x)] < \deg[p_h(x)]$. But then $r(x) = p_1(x) - p_h(x)q(x) = p_1(x) - \{p_1(x)v(x) + p_2(x)u(x)\}q(x) = p_1(x)\{1 - v(x)q(x)\} - p_2(x)\{u(x)q(x)\}$, which is of the form (F), in contradiction with the choice of $p_h(x)$. $\square$

Corollary 3.2.5. A necessary and sufficient condition that two polynomials $p_1(x)$ and $p_2(x)$ in $J[x]$, J a field, be relatively prime is that there exist two polynomials $v(x)$, $u(x)$ such that

$$p_1(x)v(x) + p_2(x)u(x) = 1$$

The polynomials $u(x)$ and $v(x)$ of Theorem 3.2.4 are not unique. In fact, if $u(x) = f(x)$ and $v(x) = g(x)$ fulfill the requirements of the theorem, so do

$$u(x) = f(x) - t(x)p_1(x), \quad v(x) = g(x) + t(x)p_2(x)$$

where $t(x)$ is an arbitrary polynomial in $J[x]$. (Verify this by direct substitution.) Therefore, $u(x)$ and $v(x)$ can be chosen so as to be of arbitrarily high degree. However, there are lower limits to their degrees.

Theorem 3.2.6. Let J be a field, and consider the polynomials $p_1(x)$ and $p_2(x)$ in $J[x]$. If $p_h(x) = gcd[p_1(x), p_2(x)]$, then there exist two *unique* polynomials $f(x)$ and $g(x)$ in $J[x]$, whose degrees are less than those of $p_1(x)$ and $p_2(x)$, respectively, such that

$$p_1(x)g(x) + p_2(x)f(x) = p_h(x)$$

Proof. For a constructive proof, see the extended Euclidean algorithm below. □

All the above results are also valid for polynomials with coefficients from an integral domain with unit element, provided **PDF** can be applied.

XEA-P (Extended Euclidean Algorithm for Polynomials over a Field)

Input: $p_1(x)$, $p_2(x) \in J[x]$, $p_2(x) \neq 0$, $m = \deg[p_1(x)] \geq \deg[p_2(x)] = n$; J is a field.

Output: $p_h(x)$, $f(x)$, $g(x) \in J[x]$, with $\deg[f(x)] < \deg[p_1(x)] - \deg \cdot [p_h(x)]$, $\deg[g(x)] < \deg[p_2(x)] - \deg[p_h(x)]$ and such that $p_h(x) = gcd[p_1(x), p_2(x)] = p_1(x)g(x) + p_2(x)f(x)$.

1. [Initialize.] $[p_0(x), p_1(x)] := [p_1(x), p_2(x)]$; $[g_0(x), g_1(x)] := (1, 0)$; $[f_0(x), f_1(x)] := (0, 1)$.
2. [Main loop.] While $p_1(x) \neq 0$, do $\{q(x) := \textbf{PDF}[p_0(x), p_1(x)]$; $[p_0(x), p_1(x)] := [p_1(x), p_0(x) - p_1(x)q(x)]$; $[g_0(x), g_1(x)] := [g_1(x), g_0(x) - g_1(x)q(x)]$; $[f_0(x), f_1(x)] := [f_1(x), f_0(x) - f_1(x)q(x)]\}$.
3. [Exit.] Return $[p_h(x), g(x), f(x)] := [p_0(x), g_0(x), f_0(x)]$.

Computing-Time Analysis of XEA-P. Clearly, the computing time of this algorithm is dominated by the execution time of step 2.

Since we are in a field, the first execution of algorithm **PDF** takes times $0[n(m - n + 1)]$, where $m = \deg[p_1(x)]$, $n = \deg[p_2(x)]$, $m \geq n$, and $m - n = \deg[q(x)]$; moreover, the first execution of each of the polynomial multiplications $p_1(x)q(x)$, $g_1(x)q(x)$, and $f_1(x)q(x)$ is also performed in time $0[n(m - n + 1)]$, and this clearly dominates the execution time for each of the corresponding polynomial subtractions. Therefore, the time for the first execution of step 2 is $0[n(m - n + 1)]$, which dominates the times for all of its subsequent executions (justify this).

So, in the worst case we can say that each execution of step 2 is performed in time $0[n(m - n + 1)]$, and since there are at most n such executions of this step we have

$$t_{\text{XEA-P}}[p_1(x), p_2(x)] = 0[n^2(m - n + 1)]$$

Example. Consider the field $\mathbf{Z}_{11}$ and the polynomials $p_1(x) = 7x^5 + 4x^3 + 2x + 1$ and $p_2(x) = 5x^3 + 2$ over this field. Applying **XEA-P** to $p_1(x)$ and $p_2(x)$, we obtain the following table (working with the nonnegative set of integers):

Iteration	$q(x)$	$p_0(x)$	$p_1(x)$	$g_0(x)$	$g_1(x)$	$f_0(x)$	$f_1(x)$
0	—	$7x^5 + 4x^3 + 2x + 1$	$5x^3 + 2$	1	0	0	1
1	$8x^2 + 3$	$5x^3 + 2$	$6x^2 + 2x + 6$	0	1	1	$3x^2 + 8$
2	$10x + 4$	$6x^2 + 2x + 6$	$9x$	1	$x + 7$	$3x^2 + 8$	$3x^3 + 10x^2 + 8x + 2$
3	$8x + 10$	$9x$	6	$x + 7$	$3x^2 + 8$	$3x^3 + 10x^2 + 8x + 2$	$9x^4 + 4x^2 + 3x + 10$
4	$7x$	6	0	$3x^2 + 8$	—	$9x^4 + 4x^2 + 3x + 10$	—

Check: $6 = (7x^5 + 4x^3 + 2x + 1)(3x^2 + 8) + (5x^3 + 2)(9x^4 + 4x^2 + 3x + 10)$.

This computation shows that 6 is a greatest common divisor of the two original polynomials. As we have indicated, since every nonzero element of the field is a unit, it follows that any nonzero multiple of a greatest common divisor is also a greatest common divisor. It is, therefore, conventional to divide in this case the result by its leading coefficient and the obtained monic polynomial is called *the* greatest common divisor of the two given polynomials. Accordingly, the greatest common divisor computed in the example above, is taken to be 1, not 6.

If $m(x)$ is an irreducible polynomial in $\mathbf{Z}_p[x]$, p a prime number, we can use the above algorithm, just as we did with the integers, in order to compute the inverse of the polynomial $p(x) \neq 0$ in $\mathbf{Z}_p[x]_{m(x)}$, where $\deg[p(x)] < \deg[m(x)]$. We simply apply **XEA-P** to $m(x)$ and $p(x)$ and obtain $f(x)$ and $g(x)$ such that

$$m(x)f(x) + p(x)g(x) = 1$$

Then, evaluating the last expression at α, where $\alpha = [x]_{m(x)}$ and $m(\alpha) = 0$, we obtain $p(\alpha)g(\alpha) = 1$ in $\mathbf{Z}_p[x]_{m(x)}$, which implies that $p(\alpha)$ has an inverse in $\mathbf{Z}_p[x]_{m(x)}$.

Let us now consider another example, and let us concentrate our attention on the growth of coefficients of the members of the polynomial remainder sequence.

Example. Consider the polynomials $p_1(x) = x^3 - 7x + 7$ and $p_2(x) = 3x^2 - 7$. Applying the Euclidean algorithm over the rationals, we obtain the following sequences:

$$\begin{aligned}
p_1(x) &= x^3 - 7x + 7 \\
p_2(x) &= 3x^2 - 7 \cdots\cdots\cdots q_1(x) = (1/3)x \\
p_3(x) &= (-14/3)x + 7 \cdots q_2(x) = (-9/14)x - 27/28 \\
p_4(x) &= -1/4 \cdots\cdots\cdots q_3(x) = (56/3)x - 28 \\
p_5(x) &= 0
\end{aligned}$$

As before, $1 = gcd[p_1(x), p_2(x)]$.

The coefficient growth in the polynomial remainder sequence can be minimized if each term is made monic as soon as it is obtained. If we do this, we have

$$\begin{aligned}
p_1(x) &= x^3 - 7x + 7 \\
p_2(x) &= x^2 - 7/3 \cdots\cdot q_1(x) = x \\
p_3(x) &= x - 3/2 \cdots\cdot q_2(x) = x + 3/2 \\
p_4(x) &= 1 \cdots\cdots\cdots q_3(x) = x - 3/2 \\
p_5(x) &= 0
\end{aligned}$$

Indeed, we see that the objective has been achieved, but at the cost of requiring integer *gcd* computations at each step in order to reduce the fractions to lowest terms.

From this example we see that it is not reasonable to use rational arithmetic to compute the polynomial remainder sequence; on one hand, the number of integer *gcd* computations required to keep these coefficients reduced to lowest terms is just too great, while on the other hand, not reducing them leads to horrendous expression growth.

3.2.3 Irreducible Factors of Polynomials

Irreducible polynomials play the same role as prime numbers do in the factorization theory of integers; therefore, we examine some facts about them. We need to know which polynomials are irreducible in $J[x]$, when $J = \mathbf{C}$, $J = \mathbf{R}$, $J = \mathbf{Q}$, and $J = \mathbf{Z}$; the case $J = \mathbf{Z}_p$, p a prime, will be examined separately in Section 3.3 and in Chapter 6. (In Chapter 6 we will also examine how to compute the irreducible factors of a given polynomial with integer coefficients—quite a difficult task.)

We know that the field **C** was invented to contain roots of irreducible polynomials in **R**[x]; we have seen that $i = \sqrt{-1}$ is a root of $x^2 + 1$. From the following theorem we conclude that the only irreducible nonconstant polynomials in **C**[x] are those of degree 1.

Fundamental Theorem of Algebra. Every polynomial $p(x)$ in **C**[x] of degree ≥ 1 has a root in **C**.

Proof. There are various proofs of this theorem, due to various famous mathematicians. These proofs can be found in most texts on algebra [e.g., Childs (1979)]. □

With the help of the fundamental theorem of algebra, we can now easily determine the irreducible polynomials in **R**[x]. However, note that knowing which polynomials in **R**[x] (or in **C**[x]) are irreducible does not make it at all easy to factor a given polynomial.

Theorem 3.2.7. A nonconstant polynomial $p(x)$ in **R**[x] is irreducible if and only if either **(a)** $p(x)$ has degree 1, or **(b)** $p(x) = ax^2 + bx + c$ with $b^2 - 4ac < 0$.

Proof. Obviously, any polynomial of degree 1 is irreducible.
Suppose now that $p(x) = ax^2 + bx + c$. Then $p(x)$ factors in **C**[x] into

$$p(x) = a\left[x + \frac{b + \sqrt{(b^2 - 4ac)}}{2a}\right]\left[x + \frac{b - \sqrt{(b^2 - 4ac)}}{2a}\right]$$

We know that if $b^2 - 4ac < 0$, then the roots of $p(x)$ are complex, and so $p(x)$ is irreducible in **R**[x]. We have, therefore, proved that the two kinds of polynomials we have claimed to be irreducible in **R**[x] are, indeed, irreducible.

We now prove the theorem in the other direction. Suppose that $p(x)$ is an irreducible polynomial in **R**[x], $\deg[p(x)] > 1$. Then it follows that $p(x)$ has no real roots. However, from the fundamental theorem of algebra, it does have a complex root, $\rho = a + ib$, where a, b are real, $b \neq 0$. Form, now, the second-degree polynomial $t(x) = (x - \rho)(x - \bar{\rho})$, where $\bar{\rho} = a - ib$, and divide $p(x)$ by $t(x)$ to obtain $p(x) = t(x)q(x) + r(x)$, with $q(x), t(x)$ in **R**[x], $\deg[r(x)] < \deg[t(x)]$. Clearly, $r(x)$ is a polynomial of degree ≤ 1; that is, $r(x) = cx + d$. Think of $p(x) = t(x)q(x) + r(x)$ as an equality of functions on **C**, and set $x = \rho$. We then obtain $0 = p(\rho) = t(\rho)q(\rho) + r(\rho)$; $t(\rho) = 0$ by the way it was constructed, and so $r(\rho) = 0$. However, given the fact that ρ is not real, the equality $r(\rho) = 0$ can be true only if $c = d = 0$, which implies that $r(x) = 0$. We, therefore, have $p(x) = t(x)q(x)$, which means that $p(x)$ is not irreducible unless $q(x)$ is a constant. However, if $q(x)$ is a constant, then $p(x)$ is an irreducible polynomial of degree 2 and has already been accounted for. $\square$

We next state some results that hold for polynomials over a field.

Theorem 3.2.8. Let J be a field and $p_1(x)$, $p_2(x)$ in $J[x]$. If an irreducible polynomial $m(x)$ in $J[x]$ divides the product $p_1(x)p_2(x)$, then $m(x)$ must divide either $p_1(x)$ or $p_2(x)$.

Proof. If either $p_1(x)$ or $p_2(x)$ is identically zero, the result is obvious. If $p_1(x)p_2(x) \neq 0$, we will assume that $m(x)$ does not divide $p_1(x)$ and will show that $m(x) | p_2(x)$. By hypothesis we have $gcd[m(x), p_1(x)] = 1$ which by Corollary 3.2.5 implies that there exist polynomials $v(x)$ and $u(x)$ such that

$$m(x)u(x) + p_1(x)v(x) = 1$$

Multiplying by $p_2(x)$, we have

$$p_2(x)m(x)u(x) + p_2(x)p_1(x)v(x) = p_2(x)$$

Since $m(x)$ divides the left side of the equation we conclude that $m(x) | p_2(x)$. $\square$

Theorem 3.2.9 (Prime Factorization Theorem for Polynomials). Let J be a field and $p(x)$ in $J[x]$, $\deg[p(x)] > 0$. Then the polynomial $p(x)$ can be uniquely factored into a product of irreducible monic polynomials over $J[x]$; that is, $p(x) = cp_1(x)p_2(x) \cdots p_k(x)$, $p_i(x)$ in $J[x]$, $i = 1, 2, \ldots, k$ and c in J. The factoring is unique except for the order.

Proof. The proof is by induction on the degree of $p(x)$. Obviously, when $\deg[p(x)] = 1$, the theorem is true, since $p(x)$ is irreducible. Suppose now that $p(x)$ has two prime factorization forms; that is, $p(x) = cp_1(x)p_2(x) \cdots p_k(x) = dq_1(x)q_2(x) \cdots q_j(x)$. According to Theorem 3.2.8, $p_1(x)$ divides some $q_i(x)$. Since both $p_1(x)$ and $q_i(x)$ are monic and irreducible we have $p_1(x) = q_i(x)$. Therefore, we can write

$$p_1'(x) = \frac{p(x)}{p_1(x)} = cp_2(x) \cdots p_k(x) = dq_1(x) \cdots q_{i-1}(x)q_{i+1}(x) \cdots q_j(x)$$

By induction hypothesis, the prime factorization of $p_1'(x)$ is unique except for factor ordering; that is, every $p_i(x)$ is equal to some $q_k(x)$, and conversely. Hence, the factorization of $p(x)$ must also be unique except for factor ordering. $\square$

Just as we did with integers, we can write the factorization of $p(x)$ in $\mathbf{R}[x]$ as

$$p(x) = [p_1(x)]^{e_1}[p_2(x)]^{e_2} \cdots [p_k(x)]^{e_k}$$

If any e_i is bigger than one, we will say that $p(x)$ has a *multiple factor*. For instance, $p(x) = (x-1)^3(x+1)$ has a multiple factor, whereas $p(x) = (x-1)(x+1)$ does not. In the former case we say that $p(x)$ has a *multiple* root in J, whereas in the latter case we say that $p(x)$ has only *simple* roots. We see, therefore, that factoring in $\mathbf{R}[x]$ or in $\mathbf{C}[x]$ is equivalent to finding the roots of a polynomial. We have the following.

Theorem 3.2.10 (Gauss). Let J be an integral domain and $p(x)$ in $J[x]$, $\deg[p(x)] > 0$. Then the polynomial $p(x)$ can be uniquely factored into a product of irreducible monic polynomials over $J[x]$ provided every element in J can be uniquely factored into a product of irreducible elements.

Proof. The proof is rather lengthy and is omitted; for details, see Sims (1984), pp. 229–234. □

As a consequence of Theorem 3.2.10 we have that $\mathbf{Z}[x]$ is a *unique factorization domain*, even though it is not a Euclidean domain.

For an example of a domain that is not a unique factorization domain, consider the integral domain that contains the numbers $a + b\sqrt{-5}$, a, b integers. (Verify that it is, indeed, an integral domain.) The number 21 has two prime factorizations: $21 = 3 \cdot 7 = (1 + 2\sqrt{-5})(1 - 2\sqrt{-5})$.

The following is a general result.

Theorem 3.2.11. If J is a Euclidean domain, then J is a unique factorization domain; that is, every nonzero element of J either is a unit or can be represented as a finite product of irreducible elements.

Proof. The proof is rather lengthy and is omitted; for details, see Sims (1984), pp. 214–221. □

We continue our discussion of irreducible polynomials over $\mathbf{Q}[x]$. Unlike $\mathbf{R}[x]$ or $\mathbf{C}[x]$, where we could describe explicitly all irreducible polynomials, in $\mathbf{Q}[x]$ we can describe only certain criteria that imply irreducibility (for reasons that will become clear below). The main point of our discussion is that factoring in $\mathbf{Q}[x]$ is "the same" as factoring in $\mathbf{Z}[x]$.

Let $p(x) = c_n x^n + c_{n-1} x^{n-1} + \cdots + c_1 x + c_0$ be a polynomial with rational coefficients. We can multiply $p(x)$ by t, the least common multiple of the denominators of the coefficients, to obtain a polynomial $tp(x) = s(x)$ with integer coefficients. Since $p(x)$ and $s(x)$ are associates, $s(x)$ will be irreducible in $\mathbf{Q}[x]$ if and only if $p(x)$ is. Therefore, in studying polynomials in $\mathbf{Q}[x]$, we can always assume that they have integer coefficients.

We say that the polynomial $p(x)$ in $\mathbf{Q}[x]$ is *primitive* if the coefficients of $p(x)$ are integers and their greatest common divisor is 1. Then any polynomial in $\mathbf{Q}[x]$ with integer coefficients is an associate of a primitive polynomial. [To see this, let $p(x)$ be a polynomial in $\mathbf{Q}[x]$ with integer coefficients, and let d be the greatest common divisor of its coefficients. Then $(1/d)p(x)$ is still a polynomial with integer coefficients, where now the greatest common divisor of its coefficients is one. Hence, $p(x)$ and $(1/d)p(x)$ are associates in $\mathbf{Q}[x]$.]

Theorem 3.2.12. The product of two primitive polynomials in $\mathbf{Q}[x]$ is again a primitive polynomial.

Proof. Clearly, the product of two polynomials with integer coefficients is again a polynomial with integer coefficients. Let $p(x)$ and $q(x)$ be two primitive polynomials. By the definition of primitivity, we have that for any prime p.

$$p(x) \neq 0 \pmod{p}, \qquad q(x) \neq 0 \pmod{p}$$

Therefore,

$$p(x)q(x) \neq 0 \pmod{p}$$

for any prime p, which implies no prime divides all the coefficients of $p(x)q(x)$. Hence, the greatest common divisor of the coefficients of $p(x)q(x)$ is 1 and $p(x)q(x)$ is primitive. □

Theorem 3.2.13 (Gauss). Let $p(x)$ be a polynomial in $\mathbf{Q}[x]$ with integer coefficients. If $p(x) = q(x)r(x)$ in $\mathbf{Q}[x]$, then $p(x) = q_1(x)r_1(x)$, where $q_1(x)$ and $r_1(x)$ are polynomials with integer coefficients and are associates of $q(x)$ and $r(x)$, respectively.

Proof. Assume without loss of generality that $p(x)$ is primitive and $p(x) = q(x)r(x)$ in $\mathbf{Q}[x]$. Then, from the preceding discussion we see that there are rational numbers a and b such that $a \cdot q(x)$ and $b \cdot r(x)$ are primitive. By the previous theorem we have that $a \cdot b \cdot q(x)r(x) = a \cdot b \cdot p(x)$ is primitive. But so is $p(x)$, and, using the fact that if r is a rational number such that $r \cdot p(x)$ and $p(x)$ are both primitive polynomials, then $r = 1$ or -1, we obtain $p(x) = \pm a \cdot b \cdot q(x)r(x)$. To complete the proof, we then set $q_1(x) = \pm a \cdot q(x)$, and $r_1(x) = b \cdot r(x)$. □

We say that a polynomial in $\mathbf{Z}[x]$ is *irreducible* if it does not factor into the product of two polynomials with degrees ≥ 1 with integer coefficients. From Theorem 3.2.13 we see that a polynomial is irreducible in $\mathbf{Z}[x]$ if and only if it is irreducible as a polynomial in $\mathbf{Q}[x]$. The following theorem helps us decide this.

Theorem 3.2.14. If $p(x) = c_n x^n + \cdots + c_1 x + c_0$ is a polynomial in $\mathbf{Z}[x]$, and r/s is a root, with $(r, s) = 1$, then $s | c_n$ and $r | c_0$.

Proof. Since r/s is a root of $p(x)$ we have $c_n(r^n/s^n) + \cdots + c_1(r/s) + c_0 = 0$. Multiplying through by s^n, we obtain $c_n r^n + c_{n-1} r^{n-1} s + \cdots + c_1 r s^{n-1} + c_0 s^n = 0$, from which we conclude that $c_n r^n = \lambda s$, for some λ in $\mathbf{Z}$. It then follows that $s | c_n r^n$, and since $(r, s) = 1$ we have $s | c_n$. Likewise, we have $c_0 s^n = \mu r$, for some μ in $\mathbf{Z}$, and since $(r, s) = 1$ it follows that $r | c_0$. □

Example. The only possible rational roots of the polynomial $x^2 - 6x + 8$ are $x = 1, -1, 2, -2, 4, -4, 8, -8$, since these are the only divisors of 8; indeed, the two roots are 2 and 4.

Note that if r/s is a root of $p(x) = 0$, then $sx - r$ is a linear factor of $p(x)$. A useful criterion for detection of an irreducible polynomial in $\mathbf{Z}[x]$ is the following.

Theorem 3.2.15 (Eisenstein's Criterion, 1850). Let $p(x) = c_0 x^n + c_1 x^{n-1} + c_2 x^{n-2} + \cdots + c_{n-1} x + c_n$ be a polynomial in $\mathbf{Z}[x]$. If there exists a prime number p such that p does *not* divide c_0, p *does* divide the other integer coefficients $c_1, c_2, \ldots, c_n$, but c_n is not divisible by p^2, then $p(x)$ is irreducible.

Proof. We will prove it by contradiction. Suppose $p(x) = (a_0 x^j + a_1 x^{j-1} + a_2 x^{j-2} + \cdots + a_{j-1} x + a_j) \cdot (b_0 x^k + b_1 x^{k-1} + b_2 x^{k-2} + \cdots + b_{k-1} x + b_k) = a_j b_k + (a_{j-1} b_k + a_j b_{k-1}) x + \cdots + a_0 b_0 x^{j+k}$, where $j + k = n$; moreover, set $a_0 = b_0 = 1$. Then, since $a_j b_k = c_n$, one of the numbers a_j, b_k would be divisible by p; say, for example, $p | a_j$ and $(b_k, p) = 1$. In addition, since $p | c_{n-1}$, where $c_{n-1} = a_{j-1} b_k + a_j b_{k-1}$, we would have $p | a_j b_{k-1}$ and $p | a_{j-1} b_k$; however, since $(b_k, p) = 1$, we would have $p | a_{j-1}$. In the same way we find that p would divide $a_{j-2}, a_{j-3}, \ldots, a_1$, and also a_0, which is equal to 1. However, this is the contradiction and, hence, $p(x)$ is irreducible. □

Theorem 3.2.15 is also true when the coefficients of $p(x)$ belong to an integral domain that is a unique factorization domain.

Example. According to Theorem 3.2.15, the polynomial $x^n - 2$ is irreducible for any n.

Theorem 3.2.15 shows that there are irreducible polynomials in $\mathbf{Q}[x]$ of any degree. Also note that there are polynomials for which Eisenstein's criterion does not apply; for example, for the polynomial $p(x) = x^2 - 6x + 5$ Eisenstein's criterion is completely useless, but $p(x)$ may still be irreducible.

Example. In $\mathbf{Q}[x]$ the polynomial $p(x) = x^3 - 2$ is irreducible. It has a root in $\mathbf{R}$, namely, $2^{1/3}$, and

$$x^3 - 2 = (x - 2^{1/3})(x^2 + 2^{1/3}x + 4^{1/3})$$

where the second term has two complex roots.

Another irreducibility test is to reduce the polynomial modulo m; that is, compute $p'(x) \equiv p(x) \pmod{m}$, where m does not divide the leading coefficient of $p(x)$, and test $p'(x)$ for irreducibility. [Note that $\deg[p'(x)] = \deg[p(x)]$.] Testing irreducibility for $p'(x)$ in $\mathbf{Z}_m[x]$ is a finite problem (since there are only a finite number of possible divisors); in Chapter 6 we discuss in detail a method for factoring polynomials in finite fields.

We have the following.

Theorem 3.2.16. If $p'(x) \equiv p(x) \pmod{m}$ for some m not dividing the leading coefficient of $p(x)$, and the polynomial $p'(x)$ is irreducible in $\mathbf{Z}_m[x]$, then $p(x)$ is irreducible in $\mathbf{Q}[x]$.

Proof. Suppose that $p(x) = q(x)r(x)$, where $q(x)$ and $r(x)$ have integer coefficients and are primitive polynomials. Then for some m not dividing the leading coefficient of $p(x)$ we have $p'(x) = q'(x)r'(x)$ in $\mathbf{Z}_m[x]$, and so $p'(x)$ would factor. □

3.2.4 Squarefree Factorization of Polynomials

A polynomial $p(x)$ is called *squarefree* in case there is no polynomial $q(x)$ of positive degree such that $q^2(x) | p(x)$. The process of finding the squarefree factors of a given polynomial has many uses in mathematics. Among its applications are polynomial factorization, partial fraction decomposition and integration of rational functions; moreover, as we shall see in Chapter 7, the solution of a polynomial equation with multiple roots can be reduced to the solution of one or more equations that have only simple roots and that are the squarefree factors of the original equation.

If J is any domain of numbers, then we define $p'(x) = D[p(x)]$, the derivative of $p(x)$ in $J[x]$ by the following two rules: (1) for a in J, and $n \geq 0$, $D(ax^n) = anx^{n-1}$; it should be noted that $D(ax^n) = 0$ if $J = \mathbf{Z}_n$ and (2) $D[p(x) + q(x)] = D[p(x)] + D[q(x)]$. Recall that for the derivative we have $D[p(x) \cdot q(x)] = p(x)D[q(x)] + D[p(x)]q(x)$, the familiar product rule.

Theorem 3.2.17. Let J be a *unique factorization domain* of characteristic zero, and let $p(x)$ be a primitive, nonconstant polynomial in $J[x]$. Let $p(x) = [p_1(x)]^{e_1}[p_2(x)]^{e_2} \cdots [p_n(x)]^{e_n}$ be the unique factorization of $p(x)$ into irreducible factors, and let $p'(x)$ be its derivative. Then

$$\gcd[p(x), p'(x)] = [p_1(x)]^{e_1-1}[p_2(x)]^{e_2-1} \cdots [p_n(x)]^{e_n-1}$$

Proof. Let $q(x) = \Pi_{2 \leq i \leq n}[p_i(x)]^{e_i}$ and $r(x) = gcd[p(x), p'(x)]$. Then $p(x) = q(x)[p_1(x)]^{e_1}$ and $p'(x) = [p_1(x)]^{e_1}q'(x) + e_1[p_1(x)]^{e_1-1}p_1'(x)q(x)$, from which it follows that $[p_1(x)]^{e_1-1}|r(x)$. By contradiction we will show that $[p_1(x)]^{e_1}$ does not divide $r(x)$. Suppose that $[p_1(x)]^{e_1}|r(x)$; then $[p_1(x)]^{e_1}|p'(x)$, from which we see that $[p_1(x)]^{e_1}|e_1[p_1(x)]^{e_1-1}p_1'(x)q(x)$. After cancellations in the last relation we obtain $p_1(x)|e_1p_1'(x)q(x)$; however, since the $p_i(x)$ values are relatively prime we have that $gcd[p_1(x), q(x)] = 1$ and, therefore, it must be the case that $p_1(x)|e_1p_1'(x)$. This is the contradiction required, since $p_1(x)|e_1p_1'(x)$ implies that $\deg[p_1(x)] \leq \deg[p_1'(x)]$. So, the order of $p_1(x)$ in $r(x)$ is $e_1 - 1$, and by symmetry, we obtain $[p_1(x)]^{e_1-1} \cdots [p_n(x)]^{e_n-1}$, which is what we wanted to prove. □

From Theorem 3.2.17 we conclude that if $gcd[p(x), p'(x)] = 1$, then $p(x)$ has no multiple factors, and conversely. We also have the following.

Corollary 3.2.18. The simple roots of a polynomial are not roots of its derivative.

Corollary 3.2.19. Let J be a field and $p(x)$ be an irreducible polynomial in $J[x]$ that divides $s(x)$ in $J[x]$. Then $[p(x)]^2|s(x)$ if and only if $p(x)|s'(x)$.

Proof. Since $p(x)|s(x)$, we can write $s(x) = p(x)q(x)$, and hence, $s'(x) = p'(x)q(x) + p(x)q'(x)$. Hence, if $[p(x)]^2|s(x)$, then $p(x)|q(x)$, and, clearly, $p(x)|s'(x)$. Conversely, if $p(x)|s'(x)$, then $p(x)|p'(x)q(x)$ and, by Theorem 3.2.8, $p(x)$ divides either $p'(x)$ or $q(x)$. However, $\deg[p'(x)] < \deg[p(x)]$, and hence, $p(x)|q(x)$, which implies that $[p(x)]^2|s(x)$. □

We are now ready to discuss the squarefree factorization algorithm. Let $p(x)$ be a primitive, univariate polynomial of positive degree defined on J, a unique factorization domain. [As we have seen, taking $p(x)$ to be primitive is no restriction of the generality.] Assume that $p(x) = [p_1(x)]^{e_1}[p_2(x)]^{e_2} \cdots [p_n(x)]^{e_n}$ is the unique factorization of $p(x)$ into irreducible factors $p_i(x)$, each of positive degree, so that for all i, $e_i > 0$, and let $e = \max(e_1, \ldots, e_n)$; for $1 \leq i \leq e$, define

$$J_i = \{j : e_j = i\}, \qquad s_i(x) = \Pi_{j \in J_i} p_j(x)$$

Then, clearly,

$$p(x) = \Pi_{1 \leq i \leq e}[s_i(x)]^i$$

which is called the *squarefree factorization* of $p(x)$. [Note that some of the $s_i(x)$ values may be 1; $s_1(x)$ is the product of all linear factors corresponding to simple roots, $s_2(x)$ is the product of all those corresponding to double roots, and so on.] The $s_i(x)$ values are the *squarefree factors* of $p(x)$, and

Theorem 3.2.17 helps us to determine them, by observing that

$$r(x) = gcd[p(x), p'(x)] = \Pi_{1 \leq i \leq n}[p_i(x)]^{e_i-1} = \Pi_{1 \leq i \leq e}[s_i(x)]^{i-1}$$

[$s_1(x)$ does not appear now]. Then the greatest squarefree divisor of $p(x)$ is

$$t(x) = p(x)/r(x) = \Pi_{1 \leq i \leq n} p_i(x) = \Pi_{1 \leq i \leq e} s_i(x)$$

and, hence,

$$v(x) = gcd[r(x), t(x)] = \Pi_{2 \leq i \leq e} s_i(x)$$

Therefore, $s_1(x) = t(x)/v(x)$; that is, the first squarefree factor of $p(x)$ can be computed by differentiation, *gcd* calculations, and divisions. Repeating the process with $r(x)$ in place of $p(x)$, we can compute $s_2(x)$ as the first squarefree factor of $r(x)$ and, eventually, obtain all the squarefree factors of $p(x)$. So, we have the following algorithm:

PSQFF (Polynomial Squarefree Factorization)

Input: $p(x)$ a primitive, univariate polynomial of positive degree over a unique factorization domain J of characteristic zero.

Output: Polynomials $s_i(x)$ and e such that $p(x) = \Pi_{1 \leq i \leq e}[s_i(x)]^i$ is the squarefree factorization of $p(x)$.

1. [Initialize.] $r(x) := gcd[p(x), p'(x)]$; $t(x) := p(x)/r(x)$; $j := 1$.
2. [Finished?] If $\deg[r(x)] = 0$, then do $\{e := j; s_j(x) = t(x); \text{return}\}$.
3. [Compute $s_j(x)$.] $v(x) := gcd[r(x), t(x)]$; $s_j(x) := t(x)/v(x)$.
4. [Update.] $r(x) := r(x)/v(x)$; $t(x) := v(x)$; $j := j+1$; go to 2.

Computing-Time Analysis of **PSQFF**. Clearly, the computing time of this algorithm is dominated by the *gcd* calculations that take place in step 3. If $n = \deg[p(x)]$, then n bounds the number of executions of the loop (consisting of steps 2, 3, and 4) within which lies step 3. Moreover, the time it takes to compute $r(x) := gcd[p(x), p'(x)]$ in step 1 is an upper bound on the time of each *gcd* calculation in step 3. We now have:

Case a. The polynomial $p(x)$ is in $J[x]$, where J is a field. In this case $gcd[p(x), p'(x)]$ is executed in time $O(n^2)$, and since there are n executions of step 3 we have

$$t_{\text{PSQFF}}[p(x)] = O(n^3)$$

Case b. The polynomial $p(x)$ is in $J[x]$, where $J = \mathbf{Z}$. In this case, as we will see in Chapter 5, $gcd[p(x), p'(x)]$ is executed in time $0\{n^5 L^2[|p(x)|_\infty]\}$, and again, since there are n executions of step 3, we have

$$t_{\text{PSQFF}}[p(x)] = 0\{n^6 L^2[|p(x)|_\infty]\}$$

Additional information on squarefree factorization algorithms can be found in Wang and Trager (1979) and Yun (1976). We close this section with an example where we implicitly use information from Chapter 5—namely, how to compute $gcd[p(x), p'(x)]$ in $\mathbf{Z}[x]$.

Example. Let us find the squarefree factors of the polynomial $p(x) = x^5 - x^4 - 2x^3 + 2x^2 + x - 1$. Applying algorithm **PSQFF**, we obtain in the first pass: $r(x) = x^3 - x^2 - x + 1$, $t(x) = x^2 - 1$, $v(x) = x^2 - 1$, and $s_1(x) = 1$, which indicates that there do not exist linear factors. In the second pass we have $r(x) = x - 1$, $t(x) = x^2 - 1$, and we obtain $v(x) = x - 1$ and $s_2(x) = x + 1$, which indicates that $(x + 1)^2$ is a factor of the original polynomial. In the beginning of the third and final pass we have $r(x) = 1$, $t(x) = x - 1$, and at step 2 we see that the degree of $r(x)$ is zero and hence $s_3(x) := t(x) = x - 1$, which indicates that $(x - 1)^3$ is also a factor of the original polynomial. Therefore, $p(x) = (x + 1)^2 (x - 1)^3$.

3.3
Galois Fields $GF(p^r)$

We have already encountered finite fields of order p, a prime, as, for example, a residue system modulo p. However, for many applications, we need number fields of order p^r, and in this section we learn how to construct them and how to calculate in them. At this point, the reader is asked to review the material in Section 2.3.2, in particular Theorems 2.3.17 through 2.3.21, and also Section 3.1.4.

3.3.1 Basic Facts of Finite Fields

From their definition, we know that every field is an integral domain; the converse of this proposition is not in general true, as can be seen from $\mathbf{Z}$. However, for finite fields we have the following.

Theorem 3.3.1. *Every finite integral domain is a field.*

Proof. Let J be a finite integral domain. If a, b are two elements in J, $a \neq b$ then, for all nonzero elements c in J we have, by the cancellation law,

$ac \neq bc$ (see also Theorem 2.3.8). Therefore, $cJ = J$, and for some d in J we have $cd = 1$, which implies that every nonzero element of J has a multiplicative inverse in J, and hence, J is a field. □

The next two theorems are direct consequences of theorems from Chapter 2. The value q will be specified later in Theorem 3.3.6; it *cannot* be arbitrary.

Theorem 3.3.2. If F is a field with q elements, and a is any nonzero element of F, then $a^{q-1} = 1$.

Proof. The nonzero elements of F form an abelian group of order $q-1$ under multiplication (see also Theorem 2.3.17). □

Corollary 3.3.3. If F is a field with q elements, then every a in F satisfies the equation $x^q - x = 0$.

Proof. From Theorem 3.3.2 we know that all the nonzero element of F satisfy the equation $x^{q-1} - 1 = 0$; the zero element of the field, 0, satisfies the equation $x = 0$. Therefore, all the elements of the field satisfy the equation $x(x^{q-1} - 1) = x^q - x = 0$. □

Theorem 3.3.4. Let F be a field with q elements and a any nonzero element of F. If n is the order of a, then $n|(q-1)$.

Proof. If n does not divide $q-1$, then we can obtain k and r such that $q - 1 = kn + r$, where $0 < r < n$. We then have $a^{q-1} = a^{kn+r} = a^{kn}a^r = (a^n)^k a^r = 1$, which implies that $a^r = 1$, since $a^{q-1} = a^n = 1$. This, however, is impossible since $0 < r < n$, and n is the smallest number such that $a^n = 1$. Therefore, $n|(q-1)$. □

In Section 3.1.4 we saw that, starting with $J[x] = \mathbf{Z}_2[x]$, we constructed a new field, $\mathbf{Z}_2[x]/\equiv_{m(x)}$, where $m(x) = x^2 + x + 1$, an irreducible polynomial over $\mathbf{Z}_2$. (Assume for the moment that there are irreducible polynomials over $\mathbf{Z}_p$ of degree n for each n; we will prove this fact later in this section.) This new field, which is the set of congruence classes of polynomials modulo $m(x)$, contains the four elements $\{[0], [1], [x], [1 + x]\}$, and is also denoted by $GF(2^2) = GF(4)$. Moreover, we can think of the elements of $GF(4)$ as being polynomials with coefficients in $\mathbf{Z}_p$ evaluated at $\alpha = [x]_{m(x)}$, which is considered the root of $m(x)$; that is, this new field was constructed by adding to $\mathbf{Z}_2[x]$ a single element, and hence is called a *simple field extension of* $\mathbf{Z}_2[x]$ and is denoted by $\mathbf{Z}_2[\alpha]$. We have the following.

Theorem 3.3.5. Let p be a prime number and $m(x)$ be an irreducible polynomial of degree r in the field $\mathbf{Z}_p[x]$. Then the residue class $\mathbf{Z}_p[x]/\equiv_{m(x)}$ is a field with p^r elements that contains $\mathbf{Z}_p$ and a root of $m(x)$.

Proof. The proof was actually given while discussing the example in Section 3.1.4. □

Theorem 3.3.6. Let F be a field with q elements. Then $q = p^r$, where p is prime and r is a natural number.

Proof. By definition F has an identity element for multiplication; we denote this by 1. Obviously, $1 + 1$ is in F, and we denote this element by 2. We continue in this way, $2 + 1 = 3$ in F, and so on, and after a finite number of steps we encounter an element that we have seen before.

Following the argument presented at the end of Section 2.1.3 (review it), suppose that

$$\Sigma_{1 \leq i \leq k'} 1 = \Sigma_{1 \leq i \leq k''} 1$$

where $k' < k''$; this implies that $\Sigma_{1 \leq i \leq k'' - k'} 1 = 0$. Hence, there must exist a *smallest* positive integer λ such that $\Sigma_{1 \leq i \leq \lambda} 1 = 0$; by an argument similar to that of Theorem 2.3.12, λ is a prime number p and so, F_p is a subfield of F isomorphic to $\mathbf{Z}_p$ (show this).

We define linear independence of a set of elements of F with coefficients from F_p in the obvious way (see the Appendix at the end of the book). Among all the linearly independent subsets of F, let $\{\alpha_1, \alpha_2, \ldots, \alpha_r\}$ be the one with the maximum number of elements. If α is an element of F, then $\{\alpha, \alpha_1, \alpha_2, \ldots, \alpha_r\}$ is a linearly dependent set; that is, there are nonzero coefficients $x_1, x_2, \ldots, x_r$ such that α is a linear combination of $\{\alpha_1, \alpha_2, \ldots, \alpha_r\}$. Obviously, there are p^r distinct linear combinations of $\{\alpha_1, \alpha_2, \ldots, \alpha_r\}$, and this proves the theorem. □

For the converse of this proposition, see Theorem 3.3.19.

Corollary 3.3.7. If F is a finite field, then F has characteristic p, for some prime $p > 0$, and so contains a subfield isomorphic to $\mathbf{Z}_p$.

As we recall from Section 2.3.2, a group element a is a primitive element or primitive root, if its powers $1, a, a^2, \ldots$ run through all the elements of the group (see also Theorem 2.3.20); examples were also studied in Section 2.3.2. The following theorem assures us that every finite field has a primitive root.

Theorem 3.3.8 (Primitive Element Theorem). Let F be a field with q elements. Then, there is some element a in F such that **(i)** every nonzero element of F is a power of a and **(ii)** the order of a is $q - 1$.

Proof. By Theorem 3.3.4 we know that the order of a divides $q - 1$. The idea of the proof is to study the set O of orders of elements of F. Obviously,

O is a set of integers $\leq q - 1$, and the proof is completed when we show that $q - 1$ is in O. This is so because we will then have $a^{q-1} = 1$ for some a in F, and no smaller power of a will be equal to 1. That is, we will have proved part ii of the theorem. Part i then follows easily from the fact that the powers $1, a, a^2, \ldots, a^{q-2}$ are all different, and thus they are all of the nonzero elements of F. Details can be found in Berlekamp (1968) or in Childs (1979). □

Theorem 3.3.8 gives us no formula for finding a primitive element in Z_p, p prime. In fact, no such formula exists. It is known, however, that if p is a prime of the form $p = 4q + 1$, where q is a prime, then 2 is a primitive element of Z_p. Therefore, 2 is a primitive root of Z_p for $p = 5, 13$, and so on. Primitive roots can be found quite readily in a finite field using the following method [found in Albert (1958)].

Finding a Primitive Root in a Finite Field. In the finite field $GF(q)$, an element a is a primitive root if and only if

$$a^{(q-1)/d_i} \neq 1 \pmod{q}$$

for all prime divisors $d_1, d_2, \ldots, d_r$ of $q - 1$.

Generalizing Theorem 3.3.5, we obtain the following.

Theorem 3.3.9. Let F be a field and $p(x)$ a (monic) polynomial in $F[x]$, $\deg[p(x)] \geq 1$. Then there exists a field K containing F such that in $K[x]$, $p(x)$ factors into a product of linear factors.

Proof. The proof is by induction on $\deg[p(x)] = n$ and is left as an exercise for the reader. □

The field K defined in Theorem 3.3.9 is called a *splitting field* for $p(x)$. For instance, according to the fundamental theorem of algebra, $\mathbf{C}$ is a splitting field for any polynomial in $\mathbf{Q}[x]$.

Example. In $\mathbf{Q}[x]$ the polynomial $p(x) = x^3 - 2$ is irreducible. It has a root in $\mathbf{R}$, namely, $2^{1/3}$, but $\mathbf{R}$ is not a splitting field for $p(x)$ because

$$x^3 - 2 = (x - 2^{1/3})(x^2 + 2^{1/3}x + 4^{1/3})$$

and the second term has two complex roots.

Let F be a field and K be a field containing F. Suppose that α in K is the root of some nonzero polynomial $m(x)$ in $F[x]$. Then we say that α is *algebraic over* F. (Those numbers that are not algebraic are called *transcedental*; examples of transcedental numbers are e and π.)

Theorem 3.3.10. Let α in K be algebraic over F. Then there is a unique monic irreducible polynomial $m(x)$ in $F[x]$ with α as a root, and every polynomial $p(x)$ in $F[x]$ with α as a root is a multiple of $m(x)$.

Proof. Using the well-ordering principle on the set of degrees of polynomials in $F[x]$ with α as a root, we conclude that there is a nonzero monic polynomial $m(x)$ of smallest degree having α as a root. We show that the polynomial $m(x)$ is irreducible by contradiction. That is, suppose that we have $m(x) = a(x)b(x)$, where $\deg[a(x)] < \deg[m(x)]$ and $\deg[b(x)] < \deg[m(x)]$. Then $0 = m(\alpha) = a(\alpha)b(\alpha)$, and, since K contains no zerodivisors, either $a(\alpha) = 0$ or $b(\alpha) = 0$. However, this the required contradiction since α is now a root of a polynomial with degree smaller than $\deg[m(x)]$. Therefore, $m(x)$ is irreducible. Finally, let $p(x)$ be any polynomial in $F[x]$ with α as a root. Applying the division theorem, we obtain $p(x) = m(x)q(x) + r(x)$ with $\deg[r(x)] < \deg[m(x)]$. Since $p(\alpha) = 0$, we have $r(\alpha) = 0$, the zero polynomial, and hence, $m(x) | p(x)$. □

The polynomial $m(x)$ defined in Theorem 3.3.10 is called the *minimal polynomial of α over F*, because, as we saw in the proof, it is a polynomial of minimal degree in $F[x]$ having α as a root.

In Chapter 4 we will be interested in finding minimal polynomials. The next theorem tells us that they exist and shows us how to find them.

Theorem 3.3.11. Let $K = F[\alpha]$ be a simple field extension of F, where the minimal polynomial $m(x)$ of α over F has degree r. Then, if β is any element of K, β is algebraic over F, and the minimal polynomial of β over F has degree $\leq r$.

Proof. We know that each element of $K = F[\alpha]$ is a polynomial in α of degree $\leq r - 1$. The same is also true for each power of β. To find the minimal polynomial of β over F, we look for a nonzero solution of

$$c_r \beta^r + c_{r-1} \beta^{r-1} + \cdots + c_1 \beta + c_0 = 0$$

Substituting the powers of β by their equivalent expressions that are polynomials in α, and collecting coefficients of powers of α, we obtain a system of n equations in $r + 1$ unknowns. We then use the fact from linear algebra (see the Appendix at the end of the book) that any system of r homogeneous linear equations in $r + 1$ unknowns has a nonzero solution, to find the nonzero polynomial of degree $\leq r$ in $F[x]$ with β as a root. □

Example. Let $m(x) = x^3 + x + 1$ be an irreducible polynomial over $\mathbf{Z}_2$ and consider $K = \mathbf{Z}_2[x]/\equiv_{m(x)} = \mathbf{Z}_2[\alpha] = GF(2^3)$. Then $\alpha^3 + \alpha + 1 = 0$ and K consists of polynomials in α of degree ≤ 2 with coefficients in $\mathbf{Z}_2$. What is the minimal polynomial of $\beta = \alpha + 1$ over $\mathbf{Z}_2$?

In order to find a nonzero solution of $c_3\beta^3 + c_2\beta^2 + c_1\beta^1 + c_0 = 0$, we write the powers of β as polynomials in $(\alpha + 1)$, and we have $c_3(\alpha + 1)^3 + c_2(\alpha + 1)^2 + c_1(\alpha + 1) + c_0 = 0$. Using the fact that $\alpha^3 = \alpha + 1$ (justify this), and collecting coefficients of powers of α, we obtain the system

$$c_2 + c_1 + c_0 = 0$$
$$c_1 = 0$$
$$c_3 + c_2 = 0$$

which we solve in $\mathbf{Z}_2$. The solution is $c_1 = 0$, and $c_3 = c_2 = c_0$ and, therefore, the minimal polynomial of $\beta = \alpha + 1$ over $\mathbf{Z}_2$ is $x^3 + x^2 + 1$.

For some shortcuts when performing hand computations, the reader is referred to Berlekamp (1968), pp. 112–117.

The following is an important result.

Theorem 3.3.12. Given a prime number p and a natural number r, all finite fields with $q = p^r$ elements are isomorphic.

Proof. Let F be the finite field with q elements. Then from Theorem 3.3.2 we know that the order of an arbitrary nonzero element α of F is $q - 1$; that is, $\alpha^{q-1} = 1$ for $\alpha \neq 0$. On multiplying the last equality by α, we obtain $\alpha^q - \alpha = 0$, which is valid for $\alpha = 0$ as well. Therefore, $\alpha_1, \alpha_2, \ldots, \alpha_q$, all the elements of F, are roots of the polynomial $x^q - x$. It then follows that, since $x^q - x$ is divisible by $(x - \alpha_i)$, $i = 1, 2, \ldots, q$, it must be divisible by $\Pi_{1 \leq i \leq q}(x - \alpha_i)$. Since the degrees are equal, we have $x^q - x = \Pi_{1 \leq i \leq q}(x - \alpha_i)$. By Corollary 3.3.7 we see that F is formed from $\mathbf{Z}_p$ by adjoining to it all the roots of the polynomial $x^q - x$, and, therefore, it is uniquely determined up to isomorphism. $\square$

As a result of Theorems 3.3.5 and 3.3.12 we have the following corollaries.

Corollary 3.3.13. Any finite field is isomorphic to a simple field extension of $\mathbf{Z}_p$, for some prime number p.

Corollary 3.3.14. If q is not a power of a prime number p, there is no finite field with q elements.

Proof. If F is a finite field, it will be isomorphic to $\mathbf{Z}_p[x]_{m(x)}$ for some irreducible polynomial $m(x)$ of degree r. Then $\mathbf{Z}_p[x]_{m(x)}$ has p^r elements, and so does F. $\square$

Since there is essentially only one finite field with p^r elements, it is customary to denote it by $GF(p^r)$. The following theorems are quite useful.

Theorem 3.3.15. In any field of characteristic p, we have $(x - a)^p = x^p - a^p$.

Proof. Expanding, we have $(x - a)^p = \Sigma_k \{p(p-1)\cdots(p-k+1)/k!\}(-a)^k x^{p-k}$, where $p/0! = 1$. The proof is now immediate from the fact that the coefficients are multiples of p. $\square$

A consequence of Theorem 3.3.15 is that in a field of characteristic p, no element can have an order that is a multiple of p. The above result is generalized into the following.

Theorem 3.3.16. Let F be a field of characteristic p. Then, we have for all r

$$(\Sigma_{1 \le i \le k} a_i)^{p^r} = (\Sigma_{1 \le i \le k} a_i^{p^r})$$

Proof. The proof is by induction first on r, for the special case $k = 2$, and then on k for any r. Details are left to the reader. $\square$

As a consequence of Theorems 3.3.15 and 3.3.16 we have the following.

Corollary 3.3.17. Let $q(x)$ be any polynomial in $\mathbf{Z}_p[x]$, and α one of its roots. Then α^p is also a root of $q(x)$.

Proof. From Theorem 3.3.15 it follows that $[q(x)]^p = q(x)$. (This is true only for polynomials in $\mathbf{Z}_p[x]$.) Let $q(x) = (x - \alpha_1)(x - \alpha_2)\cdots(x - \alpha_n)$, where $n = \deg[q(x)]$, be a factorization of $q(x)$ in its splitting field $K[x]$. Then, $q(x) = [q(x)]^p = (x - \alpha_1^p)(x - \alpha_2^p)\cdots(x - \alpha_n^p)$, and by unique factorization of polynomials in $K[x]$ we have that the set $\{\alpha_1^p, \alpha_2^p, \ldots, \alpha_n^p\}$ is the same as the set $\{\alpha_1, \alpha_2, \ldots, \alpha_n\}$. $\square$

For the irreducible polynomial $m(x)$ in $\mathbf{Z}_p[x]$ we have that, if α is one of its roots, then α^{p^k}, is also a root of $m(x)$, for all natural numbers k; that is, we have $m(\alpha^{p^k}) = [m(\alpha)]^{p^k} = 0$. However, since $m(x)$ has a finite degree, there must be a point after which the sequence $\alpha, \alpha^p, \alpha^{p^2}, \ldots$ must repeat itself. The following theorem tells us that $m(x)$, an irreducible polynomial in $\mathbf{Z}_p[x]$, does not have multiple roots in any extension field.

Theorem 3.3.18. Let $m(x)$ be an irreducible polynomial in $\mathbf{Z}_p[x]$ of degree r, and let K be a field containing $\mathbf{Z}_p$. If α in K is a root of $m(x)$, then all the roots of $m(x)$ are $\alpha, \alpha^p, \alpha^{p^2}, \ldots, \alpha^{p^{r-1}}$.

Proof. From the remarks preceding the theorem we see that $\alpha, \alpha^p, \alpha^{p^2}, \ldots$ are all roots of $m(x)$.

Let n be the smallest natural number for which the sequence starts repeating; that is, let $\alpha^{p^n} = \alpha$. Then $\alpha, \alpha^p, \alpha^{p^2}, \ldots, \alpha^{p^{n-1}}$ are all distinct roots of $m(x)$. [Otherwise, from $\alpha^{p^i} = \alpha^{p^j}$ we obtain $(\alpha^{p^i})^{p^{n-j}} = (\alpha^{p^j})^{p^{n-j}} = \alpha$, which implies that $\alpha^{p^{n-(j-i)}} = \alpha$, a contradiction to the minimality of n.] Therefore, $m(x)$ has at least n roots, $n \leq r$.

To complete the proof, let $m_1(x) = (x - \alpha)(x - \alpha^p) \cdots (x - \alpha^{p^{n-1}})$. Since $(\alpha^{p^{n-1}})^p = \alpha$, we have that $[m_1(x)]^p = m_1(x)$, and α is a root of $m_1(x)$. So, $m_1(x)$ is in $\mathbf{Z}_p[x]$ and by Theorem 3.3.10 $m(x)$ must divide $m_1(x)$; since $n \leq r$, we have $m(x) = m_1(x)$. □

We next prove the converse of Theorem 3.3.6.

Theorem 3.3.19. For every power of a prime $q = p^r$, $r > 0$, there exists one, and except for isomorphism exactly one, finite field with q elements. These elements are the roots of the polynomial $x^q - x$.

Proof. Consider the polynomial $p(x) = x^q - x$ in $\mathbf{Z}_p[x]$. Then by Theorem 3.3.9 there exists a splitting field K for $p(x)$; that is, in $K[x]$, $p(x)$ can be written as a product of linear factors.

Let F be the subset of K consisting of all roots of $x^q - x$ in K; that is, F consists of all elements a in K such that $a^q = a$. We will show that F is the required finite field.

For this we need to show that F contains $q = p^r$ elements, and that it is a field. By Theorem 3.2.17 we see that $\gcd[p(x), p'(x)] = 1$ [since we are in $\mathbf{Z}_p$ and the derivative of $p(x)$ is -1] and hence, there are no multiple roots of $p(x)$ in K. Therefore, there are $q = p^r$ distinct roots of $p(x)$ in K, and F contains p^r elements. To show that F is a field, we need to prove that if a, b are in F, then so is $a + b$, $a \cdot b$, $-a$, and a^{-1}. Details are left to the reader. (*Hint:* Use Theorem 3.3.16.) □

We are now in a position to show the following.

Theorem 3.3.20. In $\mathbf{Z}_p[x]$ there is an irreducible polynomial of degree r for each r.

Proof. Let F be a finite field with p^r elements. Then, by Corollary 3.3.13 F is isomorphic to $\mathbf{Z}_p[x]_{m(x)}$ for some irreducible polynomial $m(x)$ in $\mathbf{Z}_p[x]$. Therefore, $\mathbf{Z}_p[x]_{m(x)}$ also has p^r elements, and the degree of $m(x)$ is r. □

It is now clear that any irreducible polynomial $m(x)$ of degree r in $\mathbf{Z}_p[x]$ has a root in any field F with p^r elements. This is so because F is isomorphic to $\mathbf{Z}_p[x]_{m(x)}$ within which $m(x)$ has a root, namely, $[x]_{m(x)}$. Therefore, $m(x)$ has a root in F. More information is obtained from the following theorem, which will prove to be of great importance in Chapter 6.

Theorem 3.3.21. The polynomial $p(x) = x^{p^r} - x$ is the product of all irreducible polynomials in $\mathbf{Z}_p[x]$ whose degrees divide r.

Proof. Let $m(x)$ be an irreducible polynomial of degree r over $\mathbf{Z}_p$, and consider the field F obtained by adjoining the root α of $m(x)$ to $\mathbf{Z}_p$; obviously, F has p^r elements. By Theorem 3.3.10 we know that, for any polynomial $q(x)$, $q(\alpha) = 0$ if and only if $q(\alpha)$ is a multiple of $m(\alpha)$. Moreover, by Corollary 3.3.3 every element of F is a root of $p(x)$ in F; in particular, $p(\alpha) = 0$, which implies that $p(\alpha)$ is a multiple of $m(\alpha)$. Therefore, $p(x)$ is divisible by $m(x)$. However, the same is also true in case $\deg[m(x)] = d$ and $d|r$, because if $r = d \cdot e$, we then have $\alpha^{p^d} = \alpha$, and applying this e times we see that α is a root of $p(x)$.

To complete the proof, we next show that no irreducible polynomial of degree $>r$ over $\mathbf{Z}_p$ divides $p(x)$. We show this by contradiction. Assume that $\deg[m(x)] = s > r$, and that $m(x) | p(x)$. Consider the field F obtained from $\mathbf{Z}_p$ by adjoining to it the root α of $m(x)$. Then F is the set of all polynomials in α, of degree $\le s - 1$; that is, $F = \{q(\alpha) : q(\alpha) = c_{s-1}\alpha^{s-1} + \cdots + c_1 \alpha + c_0, c_i$ in $\mathbf{Z}_p$ for all $i\}$. Then by Theorem 3.3.16 and Corollary 3.3.3 we have $[q(\alpha)]^{p^r} = (c_{s-1})^{p^r}\alpha^{(s-1)p^r} + \cdots + (c_1)^{p^r}\alpha^{p^r} + (c_0)^{p^r} = c_{s-1}\alpha^{(s-1)p^r} + \cdots + c_1 \alpha^{p^r} + c_0$. Since by assumption $m(x)|p(x)$, α, which is a root of $m(x)$, must also be a root of $p(x)$, and hence, $\alpha^{p^r} = \alpha$. We, therefore, have $[q(\alpha)]^{p^r} = c_{s-1}\alpha^{s-1} + \cdots + c_1 \alpha + c_0 = q(\alpha)$, which implies that every $q(\alpha)$ in F is a root of $p(x)$. This, in turn, implies that there are p^s distinct roots of $p(x)$, where $p^s > p^r$, and this is the required contradiction. □

Example. Consider the Galois field $GF(2^4) = \mathbf{Z}_2[x]_{m(x)}$, where $m(x) = x^4 - x - 1$, and let us express $x^{16} - x$ as a product of irreducible polynomials in $\mathbf{Z}_p[x]$. (Note that in $GF(2)$ we have $+1 = -1$.) Clearly, $m(x) = x^4 - x - 1$ is

TABLE 3.3.1 The Three Different Representations of the Elements of $GF(2^4)$.

Power Representation	Polynomial Representation	Vector Representation
0	0	(0000)
$\alpha^0 = 1$	1	(0001)
α	α	(0010)
α^2	α^2	(0100)
α^3	α^3	(1000)
α^4	$\alpha + 1$	(0011)
α^5	$\alpha^2 + \alpha$	(0110)
α^6	$\alpha^3 + \alpha^2$	(1100)
α^7	$\alpha^3 + \alpha + 1$	(1011)
α^8	$\alpha^2 + 1$	(0101)
α^9	$\alpha^3 + \alpha$	(1010)
α^{10}	$\alpha^2 + \alpha + 1$	(0111)
α^{11}	$\alpha^3 + \alpha^2 + \alpha$	(1110)
α^{12}	$\alpha^3 + \alpha^2 + \alpha + 1$	(1111)
α^{13}	$\alpha^3 + \alpha^2 + 1$	(1101)
α^{14}	$\alpha^3 + 1$	(1001)

one factor with roots α, α^2, α^4, α^8 (by Theorem 3.3.18); that is, $x^4 - x - 1 = (x - \alpha)(x - \alpha^2)(x - \alpha^4)(x - \alpha^8)$ (see Table 3.3.1 to verify the last equality). Other obvious factors are the polynomials x and $(x - 1)$.

Next we take a power of α that has not been considered; namely, α^3. Then another factor will be the polynomial with roots, α^3, α^6, α^{12}, α^9; that is, the polynomial $(x - \alpha^3)(x - \alpha^6)(x - \alpha^{12})(x - \alpha^9)$. The term $(x - \alpha^9)$ appears because the last polynomial has α^{12} as a root, and hence, it will also have $(\alpha^{12})^2 = \alpha^{24} = \alpha^{15}\alpha^9 = \alpha^9$ as a root. We next consider α^5, and another factor is the polynomial $(x - \alpha^5)(x - \alpha^{10})$. Finally, the last factor is the polynomial $(x - \alpha^7)(x - \alpha^{14})(x - \alpha^{13})(x - \alpha^{11})$. The term $(x - \alpha^{13})$ appears because the last polynomial has α^{14} as a root, and hence, it will also have $(\alpha^{14})^2 = \alpha^{28} = \alpha^{15}\alpha^{13} = \alpha^{13}$ as a root. Likewise, the term $(x - \alpha^{11})$ appears because α^{13} is a root, and $(\alpha^{13})^2 = \alpha^{26} = \alpha^{15}\alpha^{11} = \alpha^{11}$. Therefore

$$x^{16} - x = x(x - 1)(x^4 - x - 1)(x^4 - x^3 - x^2 - x - 1)(x^4 - x^3 - 1)(x^2 - x - 1)$$

3.3.2 Construction of Galois Fields $GF(2^r)$

We now present a method for constructing Galois fields $GF(2^r)$, for any r, starting from the binary field $GF(2)$; $GF(2)$ is called the *ground* field, and the characteristic of $GF(2^r)$ is 2. Note that arithmetic in $GF(2)$ is exactly the same as arithmetic in $\mathbf{Z}_2$, and we already know how to work with elements in that field, as well as with polynomials over $\mathbf{Z}_2$. Also, we say that an irreducible polynomial $p(x)$ of degree r is *c-primitive* if the smallest integer n for which $p(x)|(x^n - 1)$ is $n = 2^r - 1$; authors on coding theory use the term *primitive* for such polynomials, but, as you recall, we reserve that term for polynomials whose coefficients are relatively prime. c-Primitive polynomials are not easy to detect, and tables are given for them.

We start with two elements 0 and 1, from $GF(2)$ and the symbol α. We define multiplication, "$\cdot$", involving elements of $GF(2)$ and α as follows: $0 \cdot \alpha = \alpha \cdot 0 = 0$, $1 \cdot \alpha = \alpha \cdot 1 = \alpha$; we then set $\alpha^2 = \alpha \cdot \alpha$, $\alpha^3 = \alpha^2 \cdot \alpha = \alpha \cdot \alpha \cdot \alpha$, and so on. Note that from the preceding definition we have $0 \cdot \alpha^k = \alpha^k \cdot 0 = 0$, $1 \cdot \alpha^k = \alpha^k \cdot 1 = \alpha^k$, and $\alpha^j \cdot \alpha^k = \alpha^{j+k}$.

So far we have constructed the set $F' = \{0, 1, \alpha, \alpha^2, \ldots, \alpha^k, \ldots\}$ on the elements of which we have defined a multiplication operation. Next we impose a condition on the element α, so that F' contains only 2^r elements and the multiplication operation is closed in F'. Let $p(x)$ be a c-primitive polynomial of degree r over $GF(2)$, and assume that $p(\alpha) = 0$. Since $p(x)|(x^{2^r-1} - 1)$, we have $x^{2^r-1} - 1 = p(x)q(x)$, for some $q(x)$ over $GF(2)$, and we obtain $\alpha^{2^r-1} - 1 = p(\alpha)q(\alpha) = 0$, or $\alpha^{2^r-1} = 1$. Therefore, under the condition $p(\alpha) = 0$, the set F' becomes a new set F, which is finite and contains the elements $\{0, 1, \alpha, \alpha^2, \ldots, \alpha^{2^r-2}\}$, which are all distinct; that is we have

$$F = \{0, 1, \alpha, \alpha^2, \ldots, \alpha^{2^r-2}\}.$$

As before, F^* denotes the nonzero elements of F.

We now show that F^* is a (multiplicative) group under multiplication, an operation that is closed in F^*. First we see that the element 1 is a unit element. Next take α^i and α^j, two elements from F^*, and consider their product $\alpha^i \cdot \alpha^j = \alpha^{i+j}$. If $i+j < 2^r - 1$, then it is an element of F^* and we have nothing to do; however, if $i+j \geq 2^r - 1$, then we have $i+j = (2^r - 1) + v$, where $0 \leq v < 2^r - 1$, and $\alpha^i \cdot \alpha^j = \alpha^{i+j} = \alpha^{(2^r-1)+v} = \alpha^v$, which is also an element of F^*. Moreover, each element α^i in F^*, $0 < i < 2^r - 1$ has a multiplicative inverse, which is $\alpha^{(2^r-1)-i}$. Therefore, under multiplication F^* is a commutative group of order $2^r - 1$.

We next define an addition operation, "$+$", on F, so that F forms a commutative group under "$+$", and, in the process of doing so, we also develop another way for representing the elements of F.

Observe that if we divide x^i by $p(x)$, $0 \leq i < 2^r - 1$, we obtain $x^i = p(x)q_i(x) + v_i(x)$, $\deg[v_i(x)] < r$ over $GF(2)$; that is, $v_i(x) = v_{ir-1}x^{r-1} + \cdots + v_{i2}x^2 + v_{i1}x + v_{i0}$. Since $\gcd[x^i, p(x)] = 1$, it follows that $p(x)$ does not divide x^i and, therefore, for any i we have $v_i(x) \neq 0$. Moreover, for $i \neq j$ we have $v_i(x) \neq v_j(x)$. [This is seen by contradiction. Assume that for $i \neq j$ we have $v_i(x) = v_j(x)$. Then from the expressions $x^i = p(x)q_i(x) + v_i(x)$ and $x^j = p(x)q_j(x) + v_j(x)$ we obtain $x^i - x^j = p(x)[q_i(x) - q_j(x)] + v_i(x) - v_j(x) = p(x)[q_i(x) - q_j(x)]$. The last equality implies that $p(x)|(x^i - x^j) = x^i(1 - x^{j-i})$, assuming, of course, that $j > i$. Since $p(x)$ does not divide x^i, it follows that $p(x)|(1 - x^{j-i})$, a polynomial of degree $< 2^r - 1$, and this is the required contradiction since $p(x)$ is c-primitive.] Therefore, for $i = 0, 1, 2, \ldots, 2^r - 2$ we obtain $2^r - 1$ distinct nonzero polynomials $v_i(x)$ of degree $\leq r - 1$. Substituting x by α in the expression $x^i = p(x)q_i(x) + v_i(x)$, we now have

$$\alpha^i = v_i(\alpha) = v_{ir-1}\alpha^{r-1} + \cdots + v_{i2}\alpha^2 + v_{i1}\alpha + v_{i0}$$

So, from this relation we see that every element of F^* is *uniquely represented by a polynomial* in α over $GF(2)$ of degree $\leq r - 1$. The zero element of F is represented by the zero polynomial. Consequently, addition in F can be considered as addition of polynomials, and we know how to do this over $GF(2)$. We can now easily verify that F is a commutative group under addition, and combined with our previous result, we have that F is a field.

So far, we have developed two representations of $GF(2^r)$: the power representation (convenient for multiplications) and the polynomial representation (convenient for additions); see also Table 3.3.1. There also exists a third one derived by representing a polynomial as a tuple of its coefficients. Defining the r-tuple

$$\mathbf{v} = (v_{r-1}, v_{r-2}, \ldots, v_2, v_1, v_0)$$

as a *vector* over $GF(2)$, we obtain the vector representation of $GF(2^r)$ in the obvious way. Note that there are 2^r distinct vectors of length r. Vector addition is done componentwise and is itself another vector; hence, it is very convenient for addition of elements of $GF(2^r)$. Multiplication of a vector $\mathbf{v}$ times a scalar s is defined by

$$s(v_{r-1}, v_{r-2}, \ldots, v_2, v_1, v_0) = (sv_{r-1}, sv_{r-2}, \ldots, sv_2, sv_1, sv_0)$$

Moreover, we have the zero vector. The set of all binary vectors of length r with the operations and properties defined above form a *vector space* over $GF(2)$ and is denoted by $\mathbf{V}_r$.

We define the *inner product* of two vectors $\mathbf{u} = (u_{r-1}, u_{r-2}, \ldots, u_2, u_1, u_0)$ and $\mathbf{v} = (v_{r-1}, v_{r-2}, \ldots, v_2, v_1, v_0)$ in $\mathbf{V}_r$ as $\mathbf{u} \cdot \mathbf{v} = (u_{r-1}v_{r-1}, u_{r-2}v_{r-2}, \ldots, u_2v_2, u_1v_1, u_0v_0)$. If the inner product of two vectors equals zero, the vectors are called *orthogonal*.

Example. Using the c-primitive polynomial $p(x) = x^4 + x + 1$, let us construct $GF(2^4)$. The elements are obtained with repeated applicatiion of the equality $\alpha^4 = \alpha + 1$, where $p(\alpha) = 0$; they are presented in Table 3.3.1 in the three representations that we discussed above. For example, we have $\alpha^5 = \alpha^4 \cdot \alpha = (\alpha + 1)\alpha = \alpha^2 + \alpha$, and so forth. To multiply two elements, we simply add their exponents and use the fact that $\alpha^{15} = 1$; for instance, $\alpha^8 \cdot \alpha^9 = \alpha^{17} = \alpha^2$. Moreover, $\alpha^5/\alpha^8 = \alpha^5 \cdot \alpha^{(15-8)} = \alpha^5 \cdot \alpha^7 = \alpha^{12}$. To add, we use either the polynomial or the vector representation of the elements and do componentwise addition.

3.3.3 Circuits for Polynomial Arithmetic in $GF(2^r)$

The material in this section is essential for the implementation of error-correcting and error-detecting codes discussed in Chapter 4 (we follow Afrati, 1985); however, the topic is rather technical and, if the reader is not familiar with shift registers, extra effort might be required for a good understanding (e.g., Taub, 1982).

The circuits that we will discuss basically consist of the elements shown in Figure 3.3.1. The adder outputs the sum of the two values that are presented at its input gates, whereas the multiplier outputs the product of the value presented at the input gate times the constant a. The storing element "holds" the value presented at its input gate and outputs it later.

Our circuits are very similar to shift registers. Shift registers operate with a shift signal that is usually provided by a clock pulse; this signal will *not* be included in the diagrams that follow. In shift registers the storage elements

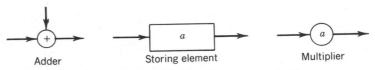

Figure 3.3.1. Building blocks of the circuits.

are simply delay devices (D-type flip-flops), where the value presented at the output gate is the value presented at the input gate exactly one clock pulse before. Each delay element is considered a stage of the shift register. Moreover, since we are dealing with elements of $GF(2)$, the adder is just an exclusive OR gate and the multiplier is just a connection, if the constant is 1, or simply no connection, if the constant is zero.

Input and output in a shift register are performed serially. When the input or output is a polynomial, then only the coefficients [which are in $GF(2)$] are presented at the input or output gate, one element per clock pulse. Note that *the coefficients of the terms of highest power are transmitted first*. (This is so because in division we first operate on the coefficients of the terms of highest power of the divident.) For instance, the polynomial

$$p(x) = c_n x^n + c_{n-1} x^{n-1} + \cdots + c_0$$

enters, or exits from, a shift register as a sequence of elements of $GF(2)$ with c_n first, c_{n-1} a clock pulse later, and so on.

Below we present circuits for the multiplication or division of any polynomial by another given polynomial. The circuit presented in Figure 3.3.2 multiplies any polynomial

$$a(x) = a_k x^k + a_{k-1} x^{k-1} + \cdots + a_0$$

that will appear at its input times the given polynomial

$$h(x) = h_r x^k + h_{r-1} x^{r-1} + \cdots + h_0$$

It is assumed that, initially, the delay elements all contain "0"; in addition, the coefficients of $a(x)$, which are entered serially, are followed by r "0."

Obviously, the product to be computed is

$$\begin{aligned} a(x)h(x) &= a_k h_r x^{k+r} + (a_{k-1} h_r + a_k h_{r-1}) x^{k+r-1} \\ &\quad + (a_{k-2} h_r + a_{k-1} h_{r-1} + a_k h_{r-2}) x^{k+r-2} \\ &\quad + \cdots + (a_0 h_2 + a_1 h_1 + a_2 h_0) x^2 + (a_0 h_1 + a_1 h_0) x + a_0 h_0 . \end{aligned}$$

As we see from Figure 3.3.2, when the first coefficient a_k of $a(x)$ appears at the input, the first coefficient $a_k h_r$ of $a(x)h(x)$ appears at the output. At this point the delay elements all contain "0." After a clock pulse a_{k-1} appears at the input, a_k is contained in the first storage element, and the rest of the storage elements contain "0"; the output is $a_{k-1} h_r + a_k h_{r-1}$, which is the correct second coefficient of $a(x)h(x)$. Likewise, after two clock pulses, a_{k-2} appears at the input, the storage elements contain $a_{k-1}, a_k, 0, \ldots, 0, 0, 0$, and at the output appears the correct third coefficient of $a(x)h(x)$. This

148   POLYNOMIALS

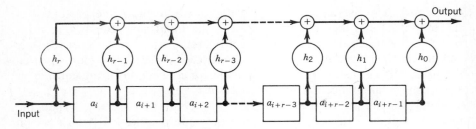

Figure 3.3.2.   A circuit for polynomial multiplication.

process continues in the same way. After $r + k - 1$ clock pulses the shift register contains $0, 0, 0, \ldots, 0, a_0, a_1$ and the one before last coefficient of $a(x)h(x)$ appears at the output, namely, $a_0h_1 + a_1h_0$. After $r + k$ clock pulses the shift register contains $0, 0, 0, \ldots, 0, 0, a_0$, and the output is a_0h_0, the last coefficient of the product $a(x)h(x)$.

Another circuit for multiplication is shown in Figure 3.3.3. The coefficients of the product are formed in the storage elements of the shift register. When the first coefficient appears at the input gate, at the output appears $a_k h_r$, and the storage elements all contain "0." After a clock pulse the shift register contains $a_k h_0, a_k h_1, \ldots, a_k h_{r-1}$, the input is a_{k-1}, and the output is $a_{k-1}h_r + a_k h_{r-1}$, which is the correct second coefficient of $a(x)h(x)$. At the next clock pulse the shift register contains $a_{k-1}h_0, a_k h_0 + a_{k-1}h_1, a_k h_1 + a_{k-1}h_2, \ldots, a_k h_{r-2} + a_{k-1}h_{r-1}$, the input is a_{k-2}, and the output is $a_{k-2}h_r + a_{k-1}h_{r-1} + a_k h_{r-2}$, the correct third coefficient of $a(x)h(x)$. The process continues in this way.

This last circuit can be seen in yet another way. Namely, the set of the r storage elements form a shift register that can store a polynomial. In the beginning this polynomial is 0. When a_k enters, then $a_k h(x)$ is added to the contents of the register. Delay by one clock pulse means multiplication by x, and at the output we receive the first coefficient. The appearance of a_{k-1} at the input gate adds $a_{k-1}h(x)$ to the contents of the register, and the unit delay multiplies by x and we obtain the second coefficient at the output, and so on.

Circuits of the type presented in Figure 3.3.3 can have more than one input gates. For example, the circuit shown in Figure 3.3.4 has two inputs,

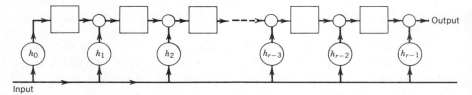

Figure 3.3.3.   Another circuit for polynomial multiplication.

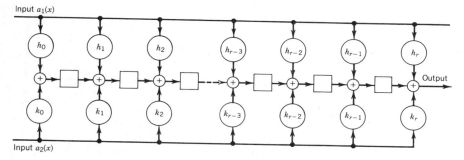

Figure 3.3.4. Polynomial multiplier with two inputs.

$a_1(x)$ and $a_2(x)$, and the output is

$$b(x) = a_1(x)h(x) + a_2(x)k(x)$$

where

$$h(x) = h_r x^r + h_{r-1} x^{r-1} + \cdots + h_0, \qquad k(x) = k_r x^r + k_{r-1} x^{r-1} + \cdots + k_0$$

Example. The circuits shown in Figure 3.3.5 multiply the input polynomial times $h(x) = x^6 + x^5 + x^4 + x^3 + 1$ over $GF(2)$. The reader should study the operation of these circuits step by step and compare the results with those obtained through hand computations.

A circuit for dividing any polynomial, say, $d(x) = d_n x^n + d_{n-1} x^{n-1} + \cdots + d_0$, by the polynomial $g(x) = g_r x^r + g_{r-1} x^{r-1} + \cdots + g_0$ is provided in Figure 3.3.6. The register contains all "0" at the beginning. The output is "0" for the first r clock pulses. Then the first nonzero output appears and it is $d_n g_r^{-1}$, the first coefficient of the quotient. For each coefficient q_j of the quotient, the polynomial $q_j g(x)$ must be subtracted from the dividend. [Recall that in $GF(2)$, addition is the same as subtraction.] This is achieved with the help of the feedback connections. After n clock pulses the whole

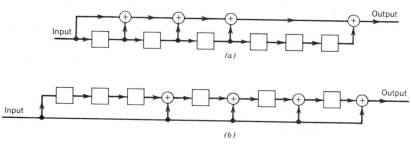

Figure 3.3.5. Circuits for multiplying times $x^6 + x^5 + x^4 + x^3 + 1$.

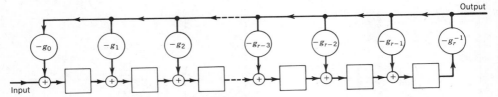

Figure 3.3.6. Circuit for polynomial division.

quotient has appeared at the output and the remainder is in the storage elements of the register. This is better understood if we consider the circuit of Figure 3.3.7, take as dividend the polynomial $x^{13} + x^{11} + x^{10} + x^7 + x^4 + x^3 + x + 1$, and try out the division.

The preceding circuits can be easily adopted for polynomial arithmetic modulo $g(x)$. The register shown in Figure 3.3.6 can store elements of $GF(2)$, which can be considered the coefficients of a polynomial

$$b(x) = b_{r-1}x^{r-1} + b_{r-2}x^{r-2} + \cdots + b_0$$

of degree $\leq r - 1$. If the contents of the register shift once to the right (or after one clock pulse), the stored polynomial becomes

$$b'(x) = b_{r-2}x^{r-2} + b_{r-3}x^{r-3} + \cdots + b_0 x - b_{r-1}\{g_r^{-1}g(x) - x^r\}$$

where the last term is the result of the feedback connections. This can be also written as

$$b'(x) = xb(x) - b_{r-1}g_r^{-1}g(x)$$

which is the same as

$$b'(x) = xb(x) \ [\text{mod } g(x)]$$

Let us consider an example to make things clear. Consider $g(x) = x^4 + x + 1$ and the field $GF(2)$. The polynomial $g(x)$ is c-primitive, and its root α is a primitive element of $GF(2^4)$; that is, all the elements of $GF(2^4)$ are obtained from the first 16 powers of α. The corresponding register is shown in Figure 3.3.8. If we place "1" in the first, starting from left, storage element, and "0" in the others, then successive shifts will give us the

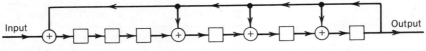

Figure 3.3.7. Circuit for division by $x^6 + x^5 + x^4 + x^3 + 1$.

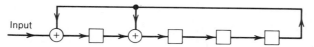

Figure 3.3.8. A circuit that counts in $GF(2^4)$.

representations of the various powers of α in exactly the same form as they are found in Table 3.3.1. Note, for example, that the "1" coming out of the last storage element corresponds to α^4, which is replaced by $\alpha + 1$ with the help of the feedback connections.

Multiplication of two elements in $GF(2^4)$ can be performed, using the register of Figure 3.3.8 as an accumulator, in exactly the same way as is done in digital computers.

Example. Using the register illustrated in Figure 3.3.8, let us multiply the elements α^{10} and α^7 of $GF(2^4)$ as they are represented in Table 3.3.1; that is, $\alpha^{10} = \alpha^2 + \alpha + 1 = (0, 1, 1, 1)$, and $\alpha^7 = \alpha^3 + \alpha + 1 = (1, 0, 1, 1)$. The content of the accumulator is shown below after each operation. Note that a circuit is used that can add vectors to the vector contained in the accumulator; moreover, the vector representation of the elements is without the commas and the coefficients of α^0 appears at the left.

```
                                    Accumulator
1 1 0 1                             0000
      └──── Add 1 · (1110)          1110
           Shift                    0111
      ──── Add 0 · (1110)           0111
           Shift                    1111
      ──── Add 1 · (1110)           0001
           Shift                    1100
      ──── Add 1 · (1110)           0010
```

which is the answer.

Exercises

Section 3.1.1

1. If we are doing polynomial arithmetic modulo 7, what is $3x^2 + 5x + 1$ minus $6x + 4$? What is the product of the above polynomials?

2. Is the product of monic polynomials monic? What is the degree of the product of polynomials of respective degrees m and n? What is the degree of the sum of polynomials of respective degrees m and n?

3. For which, if any, primes p do the polynomials $x^5 + 2x + 1$ and $x^8 + 8x^2 + 1$ agree as functions over $\mathbf{Z}_p$?

4. (a) Find the roots of $x^3 + 3x + 3$ in $\mathbf{Z}_5$.
   (b) Using **PDF**, divide $2x^4 + 3x^3 + x + 4$ by $x^2 + 2$ in $\mathbf{Z}_5$.

Section 3.1.2

1. Given the polynomial $p(x)$, derive the computing-time bound for computing the polynomial $p(\alpha + x)$, $\alpha \geq 1$. [*Hint*: See also the paper by Akritas and Danielopoulos (1980).]

2. If $p(x) = x^3 - 7x + 7$, compute $p(3 + x)$ using the scheme mentioned in this section.

Section 3.1.3

1. What is the computing time of Lagrange's interpolation?

2. Use Lagrange interpolation to find the polynomial $p(x)$ in $\mathbf{R}[x]$ of degree at most 2 such that $p(1) = -8$, $p(3) = 2$, and $p(4) = 13$.

3. Find a polynomial $p(x)$ in $\mathbf{Z}_{11}[x]$ such that $p(1) = 8 \pmod{11}$, $p(3) = 1 \pmod{11}$, and $p(7) = 4 \pmod{11}$.

Section 3.1.4

1. Show that $p_1(x) \equiv p_2(x)$ iff $m(x) | [p_1(x) - p_2(x)]$ defines an equivalence relation on $J[x]$, J a field.

2. Complete the proof of Theorem 3.1.9.

3. Using the evaluation–interpolation scheme, develop an algorithm for *trial* division of two polynomials over a field; that is, if $p_1(x) | p_2(x)$, return the quotient; otherwise, state that $p_1(x)$ does not divide $p_2(x)$. [*Hint*: The main point here is that if $\deg[p_1(x)] = n$ and $\deg[p_2(x)] = m$, $m \geq n \geq 0$, then for the quotient $q(x)$ we have $\deg[q(x)] = m - n$. So, if the degree of the polynomial obtained from interpolation is $\neq m - n$, we can clearly state that $p_1(x)$ does not divide $p_2(x)$.]

Section 3.2.1

1. Does $x^2 + x + 1$ divide $x^6 - 1$ over the integers?

2. Find all m such that $x^3 + 5$ divides $x^5 + x^3 + x^2 - 15$ in $\mathbf{Z}_m[x]$.

3. Compute the greatest common divisor of $p_1(x) = x^3 - 7x + 7$ and $p_2(x) = 3x^2 - 7$ over the reals.

Section 3.2.2

1. (Continued fractions for polynomial approximation.) Let $c(x)$ and $d(x)$ be two polynomials over a field, with $\deg[c(x)] > \deg[d(x)]$, and let $a_1(x), a_2(x), \ldots$ be the quotient polynomials when Euclid's algorithm is applied to $c(x)$ and $d(x)$; moreover, let $p_0(x) = q_{-1}(x) = 0$, and $p_{-1}(x) = q_0(x) = 1$. We wish to show that the convergents $p_n(x)/q_n(x)$ of the continued fraction $[a_1(x), a_2(x), \ldots]$ are the "best" approximations of low degree to the rational function $d(x)/c(x)$, where the convergents are defined in a way analogous to that of Section 2.2.4. (The reader should review it.)

    Prove that if $p(x)$ and $q(x)$ are two polynomials such that $\deg[q(x)] < \deg[q_n(x)]$ and $\deg[p(x)c(x) - q(x)d(x)] \leq \deg[p_{n-1}(x)c(x) - q_{n-1}(x)d(x)]$, for some $n \geq 1$, then $p(x) = cp_{n-1}(x)$ and $q(x) = cq_{n-1}(x)$, for some constant c. Each $q_n(x)$ is the "best" polynomial in the sense that no nonzero polynomial $q(x)$ of smaller degree can make the quantity $p(x)c(x) - q(x)d(x)$, for any polynomial $p(x)$, achieve a degree as small as $p_n(x)c(x) - q_n(x)d(x)$. [Hint: Use **XEA-P**; see also Knuth (1981), p. 622.]

2. Prove that for $p_1(x)$, $p_2(x)$, $p_3(x)$ in $J[x]$, J a field, if $gcd[p_1(x), p_2(x)] = 1$ and $p_3(x)|p_1(x)$, then $gcd[p_3(x), p_2(x)] = 1$.

3. Prove that for $p_1(x)$, $p_2(x)$, $p_3(x)$ in $J[x]$, J a field, if $gcd[p_1(x), p_2(x)] = 1$, then $gcd[p_1(x)p_3(x), p_2(x)] = gcd[p_3(x), p_2(x)]$.

4. (a) Apply **XEA-P** to the polynomials $x^2 + 1$ and $x^5 + 1$ in $\mathbf{Z}_2[x]$.
    (b) In $\mathbf{Q}[x]$ solve $p(x)(x^2 - 3x + 2) + q(x)(x^2 + x + 1) = 1$.
    (c) In $\mathbf{Q}[x]$ find the gcd of $x^{11} - 1$ and $x^9 - 1$. [Hint: Use the identity $x^m - 1 \pmod{x^n - 1} = x^{m(\bmod n)} - 1$.]

Section 3.2.3

1. Prove that in $\mathbf{R}[x]$, no polynomial of odd degree is irreducible and that in $\mathbf{C}[x]$ no polynomial of degree 2 is irreducible.

2. Given that $p(x) = x^3 - 2x^2 + x - 2$ has roots $-i$ and $+i$, factor $p(x)$ into a product of irreducible polynomials in $\mathbf{Z}[x]$.

3. Find, if they exist, all factors in $\mathbf{Z}[x]$ of degree 1 in the following polynomials:
    (a) $x^3 - 2x^2 - 5x + 6$.
    (b) $3x^4 + 4x^3 - 6x^2 - 9x - 10$.

Section 3.2.4

1. Apply **PSQFF** to compute the squarefree factors of $p(x) = x^{12} - 36x^{10} + 510x^8 - 3580x^6 + 12825x^4 - 21384x^2 + 11664$. [Hint: The above polynomial is the product of $x - 1$, $x + 1$, $(x - 2)^2$, $(x + 2)^2$, $(x - 3)^3$, $(x + 3)^3$.]

Section 3.3.1

1. What are the orders of the nonzero elements of $GF(7)$?
2. Prove Theorem 3.3.9.
3. If p is a prime number show that $x^{p^m} - x$ divides $x^{p^n} - x$ in $\mathbf{Z}_p[x]$ if and only if m divides n.
4. Show that if $\alpha \in GF(q)$ with $\alpha^r = 1$, then $\alpha^d = 1$, where $d = \gcd(r, q - 1)$.
5. If $m(x) = x^2 + 1$, find a primitive element β of $\mathbf{Z}_3[x]_{m(x)}$. Also find the minimal polynomial of β in $\mathbf{Z}_3[x]$.
6. Prove Theorem 3.3.16.
7. Find all the irreducible factors of $x^8 - x$ in $\mathbf{Z}_2[x]$.

Section 3.3.2

1. Let α_i, $i = 0, 1, \ldots, 7$ be the eight elements of $GF(8)$, defined by the irreducible polynomial $x^3 + x + 1$ over $\mathbf{Z}_2$. Develop the operation tables for addition and multiplication of elements of $GF(8)$.

References

Afrati, F.: *Introduction to the theory of information* (in Greek). National Technical University of Athens, Greece, 1985.

Akritas, A. G., and S. D. Danielopoulos: On the complexity of algorithms for the translation of polynomials. *Computing* **24**, 51–60, 1980.

Albert, A. A.: *Fundamental concepts of higher algebra*. University of Chicago Press, Chicago, 1958.

Berlekamp, E. R.: *Algebraic coding theory*. McGraw-Hill, New York, 1968.

Cajori, F.: Horner's method of approximation anticipated by Ruffini. *American Mathematical Society Bulletin* **17**, 409–414, 1911.

Childs, L.: *A concrete introduction to higher algebra*. Springer Verlag, New York, 1979.

Knuth, D. E.: *The art of computer programming*. Vol. 2, 2nd ed. *Seminumerical algorithms*. Addison-Wesley, Reading, MA, 1981.

Netto, E.: *Vorlesungen über Algebra*. Erster Band. Teubner, Leipzig, 1896.

Sims, C. C.: *Abstract algebra, a computational approach*. Wiley, New York, 1984.

Taub, H.: *Digital circuits and microprocessors*. McGraw-Hill, New York, 1982.

Wang, S. P., and B. T. Trager: New algorithms for polynomial square-free decompositions over the integers. *SIAM Journal on Computing* **8**, 300–305, 1979.

Yun, D. Y. Y.: On square-free decomposition algorithms. *Proceedings of the 1976 ACM SYMSAC* (Yorktown Heights, NY), 26–35, 1976.

Part III

Applications and Advanced Topics

This part of the book is devoted to applications of the material presented in Parts I and II and advanced topics. The following topics are covered:

Chapter 4: Error-correcting codes and cryptography.
Chapter 5: Greatest common divisors of polynomials over the integers and polynomial remainder sequences.
Chapter 6: Factorization of polynomials over the integers.
Chapter 7: Isolation and approximation of the real roots of polynomial equations.

The reader should keep in mind that Section 7.2 depends on Section 5.2.

Part III

Applications and Advanced Topics

4

Error-Correcting Codes and Cryptography

4.1 Error-Correcting Codes—General Concepts

4.2 Cryptography—General Concepts

Exercises

Programming Exercises

References

Coding and decoding information for the purpose of maintaining secrecy is known as *cryptology*. However, there are many other types of code where secrecy is not of interest. Examples are zip codes used by post offices, ASCII or EBCDIC codes used to convert alphabetic characters into binary form for computer representation, and the Universal Product Code, the series of black vertical lines containing product information found on many items.

In this chapter we will examine codes intended to protect the information being transmitted against corruption in the information itself and against unauthorized receivers.

4.1
Error-Correcting Codes—General Concepts

While transmitting data over a channel, errors can be introduced through a variety of means such as hardware failure, "noise," or "glitches" to which sensitive and complex electronic equipment is subject. To protect against errors, we can code the information to be sent and decode it at the other end in such a way as to maximize the probability of correcting, or at least detecting, such errors. Successful coding schemes often rely on abstract algebra.

Algebraic coding theory is now over 30 years old. It originally arose in response to the celebrated coding theorem of Shannon (1948, 1949), which bounded the error performance obtainable on a discrete channel but gave little indication of how that performance might be achieved. That is, Shannon's theorem guarantees the existance of codes that can transmit information at rates close to capacity with an arbitrarily small probability of error.

In this section we describe two families of codes: the first is able to detect and correct a single error and was developed by Hamming in 1950, whereas the second is able to detect and correct two errors and was developed 10 years later by Bose, Chaudhuri, and Hocquenghem (BCH for short) (Bose and Chaudhuri, 1960; Hocquenghem, 1959). The main tools in Hamming codes are linear algebra (the reader should review the Appendix at the end of the book) and the language of cosets, whereas finite fields play an important role in the BCH codes. (In our discussion, we follow Afrati, 1985; see also Berlekamp, 1968; Blake, 1979; Levinston, 1970; MacWilliams, 1977; Peterson et al., 1972; and Wakerly, 1978.)

In Figure 4.1.1 we describe our model of a communication system. Messages go through the system starting from the sender (also known as *source* or *transmitter*). The channel could be air, or wires.

We consider only sources with a finite number of discrete signals as opposed to continuous sources such as ratio. More precisely, we are dealing

ERROR-CORRECTING CODES—GENERAL CONCEPTS

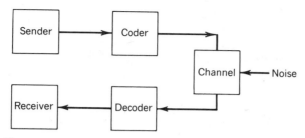

Figure 4.1.1. The model of a communication system.

with *binary codes*, where we assume that any information we want to transmit can be represented as a sequence of binary digits; this is the *message*. Moreover, we assume that our channel is a *binary symmetric channel* (BSC) of the form shown in Figure 4.1.2.

That is, if p is the probability that a binary signal is received correctly, then $1 - p$ is the probability of incorrect reception. We assume that errors occurring during transmission are independent and such that only one binary digit is altered, whereas the surrounding digits remain unaffected. (In reality this is not so, and attention has been given to codes that deal with channels where errors occur in *bursts* or *clusters*.) Finally, we assume that it is just as likely for a "0" to be converted to "1" as it is for a "1" to be converted to "0," and that the probability of either event is relatively small, say, .01. (Actually, ones are converted more frequently than zeros.)

Although we are unable to prevent the channel from causing such errors, we can take certain measures to protect ourselves from them. The idea is to annex r *check* digits to the set of k *information* digits that we wish to transmit, and then transmit the entire block of $n = k + r$ digits. The assumption, of course, is that the channel will change relatively few of these n transmitted digits, so that the user, based on the r check digits, will have sufficient information to detect and correct the channel errors. (This is an example of *redundancy*, that is, transmission of more than the k message digits in order to improve the reliability of the transmission process.) Of course, there must be a rule for selecting the r check digits, and the process is called *encoding*. We call $R = k/n$ the *information rate* of the code.

Definition 4.1.1. The k-dimensional subspace C of the vector space $\mathbf{V}_n$ of all binary n-tuples over $GF(2)$ is called an (n, k) *code*.

$$0 \xrightarrow{p} 0, \qquad 0 \xrightarrow{(1-p)} 1,$$
$$1 \xrightarrow{(1-p)} 0, \qquad 1 \xrightarrow{p} 1.$$
$$(a) \qquad\qquad (b)$$

Figure 4.1.2. The two forms of a binary symmetric channel.

Any sequence of n digits that can be transmitted is called a *codeword*. Obviously, there are 2^n sequences, or vectors, of length n, but only 2^k of those are codewords because the r check digits within each codeword are completely determined from the k message digits; we refer to the set of these 2^k vectors of length n as the *code*. For any codeword that is transmitted, and given the fact that we are dealing with a noisy channel, any one of the 2^n vectors of length n may be received; then, the user has to decide which of the 2^k possible codewords was transmitted.

Definition 4.1.2. If $\mathbf{u}$ is the transmitted codeword and $\mathbf{v}$ is received, then $\mathbf{e} = \mathbf{u} - \mathbf{v} = (e_1, e_2, \ldots, e_n)$ is called an *error word*.

On receiving $\mathbf{v}$, the user decides which codeword was transmitted, by deciding the most likely error. If all codewords of length n are equally likely, then this procedure minimizes the probability for incorrect decoding. *Maximum-likelihood decoding* simply assumes the most probable situation, namely, that the smallest number of errors occurred.

Definition 4.1.3. The *Hamming distance* $d(\mathbf{u}, \mathbf{v})$ between two vectors of length n is the number of coordinates in which $\mathbf{u}$ and $\mathbf{v}$ differ. The *Hamming weight* $w(\mathbf{u})$ of a vector of length n is the number of nonzero coordinates u_i, and obviously, $w(\mathbf{u}) = d(\mathbf{u}, \mathbf{0})$.

For example, if $\mathbf{u} = (1, 0, 1, 1, 0)$ and $\mathbf{v} = (1, 1, 0, 0, 1)$, then $d(\mathbf{u}, \mathbf{v}) = 4$. It can be shown that the Hamming distance is a *metric* on the vector space V_n over $GF(2)$ (see Exercise 1 for this section); that is, for any $\mathbf{u}_1, \mathbf{u}_2$ in V_n the following properties are satisfied:

a. $d(\mathbf{u}_1, \mathbf{u}_2) \geq 0$, and $d(\mathbf{u}_1, \mathbf{u}_2) = 0$ iff $\mathbf{u}_1 = \mathbf{u}_2$.
b. $d(\mathbf{u}_1, \mathbf{u}_2) = d(\mathbf{u}_2, \mathbf{u}_1)$.
c. $d(\mathbf{u}_1, \mathbf{u}_3) \leq d(\mathbf{u}_1, \mathbf{u}_2) + d(\mathbf{u}_2, \mathbf{u}_3)$ (triangular inequality).

Likewise, it can be shown that the Hamming weight is a *norm* on the vector space V_n over $GF(2)$; that is, for any $\mathbf{u}, \mathbf{v}$ in V_n and λ in $\mathbf{R}$ we have the following:

a. $w(\mathbf{u}) \geq 0$, and $w(\mathbf{u}) = 0$ iff $\mathbf{u} = \mathbf{0}$.
b. $w(\mathbf{u} + \mathbf{v}) \leq w(\mathbf{u}) + w(\mathbf{v})$.
c. $w(\lambda \mathbf{u}) = |\lambda| w(\mathbf{u})$.

For an understanding of the next theorem, some knowledge of probability theory is required.

Theorem 4.1.4. Let C be a code for a binary symmetric channel, and suppose that C contains s codewords of length n and of equal probability.

Then the *nearest-neighbor decoding* is the maximum likelihood decoding if the probability p for correct transmission is $> \frac{1}{2}$.

Proof. Let $d(\mathbf{u}, \mathbf{v})$ be the distance between a codeword $\mathbf{u}$ and the received vector $\mathbf{v}$. Then, in order for the user to receive the vector $\mathbf{v}$, when $\mathbf{u}$ was transmitted, $d(\mathbf{u}, \mathbf{v})$ errors must have occurred. The probability of this event is

$$\text{prob}(\mathbf{v}|\mathbf{u}) = (1-p)^{d(\mathbf{u},\mathbf{v})} p^{n-d(\mathbf{u},\mathbf{v})}$$

where p is the probability of correct transmission. Set $d_i = d(\mathbf{u}_i, \mathbf{v})$, $i = 1, 2$, and let us compare the probabilities $\text{prob}(\mathbf{v}|\mathbf{u}_1)$ and $\text{prob}(\mathbf{v}|\mathbf{u}_2)$ for two codewords $\mathbf{u}_1$ and $\mathbf{u}_2$. Obviously,

$$\frac{\text{prob}(\mathbf{v}|\mathbf{u}_1)}{\text{prob}(\mathbf{v}|\mathbf{u}_2)} = \frac{(1-p)^{d_1} p^{n-d_1}}{(1-p)^{d_2} p^{n-d_2}} = \left(\frac{p}{1-p}\right)^{d_2-d_1}$$

and, since we always assume $p > \frac{1}{2}$, it follows that

$$\text{prob}(\mathbf{v}|\mathbf{u}_1) > \text{prob}(\mathbf{v}|\mathbf{u}_2) \text{ if and only if } d_1 < d_2.$$

Therefore, $\text{prob}(\mathbf{v}|\mathbf{u})$ is minimized when $\mathbf{u}$ is the codeword closest to $\mathbf{v}$. If there are several codewords with this property, we can randomly choose one of them. □

Example. Suppose that the code contains the following four codewords:

$$\mathbf{u}_1 = (0, 0, 0, 0, 0)$$
$$\mathbf{u}_2 = (1, 0, 0, 1, 1)$$
$$\mathbf{u}_3 = (1, 1, 1, 0, 0)$$
$$\mathbf{u}_4 = (0, 1, 1, 1, 1)$$

If the received vector is $\mathbf{v} = (0, 1, 0, 1, 1)$, then it will be decoded as $\mathbf{u}_4$, because $d(\mathbf{u}_4, \mathbf{v}) = 1$ and $d(\mathbf{u}_i, \mathbf{v}) > 1$ for $i = 1, 2, 3$. Likewise, the received vector $\mathbf{w} = (0, 0, 1, 1, 0)$ could be decoded as either $\mathbf{u}_1$ or $\mathbf{u}_4$ because $d(\mathbf{u}_1, \mathbf{w}) = d(\mathbf{u}_4, \mathbf{w}) = 2$, and $d(\mathbf{u}_2, \mathbf{w}) = d(\mathbf{u}_3, \mathbf{w}) = 3$.

Since every vector $\mathbf{v}$ presented to the receiver is decoded to its nearest codeword (nearest with respect to the Hamming distance), it seems that a good code is one whose codewords lie "far" apart from each other. This is so because if the distance between codewords is great, then many errors would have to occur before the codeword $\mathbf{u}$ is converted to a (received) vector $\mathbf{v}$ that will be closer to another codeword $\mathbf{u}'$ than it is to $\mathbf{u}$. We have the following.

Lemma 4.1.5. Let C be a binary code with s codewords $\mathbf{u}_1, \mathbf{u}_2, \ldots, \mathbf{u}_s$ of length n. Then the code can correct all the combinations of t errors or less if and only if
$$d(\mathbf{u}_i, \mathbf{u}_j) \geq 2t + 1$$
for all $i \neq j$.

Proof. We prove it in one direction and leave the rest to the reader. Assume that the distance between any two codewords is at least $2t + 1$. If the transmitted codeword is $\mathbf{u}$ and the received word is $\mathbf{v}$ with $t' \leq t$ errors, that is, $d(\mathbf{u}, \mathbf{v}) = t'$, then, because of the triangular inequality of the metric, the distance of $\mathbf{v}$ from any other codeword $\mathbf{u}'$ will be
$$d(\mathbf{u}', \mathbf{v}) \geq d(\mathbf{u}', \mathbf{u}) - d(\mathbf{u}, \mathbf{v}) \geq 2t + 1 - t = t + 1 > t > t' = d(\mathbf{u}, \mathbf{v})$$
that is, it is always greater than $d(\mathbf{u}, \mathbf{v})$ and if $t' \leq t$ $\mathbf{v}$ will be decoded to the correct codeword $\mathbf{u}$. □

Definition 4.1.6. The minimum distance $d_{\min}$ of a binary code C is given by $d_{\min} = \min d(\mathbf{u}, \mathbf{v})$, where the minimum is taken over all vectors $\mathbf{u}, \mathbf{v}$ in C, $\mathbf{u} \neq \mathbf{v}$.

If we set $d = d_{\min}$, then another characterization of a code, analogous to Definition 4.1.1, is (n, k, d). Moreover, we have the following.

Theorem 4.1.7. The minimum distance of a binary code C is equal to the least weight of all nonzero codewords.

Proof. The proof is immediate from the fact that $d(\mathbf{u}, \mathbf{v}) = d(\mathbf{u} - \mathbf{v}, \mathbf{0}) = w(\mathbf{u} - \mathbf{v})$. □

Definition 4.1.8. Let $\mathbf{V}_n$ be the vector space of all binary n-tuples over $GF(2)$. Then the set $S_r(\mathbf{y}) = \{\mathbf{x} \in \mathbf{V}_n : d(\mathbf{x}, \mathbf{y}) \leq r\}$ is called the *sphere of radius r* about $\mathbf{y} \in \mathbf{V}_n$.

Lemma 4.1.9. A binary code C with minimum distance $d_{\min}$ can correct all combinations from 1 to t errors and can detect all combinations from $t + 1$ to $t + j$ errors, $0 < j \leq s$, if and only if $d_{\min} \geq 2t + s$.

Proof. The first part is proved by Lemma 4.1.5. We leave it as an exercise for the reader to show that any combination of $t + s$ errors can be detected. □

A geometric justification for Lemmas 4.1.5 and 4.1.9 is given by Figure 4.1.3. For instance, if $d(\mathbf{w}_i, \mathbf{w}_j) = 9$ as in Figure 4.1.3a, then a change of

ERROR-CORRECTING CODES—GENERAL CONCEPTS

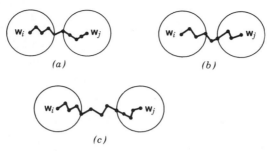

Figure 4.1.3. The relationship between distance and correcting ability of a code.

four digits, or less, in the codeword $\mathbf{w}_i$ will have as a result that the received vector will be closer to $\mathbf{w}_i$ than to $\mathbf{w}_j$. If, however, $d(\mathbf{w}_i, \mathbf{w}_j) = 8$ as in Figure 4.1.3b, then errors in three digits, or less, will be corrected, but four errors result in the received vector being equidistant from both $\mathbf{w}_i$ and $\mathbf{w}_j$ and so on.

Another geometric representation of binary codes is to view the vectors of length n as the vertices of an n-dimensional hypercube (see Figure 4.1.4). A change in a digit of a binary vector corresponds to a movement to a neighboring vertex, where the movement is along the edge parallel to the direction corresponding to the position of the digit that changed. To form a code with minimum distance d, we choose those vertices of the hypercube that cannot be connected with a path consisting of fewer than d edges. In Figure 4.1.4 the code consisting of the codewords $(0, 0, 1)$, $(0, 1, 0)$ and $(1, 1, 1)$ has $d_{\min} = 2$.

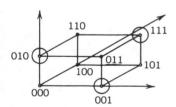

Figure 4.1.4. Binary codewords as vertices of a hypercube.

It is of interest to know how many codewords a code can have, when it is able to correct all the combinations of t errors or less. An upper bound on the number of codewords is given by the following.

Theorem 4.1.10 (Hamming's upper bound on the number of codewords). If the binary code C, which has s codewords of length n, is able to correct all combinations of t errors or less, then

$$s \leq \frac{2^n}{\sum_{0 \leq i \leq t} \binom{n}{i}}$$

Proof. Let $\mathbf{u}_1, \mathbf{u}_2, \ldots, \mathbf{u}_s$ be the codewords of length n in C. For each codeword consider the sphere $S_t(\mathbf{u}_i)$ with radius t around $\mathbf{u}_i$ (Definition 4.1.8). Since the code corrects up to t errors, the spheres do not intersect. Since the sphere around $\mathbf{u}_i$ contains all the vectors that differ from $\mathbf{u}_i$ in $0, 1, 2, \ldots, t$ positions, we have that the number of these vectors is

$$1 + n + \binom{n}{2} + \cdots + \binom{n}{t} = \Sigma_{0 \le i \le t} \binom{n}{i}$$

Therefore, in all s spheres we have

$$s \Sigma_{0 \le i \le t} \binom{n}{i}$$

vectors, and the proof is completed if we consider that the total number of vectors inside all the spheres cannot exceed 2^n, the total number of binary vectors of length n. □

When equality holds in Theorem 4.1.10, then the code is called *perfect*. Also note that, given t, n, and s as defined by that Theorem, it does not necessarily follow that there exists a code with s codewords of length n that can correct t errors. As an example, consider $t = 1$, and $n = 4$; then, by Theorem 4.1.10, $s = 3$, but there does not exist a code with these parameters (try it). When such codes exist, another characterization of them is (n, s, t).

4.1.1 Hamming Codes

Let us summarize what we have seen so far. We have considered a binary symmetric channel and a code consisting of 2^k codewords of length n, where the information rate is $R = k/n$. In order to have high information rate we would like to have as many codewords as possible. At the same time we would like to correct as many errors as possible. However, these two things are incompatable, because the more codewords we have the closer they will be to each other and, by Lemma 4.1.5, the smaller will be the correcting ability of the code. For instance, if $n = 5$ and we want to correct up to one error, we can find a code with four codewords, namely, $(0,0,0,0,0)$, $(1,0,0,1,1)$, $(1,1,1,0,0)$, $(0,1,1,1,1)$. However, if $n = 5$ and we want to correct two errors, then we cannot find more than two words, $(0,0,0,0,0)$, $(1,1,1,1,1)$. For a code of fixed length n, Theorem 4.1.10 indicates that as the demand for correcting ability increases the maximum possible number of codewords decreases.

So far, one problem we have completely ignored is how easily can we decode. The nearest-neighbor decoder that we studied earlier decodes the received vector $\mathbf{v}$ to that codeword $\mathbf{u}$, for which the distance $d(\mathbf{v}, \mathbf{u})$ is minimized. However, in order to do so, the decoder must keep a table of all the codewords, and when the vector $\mathbf{v}$ arrives, it is decided to which codeword $\mathbf{u}$ is nearest. However, such a decoder requires enormous memory, and the long decoding time would render real-time decoding infeasible in most cases.

Below we develop the theory of linear (the term will be defined below) binary codes that have many desirable characteristics as far as the complexity of coding and decoding is concerned.

One of the simplest examples of linear binary codes are the *repetition codes* $(n, 1)$; that is, $k = 1$ and $r = n - k$. In other words, the code is a one-dimensional subspace of all binary n-tuples and contains two codewords, $(0, 0, 0, \ldots, 0)$, the vector of n zeros, and $(1, 1, 1, 1, \ldots 1)$, the vector of n ones. On receiving a vector of length n the user can decide which codeword was transmitted using various rules; for example, if the number of received zeros/ones is greater than the number of the received ones/zeros, the user may decide that the all-zero/one codeword was transmitted. This scheme has great error-detection and error-correction capabilities, but it also has very low information rate.

At the other extreme are the *single-parity-check* codes $(n, n - 1)$ which contain only one check bit, $r = 1$. That is, these codes have very high information rate but no error-correction capability; they can only detect an *odd* number of errors. (In fact, any odd number of errors is indistinguishable from a single error.) The check bit, which is also called a *parity bit*, is taken to be the modulo 2 sum of the $n - 1$ message digits. In an *even/odd parity check*, the last digit is chosen so as to make the total number of ones an even/odd number. For example, in an even parity check code, the codeword 1101 is 11011. Note that any even number of channel errors cannot be detected. (Such codes are used in storing data on magnetic tapes.)

We are interested in finding codes with moderate information rates and moderate error-correction capabilities.

Let us examine the single-parity-check code described above. We observe that the $n - 1$ information bits can be freely chosen, whereas the nth bit, the check bit, is a linear combination of the first $n - 1$, and was added to help us detect possible transmission errors. If we denote the information bits by m_i and the check bits by c_i, then in the single-parity-check code we have

$$c_1 = m_1 + m_2 + \cdots + m_{n-1}$$

Generalizing this idea, we have the following. Let $m_1, m_2, \ldots, m_k$ be the k information bits, and let us suppose that we add r check bits to form a codeword of length $n = k + r$. (This is the encoding procedure.) Whereas each of the k information bits is formed independently from the preceding ones, the r check bits are linear combinations of the information bits; that is, we have

$$c_1 = h_{11}m_1 + h_{12}m_2 + \cdots + h_{1k}m_k$$
$$c_2 = h_{21}m_1 + h_{22}m_2 + \cdots + h_{2k}m_k$$
$$\cdots$$
$$c_r = h_{r1}m_1 + h_{r2}m_2 + \cdots + h_{rk}m_k$$

or, since addition is the same as subtraction in $GF(2)$,

$$h_{11}m_1 + h_{12}m_2 + \cdots + h_{1k}m_k + c_1 \qquad\qquad = 0$$
$$h_{21}m_1 + h_{22}m_2 + \cdots + h_{2k}m_k \qquad + c_2 \qquad = 0$$
$$\cdots\cdots$$
$$h_{r1}m_1 + h_{r2}m_2 + \cdots + h_{rk}m_k \qquad\qquad + c_r = 0$$

This system of equations can be also written as

$$\underset{\text{(the matrix }\mathbf{H}\text{)}}{\underbrace{\begin{bmatrix} h_{11} & h_{12} & \cdots & h_{1k} & 1 & 0 & \cdots & 0 \\ h_{21} & h_{22} & \cdots & h_{2k} & 0 & 1 & \cdots & 0 \\ & & \cdot & & & & & \\ & & \cdot & & & & & \\ h_{rl} & h_{r2} & \cdots & h_{rk} & 0 & 0 & \cdots & 1 \end{bmatrix}}_{n}} \begin{bmatrix} m_1 \\ m_2 \\ \cdot \\ \cdot \\ m_k \\ c_1 \\ c_2 \\ \cdot \\ \cdot \\ c_r \end{bmatrix} = \begin{bmatrix} 0 \\ 0 \\ \cdot \\ \cdot \\ 0 \end{bmatrix}$$

or as

$$\mathbf{H}\mathbf{v}^T = \mathbf{0}^T, \quad \text{or} \quad \mathbf{v}\mathbf{H}^T = \mathbf{0} \qquad (*)$$

where $\mathbf{v} = (m_1, m_2, \ldots, m_k, c_1, c_2, \ldots, c_r)$ is the codeword defined by the preceding system of equations and $\mathbf{v}^T$ is the transposed vector. More precisely, a vector $\mathbf{v}$ is a codeword, of a code defined by $\mathbf{H}$, if and only if $\mathbf{v}$ satisfies $(*)$. The above equations are called *parity check equations*, and the $r \times n$ matrix $\mathbf{H}$ is called *parity check matrix*.

If $\mathbf{v}_1$ and $\mathbf{v}_2$ are two codewords, then we have

$$\mathbf{H}(\mathbf{v}_1^T + \mathbf{v}_2^T) = \mathbf{H}\mathbf{v}_1^T + \mathbf{H}\mathbf{v}_2^T = \mathbf{0}^T + \mathbf{0}^T = \mathbf{0}^T$$

from which we conclude that $\mathbf{v}_1 + \mathbf{v}_2$ is also a codeword. Therefore, the codewords form a subspace of the vector space $\mathbf{V}_n$. We say that a code is *linear* if and only if the codewords form a subspace of $\mathbf{V}_n$.

Since the set of codewords forms a subspace, the codewords can be expressed as a linear combination of k linearly independent vectors that constitute the basis of the subspace. These k linearly independent vectors can be considered as rows $\mathbf{g}_i$ of a $k \times n$ matrix $\mathbf{G}$, called the *generator matrix* of the code. Each $\mathbf{g}_i$ is itself a codeword, which implies that

$$\mathbf{g}_i \mathbf{H}^T = \mathbf{0}$$

from which we conclude that

$$\mathbf{GH}^T = \mathbf{0}$$

Example. Suppose that for a code with $n = 7$ and $k = 4$ we have the following parity check equations:

$$c_1 = m_1 + m_2 + m_3$$
$$c_2 = m_1 + m_2 + m_4$$
$$c_3 = m_1 + m_3 + m_4$$

Then the parity check matrix **H** is.

$$\mathbf{H} = \begin{bmatrix} 1 & 1 & 1 & 0 & 1 & 0 & 0 \\ 1 & 1 & 0 & 1 & 0 & 1 & 0 \\ 1 & 0 & 1 & 1 & 0 & 0 & 1 \end{bmatrix}$$

The generator matrix **G** of the code is (see also Theorem 4.1.11 below)

$$\mathbf{G} = \begin{bmatrix} 1 & 0 & 0 & 0 & 1 & 1 & 1 \\ 0 & 1 & 0 & 0 & 1 & 1 & 0 \\ 0 & 0 & 1 & 0 & 1 & 0 & 1 \\ 0 & 0 & 0 & 1 & 0 & 1 & 1 \end{bmatrix}$$

The codewords are the vectors derived from the 2^4 linear combinations of the rows of the matrix **G**.

Obviously, the vector space derived from the rows of a matrix does not change if we perform any one of the following elementary row operations:

1. Interchange any two rows.
2. Multiply any row times a nonzero element.
3. Replace a row by the sum of that row with any other row.

(In the binary case the second operation is meaningless, because the only nonzero element is 1.)

Since we are interested in studying those properties of linear codes that are relevant to their error-correcting capability, we observe that two codes differing only in the placement order of the digits have the same error probability, and such codes are called *equivalent*. More precisely, if **V** is the vector space of the rows of a matrix **G**, then **V'** is a code equivalent to **V** if and only if **V'** is the vector space of the rows of a matrix **G'** that is obtained after a reordering of the columns of **G**. Therefore, reordering of the columns of a generator matrix leads us to a generator matrix of an equivalent code. Moreover, any one of the three elementary row operations mentioned above when performed on the rows of **G** leads us to a matrix **G''** with the same vector space, and therefore, **G** and **G″** are generators of the same code. Combining the two cases, we see that if a matrix **G′** is obtained from a matrix **G** after a series of elementary row operations and column reorderings of the latter, then **G** and **G′** are matrix generators of equivalent codes.

We usually want the generator matrix in the following form:

$$\mathbf{G} = \begin{bmatrix} 1 & 0 & 0 & \cdots & 0 & p_{11} & \cdots & p_{1,n-k} \\ 0 & 1 & 0 & \cdots & 0 & p_{21} & \cdots & p_{2,n-k} \\ & & & \cdots & & & & \\ 0 & 0 & 0 & \cdots & 1 & p_{k1} & \cdots & p_{k,n-k} \end{bmatrix}$$

that is, $\mathbf{G} = [\mathbf{I}_k \mathbf{P}]$, where $\mathbf{I}_k$ is the $k \times k$ unit matrix and $\mathbf{P}$ is any $k \times (n-k)$ matrix. Using elementary row operations, we can always bring a generator matrix in the above form and this is best illustrated in the example below.

Example. Suppose that we have the generator matrix

$$\mathbf{G} = \begin{bmatrix} 1 & 1 & 0 & 1 \\ 1 & 1 & 1 & 0 \\ 0 & 1 & 1 & 0 \end{bmatrix}$$

At every step we perform the necessary operations to bring one of the columns into the desired form; we begin from left to right.

For the first column: add the first row to the second and replace the second row with the sum

$$\begin{bmatrix} 1 & 1 & 0 & 1 \\ 0 & 0 & 1 & 1 \\ 0 & 1 & 1 & 0 \end{bmatrix}$$

For the second column: interchange the second and third rows

$$\begin{bmatrix} 1 & 1 & 0 & 1 \\ 0 & 1 & 1 & 0 \\ 0 & 0 & 1 & 1 \end{bmatrix}$$

and replace the first row by the sum of the first and second rows

$$\begin{bmatrix} 1 & 0 & 1 & 1 \\ 0 & 1 & 1 & 0 \\ 0 & 0 & 1 & 1 \end{bmatrix}$$

For the third column: replace the first and the second rows by their respective sums with the third row

$$\begin{bmatrix} 1 & 0 & 0 & 0 \\ 0 & 1 & 0 & 1 \\ 0 & 0 & 1 & 1 \end{bmatrix}$$

We see, therefore, that with repeated application of elementary operations we can always bring a generator matrix $\mathbf{G}$ into the form $\mathbf{G} = [\mathbf{I}_k \mathbf{P}]$.

Now, let $\mathbf{a} = (a_1, a_2, \ldots, a_k)$ be a vector of dimension k, and let us consider the product of $\mathbf{a}$ times the generator matrix $\mathbf{G} = [\mathbf{I}_k \mathbf{P}]$

$$\mathbf{b} = \mathbf{aG} = (a_1 \, a_2, \ldots, a_k, c_1, c_2, \ldots, c_{n-k})$$

where

$$c_j = \Sigma_{1 \leq i \leq k} a_i p_{ij} \qquad (**)$$

We observe that if we see the vector $\mathbf{a}$ as the vector with the information bits, then, in the codeword $\mathbf{b}$ of the code obtained from $\mathbf{G}$, the first k information bits and the rest $n - k$ components are linear combinations of the first k components. Therefore, coding is extremely simple; it suffices to multiply the information vector times the generator matrix. An *ordered* code is one consisting of codewords in each one of which the first k bits are information bits and the rest are check bits. It follows from the above that every linear code is equivalent to an ordered code. We say that a linear code with codewords of length n has dimension k if and only if it has k information bits, and, as we have already seen, we denote this code by (n, k).

Summarizing what we have said above, we see that a binary linear code of length n and dimension k is a k-dimensional subspace of the n-dimensional vector space over $GF(2)$. Therefore, the code is completely defined by the generator matrix $\mathbf{G}$, whose rows are the basis vectors of the subspace. Alternatively, a linear code C can be completely defined by the parity check matrix $\mathbf{H}$ that satisfies the equations $\mathbf{Hv}^T = \mathbf{0}^T$, or $\mathbf{vH}^T = \mathbf{0}$, for every codeword $\mathbf{v}$. We say that the vector space that defines the code is *orthogonal* to the vector space defined by the rows of $\mathbf{H}$. The vector space defined by the rows of the matrix $\mathbf{H}$ can be also considered as defining a linear code C'; $\mathbf{H}$ is now the generator matrix and $\mathbf{G}$ is the parity check matrix. The codes C and C' are called *dual codes*, and if C is a (n, k) code, then C' is a $(n, n-k)$ code.

Theorem 4.1.11. If V is a code with generator matrix $\mathbf{G} = [\mathbf{I}_k, \mathbf{P}]$, where $\mathbf{I}_k$ is the $k \times k$ unit matrix and $\mathbf{P}$ is any $k \times (n-k)$ matrix, then the parity check matrix of the code is $\mathbf{H} = [\mathbf{P}^T \mathbf{I}_{n-k}]$.

Proof. If $\mathbf{v} = (a_1, a_2, \ldots, a_k, c_1, c_2, \ldots, c_{n-k})$ is a codeword, then

$$\mathbf{vH}^T = \mathbf{0} = (\Sigma_i a_i p_{i1} + c_1, \Sigma_i a_i p_{i2} + c_2, \ldots, \Sigma_i a_i p_{i,n-k} + c_{n-k})$$

which is exactly the same as (**). (The relation between **G** and **H** can be understood if we observe that, in both cases, p_{ij} is the coefficient of the ith information bit in the sum yielding the jth parity check bit c_j.)

Example. From the generator matrix

$$\mathbf{G} = \begin{bmatrix} 1 & 0 & 0 & 1 & 1 \\ 0 & 1 & 0 & 1 & 0 \\ 0 & 0 & 1 & 0 & 1 \end{bmatrix} = [\mathbf{I}_3 \mathbf{P}]$$

we find the parity check matrix

$$\mathbf{H} = [\mathbf{P}^T \mathbf{I}_2] = \begin{bmatrix} 1 & 1 & 0 & 1 & 0 \\ 1 & 0 & 1 & 0 & 1 \end{bmatrix}$$

The codewords are

$$00000 \quad 10011 \quad 01010 \quad 11001 \quad 00101 \quad 10110 \quad 01111 \quad 11100$$

Theorem 4.1.12. A code with parity check matrix **H** has minimum weight (and therefore, minimum distance) w if and only if every set of $w-1$ or fewer columns of **H** is linearly independent.

Proof. We prove the theorem in one direction and leave the rest to the reader. A vector $\mathbf{v} = (a_1, a_2, \ldots, a_n)$ is a codeword if and only if $\mathbf{vH}^T = \mathbf{0}$, or equivalently, if and only if

$$\Sigma_{1 \leq i \leq n} a_i \mathbf{h}_i = \mathbf{0}$$

where $\mathbf{h}_i$ is the ith column of the matrix **H**. So, if w is the minimum weight of the code, then there does not exist a linear combination of $w-1$ or fewer columns of **H** equaling 0. □

From Theorem 4.1.12 we conclude that a code can have minimum distance 3 if and only if every $3 - 1 = 2$ columns of the parity check matrix **H** are linearly independent. However, two vectors in $\mathbf{V}_{n-k}$ over $GF(2)$ are linearly independent if they are simply different. Therefore, we can construct a code with minimum distance 3 (i.e., the code can correct *one* error) if we consider as parity check matrix **H**, a matrix with columns all the nonzero elements of the space $\mathbf{V}_{n-k}$. So we can construct the so called Hamming codes $(2^m - 1, 2^m - m - 1, 3)$; that is, $n = 2^m - 1$, $k = 2^m - m - 1$, and $d_{\min} = 3$.

ERROR-CORRECTING CODES—GENERAL CONCEPTS   171

Definition 4.1.13. Let **H** be the parity check matrix of a linear (n, k) code. If **v** is the received vector, then the vector $\mathbf{s} = \mathbf{v}\mathbf{H}^T$, of length $n - k$, is called the *syndrome* of **v**.

From this definition we see that a vector is a codeword if and only if the syndrome is **0**. With this in mind, the decoding procedure for the Hamming codes is very simple. If **v** is the received and **u** the transmitted vector, where $\mathbf{e} = \mathbf{v} - \mathbf{u}$, we work as follows. We first compute the syndrome of **v**, which is

$$\mathbf{s} = \mathbf{v}\mathbf{H}^T = (\mathbf{u} + \mathbf{e})\mathbf{H}^T = \mathbf{e}\mathbf{H}^T$$

If *one* error occurred, **e** is a vector with weight = 1; that is, $n - 1$ components are 0 and we have a "1" at the ith position, where the error occurred. Therefore, the product $\mathbf{e}\mathbf{H}^T$ will be the ith column of **H**, and the transmitted vector is obtained if we add to **v** the vector **e**, which is a vector of zeros except for the ith position that is "1."

Example. Let us consider the $(7, 4)$ Hamming code with parity matrix

$$\mathbf{H} = \begin{bmatrix} 1 & 0 & 0 & 1 & 1 & 0 & 1 \\ 0 & 1 & 0 & 1 & 0 & 1 & 1 \\ 0 & 0 & 1 & 0 & 1 & 1 & 1 \end{bmatrix}$$

and suppose that the received vector is $\mathbf{v} = (0, 1, 1, 0, 1, 0, 0)$. We see that the syndrome of **v** is

$$\mathbf{s} = \mathbf{v}\mathbf{H}^T = (1, 1, 0)$$

which is the fourth column of **H**; hence, the error vector is $\mathbf{e} = (0, 0, 0, 1, 0, 0, 0)$, and adding this to **v**, we obtain the transmitted vector $\mathbf{u} = (0, 1, 1, 1, 1, 0, 0)$.

We will now examine another decoding procedure for linear codes; that is, we will obtain another view of maximum likelihood decoding by considering the *cosets* of the code C in $\mathbf{V}_n$.

Let C be a linear (n, k) code over $GF(2)$, where the code is regarded as a subspace of $\mathbf{V}_n$, the vector space of all binary n-tuples. The factor space $\mathbf{V}_n/C$ consists of all cosets $\mathbf{v} + C = \{\mathbf{v} + \mathbf{x}: \mathbf{x} \in C\}$ for arbitrary $\mathbf{v} \in \mathbf{V}_n$, where each coset contains 2^k vectors. There is a partition of $\mathbf{V}_n$ of the form

$$\mathbf{V}_n = C \cup \{\mathbf{v}_1 + C\} \cup \cdots \cup \{\mathbf{v}_s + C\}, \qquad s = 2^{n-k} - 1$$

If the vector **y** is received, then **y** must be an element of one of these cosets, say, $\{\mathbf{v}_i + C\}$. If the codeword $\mathbf{x}_i$ has been transmitted, then the error vector e is $e = y - x_i \in v_i + C - x_i = v_i + C$. Hence, we have the following decoding rule: If a vector y is received, then the possible error vectors e are the vectors in the coset containing y. The most likely error is the vector e' with minimum weight in the coset of y. Therefore, y is decoded as $x' = y + e'$. (A vector of least weight in a coset is called the *coset leader*. If there are several such vectors, then we arbitrarily choose one of them as coset leader.)

To employ the decoding scheme outlined above we need coset tables and below, we show how to create the Slepian decoding table.

Let C be the (n, k) linear code mentioned above, and let $\mathbf{h}_1, \mathbf{h}_2, \ldots, \mathbf{h}_r$, $r = 2^k$ be the codewords, where $\mathbf{h}_1$ is the zero vector of the code. The codewords are placed in the first row with the zero vector at the left. We then choose one of the remaining vectors of $\mathbf{V}_n$, say, $\mathbf{g}_1$, and we place it under the zero vector. (Usually $\mathbf{g}_1$ is chosen in such a way so that it has the greatest probability of reaching the receiver had the zero vector been transmitted.) Following this, we complete the second row placing under each codeword $\mathbf{h}_i$ the vector $\mathbf{g}_1 + \mathbf{h}_i$. Likewise, a second vector $\mathbf{g}_2$ of $\mathbf{V}_n$, which has not yet appeared in the table, is placed in the first column of the third row, and the row is completed in the same way. This process is continued until all the vectors of $\mathbf{V}_n$ have appeared in the table. The set of the elements of a row form the coset, and, if at each stage the vector $\mathbf{g}_i$ is chosen to be a vector of least weight that is not present in the first i columns, the element in the *first column* is guaranteed to be a coset leader (verify this in the example that follows).

Two elements $\mathbf{g}_i$ and $\mathbf{g}_j$ that belong to the same coset k have the following property: $\mathbf{g}_i + \mathbf{g}_j = \mathbf{g}_k + \mathbf{h}_i + \mathbf{g}_k + \mathbf{h}_j = \mathbf{h}_i + \mathbf{h}_j = \mathbf{h}_m$; that is, they sum up to a codeword. Moreover, each vector of the space $\mathbf{V}_n$ appears only once in the table. (This is easily seen if we assume that two vectors $\mathbf{g} = \mathbf{g}_k + \mathbf{h}_i$, and $\mathbf{g}' = \mathbf{g}_m + \mathbf{h}_j$ are the same, where $\mathbf{g}_k$ and $\mathbf{g}_m$ are the coset leaders. Then $\mathbf{g}_k + \mathbf{h}_i = \mathbf{g}_m + \mathbf{h}_j$ if and only if $\mathbf{g}_k = \mathbf{g}_m + \mathbf{h}_i + \mathbf{h}_j = \mathbf{g}_m + \mathbf{h}_m$, where $\mathbf{h}_m$ is a codeword; but then $\mathbf{g}_k$ belongs to the coset that has $\mathbf{g}_m$ as coset leader and this contradicts the way in which the table was constructed, according to which the first element in each row is a vector that has not appeared before.)

Example. Consider the $(5, 3)$ linear code whose check bits are defined by the equations

$$c_1 = m_1 + m_2$$
$$c_2 = m_1 \quad\quad + m_3$$

Then the Slepian table of the code is

```
00000  10011  01010  11001  00101  10110  01111  11100
00001  10010  01011  11000  00100  10111  01110  11101
00010  10001  01000  11011  00111  10100  01101  11110
10000  00011  11010  01001  10101  00110  11111  01100.
```

Note that there are $2^{5-3} = 4$ cosets, each with $2^3 = 8$ elements. The following two theorems show how the Slepian table described above can be used for decoding linear codes.

Theorem 4.1.14. If we use the Slepian table to decode a received vector to the codeword right above it, then the received vector $\mathbf{v}$ is correctly decoded to the transmitted vector $\mathbf{u}$ if and only if the error vector $\mathbf{e}_i = \mathbf{v} - \mathbf{u}$ is coset leader.

Proof. If $\mathbf{e}_i = \mathbf{v} - \mathbf{u}$ is the leader of the ith coset, then $\mathbf{v} = \mathbf{u} + \mathbf{e}_i$ appears in the ith coset of the Slepian table right under the codeword $\mathbf{u}$, and it will be correctly decoded. However, if $\mathbf{v} - \mathbf{u}$ is not a coset leader, the received vector $\mathbf{v}$ will appear in some coset, say, the jth coset with leader $\mathbf{g}_j$. Then $\mathbf{v}$ belongs to jth row of the table, but it is not under $\mathbf{u}$ because $\mathbf{v} \neq \mathbf{g}_j + \mathbf{u}$. □

Theorem 4.1.15. Two vectors $\mathbf{v}_1$ and $\mathbf{v}_2$ belong to the same coset if and only if their syndromes are equal.

Proof. If $\mathbf{v}_1$ and $\mathbf{v}_2$ both belong to the ith coset, then $\mathbf{v}_1 = \mathbf{g}_i + \mathbf{h}_1$ and $\mathbf{v}_2 = \mathbf{g}_i + \mathbf{h}_2$. Therefore, $\mathbf{v}_1 + \mathbf{v}_2 = \mathbf{h}_1 + \mathbf{h}_2$ and $(\mathbf{v}_1 + \mathbf{v}_2)\mathbf{H}^T = (\mathbf{h}_1 + \mathbf{h}_2)\mathbf{H}^T = \mathbf{0}$, from which we conclude that $\mathbf{v}_1\mathbf{H}^T = \mathbf{v}_2\mathbf{H}^T$.
Conversely, if $\mathbf{v}_1\mathbf{H}^T = \mathbf{v}_2\mathbf{H}^T$, then $(\mathbf{v}_1 + \mathbf{v}_2)\mathbf{H}^T = \mathbf{0}$, which implies that $\mathbf{v}_1 + \mathbf{v}_2 = \mathbf{h}_m$; and if $\mathbf{v}_1 = \mathbf{h}_1 + \mathbf{g}_i$, we have $\mathbf{v}_2 = \mathbf{h}_m + \mathbf{v}_1 = \mathbf{h}_m + \mathbf{h}_1 + \mathbf{g}_i = \mathbf{g}_i + \mathbf{h}_2$, from which we see that $\mathbf{v}_1$ and $\mathbf{v}_2$ belong to the same coset. □

We are now in a position to present a simplified decoding procedure that works as follows. We construct a decoding table which consists of two columns, one for the 2^{n-k} syndromes and another one for the corresponding coset leaders. When a vector $\mathbf{v}$ is received, we first compute its syndrome, and we then find from the table the corresponding coset leader. It is assumed that the coset leader is the error vector, and adding it to the vector $\mathbf{v}$, we obtain the transmitted vector $\mathbf{u}$.

Example. For the linear code (5, 3) described in the last example we have the following decoding table

Coset Leader	Syndrome
00000	00
00001	01
00010	10
10000	11

We see that for an (n, k) code the decoding table requires a list of syndromes and coset leaders corresponding to the 2^{n-k} cosets. This may or may not be less expensive than checking a received vector against all possible 2^k codewords and selecting the nearest one. In a (100, 80) code, for instance, there are only 2^{20} cosets as opposed to 2^{80} codewords, although 2^{20} is a very big number.

Theorem 4.1.16. Let C be an (n, k) linear code whose codewords have the same transmission probability. Then, use of the Slepian table constructed as above (i.e., the coset leader having the least weight in the coset) maximizes the probability for correct decoding.

Proof. Let $\mathbf{v}_{ij}$ be the vector in the ith row and jth column of the Slepian table. The codewords $\mathbf{v}_{0j}$ are at the 0th row and at the top of every column. Let d_{ij} be the Hamming distance between the received vector $\mathbf{v}_{ij}$ and the codeword $\mathbf{v}_{0j}$ to which $\mathbf{v}_{ij}$ was decoded. We have already seen that the probability of correct decoding is maximized when the received vector is decoded to the nearest codeword.

Suppose, now, that a vector $\mathbf{v}$ appears in the Slepian table under the codeword $\mathbf{u}$, where $d(\mathbf{v}, \mathbf{u}) = w$, and let $\mathbf{g}$ be the leader of the coset containing $\mathbf{v}$; moreover, suppose that $\mathbf{u}_1$ is the nearest codeword to $\mathbf{v}$, with $d(\mathbf{v}, \mathbf{u}_1) = w_1$. Then, $\mathbf{g} = \mathbf{v} - \mathbf{u}$ has weight w, whereas the element $\mathbf{v} - \mathbf{u}_1 = \mathbf{g} + (\mathbf{u} - \mathbf{u}_1)$ has weight w_1 and is in the same coset. Since by definition $\mathbf{g}$ has the least weight in the coset, $w_1 \geq w$, and so $\mathbf{v}$ is at least as near to $\mathbf{u}$ as it is to $\mathbf{u}_1$. □

4.1.2 BCH Codes

In this section we examine the BCH codes, which enable us to detect and correct two errors. These codes are a major improvement over the Hamming codes, which can detect two errors but can correct only one. The BCH codes were developed by Bose and Chaudhuri in 1960 and by Hocquenghem in 1959, and they form a subset of the *cyclic* codes; the latter are examined in detail below. We begin with two definitions.

Definition 4.1.17. A subspace C of the n-dimensional binary vector space $\mathbf{V}_n$ is called a *cyclic subspace* or a *cyclic code*, if for each vector $\mathbf{v} = (v_{n-1}, v_{n-2}, \ldots, v_0)$ of C, the vector $\mathbf{v}' = (v_0, v_{n-1}, v_{n-2}, \ldots, v_1)$, obtained from $\mathbf{v}$ with a "cyclic shift" of one position to the right, belongs also to C.

Definition 4.1.18. A set of polynomials is called an *ideal* if and only if it consists of all the multiples of some polynomial.

Strictly speaking, we have just defined a *principal ideal*. However, it is easily shown that if J is a field, then every ideal in the ring of polynomials $J[x]$ is a principal ideal generated by a polynomial of least degree in $J[x]$. (The remainder, on dividing any other polynomial in the ideal by the *generating* polynomial, has to be zero.) Therefore, for our discussion, we use just the term ideal. Moreover, we will use the algebra of polynomials modulo $x^n - 1$ to describe cyclic codes, as well as the fact that a polynomial of degree $n - 1$ can be represented by a vector in $\mathbf{V}_n$; the latter will always be assumed a binary vector space, even when this is not explicitly mentioned.

To see why we use the algebra of polynomials modulo $x^n - 1$, we consider the following alternate description of the (7, 4) Hamming code that we have already encountered.

The columns of the parity check matrix $\mathbf{H}$ of the (7, 4) Hamming codes are binary vectors of length 3. As such, we know that they can be thought of as polynomials of degree ≤ 2 over $GF(2)$. This suggests that the columns of $\mathbf{H}$ can be indexed by the elements of $GF(2^3)$. More precisely, if α is a root of the polynomial $x^3 + x + 1$, then the seven nonzero elements of $GF(2^3)$ are the powers α^i, and they correspond to the columns of $\mathbf{H}$. (There is a correspondence analogous to that of Table 3.3.1, and the reader should construct a similar table to verify this; see also Theorem 3.3.11.) Likewise, codewords in the (7, 4) Hamming code can be written as binary polynomials of degree ≤ 6; in this case codewords are all the polynomials $p(x)$ of degree ≤ 6 having α as a root.

Now, $m(x) = x^3 + x + 1$ is the minimal polynomial of α over $GF(2)$, and all polynomials having α as a root are precisely the elements of the ideal generated by $m(x)$ in the polynomial ring $\mathbf{Z}_2[x]$. Since α is an element of order 7 (i.e., α is a root of $x^7 - 1$), we have $m(x)|(x^7 - 1)$. Consider the quotient ring $\mathbf{Z}_2[x]_{(x^7-1)}$ and the image of the ideal generated by $m(x)$ in this quotient ring. Clearly, given any multiple $k(x)m(x)$ of $m(x)$, we divide it by $x^7 - 1$ and obtain $k(x)m(x) = q(x)(x^7 - 1) + r(x)$, deg $[r(x)] \leq 6$. Moreover, $r(x) = k(x)m(x) - q(x)(x^7 - 1)$, and since $m(x)|(x^7 - 1)$, it follows that $m(x)|r(x)$, and $r(x)$ is a multiple of $m(x)$. Therefore, the image of the ideal generated by $m(x)$ under the quotient map from $\mathbf{Z}_2[x]$ to $\mathbf{Z}_2[x]_{(x^7-1)}$ can be viewed as precisely those polynomials $p(x)$ of degree ≤ 6, which are divisible by $m(x)$. We, therefore, arrive at a description of the (7, 4) Hamming code as the ideal generated by the image of $m(x)$ in $\mathbf{Z}_2[x]_{(x^7-1)}$. Below we will generalize the above, but first we present an alternate coding/decoding scheme for the (7, 4) Hamming code:

Coding. Let us suppose that we want to transmit the information bits m_1, m_2, m_3, and m_4. We form the polynomial $m_1 x^6 + m_2 x^5 + m_3 x^4 + m_4 x^3$, and divide it by $m(x) = x^3 + x + 1$, to obtain a quotient $q(x)$ and a remainder $r(x) = r_1 x^2 + r_2 x + r_3$, where deg $[r(x)] < 3$. Then, the transmitted codeword is $m_1, m_2, m_3, m_4, r_1, r_2$, and r_3, corresponding to the polynomial $m(x)q(x) = m_1 x^6 + m_2 x^5 + m_3 x^4 + m_4 x^3 + r_1 x^2 + r_2 x + r_3$, which is zero when evaluated at α.

Decoding. Assume that at most one error occurred; then, on receiving a vector of 7 bits, the user evaluates the corresponding polynomial at α, using of course the property $\alpha^2 = \alpha + 1$. If the answer is zero, then there were no errors, whereas if the answer is α^e, the error was at the coefficient of the eth power of x.

Example. To transmit the information $(1,0,0,1)$, we divide the polynomial $x^6 + x^3$ by $x^3 + x + 1$ to obtain the remainder $x^2 + x$; the transmitted codeword is $(1,0,0,1,1,1,0)$. Note that the polynomial $x^6 + x^3 + x^2 + x$ is zero when evaluated at α. (Note that, $\alpha^3 = \alpha + 1$ and $\alpha^6 = \alpha^2 + 1$.) If, now, the received vector is $(0,0,0,1,1,1,0)$, we evaluate the corresponding polynomial at α and obtain $\alpha^2 + 1 = \alpha^6$; from that we conclude that the error was at the coefficient of x^6.

Having justified the need for algebra of polynomials modulo $x^n - 1$ to describe cyclic codes, we present two theorems that define some basic properties of ideals.

Theorem 4.1.19. Let J be an ideal in the algebra of polynomials modulo $f(x)$, and let $g(x)$ be one of the nonzero polynomials of least degree in J. Then the polynomial $s(x)$ belongs to J if and only if $g(x)|s(x)$. Moreover, $g(x)|f(x)$.

Proof. Applying the division algorithm for polynomials, we have $s(x) = g(x)q(x) + r(x)$, where $\deg[r(x)] < \deg[g(x)]$. From this we have $r(x) = s(x) + g(x)q(x)$, and since both $s(x)$ and $g(x)$ belong to J, so does $r(x)$. However, $g(x)$ is the polynomial with the least degree in J, and so $r(x)$ must be zero and $s(x)$ is a multiple of $g(x)$. Regarding $f(x)$ we have $f(x) = g(x)q(x) + r(x)$, and, since $f(x) = 0 \,[\text{mod } f(x)]$, we have $g(x)q(x) + r(x) = 0 \,[\text{mod } f(x)]$, which implies that $g(x)q(x) = r(x) \,[\text{mod } f(x)]$. With the same reasoning as above, we have that $r(x) = 0$, and $g(x)|f(x)$. □

Therefore, every ideal in the algebra of polynomials modulo $f(x)$ is generated by a polynomial $g(x)$ whose degree is smaller than the degree of any other polynomial in J, and the elements of J are multiples of $g(x)$. If $n = \deg[f(x)]$ and $r = \deg[g(x)]$, then the ideal J has dimension $n - r$. Indeed, J is a vector subspace and the vectors $g(x), xg(x), \ldots, x^{n-r-1}g(x)$ are linearly independent, because any linear combination $(a_0 + \cdots + a_{n-r-1}x^{n-r-1})g(x)$ cannot be $0\,[\text{mod } f(x)]$ since its degree is $< n$. Obviously, every $s(x)$ in J can be expressed through these basis vectors. The polynomial $g(x)$ is called the *generator* of the ideal J.

We say that the polynomial $r(x)$ is *orthogonal* to the ideal J if $r(x)s(x) = 0\,[\text{mod } f(x)]$ for every polynomial $s(x)$ in J.

Theorem 4.1.20. Consider the polynomials $f(x)$, $g(x)$, and $h(x)$, where $f(x) = g(x)h(x)$. Then in the algebra of polynomials modulo $f(x)$ the polynomial $a(x)$ is orthogonal to the ideal generated by $h(x)$ if and only if $a(x)$ belongs to the ideal generated by $g(x)$.

Proof. Suppose that $a(x)$ belongs to the ideal generated by $g(x)$ and that $b(x)$ belongs to the ideal generated by $h(x)$; then, by Theorem 4.1.19, $a(x)b(x) = q(x)g(x)h(x) = q(x)f(x) = 0\,[\text{mod } f(x)]$.

Conversely, if $a(x)$ is orthogonal to the ideal generated by $h(x)$, then $a(x)h(x) = 0 \,[\mathrm{mod}\, f(x)]$; that is, $a(x)$ is a multiple of $f(x) = g(x)h(x)$. From the latter we conclude that $a(x)$ must be a multiple of $g(x)$ and, therefore, belongs to the ideal generated by $g(x)$. □

Example. We consider the algebra of polynomials modulo $x^7 - 1$; since the coefficients are from $GF(2)$, we have $-1 = +1$ and, hence, $x^7 - 1 = x^7 + 1$. We now work with $x^7 + 1$. The polynomial $g(x) = x^3 + x^2 + 1$ divides $x^7 + 1$, and, therefore, by Theorem 4.1.19 we can consider it as the generator of an ideal with basis vectors $g(x)$, $xg(x)$, $x^2g(x)$, and $x^3g(x)$. The polynomial $a(x) = (x^2 + 1)g(x) = x^5 + x^4 + x^3 + 1$ belongs to this ideal. Considering the correspondence between vectors and polynomials, this ideal is nothing else but the subspace of the vector space $\mathbf{V}_7$ with basis vectors $(0, 0, 0, 1, 1, 0, 1)$, $(0, 0, 1, 1, 0, 1, 0)$, $(0, 1, 1, 0, 1, 0, 0)$, $(1, 1, 0, 1, 0, 0, 0)$, and the polynomial $a(x)$ is the vector $(0, 1, 1, 1, 0, 0, 1) = (0, 1, 1, 0, 1, 0, 0) + (0, 0, 0, 1, 1, 0, 1)$. The ideal orthogonal to the one mentioned above is the ideal generated by $h(x) = (x^7 + 1)/(x^3 + x^2 + 1) = x^4 + x^3 + x^2 + 1$, and basis vectors $h(x)$, $xh(x)$, and $x^2h(x)$.

Observe that in the algebra of polynomials modulo $(x^n - 1)$, the dimension of an ideal is $n - r$, if r is the degree of its generator polynomial. The dimension of its orthogonal ideal is r, and the degree of its generator polynomial is $n - r$. In the example above $\deg[g(x)] = 3$, $\deg[h(x)] = 7 - 3 = 4$, and the dimensions of the corresponding ideals are 4 and 3.

Theorem 4.1.21. In the algebra of polynomials modulo $(x^n - 1)$, a subspace of $\mathbf{V}_n$ is cyclic if and only if it is an ideal.

Proof. The basic observation in this case is that multiplication modulo $(x^n - 1)$ of a polynomial times x is equivalent to a cyclic shift because

$$x(a_{n-1}x^{n-1} + a_{n-2}x^{n-2} + \cdots + a_1x + a_0)$$
$$= a_{n-1}(x^n - 1) + a_{n-2}x^{n-1} + \cdots + a_1x^2 + a_0x + a_{n-1}$$
$$= a_{n-2}x^{n-1} + \cdots + a_1x^2 + a_0x + a_{n-1}\,[\mathrm{mod}\,(x^n - 1)]$$

If C is a cyclic subspace of $\mathbf{V}_n$, then from $\mathbf{v} \in C$ it follows that $x \cdot \mathbf{v}$, $x^j \cdot \mathbf{v}$, and $(c_{n-1}x^{n-1} + c_{n-2}x^{n-2} + \cdots + c_1x + c_0) \cdot \mathbf{v}$ all belong to C, which, by definition, is an ideal.

Conversely, if the subspace C of $\mathbf{V}_n$ is an ideal, then from $\mathbf{v} \in C$ it follows that $\mathbf{v}' = x \cdot \mathbf{v} \in C$, and since the coefficients of $\mathbf{v}'$ are obtained from a cyclic shift of the coefficients of $\mathbf{v}$, we conclude that C is a cyclic subspace. □

Therefore, we have that a cyclic code is completely defined from the generator polynomial, which divides $x^n - 1$. A message $a_{k-1}, a_{k-2}, \ldots, a_0$ (of maximum length k) can be encoded using a code C by computing the product $a(x)g(x)$, where $a(x)$ is the polynomial corresponding to a_{k-1}, $a_{k-2}, \ldots, a_0$, and $g(x)$, $\deg[g(x)] = r$, is the generator of the ideal C whose dimension is $n - r$.

Example. Suppose that the messages to be transmitted consist of 3 bits, and that the generator polynomial of the cyclic code is $g(x) = x^3 + x + 1 = (1, 0, 1, 1)$. Here $x^3 + x + 1$ divides $x^7 - 1$ or $x^7 + 1$. Then the possible codewords are computed as the product of $a(x)$ times $g(x)$, for each message polynomial $a(x)$. That is we have

$$000 \rightarrow 0000000 \quad 100 \rightarrow 0101001$$
$$001 \rightarrow 0001011 \quad 101 \rightarrow 0100010$$
$$010 \rightarrow 0010110 \quad 110 \rightarrow 0111010$$
$$011 \rightarrow 0011101 \quad 111 \rightarrow 0110001$$

Obviously, for this code the message could have a maximum of 4 bits of information.

Equivalently, a code can be defined as the orthogonal subspace of the ideal generated by the polynomial $h(x) = (x^n - 1)/g(x)$. Note that $g(x) \cdot h(x) \equiv 0 \,[\bmod\,(x^n - 1)]$. If $\deg[g(x)] = r$, then the dimension of the code is $n - r$. The element $a(x)$ belongs to the code if and only if $g(x)|a(x)$, or equivalently, if and only if $a(x)h(x) \equiv 0 \,[\bmod\,(x^n - 1)]$. "To prove it in one direction, consider the polynomial $a(x)$ in the code; then there is a polynomial $u(x)$ such that $a(x) = u(x)g(x)$. However, since $g(x)h(x) \equiv 0 \,[\bmod\,(x^n - 1)]$, we have $a(x)h(x) \equiv u(x)g(x)h(x) \equiv 0 \,[\bmod\,(x^n - 1)]$. We leave it as an exercise for the reader to prove it in the other direction. From $a(x)h(x) \equiv 0 \,[\bmod\,(x^n - 1)]$, we conclude that the coefficient of x^i in the product $a(x)h(x) \equiv 0 \,[\bmod\,(x^n - 1)]$ is given by $\Sigma_{0 \leq i \leq n-1} a_i h_{j-i} = 0, j = 0, 1, \ldots, n - 1$, where the indices are computed modulo n. This accounts for the fact that the matrix $\mathbf{H}$, corresponding to $h(x)$, has the order of its entries reversed. See the example below for details." For instance, in the example above, the codeword 0110001 corresponds to the polynomial $x^5 + x^4 + 1$ and $h(x) = (x^7 + 1)/(x^3 + x + 1) = x^4 + x^2 + x + 1$; clearly, $a(x)h(x) \equiv 0 \,[\bmod\,(x^7 - 1)]$. The polynomial $h(x)$ is called the (*parity*) *check polynomial* of the code generated by $g(x)$. Since $h(x)|(x^n - 1)$, it can be also used as the generator of another code C' that is the *dual* code of C. The analogy with the dual code presented before, as well as the correspondence of the polynomials $g(x)$ and $h(x)$ with the matrices $\mathbf{G}$ and $\mathbf{H}$, is easily understood with the following example.

Example. Consider the polynomial $x^7 - 1 = x^7 + 1 = (x + 1)(x^3 + x + 1) \cdot (x^3 + x^2 + 1)$. The polynomial $g(x) = x^3 + x^2 + 1$ generates a cyclic code C with $n = 7$, $k = 4$. The elements $x^3 g(x)$, $x^2 g(x)$, $xg(x)$, and $g(x)$ form the basis of the subspace of the code and, hence, the matrix **G** can be considered the generator matrix of the code.

$$x^3 g(x) = (1, 1, 0, 1, 0, 0, 0)$$
$$x^2 g(x) = (0, 1, 1, 0, 1, 0, 0)$$
$$x g(x) = (0, 0, 1, 1, 0, 1, 0)$$
$$g(x) = (0, 0, 0, 1, 1, 0, 1)$$

$$G = \begin{bmatrix} 1 & 1 & 0 & 1 & 0 & 0 & 0 \\ 0 & 1 & 1 & 0 & 1 & 0 & 0 \\ 0 & 0 & 1 & 1 & 0 & 1 & 0 \\ 0 & 0 & 0 & 1 & 1 & 0 & 1 \end{bmatrix}$$

Moreover, this code is the orthogonal subspace of the ideal generated by $h(x) = (x^7 - 1)/g(x) = (x - 1)(x^3 + x + 1) = x^4 + x^3 + x^2 + 1$; the basis vectors of this ideal are

$$x^2 h(x) = (1, 1, 1, 0, 1, 0, 0)$$
$$x h(x) = (0, 1, 1, 1, 0, 1, 0)$$
$$h(x) = (0, 0, 1, 1, 1, 0, 1)$$

Since polynomial multiplication and inner product of vectors differ, the code C is the orthogonal subspace of a matrix **H** that is formed from the vectors $x^2 h(x)$, $x h(x)$, $h(x)$ with *reversed* order of their coordinates. Therefore

$$H = \begin{bmatrix} 0 & 0 & 1 & 0 & 1 & 1 & 1 \\ 0 & 1 & 0 & 1 & 1 & 1 & 0 \\ 1 & 0 & 1 & 1 & 1 & 0 & 0 \end{bmatrix}$$

It is easily seen that GH^T is the zero matrix. This code is equivalent to the Hamming code $(7, 4)$.

Another way to define a cyclic code is through the roots of the generator polynomial $g(x)$ in $GF(2^m)$; that is, the polynomial $g(x)$ is indirectly defined. In this way a vector $a(x)$ belongs to the code if it has as roots all the roots of $g(x)$. Moreover, since $g(x)$ must divide $x^n - 1$, all the roots $\alpha_1, \alpha_2, \ldots, \alpha_r$ of $g(x)$ must be roots of $x^n - 1$, and, according to what we said in Section 3.3, the order of each root must divide n. Therefore, if we have the roots of the generator polynomial $g(x)$ in $GF(2^m)$, we can determine n, the length of the code, as the least common multiple of the orders of the roots. The following examples help to clarify this concept.

Example. The code of the previous example can be defined as follows. Every polynomial that belongs to the code must have α as root, where α is one of the primitive elements of $GF(2^3)$. The primitive elements of $GF(2^3)$

are α and α^3 (verify this), which implies that the generator polynomial of the code is

$$g(x) = (x + \alpha)(x + \alpha^2)(x + \alpha^4), \quad \text{or} \quad g(x) = (x + \alpha^3)(x + \alpha^6)(x + \alpha^5)$$

that is, the generator polynomial is either the irreducible polynomial $x^3 + x^2 + 1$ or $x^3 + x + 1$. The order of α is $2^3 - 1 = 7$, and hence, the length of the code is $n = 7$. Since $\deg[g(x)] = 3$, there are $7 - 3 = 4$ information bits.

Example. Let α be a primitive element of $GF(2^4)$. We consider the code in which every polynomial has as roots $\alpha, \alpha^2, \alpha^3, \alpha^4, \alpha^5$, and α^6; for α we have $\alpha^{15} - 1 = 0$.

However, the polynomial that has α as a root will also have α^2, α^4, and α^8 as roots; likewise, the polynomial with α^3 as a root will also have $\alpha^6, \alpha^{12}, \alpha^{24} = \alpha^{15}\alpha^9 = \alpha^9$ as roots, and, finally, the polynomial with α^5 as a root will also have α^{10} as a root. Therefore

$$g(x) = [(x + \alpha)(x + \alpha^2)(x + \alpha^4)(x + \alpha^8)]$$
$$\cdot [(x + \alpha^3)(x + \alpha^6)(x + \alpha^{12})(x + \alpha^9)][(x + \alpha^5)(x + \alpha^{10})]$$
$$= (x^4 + x + 1)(x^4 + x^3 + x^2 + x + 1)(x^2 + x + 1)$$

The polynomial $g(x)$ has degree 10, the length of the code is $2^4 - 1 = 15$, and it has $k = 15 - 10 = 5$ information bits.

Next we summarize a simple coding and decoding rule corresponding to this definition of the cyclic code:

Coding. Let $g(x)$ be the generator polynomial of the code, $\deg[g(x)] = r$, and let $a_{k-1}, a_{k-2}, \ldots, a_0$ be a message with $k = n - r$ information bits. Then, this message is regarded as a polynomial $a(x)$ over $GF(2)$ and is encoded as $a(x)g(x)$.

Decoding. A received vector $v(x)$ is divided by $g(x)$. [The remainder of this division is the syndrome of $v(x)$.] If the division is not exact and a remainder occurs, then there must have been a transmission error. To recognize the error $e(x)$, we compute the product $v(x)h(x)$ [mod $(x^n - 1)$], where $h(x)$ is the parity check polynomial of the code. According to an observation made earlier, we have $v(x)h(x) = [u(x) + e(x)]h(x) = 0 + e(x)h(x)$ [mod $(x^n - 1)$]. Then, division of the product by $h(x)$ gives us the error polynomial $e(x)$, and $u(x) = v(x) - e(x)$. Dividing $u(x)$ by $g(x)$, we obtain the transmitted message.

Example. Let $g(x) = x^3 + x^2 + 1$ be the generator polynomial of a $(7, 4)$ code that we saw in an example above. The message $(1, 1, 1, 0)$ is encoded as $(1, 0, 0, 0, 1, 1, 0)$.

Suppose now that the received vector $v(x)$ is $(1,0,0,0,0,1,1)$ (two errors); then dividing $x^6 + x + 1$ by $g(x)$, we obtain $x^2 + 1$ as remainder, which implies that error(s) occurred in transmission. The check polynomial is $h(x) = x^4 + x^3 + x^2 + 1$, and $v(x)h(x) \,[\bmod (x^7 - 1)]$ is $x^6 + x^5 + x^3 + 1 = (1,1,0,1,0,0,1)$. So, we have that $e(x)h(x) = x^6 + x^5 + x^3 + 1$, and dividing this by $h(x)$, we obtain $e(x) = x^2 + 1 = (0,0,0,0,1,0,1)$. Therefore, $u(x) = v(x) - e(x) = (1,0,0,0,1,1,0) = x^6 + x^2 + x$, and dividing this by $g(x)$, we obtain the transmitted vector $(1,1,1,0)$.

If these is only one transmission error, then $e(x) = x^i$, and we correct the ith bit.

In general, if $g(x) = g_r x^r + g_{r-1} x^{r-1} + \cdots + g_0$ is the generator polynomial of the code, then the polynomials $x^{n-r-1}g(x), x^{n-r-2}g(x), \ldots, g(x)$ are all codewords. Therefore, the rows of the following matrix are all codewords

$$\mathbf{G} = \begin{bmatrix} g_r & g_{r-1} & \cdots & g_0 & 0 & 0 & \cdots & 0 \\ 0 & g_r & \cdots & g_1 & g_0 & 0 & \cdots & 0 \\ & & & \cdots & & & & \\ 0 & 0 & \cdots & g_r & g_{r-1} & & \cdots & 0 \\ 0 & 0 & \cdots & 0 & g_r & & \cdots & g_0 \end{bmatrix}$$

Obviously, the rows of $\mathbf{G}$ are linearly independent and the rank of $\mathbf{G}$ is $n - r$, which is the dimension of the code.

We are lead to yet another representation of a cyclic code if we assume that a polynomial $f(x)$ belongs to the code if and only if it has as roots the elements $\alpha_1, \alpha_2, \ldots, \alpha_r$ of $CF(2^m)$. Then, if $f(x) = f_{n-1}x^{n-1} + f_{n-2}x^{n-2} + \cdots + f_0$, we have

$$f(\alpha_i) = f_{n-1}\alpha_i^{n-1} + f_{n-2}\alpha_i^{n-2} + \cdots + f_0 = 0$$

for $i = 1, 2, \ldots, r$. This can be also written as a matrix product

$$[f_{n-1}, f_{n-2}, \ldots, f_0][\alpha_i^{n-1}, \alpha_i^{n-2}, \ldots, \alpha_i, 1]^T = \mathbf{0}$$

This condition is equivalent to α_i being a root of $f(x)$. However, from our discussion of finite fields, this is equivalent to $f(x)$ being divisible by the minimal polynomial $m_i(x)$ of α. For all α_i, this above condition can be written as

$$\begin{bmatrix} \alpha_1^{n-1} & \alpha_1^{n-2} & \cdots & \alpha_1 & 1 \\ \alpha_2^{n-1} & \alpha_2^{n-2} & \cdots & \alpha_2 & 1 \\ & & \cdots & & \\ \alpha_{r-1}^{n-1} & \alpha_{r-1}^{n-2} & \cdots & \alpha_{r-1} & 1 \\ \alpha_r^{n-1} & \alpha_r^{n-2} & \cdots & \alpha_r & 1 \end{bmatrix} \begin{bmatrix} f_{n-1} \\ f_{n-2} \\ \\ f_1 \\ f_0 \end{bmatrix} = \mathbf{0}^T,$$

from which we conclude that the parity check matrix is

$$\mathbf{H} = \begin{bmatrix} \alpha_1^{n-1} & \alpha_1^{n-2} & \cdots & \alpha_1^1 & \alpha_1^0 \\ \alpha_2^{n-1} & \alpha_2^{n-2} & \cdots & \alpha_2^1 & \alpha_2^0 \\ & & \cdots & & \\ \alpha_{r-1}^{n-1} & \alpha_{r-1}^{n-2} & \cdots & \alpha_{r-1}^1 & \alpha_{r-1}^0 \\ \alpha_r^{n-1} & \alpha_r^{n-2} & \cdots & \alpha_r^1 & \alpha_r^0 \end{bmatrix}$$

The ith row of the matrix $\mathbf{H}$ denotes that every polynomial of the code has α_i as a root. Let us consider for the moment only the ith row of the matrix. Then, all the polynomials having α_i as a root, constitute the space orthogonal to the space of the matrix

$$[\alpha_i^{n-1}, \alpha_i^{n-2}, \ldots, \alpha_i^1, \alpha_i^0], \qquad (***)$$

and since these are exactly the polynomials divisible by $m_i(x)$, the minimal polynomial of α_i, where $\deg[m_i(x)] = m_i$, they form an ideal. [The matrix form of (***) is easily seen if each power of α_i is replaced by a column corresponding to its vector representation as in Table 3.3.1.] Since the orthogonal space of the matrix has dimension $n - m_i$, the dimension of the space of the matrix rows is m_i. Note that the coefficients of the polynomials representing the codewords are in $GF(2)$, whereas the elements α_i are in $GF(2^{m_i})$. However, we have mentioned that the elements of $GF(2^{m_i})$ can be considered as vectors with m_i components from $GF(2)$, and, hence, the dimension of the space of the matrix (***) is m_i over $GF(2)$. See also the example below.

If α_i and α_j have the same minimal polynomial, then their orthogonal spaces are the same, and, hence, the spaces of the rows of the corresponding matrix of the form (***)—considered as spaces over $GF(2)$—will be the same. So, in order to construct the matrix $\mathbf{H}$, *we need to know only one root for each irreducible factor of $g(x)$.*

Example. We want the binary cyclic code, for which the polynomial $f(x)$ is a codeword if and only if it has as roots the elements $\alpha, \alpha^2, \ldots, \alpha^6$, where α is the primitive element of $GF(2^4)$. Now the minimal polynomial of α, $m_1(x)$ has $\alpha^2, \alpha^4, \alpha^8$ also as roots. Likewise, $m_2(x)$, the minimal polynomial of α^3, has $\alpha^6, \alpha^{12}, \alpha^9$ also as roots, and $m_3(x)$, the minimal polynomial of α^5, has α^{10} as root, too. Therefore

$$g(x) = m_1(x) m_2(x) m_3(x)$$

and it suffices to ask that every polynomial $f(x)$ has as roots α, α^3, and α^5. So, the code is the orthogonal space of the matrix

$$H = \begin{bmatrix} \alpha^{14} & \alpha^{13} & \alpha^{12} & \cdots & \alpha^{1} & \alpha^{0}=1 \\ (\alpha^3)^{14}=\alpha^{12} & (\alpha^3)^{13}=\alpha^{9} & (\alpha^3)^{12}=\alpha^{6} & \cdots & (\alpha^3)^{1}=\alpha^{3} & (\alpha^3)^{0}=1 \\ (\alpha^5)^{14}=\alpha^{10} & (\alpha^5)^{13}=\alpha^{5} & (\alpha^5)^{12}=1 & \cdots & (\alpha^5)^{1}=\alpha^{5} & (\alpha^5)^{0}=1 \end{bmatrix}$$

or, taking into consideration the vector representation of each power of α (see Table 3.3.1) the **H** matrix becomes

$$H = \begin{bmatrix}
1 & 1 & 1 & 1 & 0 & 1 & 0 & 1 & 1 & 0 & 0 & 1 & 0 & 0 & 0 \\
0 & 1 & 1 & 1 & 1 & 0 & 1 & 0 & 1 & 1 & 0 & 0 & 1 & 0 & 0 \\
0 & 0 & 1 & 1 & 1 & 1 & 0 & 1 & 0 & 1 & 1 & 0 & 0 & 1 & 0 \\
1 & 1 & 1 & 0 & 1 & 0 & 1 & 1 & 0 & 0 & 1 & 0 & 0 & 0 & 1 \\
1 & 1 & 1 & 1 & 0 & 1 & 1 & 1 & 1 & 0 & 1 & 1 & 1 & 1 & 0 \\
1 & 0 & 1 & 0 & 0 & 1 & 0 & 1 & 0 & 0 & 1 & 0 & 1 & 0 & 0 \\
1 & 1 & 0 & 0 & 0 & 1 & 1 & 0 & 0 & 0 & 1 & 1 & 0 & 0 & 0 \\
1 & 0 & 0 & 0 & 1 & 1 & 0 & 0 & 0 & 1 & 1 & 0 & 0 & 0 & 1 \\
0 & 0 & 0 & 0 & 0 & 0 & 0 & 0 & 0 & 0 & 0 & 0 & 0 & 0 & 0 \\
1 & 1 & 0 & 1 & 1 & 0 & 1 & 1 & 0 & 1 & 1 & 0 & 1 & 1 & 0 \\
1 & 1 & 0 & 1 & 1 & 0 & 1 & 1 & 0 & 1 & 1 & 0 & 1 & 1 & 0 \\
1 & 0 & 1 & 1 & 0 & 1 & 1 & 0 & 1 & 1 & 0 & 1 & 1 & 0 & 1
\end{bmatrix}$$

The irreducible factors of $x^{15} - 1$ are

$$x^{15} - 1 = (x-1)(x^2 + x + 1)(x^4 + x^3 + x^2 + x + 1)(x^4 + x^3 + 1)(x^4 + x + 1)$$

and we can easily see that $m_1(x) = x^4 + x + 1$, where $m_1(\alpha) = 0$, $m_2(x) = x^4 + x^3 + x^2 + x + 1$, with $m_2(\alpha^3) = 0$, and $m_3(x) = x^2 + x + 1$, where $m_3(\alpha^5) = 0$. (At this point the reader should note the need for efficient procedures to factor polynomials with coefficients from a finite field; this topic is discussed in Chapter 6.)

We will now show that the Hamming codes $(2^m - 1, 2^m - m - 1, 3)$, which we have already encountered, are equivalent to cyclic codes.

Let α be a primitive element of $GF(2^m)$, and consider the code with parity check matrix

$$H = (\alpha^{q-2}, \alpha^{q-3}, \ldots, \alpha, 1), \quad q = 2^m$$

If the powers of α are represented as vectors (columns) of length m over $GF(2)$, then every nonzero vector of length m appears only once as a column of **H**. Therefore, **H**, indeed, describes a Hamming code. If can be described as a cyclic code if we say that the polynomial $f(x)$ is a codeword if and only if α is a root of $f(x)$. The minimal polynomial of α is a c-primitive polynomial, and because of this, the generator polynomial $g(x)$ is c-primitive. Likewise, every code generated by a c-primitive polynomial is a Hamming code.

Example. Let $m = 4$ and α be a root of the c-primitive polynomial $x^4 + x + 1$. Then the parity check matrix of the Hamming code $(15, 11)$ generated from this polynomial is

$$\mathbf{H} = (\alpha^{14}, \alpha^{13}, \ldots, \alpha^2, \alpha, 1) =$$

$$\begin{bmatrix} 1 & 1 & 1 & 1 & 0 & 1 & 0 & 1 & 1 & 0 & 0 & 1 & 0 & 0 & 0 \\ 0 & 1 & 1 & 1 & 1 & 0 & 1 & 0 & 1 & 1 & 0 & 0 & 1 & 0 & 0 \\ 0 & 0 & 1 & 1 & 1 & 1 & 0 & 1 & 0 & 1 & 1 & 0 & 0 & 1 & 0 \\ 1 & 1 & 1 & 0 & 1 & 0 & 1 & 1 & 0 & 0 & 1 & 0 & 0 & 0 & 1 \end{bmatrix}$$

We next describe two coding procedures for cyclic codes, where the simplicity of such codes can be easily seen. The common characteristic of both methods is that the register accepts the k information bits from the source and, without delay, forwards them to the transmission channel. Each such block of information bits is followed by $n - k$ check bits. Obviously, while the check bits are being prepared, the register cannot accept the next block of information bits, and so, either the source must be able to stop and restart transmitting, or a temporary buffer must be used.

The first coding procedure to be described uses a register with k-stages, whereas the second uses a register with $n - k$ stages. If there are more check bits than information bits, the first method is preferred, whereas otherwise, the second method is more economical. Both procedures produce the *same* codewords.

The coding of a cyclic (n, k) code generated by the polynomial $g(x)$, $\deg[g(x)] = n - k$, is achieved with the circuit shown in Figure 4.1.5, where a register with k stages is used; the connections of the register correspond to the polynomial $h(x) = (x^n - 1)/g(x)$. The information bits are originally placed in the k storage elements, and afterward the successive shifts take place. The first k bits that come out of the register are the k information bits, and the rest $n - k$ are the check bits, for a total of n bits per codeword. (See Exercise 5 for this section.)

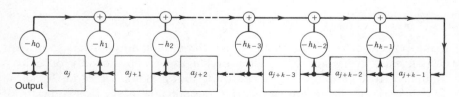

Figure 4.1.5. Coding of a cyclic (n, k) code.

Using a register with $n - k$ stages, coding proceeds as follows. A codeword can be formed multiplying any polynomial of degree $\leq k - 1$, whose coefficients are the information bits, times the polynomial $g(x)$, which generates the code. This can be achieved using either the circuit shown in Figure 3.3.2 or that in Figure 3.3.3. The information bits are now altered and can be obtained from a codeword dividing the corresponding polynomial by $g(x)$; the circuit in Figure 3.3.6 can be used for this purpose. (See Exercise 6 for this section.)

The procedure for error detection is also quite simple for cyclic codes. If $v(x)$ is the received polynomial, it suffices to check if the polynomial–syndrome $r(x) = g(x)v(x)$ is zero or not (see Exercise 7 for this section). Therefore, for cyclic codes, coding and error detection is performed in linear time with regard to the length of the code.

Finally, we discuss the error-correcting capabilities of cyclic codes arriving thus at the BCH codes. We say that an error vector of length n is a *burst of errors* of length b if and only if all its nonzero components are not spread out over the length n, but are concentrated in an interval of length b. We will examine some theorems concerning the correcting ability of cyclic codes with regard to bursts of errors.

Theorem 4.1.22. In a cyclic (n, k) code no codeword is a burst of errors of length $n - k$ or less. So, every cyclic code can detect any burst of errors of length $n - k$ or less.

Proof. Let $r(x)$ be a burst of errors of length $n - k$ or less, and suppose that the (n, k) cyclic code we discuss is generated by the polynomial $g(x)$ of degree $n - k$. Moreover, let us assume that the first nonzero coefficient of $r(x)$ is the coefficient of x^j. Then, $r(x) = x^j r_0(x)$, where $\deg[r_0(x)] < n - k$, since $r(x)$ is a burst of errors of length $n - k$ or less. Now, since $g(x)|(x^n - 1)$, $g(x)$ is not divisible by x and, hence, x^j and $g(x)$ are relatively prime. Moreover, since $g(x)$ must divide $r(x)$, it follows that if $r(x)$ were a codeword, it would necessarily divide $r_0(x)$, but this is a contradiction given the fact that $\deg[r_0(x)] < n - k$. So, $r(x)$ cannot be a codeword. □

The next theorem sets a lower bound on the minimum distance of every cyclic code. In the case of the BCH codes, to be defined below, the generating polynomials are so chosen as to guarantee a relatively large minimum distance.

Theorem 4.1.23. Let $g(x)$ be the generating polynomial of a binary cyclic code of length n, and let $\alpha^{e_1}, \alpha^{e_2}, \ldots \alpha^{e_{n-k}}$ be the roots of $g(x)$, where α is an element of order n in a Galois field. Then, the minimum distance of the code is greater than the maximum number of consecutive integers in the set $e = (e_1, e_2, \ldots, e_{n-k})$.

186 ERROR-CORRECTING CODES AND CRYPTOGRAPHY

Proof. Suppose that $m_0, m_0 + 1, \ldots, m_0 + d_0 - 2$ is the biggest set of consecutive integers in the set e. We have previously mentioned that a cyclic code with roots $\alpha^{e_1}, \alpha^{e_2}, \ldots, \alpha^{e_{n-k}}$ is the orthogonal space to the matrix

$$\mathbf{H} = \begin{bmatrix} (\alpha^{e_1})^{n-1} & (\alpha^{e_1})^{n-2} & \cdots & \alpha^{e_1} & 1 \\ (\alpha^{e_2})^{n-1} & (\alpha^{e_2})^{n+2} & \cdots & \alpha^{e_2} & 1 \\ & & \cdots & & \\ (\alpha^{e}_{n-k})^{n-1} & (\alpha^{e}_{n-k})^{n-2} & \cdots & \alpha^{e}_{n-k} & 1 \end{bmatrix}$$

Now, if we prove that no linear combination of $d_0 - 1$ columns of the submatrix

$$\mathbf{S} = \begin{bmatrix} (\alpha^{m_0})^{m-1} & (\alpha^{m_0})^{m-2} & \cdots & \alpha^{m_0} & 1 \\ (\alpha^{m_0+1})^{n-1} & (\alpha^{m_0+1})^{n-2} & \cdots & \alpha^{m_0+1} & 1 \\ & & \cdots & & \\ (\alpha^{m_0+d_0-2})^{n-1} & (\alpha^{m_0+d_0-2})^{n-2} & \cdots & \alpha^{m_0+d_0-2} & 1 \end{bmatrix}$$

is zero, then, clearly, the same will be true for the columns of $\mathbf{H}$ and, by Theorem 4.1.12, the minimum distance of the code will be d_0 or greater.

That the above matrix $\mathbf{S}$ has the desired property can be proved by showing that the determinant of the matrix obtained by taking any set of $d_0 - 1$ columns of $\mathbf{S}$ is nonzero. Indeed, such a determinant can be written as

$$\det \begin{bmatrix} (\alpha^{m_0})^{j_{d_0-1}} & (\alpha^{m_0})^{j_{d_0-2}} & \cdots & (\alpha^{m_0})^{j_1} \\ (\alpha^{m_0+1})^{j_{d_0-1}} & (\alpha^{m_0+1})^{j_{d_0-2}} & \cdots & (\alpha^{m_0+1})^{j_1} \\ & \cdots & & \\ & \cdots & & \\ (\alpha^{m_0+d_0-2})^{j_{d_0-1}} & (\alpha^{m_0+d_0-2})^{j_{d_0-2}} & \cdots & (\alpha^{m_0+d_0-2})^{j_1} \end{bmatrix}$$

which, in turn, can be written as

$$\alpha^{m_0(j_1+j_2+\cdots+j_{d_0-1})} \det \begin{bmatrix} 1 & 1 & \cdots & 1 \\ x_1 & x_2 & \cdots & x_s \\ x_1^2 & x_2^2 & \cdots & x_s^2 \\ & \cdots & & \\ x_1^{s-1} & x_2^{s-1} & \cdots & x_s^{s-1} \end{bmatrix}$$

where $s = d_0 - 1$ and $x_i = \alpha^{j_{d_0-i}}$. This last determinant, however, is the Vandermonde matrix and is equal to $\prod_{i>j}(x_i - x_j)$; it becomes zero only when $x_i = x_j$ for some i, j, but this is not possible in our case. □

Definition 4.1.24. Let α be an element of $GF(2^m)$. Then for given m_0 and d_0, the code generated from the polynomial $g(x)$ is a BCH code if and only if $g(x)$ is the minimal polynomial over $GF(2)$ having as roots the elements $\alpha^i, i = m_0, m_0 + 1, \ldots, m_0 + d_0 - 2$.

The length n of such a code is the least common multiple (*lcm*) of the orders of its roots. This is equivalent to n being equal to the order e of α (except for the trivial case where only one root α^{m_0} is given). The latter is so because we have (since n is the *lcm* of the orders of the roots)

$$(\alpha^{m_0})^n = 1, \quad (\alpha^{m_0+1})^n = 1$$

from which we conclude that $\alpha^n = 1$, and that the order e of α divides n, and hence, $e \leq n$. On the other hand, if $\alpha^e = 1$, we have $(\alpha^j)^e = 1$ for every j; therefore, the order of every element α^j divides e, from which we conclude that $e = n$, since n, the *lcm*, cannot be greater than e.

The number of information bits as well as the number of check bits can be found from the general way we presented for cyclic codes. From Theorem 4.1.23 we have that the minimum distance of these codes is greater than d_0.

The most important BCH codes are formed by taking α to be a primitive element of $GF(2^m)$ and setting $m_0 = 1$ and $d_0 = 2t_0 + 1$. Then the roots of the polynomial are $\alpha, \alpha^2, \alpha^3, \ldots, \alpha^{2t_0}$. However, since every even power of α has the same minimal polynomial as some previous odd power of α (e.g., α^{10} has the same minimal polynomial as α^5), we can say that $g(x)$ is the polynomial with roots $\alpha, \alpha^3, \ldots, \alpha^{2t_0-1}$; let $m_1(x), m_3(x), \ldots, m_{2t_0-1}(x)$ be the corresponding minimal polynomials. Then, we have

$$g(x) = lcm[m_1(x), m_3(x), \ldots, m_{2t_0-1}(x)]$$

From our previous observations we can show that $\deg[m_i(x)] < m$, where m defines $GF(2^m)$, which implies that $\deg[g(x)] < mt_0$, and therefore, the code has fewer than mt_0 check bits. The above remarks are summarized in the following theorem.

Theorem 4.1.25. For every pair of positive integers m and $t_0 < n/2$ there is a binary BCH code of length $n = 2^m - 1$, which corrects all the combinations of t_0 errors or less and does not have more than mt_0 check bits.

Example. Let us consider the field $GF(2^4)$ and α a primitive element in this field; that is, $n = 2^4 - 1 = 15$. Moreover, let us consider the BCH code consisting of all the polynomials having as roots α and α^3. The minimal polynomials for α and α^3 are $m_1(x) = x^4 + x + 1$, and $m_2(x) = x^4 + x^3 + x^2 + x + 1$, respectively. They are both divisors of $x^{15} - 1$. The cyclic code C

over $GF(2)$ is defined by the generator polynomial $g(x) = m_1(x)m_2(x)$ of degree 8, and parity check matrix

$$\mathbf{H} = \begin{bmatrix} \alpha^{14} & \alpha^{13} & \alpha^{12} & \cdots & \alpha^1 & a^0 = 1 \\ (\alpha^3)^{14} = \alpha^{12} & (\alpha^3)^{13} = \alpha^9 & (\alpha^3)^{12} = \alpha^6 & \cdots & (\alpha^3)^1 = \alpha^3 & (\alpha^3)^0 = 1 \end{bmatrix}.$$

From the above we know that the minimum distance of the code is ≥ 5 ($t_0 = 2$ in this case), which implies that the code can correct up to two errors; C is a (15, 7) code. We have that $\mathbf{v} \in C$ if and only if $S(\mathbf{v})$, the syndrome of $\mathbf{v}$ is $\mathbf{0}$; that is, $\mathbf{v} \in C$ if and only if $\mathbf{Hv}^T = \mathbf{0}^T$, or if and only if $S_1 = S_3 = 0$, where $S_1 = \Sigma_{1 \leq i \leq 14} v_i \alpha^i$, and $S_3 = \Sigma_{1 \leq i \leq 14} v_i \alpha^{3i}$ are the components of the syndrome $S(\mathbf{v}) = (S_1, S_3)^T$ of $\mathbf{v}$ with respect to $\mathbf{H}$. If we use the vector representation for the elements of $GF(2^4)$, we obtain an 8×15 matrix $\mathbf{H}$, the like of which we have seen in a previous example.

Suppose that the received vector $\mathbf{v} = (v_0, v_2, \ldots, v_{14})$ is a vector with at most two errors. Let $e = x^{e_1} + x^{e_2}, 0 \leq e_1, e_2 \leq 14, e_1 \neq e_2$ be the error vector. We have

$$S_1 = \alpha^{e_1} + \alpha^{e_2}, \qquad S_3 = \alpha^{3e_1} + \alpha^{3e_2}.$$

Moreover, if we let $x_1 = \alpha^{e_1}$, and $x_2 = \alpha^{e_2}$, then $S_1 = x_1 + x_2$, and $S_3 = x_1^3 + x_2^3$, and after some manipulations we see that if two that if two errors occurred, then $1/x_1$ and $1/x_2$ are two roots, in $GF(2^4)$, of the *error-locator polynomial*

$$\sigma = 1 + S_1 x + \left(S_1^2 + \frac{S_3}{S_1}\right) x^2.$$

If only one error occurred, then $S_1 = x_1, S_3 = x_1^3$ and $S_1^3 + S_3 = 0$. Hence, $\sigma = 1 + S_1 x$. And, finally, if no error occurred, then $S_1 = S_3 = 0$.

From the above example we see that in order to be able to correct two errors, we have to double, in a specific way, the number of rows of the matrix $\mathbf{H}$, used in Hamming codes. In the literature there exist more general decoding procedures for correcting any number of errors; see for example Berlekamp, 1968; Childs, 1979; Mackiw, 1985; MacWilliams, 1977; and Peterson et al., 1972.

4.2
Cryptography—General Concepts

Cryptology is the art of designing and breaking secrecy systems; its designing part is called *cryptography*, whereas the breaking part is called *cryptanalysis*. The whole process can be paralleled with the basic process of coding and decoding. Again we consider a transmitter who wants to

CRYPTOGRAPHY—GENERAL CONCEPTS

communicate "sensitive" information to a receiver over an unreliable communication channel. The unreliability of the channel can be caused by an unauthorized interceptor having access to the channel and whose (not always exclusive) goals are to (1) violate the secrecy of communications, (2) confound the receiver with a corrupted message, and (3) deceive either the transmitter or the receiver, or both, about the identity of the opposite party.

The first threat above is the most widely known realization of the cryptographic problem. Protection from an interceptor who pursues the latter two goals has come into prominance only recently, and it comprises *authentication* and *integrity*. For instance, the authentication problem arises in the login procedure of multiuser computer systems, whereas the integrity problem arises in electronic funds transfer. Threat to authorization occurs when C' in Figure 4.2.1 is composed by the interceptor, whereas the receiver believes it originated from the transmitter. Threat to integrity occurs when $M \neq M'$ and neither the transmitter nor the receiver is able to detect the change of C.

The problem of secret writing is a very old one, and many attempts to secure sensitive messages, particularly military and diplomatic messages, have endowed the field with a rich history [of great interest is the book by Kahn (1967)]. Whereas in coding the aim is fast and correct transmission of a message over a noisy channel, in cryptology the aim is safe transmission by altering the original message in such a way that an unauthorized interceptor would be unable to understand the message. This is achieved with the help of a cryptosystem, like the one shown in Figure 4.2.1.

A cryptographic system (*cryptosystem* or *cipher*) consists of transforming a message M, called the *plaintext*, using an *encryption* or *enciphering* device in such a way that only an authorized recipient can invert the transformation and restore the message. The enciphering device invokes an encrypting function E that, besides the plaintext M, also requires K, the encryption *key*, which is a parameter particular to one transformation. (A more precise definition of the key is given below.) The encrypting function is defined by an algorithm and the result of the enciphering process, $E(K, M) = C$, is

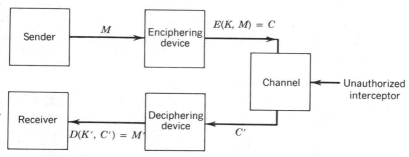

Figure 4.2.1. A cryptosystem for transmitting a message M.

called *ciphertext* or *cryptogram*; C is transmitted over the insecure channel, where the interceptor may examine it, store it, operate on it, and finally change it to C'. The *decryption* or *deciphering* device invokes a decrypting algorithm D that takes as arguments the alleged ciphertext C' and the decrypting key K'. The key and the ciphertext must determine the plaintext uniquely. Note that K, M are not necessarily equal to K', M', where $M' = D(K', C')$, even though in classical cryptography $K = K'$. We refer to systems where $K = K'$ as *single key* or *symmetric cryptosystems* to distinguish them from the newer *public key* or *asymmetric cryptosystems* where two different keys are used.

Codes and ciphers are the only two possibilities available in the enciphering/deciphering process. Using codes to encipher a message means to replace some or all of the words and phrases by codewords and codephrases obtained from a special book that resembles a dictionary; in fact, the word *code* refers to such a cryptosystem, even though the terms *secret code* and *code breaking* are used to cover all aspects of secret writing. Stated in other words, the code must have established the semantic content of every possible message to be sent through the channel, and both the transmitter and the receiver must have a codebook; provided the codebook is well protected, messages are extremely hard to break (if at all possible). However, communication is not possible if a phrase is not included in the codebook. On the contrary, using a cipher enables the communication of arbitrary messages because the cipher is an algorithm that assigns new symbols of ciphertext to symbols, or groups of symbols, of the plaintext. Ciphers are of great interest because the relationship between ciphertext and plaintext is not arbitrary and may be revealed by cryptanalysis.

4.2.1 Symmetric (Single-Key) Cryptosystems

In classical cryptography two basic transformations of plaintext messages exist along with their combination:

1. *Transposition* or *permutation ciphers* rearrange a group of characters according to some rule, without changing their identity; that is, if the message M is composed of m blocks $M = B_1 B_2 \cdots B_m$, where each block B_i contains n characters, $B_i = b_{i,1} b_{i,2} \cdots b_{i,n}$, $i = 1, 2, \ldots, m$, then the ciphertext is $C = C_1 C_2 \cdots C_m$, with $C_i = b_{i,\pi(1)} b_{i,\pi(2)} \cdots b_{i,\pi(n)}$, for each $i = 1, 2, \ldots, m$, where π is a fixed permutation of the integers $1, 2, \ldots, n$.
2. *Substitution ciphers* replace plaintext characters by corresponding characters from a ciphertext alphabet (the key specifies the mapping); that is, if the message is $M = a_1 a_2 \cdots a_n$, then the ciphertext $C = f_1(a_1) f_2(a_2) \cdots f_n(a_n)$ is defined with respect to the n mappings f_i, $i = 1, 2, \ldots, n$, from the plaintext alphabet into the ciphertext alphabet.
3. Of course, by combining (1) and (2) we obtain the substitution/permutation (S/P) cipher.

Example (Transposition Ciphers). Suppose that we want to encipher the message "computer algebra." The first example of a transposition cipher is to *write the text backward*, in the traditional way of groups of five characters. So we have

arbeg laret upmoc

Another example of transposition cipher is the *rail fence cipher* where we write the text alternating in two rows and then read it row by row; that is, we have

c m u e a g b a
o p t r l e r

and the ciphertext is "cmuea gbaop trler." Ways to break the transposition ciphers can be found in Kahn (1967), pp. 225–226 and in Sinkov (1968), Chapter 5.

If only one cipher alphabet is used, the cryptosystem is called *monoalphabetic*. Cryptosystems, in which a ciphertext letter may represent more than one plaintext letter are called *polyalphabetic*. We also distinguish between *stream* and *block* ciphers. Stream ciphers treat the plaintext as a sequence of symbols to be encrypted, whereas block ciphers divide messages into blocks of equal length and derive ciphertext by operating on blocks of plaintext characters. In a stream cipher, the basic allowable operation on a message is the substitution of one symbol for another, whereas with a block cipher, besides substitution we have permutation. Although stream ciphers retain their importance for many applications, block ciphers have had the greatest recent impact on cryptography. We examine stream ciphers first.

Definition 4.2.1. Let A and B be the plaintext and ciphertext alphabet, respectively. A *key* is an injective (one-to-one) mapping from A into B. The key is *fixed* if the mapping is the same for each element of A; otherwise, the key is *variable*.

A fixed key $f: A \to B$ can be extended to a string of elements of A, that is to words with letters in A, by considering $a_1 a_2 \cdots a_m \to f(a_1) f(a_2) \cdots f(a_m)$; likewise, a variable key $f_1, f_2, \ldots$ can be extended by $a_1 a_2 \cdots a_m \to f_1(a_1) f_2(a_2) \cdots f_m(a_m)$. The alphabets mentioned in the definition above may be letters of the English alphabet, or n-tuples of letters, or elements of $\mathbf{Z}_m$ ($m = 2$, or $m = 26$ or $m = 96$ as in some commercial systems). It is left as an exercise for the reader to verify that the mapping $a \to an + k \pmod{m}$ from $\mathbf{Z}_m$ into itself is injective (one-to-one) if and only if $\gcd(m, n) = 1$.

Definition 4.2.2. The mapping $a \to an + k \pmod{m}$, from $\mathbf{Z}_m$ into itself, with fixed n, k in $\mathbf{Z}_m$ and $\gcd(m, n) = 1$, is called a *modular cipher*.

For $n = 1$, $k = 3$, and $m = 26$ we obtain the Caesar cipher, a code of historical interest since, according to legend, it was used by the Roman emperor Gaius Julius Caesar; observe that the Caesar cipher is a cyclic shift of the alphabet by three letters.

Example. We identify a with 0, b with 1, c with 2, and so on:

a	b	c	d	e	f	g	h	i	j	k	l	m	n
0	1	2	3	4	5	6	7	8	9	10	11	12	13

o	p	q	r	s	t	u	v	w	x	y	z
14	15	16	17	18	19	20	21	22	23	24	25

Using the Caesar cipher, $a \to a + 3$, and writing the text in the traditional way of groups of five, the message "computer algebra," which is equivalent to 2, 14, 12, 15, 20, 19, 4, 17, 0, 11, 6, 4, 1, 17, 0, is enciphered as 5, 17, 15, 18, 23, 22, 7, 20, 3, 14, 9, 7, 4, 20, 3, or, equivalently, as "frpsx whudo jheud."

Monoalphabetic substitution ciphers use only one key and can thus be easily broken by observing the frequency distribution of the characters, which is preserved in the ciphertext. This task is easy because there are tables with various letter frequencies, such as initial letter frequencies, terminal letter frequencies, frequencies of digraphs (i.e., the frequency of a being followed by b) and so on. The table shown in Figure 4.2.2 was generated by Sinkov from a sample of 1000 letters (see the appendixes of Sinkov's book).

Monoalphabetic ciphers can be strengthened if we use polyalphabetic substitution ciphers that conceal the letter frequencies by using multiple substitution. Here more than one alphabet is used in enciphering the message, and the key includes a specification of which substitution is to be

Letter	% Frequency	Letter	% Frequency	Letter	% Frequency
a	7.3	j	0.2	s	6.3
b	0.9	k	0.3	t	9.3
c	3.0	l	3.5	u	2.7
d	4.4	m	2.5	v	1.3
e	13.0	n	7.8	w	1.6
f	2.8	o	7.4	x	0.5
g	1.6	p	2.7	y	1.9
h	3.5	q	0.3	z	0.1
i	7.4	r	7.7		

Figure 4.2.2. Relative frequencies of the letters in the English language.

used for each symbol. These ciphers are also known as *Vigenere ciphers*, due to the sixteenth-century French cryptologist Blaise de Vigenere.

More precisely, a polyalphabetic substitution cipher with period p consists of p cipher alphabets B_i and the mappings $f_i: A \to B_i$, $i = 1, 2, \ldots, p$, defined by the key. The key often is a word $K = k_1 k_2 \cdots k_p$ and $f_i(a) \equiv a + k_i$ (mod 26). A plaintext message $M = a_1 a_2 \cdots a_p a_{p+1} \cdots a_{2p} \cdots$ is enciphered as $f_1(a_1) f_2(a_2) \cdots f_p(a_p) f_1(a_{p+1}) \cdots f_p(a_{2p}) \cdots$ by repeating the sequence of mappings $f_1, f_2, \ldots, f_p$ every p characters. A Vigenere square, with the letters a, b, ..., z of the alphabet as first row and column, can be used for enciphering; that is, we have

$$
\begin{array}{l}
a\,b\cdots y\,z \\
b\,c\cdots z\,a \\
\cdots\cdots \\
y\,z\cdots w\,x \\
z\,a\cdots x\,y
\end{array}
$$

The enciphering process can be simplified if we consider the fact that not all rows (alphabets) of the Vigenere square are used. For example, suppose that "alkis" is the key, in which case the period is $p = 5$; then the mappings f_i are given by the following simple tableau, where the key appears on the left column.

a	b	c	d	e	f	g	h	i	j	k	l	m	n	o	p	q	r	s	t	u	v	w	x	y	z
a	b	c	d	e	f	g	h	i	j	k	l	m	n	o	p	q	r	s	t	u	v	w	x	y	z
l	m	n	o	p	q	r	s	t	u	v	w	x	y	z	a	b	c	d	e	f	g	h	i	j	k
k	l	m	n	o	p	q	r	s	t	u	v	w	x	y	z	a	b	c	d	e	f	g	h	i	j
i	j	k	l	m	n	o	p	q	r	s	t	u	v	w	x	y	z	a	b	c	d	e	f	g	h
s	t	u	v	w	x	y	z	a	b	c	d	e	f	g	h	i	j	k	l	m	n	o	p	q	r

In this case the message "computer algebra" is enciphered as "czwxm tpbid gplzs," where the ith plaintext character is mapped to the character found in its column in the i (mod 5)th cipher alphabet; for instance, the character m of the plaintext is mapped to the letter w, which is found in the column under m in the alphabet starting from k.

From the last example we see that the key "alkis" corresponds to the sequence of numbers 0, 11, 10, 8, 18, and hence, the enciphering could be equally well performed by repeatedly writing, under the plaintext, the numbers 0, 11, 10, 8, 18, 0, 11, 10, 8, 18, 0, 11, 10, 8, 18, ... and adding them modulo 26 to the numbers corresponding to the letters of the plaintext.

The deciphering process is as follows. The number corresponding to the ith character of the plaintext is obtained by adding modulo 26 the number corresponding to the ith character of the ciphertext to the m-complement of the number corresponding to the i (mod p)th character of the key. (Recall that the receiver also knows the key.) For example, if $m = 26$, the received message is "czwxm tpbid gplzs," the key is "alkis" or 0, 11, 10, 8, 18, and we wish to compute the *second* character of the plaintext, then we add modulo 26 the number 25, which coresponds to the letter z of the ciphertext, to the number 15, which is the 26-complement of the second character of the key $(26 - 11 = 15)$ obtaining 14 or "o."

Vigenere ciphers were considered unbreakable in their days, but in 1863 a Prussian military officer named F. W. Kasiski published a simple number-theoretic means for searching for the key length. In the polyalphabetic ciphers, a shift is applied to the alphabet just as in the Caesar ciphers, but the length of the shift varies, usually in a periodic way. The changing shifts even out the letter frequencies, defeating the kind of analysis used to break the Caesar ciphers; however, the characteristic frequencies are retained in subsequences of the enciphered message corresponding to repetitions in the key sequence. If we determine the length of the key's period, letters can be identified by frequency analysis.

The period of the key can be discovered by looking for repeated blocks in the ciphertext. Some of these are accidental, but a large proportion results from matches between repeated words or subwords of the plaintext and repeated blocks in the key sequence. When this happens, the distance between repetitions will be a multiple of the key period. For example, the plaintext "send more men and more arms" when enciphered using the key "bhenf" results in the ciphertext "tlrqr pyizj ohrqr pyinw nz" (if we leave unpadded the last block). The distance, in this case, between the two occurrences is 10, indicating that the key length is 10 or 5.

Other variants of the polyalphabetic substitution ciphers are the *auto key ciphers*, where the message itself (plaintext or ciphertext) is used as the key; a short "seed" key, usually a single letter, is used to start the cipher. These variants were proposed in 1550 by Cardano and were developed by Vigenere; if we work in $\mathbf{Z}_m$, then the auto key ciphers are defined by $a_i \rightarrow b_i \equiv na_i + ca_{i-1}$ (mod m), with c, a_0 given, or $a_i \rightarrow b_i \equiv na_i + cb_{i-1}$ (mod m), with c, b_0 given, where, of course, we choose n so that $\gcd(m, n) = 1$.

Example. If we number the letters from a to z as in the previous example (i.e., $m = 26$), then the message "computer algebra," which is equivalent to $2, 14, 12, 15, 20, 19, 4, 17, 0, 11, 6, 4, 1, 17, 0 = a_1 a_2 \cdots a_{15}$ can be enciphered in the following two ways:

1. Using the transformation $a_i \rightarrow b_i \equiv na_i + ca_{i-1}$ (mod m), with $n = 1$, $c = 1$, and $a_0 = 2$, we obtain $b_1 = 4$, $b_2 = 16$, $(14 + 2)$, and so on and finally obtain the ciphertext $4, 16, 0, 1, 9, 13, 23, 21, 17, 11, 17, 10, 5, 18, 17$.

2. Using the transformation $a_i \rightarrow b_i \equiv na_i + cb_{i-1}$ (mod m), with $n = 1$, $c = 1$, and $b_0 = 2$, we obtain $b_1 = 4$, $b_2 = 18$, $(14 + 4)$, and so on and finally obtain the ciphertext $4, 18, 4, 19, 13, 6, 10, 1, 1, 12, 18, 22, 23, 14, 14$.

Another variant is the *running key Vigenere cipher*, which uses nonrepeating text for a key. Originally these ciphers were also considered unbreakable, but in 1883 Kerchoffs developed a general solution for polyalphabetic substitution with no restriction on the type or length of the key [details can be found in Kahn (1967), pp. 236–237]; a more general solution to the problem was given by Friedman in 1918. An interesting presentation of these ideas can be found in Gass, 1986.

The most important variant of the Vigenere ciphers was proposed in 1918 by the American engineer G.S. Vernam working for the AT & T (American Telephone & Telegraph) teletype system. Messages were then transmitted using a binary code, and Vernam suggested to add modulo 2 a random sequences of marks and spaces so that all frequency information, intersymbol correlation, periodicity, and similar are lost. The major drawback of this procedure is that it requires exchanging impractical amounts of key in advance of communications (i.e., one key character for each message character). Vernam originally thought to use either a short key or a linear combination of short keys, but both approaches were proved vulnerable. On one hand using a short key, is vulnerable to Kasiski-type solution, whereas, on the other hand, it was proved by Major J. O. Mauborgne of the U.S. Army Signal Corps that using a linear combination of short keys can be successfully analyzed by techniques essentially the same as those used against running key systems.

Both Friedman and Mauborgne reached the conclusion that for an unconditionally secure cryptosystem the key must be as long as the message and incoherent (i.e., the uncertainty of each key character must be at least as great as the average information content per plaintext character). If we assume that M is an upper bound on the length of all possible messages, then we choose the key sequence to be at least as large as M; all key sequences are then equally probable. Working modulo 26, if the plaintext is $a_1 a_2 \cdots a_M$, and each a_i is represented by one of the integers from 0 to 25, the key sequence will be $k_1 k_2 \cdots k_M$, where k_i is a randomly chosen number between 0 and 25; then each ciphertext character is $c_i \equiv a_i + k_i \pmod{26}$. This scheme is called a *one-time pad*, and its name is derived from the fact that the enciphering/deciphering process utilizes written pads of random numbers to obtain the key sequence, and each key sequence is used only once. As mentioned above, the major drawback of this scheme is that it requires exchanging impractical amounts of key in advance of communications. However, one-time pads are clearly unbreakable; the randomness of the key means that any two message sequences of the same length are equally likely to have produced the ciphertext. As a result of the above, one-time pads are used to transmit highly sensitive information—as, for example, in the hotline between Washington and Moscow.

In summary, we see that security increases as we move from Caesar ciphers to one-time pads, and relates to the increase in the length of the key. The key is a single number in the Caesar ciphers, whereas in the one-time pads the key is, potentially, an infinite sequence. In an unbreakable system all messages are equiprobable, and hence, the cipher is unresolved.

So far we dealt mainly with stream ciphers where each message symbol was individually treated. We will now examine *polygraphic systems*; these are block ciphers that treat a group of plaintext symbols at a time. Block ciphers were developed with the intension to deny the cryptanalyst raw frequencies of occurring statistics; they usually operate on pairs, *digraphs*, triples, *trigraphs*, and in general on *polygraphs*. Manual systems were restricted to digraphs.

The best known manual digraph system is due to the English scientist Playfair. According to this scheme, a mixed alphabetic sequence is written in a 5×5 square. (The letter "*j*" is omitted since it occurs infrequently and can be filled in by the context.) So, for example, we can have

$$\begin{array}{ccccc} a & l & k & i & s \\ b & y & t & q & x \\ r & n & c & u & g \\ d & w & h & m & o \\ z & p & f & v & e \end{array}$$

and the message "computer algebra" is enciphered as "ghwvc qzglk osrda," where the pair "co" is mapped to the pair "gh," with g being in the same row as c (note the parallelogram defined by cgoh), and so on. From this example the reader can easily derive the rules for enciphering a text having two letters in the same row or column. [These rules are explicitly stated in Sinkov (1968), p. 114.] Using digraph frequency counts, Mauborgne cryptanalyzed this scheme in 1914.

The cornerstone of modern mathematical cryptography was laid by Hill (1929, 1931). Hill recognized that nearly all the existing cryptosystems could be formulated in the single model of linear transformations on a message space. He identified a message *n*-tuple with an *n*-tuple of integers and equated the operations of encryption and decryption with a pair of inverse linear transformations. Generalizing the notion of the modular cipher, we have the following.

Definition 4.2.4. Let **K** be an $n \times n$ matrix and let **a** and **d** be *n*-dimensional vectors with elements from $\mathbf{Z}_m$. A *Hill cipher* is a mapping of the form $\mathbf{a} \to \mathbf{Ka}^T + \mathbf{d}$, which is injective if and only if $gcd[\det(\mathbf{K}), m] = 1$; all operations are performed modulo m.

To encipher, we subdivide the plaintext into blocks of n letters each, replace each letter by its corresponding element in $\mathbf{Z}_m$, form the transposed column vectors, and apply the given linear transformation to each block **a**. In this context the *involutory* (or self-inverse) matrices **K** are employed, which are their own inverse; they are defined by $\mathbf{K}^2 = \mathbf{I}$ or $\mathbf{K} = \mathbf{K}^{-1}$. (The operations are performed modulo m.) Observe that since $\mathbf{K}^2 = \mathbf{I}$, we have $\det^2(\mathbf{K}) = 1$, which implies that $\det(\mathbf{K}) = \pm 1$. Trying to determine the involutory matrices when the dimension $n = 2$ and $m = 26$, we see that we can obtain only seven such matrices when $\det(\mathbf{K}) = -1$; clearly, the case $\det(\mathbf{K}) = +1$ is much more interesting (see also Exercise 5 for this section). There is a combined total of 736 involutory matrices obtained in this case, and a cryptogram can be deciphered even if no word of minimal length is known from the plaintext. All the involutory matrices are determined and tried. For larger n it is no longer possible to decipher by this trial-and-error method.

Example. Consider the message "computer algebra," where the letters $a, b, \ldots, z$ of the plaintext are assigned the numbers $0, 1, \ldots, 25$, and the involutory matrix

$$\mathbf{K} = \begin{bmatrix} 2 & 7 \\ 7 & 24 \end{bmatrix}$$

the vector $\mathbf{d}$ is taken to be zero. The ciphertext becomes "ymzcr yxuze oirza," with the last plaintext character left as it is. The pair "co" that is equivalent to the vector $(2, 14)$ is mapped to the vector $\mathbf{K}(2, 14)^T = (4 + 98, 14 + 336) \equiv (24, 12) \pmod{26}$, which corresponds to the pair "ym", and so on.

Replacing the constant matrix $\mathbf{K}$ by a matrix having variable elements, we obtain a variation to the above scheme. Details of this approach can be found in Lidl and Pilz (1984); also, another interesting scheme is described in Krishnamurthy and Ramachandran (1980).

Before we discuss modern cryptosystems, we need certain concepts relating to the security of such systems. In assessing the security of any system it is essential to assume the worst case; that is, the opponent may have access to other information besides the cipher. Accordingly, we have:

Plaintext-only analysis. In this attack on the system, the opponent has access only to the intercepted ciphertext.

Known plaintext analysis. Now the opponent has several plaintext–ciphertext pairs from which to work.

Chosen plaintext analysis. The opponent can submit unlimited portions of plaintext to the system and can observe the corresponding ciphertext. (This is the most severe attack on a system.)

From the previous discussion we see that only one-time pads are *unconditionally secure*; this means that no matter how much computational power the enemy employs, they cannot break the system. However, in order to avoid the drawbacks of one-time pads, a compromise has been reached;

namely, an *unconditionally insecure* system is considered in practice, knowing that the interceptor can successfully analyze the ciphertext with an impossibly powerful computer. The basic idea here is to limit the power of the opponent to feasible computation. To understand the meaning of this statement, we need some basic facts of computational complexity (Lewis and Papadimitriou, 1978).

Problems in mathematics can be first divided into two categories: *solvable* and *proven unsolvable* problems. An example of a proven unsolvable problem is the Turing machine "halting" problem, which is equivalent to the following: "A barber shaves all those who do not shave themselves; does he shave himself or not?" (*Answer*: If he does, then he does not, and if he does not, then he does; that is, there is no answer to this problem, mainly because of the self-reference.)

The solvable problems can be further subdivided into the following general categories:

Provably difficult problems, which have only exponential computing time, that is, of the form $O(2^n)$. An example of such a problem is Presburger arithmetic with real numbers. Here we want to find out whether a formula is true, under the assumptions that we cannot multiply two variables (we can only multiply a variable times a scalar), and that we have only the quantifier $\exists$.

P-Problems, which have a polynomial computing time, specifically, of the form $O(n^k)$. An example of such a problem is Euler's circuit problem from graph theory, where we want to find in a graph a path that crosses each line once.

NP-Problems (*nondeterministic polynomial*), for which only exponential algorithms are known, but it has not been proved that polynomial time algorithms do not exist. Obviously, the set of *P*-problems is a subset of the set of *NP*-problems, but the question as to whether $P = NP$ is the biggest open problem in theoretical computer science. A characteristic property of the *NP*-problems is the following: whereas it is very difficult to find a solution for them, once such a solution is found, it is very easy to verify it in polynomial time; this property will be used below.

NP-Complete problems, which form a small subset of the *NP*-problems characterized by the property that if any one of them is solved by an efficient algorithm, then all the other problems in the *NP* class could be solved efficiently. An example is Hamilton's circuit problem, also from graph theory, where we want to find in a graph a path that crosses each point once.

Let us now use the above ideas in modern cryptology (Feistel, 1973 and Lempel, 1979). We are interested in block ciphers that operate on groups of bits. To see how such an operation is performed by an electronic device, let us consider the case where we have only 3 bits (with 3 bits we can represent a total of eight items). The device is called a substitution box, or S box, and is shown in Figure 4.2.3.

In Figure 4.2.3 we see that the substitution device consists of two switches. The first converts a sequence of 3 bits into its corresponding value to the base 8, thereby energizing any one of the eight output lines. These eight lines can be connected to the second switch in any one of $8! = 40,320$ ways. We are at liberty to decide which one of these 40,320 wire permutations is to be made between the first and the second switch. The role of the second switch is to convert the input, which is presented as one digit to the base 8, back into a 3-bit output.

Suppose now that we increase the number of input bits to the S box from 3 to 5, so that we can represent all the letters of the English alphabet; that is, a corresponds to 00000, b to 00001, and so on and finally, z to 11001. Then there are 32! possible connection patterns between the two switches. However, the cipher produced is still very weak; it could not resist letter-frequency analysis. The problem is that, in spite of the large number of possible connection patterns, there are only 32 possible inputs and outputs. Therefore, what is required is a large number of inputs and outputs, so large that it is impractical for any opponent to recover it. For instance, if we had a box with 128 inputs and outputs, an analyst would have to cope with 2^{128} possible input/output blocks, a number so large that frequency analysis would no longer be feasible. Of course, the drawback of such a scheme is that it would require 2^{128} connections between the switches (in the P box), a technological impossibility (at present ?). So a compromise has to be reached.

Clearly, one device with many inputs and outputs is the P box itself; in Figure 4.2.3 it has eight inputs and outputs. Despite the fact that there are 40,320 possible interconnections, it is very easy to find the actual ones by simply feeding the device with inputs where only one bit is 1 and the rest are

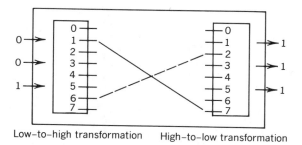

Low-to-high transformation   High-to-low transformation

Figure 4.2.3. A substitution box (S box) for block ciphers; the connections between the two switches can be considered a permutation box (P box).

0, and observing where the 1 is coming out. In our example we can find out the wiring with only seven trials. Note that the S box is a *nonlinear* device, whereas the P box is a *linear* device.

To improve on this scheme, we have to introduce the *product-cipher systems*, which combine P and S boxes. Product-cipher systems were first proposed by Shannon (1948, 1949), and they are made up of layers of P and S boxes. Assume that the P boxes have 15 inputs and outputs and that the S boxes have only 3; then each P box is followed by a layer of five S boxes and the operation of a product-cipher system is as follows, provided the input consists of 14 zeros and 1 one: the first box, P, moves the *single* 1 to some box S which, being a nonlinear device, may convert the one to as many as 3 ones. These ones are then fed into the next P box, which shuffles them and feeds them to separate S boxes, and the process is repeated. At the end the output contains a balanced number of zeros and ones. These ideas are illustrated in Figure 4.2.4.

Figure 4.2.4 illustrates the principle on which the IBM system LUCIFER is based. In LUCIFER the P boxes have either 64 or 128 inputs and outputs and the S boxes, only 4. Of course, the goal of the designer is to make it very difficult for an opponent to trace the pattern back and thus reconstruct the permutation keys on S and P. The Lucifer system is a high-quality block cipher; however, now it is not considered to be a secure system.

In 1977 the National Bureau of Standards issued a Data Encryption Standard (DES) with the intention to provide a cryptographic standard to be used for secure data transmission for all but national security-related data. The DES algorithm is a block cipher developed by IBM based on an S/P scheme as in the LUCIFER system described above. DES enciphers 64-bit blocks of plaintext using a 64-bit key (56 bits of key and 8 parity bits). Encryption is by 16 seperate rounds of encipherment, each round a product-cipher under the control of a 48 bit key. That is, each round uses a distinct 48 bit key $K_i, i = 1, 2, \ldots, 16$ obtained by a scheduling algorithm from the external key K.

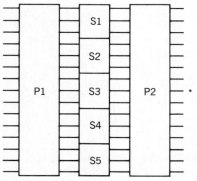

Figure 4.2.4. A product-cipher system where the P boxes have 15 inputs and outputs and the S boxes have only 3.

Although from a complexity theoretical point of view DES appears to be secure, the key size of the standard has been criticized (Diffie and Hellman, 1977). The problem is that the 56-bit key permits a known plaintext analysis by an opponent with massive, but feasible, computational resources. This is done by brute force. Suppose, for instance, that a known message M is enciphered using DES with key K, and that $E(K, M) = C$. To discover K, the cryptanalyst deciphers C with each of the 2^{56} possible keys. On discovering M, the cryptanalyst stops and announces K. Although this exhaustive search sounds prohibitive, a determined opponent can build a special purpose computer, which enables a million parallel searches to take place in just a few hours (<10 hours). However, it seems that increasing the key size from 56 to 128 rules out an exhaustive search.

Dissatisfaction with DES gave momentum for further research (Diffie and Hellman, 1976) and lead to the discovery of asymmetric encryption systems, also known as *public key cryptosystems*. Note that, up to now, all systems were symmetric, in the sense that both the transmitter and the receiver had in posession the same key that must be secured.

4.2.2 Asymmetric (Public-Key) Cryptosystems

The concept of public key cryptosystems was first introduced in 1976 by Diffie and Hellman. According to these schemes, each user places in a public directory an encryption procedure E (to be used by other subscribers for the encryption of messages addressed to that user), but keeps the details of the corresponding decryption procedure D secret. (Of interest are the articles by Ecker, 1982; Hellman, 1979; and Simmons, 1979.) Obviously, these methods are based on the observation that the encipherment procedure and the corresponding key do not have to be the same as the decipherment procedure and the corresponding key. For such systems to be viable, there must be a simple method by which the precedures E and D can be produced by each user. The desirable features of the procedures E and D must be:

1. If M is the plaintext message, E and D must be such that

$$D[E(M)] = M$$

   that is, deciphering the enciphered text M yields M.
2. Both E and D can be easily computed.
3. Revealing E does not lead to an easy way to compute D. [A very inefficient way to compute D would be to try $E(M) = C$ for all possible messages M.]
4. For each message M, $E[D(M)] = M$. This is useful in implementing signatures and it is explained below.

For example, if user B wants to send a private message M to user A, user B looks up E_A in the public directory under A (like a telephone directory) and transmits $C = E_A(M)$ in the open, knowing that only A can decipher it using the secret procedure D_A.

User B can also "sign" the messages sent to user A, so that A is assured of their authenticity. To sign a message M, user B first computes the message-dependent signature

$$S = D_B(M)$$

then, user B looks up E_A in the directory and transmits $C = E_A(S)$. Now, only A can recover S from C by computing $D_A(C) = S$, and then use E_B from the directory to obtain the message M; that is,

$$E_B(S) = E_B[D_B(M)] = M$$

Since only user B could have created an S that deciphers to M by E_B, user A knows that M could have been sent only by user B. For signatures, the RSA cryptosystem (named after its inventors, Rivest, Shamir, and Adleman, 1978), which is described below, is a very good technique.

As an application of the above scheme, consider the case where a test ban treaty between country A and country B is signed, and in order to monitor compliance with the treaty, it is agreed that seismic instruments be placed in each other's territory, to record any disturbances and, hence, detect underground tests. Whereas it is possible to protect the instruments (they could self-destruct in case anybody tampered with them), it is impossible to protect the channel over which the information is transmitted, since the host nation could cut the wires and send false information. Moreover, if the messages are sent in ciphered form, the host nation may suspect that, in addition to seismic data, additional unauthorized information is being transmitted.

The problems facing the two nations can be solved using a digital signature system. Nation A's seismic station contains a computer that converts the message M to $S = D_A(M)$. Nation B cannot substitute any S' for S, because chances are great that $E_A(S') = M'$ will not be meaningful. However, nation A *does* supply nation B with the procedure E_A, which B can then use to recover $E_A(S) = M$ and thus be assured that only authorized messages are being sent.

While presenting a complete description of these elegant concepts, Diffie and Hellman stopped short of presenting a practical implementation of a public key cryptosystem. They indicated, however, that such an implementation can be achieved with the help of computationally hard problems, such as the inversion of one-way functions.

A function f is called *one-way* if f is invertible and easy to compute, but it is computationally infeasible to compute f^{-1} from a complete description of

f. A function f is called *trapdoor* one-way if f^{-1} is easy to compute once certain "trapdoor" information is known (namely, the secret deciphering key), and if without such knowledge f is one-way.

The first public key cryptosystem (Merkle and Hellman, 1978) is based on the *NP knapsack or subset problem*, which can be stated as follows: Given a number N and k numbers $\{n_1, n_2, \ldots, n_k\}$ find, if it exists, a subset of $\{n_1, n_2, \ldots, n_k\}$ such that their sum is equal to N.

Example. Consider the set $\{1292, 2089, 2110, 625, 283, 1599, 3759, 1315, 250, 2460\}$ of 10 ($k = 10$) numbers and $N = 8329$. Which of those numbers sum up to N? In general, and in the worst case, we have to do 2^k trials; in our case, by trial and error, we find out that $8329 = 2110 + 3759 + 2460$.

Note the characteristic property of the *NP*-problems; namely, whereas it is difficult to find the numbers 2110, 2460, and 3759 once they are given, we can easily verify (we just add them) that they, indeed, solve our problem. Moreover, the knapsack problem is also *NP*-complete. (Obviously, as k increases to infinity, in the above example, it is impossible to solve the knapsack problem by search.)

In general, if we define the *scalar* or *dot* product of any two n-dimensional vectors $\mathbf{n} = (n_1, n_2, \ldots, n_k)$, and $\mathbf{x} = (x_1, x_2, \ldots, x_k)$ as $\mathbf{n} \cdot \mathbf{x} = n_1 x_1 + n_2 x_2 + \cdots + n_k x_k$, then the knapsack problem can be expressed as $N = \mathbf{n} \cdot \mathbf{x}$, where $x_i = 1$ if the ith component of $\mathbf{n}$, n_i is in the sum and $x_i = 0$ otherwise.

Let us now consider a special version of the knapsack problem where the given numbers form a *superincreasing sequence*; that is, each number n_i is larger than the sum of all the preceding elements. In this case we have an easy knapsack problem because it is decided in linear time whether there is a solution and, if it exists, a simple algorithm will find it.

Example. Consider the set $\{5, 10, 20, 42, 90, 205, 500\}$ of seven numbers ($k = 7$), and $N = 305$. To solve this knapsack problem is like trying to find solutions $x_i = 0$ or 1, $i = 1, 2, \ldots, 7$, to the equation

$$305 = 5x_1 + 10x_2 + 20x_3 + 42x_4 + 90x_5 + 205x_6 + 500x_7$$

This is, however, trivial since we can see by inspection that $x_7 = 0$ and $x_6 = 1$. (Since the sum of all the preceding numbers is less than 205, x_6 has to be 1, because otherwise we would never obtain a sum of 305.) The problem now becomes

$$100 = 5x_1 + 10x_2 + 20x_3 + 42x_4 + 90x_5$$

and continuing in the same way we have $x_5 = 1$, $x_4 = x_3 = 0$, $x_2 = 1$, and $x_1 = 0$. In other words, we need, in general, only k subtractions and comparisons, hence this version of the knapsack problem can be solved in polynomial time (linear).

Therefore, the basic idea of using a knapsack for a cryptosystem is to convert a simple knapsack into a complex one, a trapdoor knapsack. This is done by choosing two large integers m and w, $(m, w) = 1$, and then forming a new trapdoor knapsack $\mathbf{n} = (n_1\, n_2, \ldots, n_k)$ from a given superincreasing knapsack vector $\mathbf{n}^s = (n_1^s, n_2^s, \ldots, n_k^s)$, by letting $n_i \equiv w n_i^s \pmod{m}$, $i = 1, 2, \ldots, k$. Modular arithmetic throws the numbers in disarray, and, hence, the n_i^s values no longer form a superincreasing sequence. The knapsack vector $\mathbf{n}$ is *published* and forms the public key, whereas the vector $\mathbf{n}^s$, along with the numbers m, w, and w^{-1}, the multiplicative inverse of w (mod m), are *kept secret* by the person expecting to receive messages.

To summarize,

The receiver knows n^s, m, w, and w^{-1}, which are the secret keys, the sender knows the message M, everybody knows n, which is the public key

If $M = (x_1, x_2, \ldots, x_{bk})$ is the message in binary form, then it is enciphered by computing the b quantities $c_i = n_1 \cdot x_{(i-1)k+1} + n_2 \cdot x_{(i-1)k-2} + \cdots + n_k \cdot x_{ik}$ that are transmitted. An unauthorized interceptor is faced with the task of solving the hard knapsack problems with sums c_i and vector $\mathbf{n}$. However, multiplying c_i by w^{-1}, the intented receiver easily computes $c_i^s \equiv w^{-1} c_i \pmod{m}$, $i = 1, 2, \ldots, b$, and transforms the difficult knapsack problems with sums c_i and vector $\mathbf{n}$ into the easy knapsacks with sums c_i^s and vector $\mathbf{n}^s$.

It is not clear how secure are the systems that are based on the knapsack problem. The security of these cryptosystems is based on the worst-case analysis of the computing time of the knapsack problem. For instance, if $k = 1000$, we said that there will be needed at most 2^{1000} trials to solve the problem. However, it can also happen that a solution will be found after only a few trials, in which case the system is not as secure as we thought. Multiple encryption was initially thought to enhance the security of the trapdoor knapsacks. However, today attacks on simple and doubly iterated knapsacks are known. (To obtain a doubly iterated knapsack, we choose another pair of numbers m_1, w_1 and obtain a new vector from $\mathbf{n}$; the process can be repeated as often as we want.) Of interest is the paper by Merkle and Hellman (1981).

The operation of this type of cryptosystem is clarified below. However, it should be borne in mind that the values chosen are too small to provide any security; that is, they are too small to ensure that parts of the calculation are not feasible for an unauthorized interceptor.

Example. We choose the following correspondence between the letters of the alphabet and the binary 5-tuples.

a	00000	b	00001	c	00010
d	00011	e	00100	f	00101
g	00110	h	00111	i	01000
j	01001	k	01010	l	01011
m	01100	n	01101	o	01110
p	01111	q	10000	r	10001
s	10010	t	10011	u	10100
v	10101	w	10110	x	10111
y	11000	z	11001	,	11010
.	11011	!	11100	?	11101
"	11110	+	11111		

where + denotes space. We also choose the secret information, $\mathbf{n}^s = (1, 7, 13, 28, 52)$, $m = 111$, and $w = 55$, in which case $w^{-1} \pmod{111} = 109$; then, the public key is $\mathbf{n} = (55, 52, 49, 97, 85)$, where $n_i \equiv n_i^s w \pmod{m}$. Now, the message "computer algebra," which is the binary sequence 00010 01110 01100 01111 10100 10011 00100 10001 00000 01011 00110 00100 00001 10001 00000 ($=x_1 x_2 \ldots x_{75}$) is transmitted as the sequence of numbers 97, 198, 101, 283, 104, 237, 49, 140, 0, 234, 146, 49, 85, 140, 0, where $97 = 55 \cdot 0 + 52 \cdot 0 + 49 \cdot 0 + 97 \cdot 1 + 85 \cdot 0$, and so on. Note that in general each transmitted number is $c_i = n_1 \cdot x_{(i-1)k+1} + n_2 \cdot x_{(i-1)k+2} + \cdots + n_k \cdot x_{ik}$; that is, the original message in binary form is broken down into groups of k bits. An unauthorized interceptor knows c_i and the public key $\mathbf{n}$, but for large k will have to try an enormous number of cases to decipher it (2^k, the powerset, in the worst case).

The receiver, who possesses the superincreasing vector $\mathbf{n}^s$, only has to compute

$$c_i^s \equiv c_i w^{-1} \pmod{m}$$

and solve the easy knapsack problem $c_i^s = \mathbf{n}^s \cdot \mathbf{x}$. This is so because, if we choose $m > \Sigma n_i^s$, we have

$$c_i^s \equiv [n_1 \cdot x_{(i-1)k+1} + n_2 \cdot x_{(i-1)k+2} + \cdots + n_k \cdot x_{ik}] w^{-1} \pmod{m}$$
$$\equiv w^{-1} n_1 \cdot x_{(i-1)k+1} + w^{-1} n_2 \cdot x_{(i-1)k+2} + \cdots + w^{-1} n_k \cdot x_{ik} \pmod{m}$$
$$\equiv w^{-1} w n_1^s \cdot x_{(i-1)k+1} + w^{-1} w n_2^s \cdot x_{(i-1)k+2} + \cdots + w^{-1} w n_k^s \cdot x_{ik} \pmod{m}$$
$$\equiv n_1^s \cdot x_{(i-1)k+1} + n_2^s \cdot x_{(i-1)k+2} + \cdots + n_k^s \cdot x_{ik} \pmod{m}$$
$$= n_1^s \cdot x_{(i-1)k+1} + n_2^s \cdot x_{(i-1)k+2} + \cdots + n_k^s \cdot x_{ik}$$

So, continuing our example, we first observe that in our case $m > \Sigma n_i^s$, and the receiver computes the 15 quantities c_i^s that are 28, 48, 20, 100, 14, 81, 13, 53, 0, 87, 41, 13, 52, 53, 0 and then easily solves the corresponding knapsack problems with vector $\mathbf{n}^s = (1, 7, 13, 28, 52)$ and recovers the transmitted message; for example, 28 corresponds to the binary 5-tuple 00010 ($=c$), 48 corresponds to 01110 ($=o$), 20 corresponds to 01100 ($=m$), and so on.

A variant of the above cryptosystem is based on the *multiplicative knapsack*. Here given a number N and k relatively prime numbers $\{n_1, n_2, \ldots, n_k\}$ we want to find, if it exists, a subset of $\{n_1, n_2, \ldots, n_k\}$ such that their *product* equals N.

Finishing our discussion of the knapsack cryptosystems, it should be noted that they do not possess a straightforward digital signature feature. This has to do with the fact that not every integer c in the cryptogram range can be obtained as the sum of some subset of the n_i values; that is, the mapping is not "onto." Details can be found in the literature.

We next continue with a discussion of the RSA cryptosystem, which is based on the power function as a one way function. The success of this particular method depends on the difficulty of finding the factors of large integers. If that problem could be done in polynomial time, then the RSA cryptosystem could be compromised.

For a better understanding of the RSA cryptosystem, the reader should review Section 2.3, in general; particularly useful is the definition of the ϕ function in Section 2.3.1 and (Euler's) Theorem 2.3.14, which states that if $(a, m) = 1$, then $a^{\phi(m)} \equiv 1 \pmod{m}$. Moreover, we have already seen in Section 2.3.1 that for any integer n we have $\phi(n) = (1 - 1/p_1)(1 - 1/p_2) \cdots (1 - 1/p_k)$, where the p_i values are the primes involved in the prime-power decomposition of n. Of interest to us is the special case $n = pq$, where p, q are two distinct primes; then, from the last formula we obtain $\phi(pq) = (p - 1)(q - 1)$. Now, here is how the RSA cryptosystem works. The receiver computes the following items:

- p, q: Two large prime numbers, easily chosen and kept *secret*.
- n: The product pq, which is placed in the public directory.
- E: A random integer, also placed in the public directory. [E is used in encrypting messages to the receiver, who has made sure that E is relatively prime to $\phi(n) = (p - 1)(q - 1)$; this is easy to do because the receiver knows p and q and the *gcd* calculation is fast.]
- D: An integer used by the receiver in decoding. D is the multiplicative inverse modulo $(p - 1)(q - 1)$ of E and is also kept *secret*; that is, $DE \equiv 1 \ [\bmod (p - 1)(q - 1)]$. The inverse exists because we have $\gcd[E, (p - 1)(q - 1)] = 1$. Again, since p and q are known to the receiver, this computation is fast.

The reason for the above technical constrains on the various integers will be made clear below.

To summarize,

The receiver knows p, q, and D, which are the secret keys, the sender knows the message M, everybody knows n and E, which are the public keys

Messages M to be transmitted to the receiver are first converted into digital form by some standard nonsecret procedure. One way to do this is to use the assignment $a = 00$, $b = 01$, $c = 02$, $d = 03$, $e = 04, \ldots, z = 25$, $. = 26$, $, = 27$, $? = 28$, $; = 29$; that is, each letter is substituted by a two-digit number. For example, the message "computer algebra" would be written as "0214121520 1904170011 0604011700." Of course, if the message is long, it is broken up into blocks; the length of each block is such that its numerical value *does not exceed n*. It is important that each message block M_i be in the range $0 \leq M_i \leq n - 1$, since otherwise it would be impossible to distinguish it from any larger integer congruent to it modulo n. In practice, the modulus n is chosen to be large, on the order of 200 digits, so that block sizes of up to 10^{200} can be used.

The sender takes each message block M_i, looks at the public keys E and n, and transmits $C_i \equiv M_i^E \pmod{n}$, where $0 \leq M_i \leq n - 1$.

The receiver receives C_i and computes $(C_i)^D \pmod{n}$. Observe, however, that by definition we have $ED \equiv 1 \ [\mathrm{mod} \ \phi(n)]$, which implies that for some integer k we have $ED = 1 + k\phi(n)$. So

$$(C_i)^D \equiv (M_i)^{ED}$$
$$= (M_i)^{[1+k\phi(n)]}$$
$$\equiv M_i \pmod{n}$$

where the last relation is by (Euler's) Theorem 2.3.14. Therefore, the receiver has now recovered the original message block M_i.

The operation of this type of cryptosystem is made clear below. However, as in the case of the knapsack cryptosystem, it should be borne in mind that the values chosen are too small to provide any security; that is, they are too small to ensure that parts of the calculation are not feasible for an unauthorized interceptor.

Example. As we have mentioned above, using the assignment $a = 00$, $b = 01$, $c = 02$, $d = 03$, $e = 04, \ldots, z = 25$, $. = 26$, $, = 27$, $? = 28$, $; = 29$, the message "computer algebra" is written as "0214121520 1904170011 0604011700." Choosing $p = 3$, $q = 11$, we have $n = 3 \cdot 11 = 33$, and $\phi(n) = 20$; moreover, we choose $E = 7$, in which case $D = 3$. Now, each message block M_i consists of two-digit numbers, and the enciphered message is obtained from $C_i \equiv (M_i)^7 \pmod{33}$. Namely, we obtain the sequence of numbers 29, 20, 12, 27, 26, 13, 16, 08, 00, 11, 30, 16, 01, 08, 00, which is transmitted. All the receiver has to do to recover the transmitted message block (which in our case is just a letter) is to raise each two-digit number to the power 3 (modulo 33); this is left to the reader as an exercise. (In Section 2.3.2 we examined efficient algorithms for raising an integer to a power, as well as for computing the multiplicative inverse of an integer modulo some other number; the reader should review these algorithms.)

An unauthorized interceptor of the message in the example above would have to compute D, the multiplicative inverse of E modulo $(p-1)(q-1)$. However, in order to achieve this, the interceptor has to find out p and q from $n = pq$, which is in the public directory and is, therefore, compelled to derive a polynomial time algorithm for factoring large integers. Note that p and q are chosen so that they have an equal number of digits, and, therefore, n has twice that many. Moreover, they have to be so chosen that $p-1$ should have a large prime factor, say, f, and $f-1$ should also have a large prime factor. A similar condition holds for q. This ensures that the plaintext cannot be found by iterating the enciphering more quickly than by random search; see also Blakley et al., 1979; Herlestam, 1978; and Williams et al., 1979. (Decryption by *iteration* is the process according to which one successively reencrypts the ciphertext C until C is again obtained; that is, set C_1 to C, and compute $C_{i+1} \equiv (C_i)^E \pmod{n}$ until $C_{i+1} = C$. Then $C_i = M$.)

Another way for the interceptor to approach the problem would be to determine $\phi(n)$, since in that case D could be easily computed from $ED \equiv 1 \cdot [\bmod \phi(n)]$. However, the argument is that this approach is no easier than attempting to factor n; if one knew $\phi(n)$, then p and q could be found as follows. From $\phi(n) = (p-1)(q-1) = pq - (p+q) + 1 = n - (p+q) + 1$, knowledge of $\phi(n)$ would then yield knowledge of $p+q$. Moreover, the relation $(p-q)^2 = (p+q)^2 - 4pq = (p+q)^2 - 4n$ indicates that knowing n and $p+q$, one can easily obtain $p-q$. Then, p and q are obtained from

$$p = 1/2[(p+q) + (p-q)]$$
$$q = 1/2[(p+q) - (p-q)]$$

Therefore, any attempt to find $\phi(n)$ would be equivalent to solving the hard problem of factoring n.

Let us assume that present-day factorization algorithms can factor integers of up to 200 decimal digits in a matter of hours in the fastest machine; then, if n has about 2000 decimal digits, it can be safely concluded that it is impossible to factor it in any reasonable amount of time. However, it is perhaps too early to tell whether the RSA cryptosystem will stand the test of time.

There also exists another public key cryptosystem that is based on the difficulty of the general decoding problem for linear error-correcting codes; for more information, see Lempel (1979), pp. 297–298. Moreover, there also exist public key distribution systems. The purpose of such systems is to enable every pair of subscribers to exchange a private key securely over an insecure channel for use in a conventional cryptosystem; see Merkle, 1978 and Pohlig et al., 1978.

Exercises

Section 4.1

1. Prove that in the vector space of length n over $GF(2)$, the Hamming distance is a metric and the Hamming weight is a norm.
2. Complete the proof of Lemmas 4.1.5 and 4.1.9.
3. Show that if the minimum distance of a code is $\geq r + t + 1$, then this code can correct $\leq r$ errors and detect $r + t$ errors.
4. Construct a code consisting of eight words of length 7 such that any two distinct codewords have distance at least 4.

Section 4.1.1

1. Let
$$G = \begin{bmatrix} 1 & 0 & 1 & 1 & 1 \\ 0 & 1 & 0 & 1 & 0 \end{bmatrix}$$
be the generator matrix of a binary $(5, 2)$ code. Determine its parity check matrix, its codewords, and the coset leaders for this code.

2. With the $(7, 4)$ Hamming code, whose parity check matrix is
$$H = \begin{bmatrix} 1 & 0 & 0 & 1 & 1 & 0 & 1 \\ 0 & 1 & 0 & 1 & 0 & 1 & 1 \\ 0 & 0 & 1 & 0 & 1 & 1 & 1 \end{bmatrix}$$
code the messages $(0, 1, 1, 0)$ and $(1, 0, 1, 0)$; moreover, assuming that at most one error occurred, decode the vectors **(a)** $\mathbf{v} = (1, 1, 0, 0, 0, 0, 0)$, **(b)** $\mathbf{v} = (1, 0, 0, 1, 0, 1, 0)$ and **(c)** $\mathbf{v} = (1, 1, 0, 1, 0, 1, 1)$, which were received over a noisy channel.

3. A modification of the parity check matrix H of Exercise 2 enables us to detect the presence of two errors, as well as to correct one error.

Namely, adding a column of zeros at the beginning (leftside) and filling the top row with ones, we obtain

$$\mathbf{H} = \begin{bmatrix} 1 & 1 & 1 & 1 & 1 & 1 & 1 & 1 \\ 0 & 1 & 0 & 0 & 1 & 1 & 0 & 1 \\ 0 & 0 & 1 & 0 & 1 & 0 & 1 & 1 \\ 0 & 0 & 0 & 1 & 0 & 1 & 1 & 1 \end{bmatrix}$$

Now, if $\mathbf{v}$ is the received, $\mathbf{u}$ the transmitted vector and $\mathbf{e} = \mathbf{v} - \mathbf{u}$, then:

(a) If $\mathbf{e}$ has all entries zero, there were no errors in transmission.
(b) If $\mathbf{e}$ has one nonzero entry, one error occurred and this is determined as in Exercise 2.
(c) If $\mathbf{e}$ has two nonzero entries, two errors occurred.

In the last case above, $\mathbf{vH}^T = \mathbf{eH}^T$ is the sum of two columns of $\mathbf{H}$ but it cannot be determined which two columns of $\mathbf{H}$ make up the sum; what is certain, though, is the fact that $\mathbf{vH}^T$ is not a column of $\mathbf{H}$, since the sum of two columns of $\mathbf{H}$ always has top entry $= 0$.

Decode the vectors $\mathbf{v} = (1, 0, 1, 1, 0, 1, 1, 1)$, $\mathbf{v} = (0, 0, 0, 1, 1, 1, 1, 1)$, $\mathbf{v} = (1, 0, 1, 1, 1, 0, 0, 1)$, $\mathbf{v} = (1, 0, 1, 1, 0, 1, 0, 1)$ assuming that in each of them there are 0, 1, or 2 errors.
In the (8, 4) Hamming code do there exist received words for which the received can say with certainty that they have at least three errors?

4. Show that the $(2^m - 1, 2^m - 1 - m)$ Hamming codes are perfect.

5. Binary codes can be generalized to codes that are vector spaces over any finite field $GF(q)$. Let

$$\mathbf{G} = \begin{bmatrix} 1 & 0 & 1 & 1 \\ 0 & 1 & 2 & 1 \end{bmatrix}$$

be the generator matrix of a ternary (4, 2) code C. Find its parity check matrix.

Section 4.1.2

1. Determine all codewords of a code with generator polynomial $g(x) = x^3 + x + 1$ over $GF(2)$, if the length of the messages is 4. Does $(0, 1, 1, 1, 0, 1, 1)$ have detectable errors?

2. The polynomial $g(x) = x^6 + x^5 + x^4 + x^3 + 1$ is the generator polynomial for a cyclic code over $GF(2)$ with block length 15. Find the parity check polynomial and the generator and parity check matrix for this code. How many errors can this code correct?

EXERCISES

3. Verify that the binary polynomial $x^4 + x^3 + x^2 + x + 1$ is irreducible, and using it, construct the matrix **H** for a double-error-correcting BCH code with block length 15 having 7 information bits and eight check bits.

4. Consider a cyclic (15, 7) code with $g(x) = x^8 + x^7 + x^6 + x^4 + 1$. Is $x^{14} + x^5 + x + 1$ a codeword, and if not, what is its syndrome?

5. Show that the circuit shown in Figure 4.1.5, indeed, produces codewords of the code having $h(x)$ as check polynomial. Construct such a coding circuit for the special case of a binary Hamming code (7, 4) with $g(x) = x^3 + x + 1$.

6. [Coding of an (n, k) cyclic code using a register with $n - k$ stages.] Let $f_0(x)$ be the polynomial of degree $n - 1$ whose coefficients of the k highest powers of x, namely, $x^{n-1}, x^{n-2}, \ldots, x^{n-k}$, are the k information bits, and the other coefficients are "0." If $g(x)$ is a polynomial of degree $n - k$, then from the division algorithm we have $f_0(x) = g(x)q(x) + r(x)$, $\deg[r(x)] < n - k$. Then, $f_0(x) - r(x)$ is a polynomial of degree $\leq n - 1$ divisible by $g(x)$ and can be considered a codeword of the (n, k) cyclic code generated by the polynomial $g(x)$. Design the coding circuit, using a register with $n - k$ stages, based on the above observation.

7. Design a circuit to detect errors made by a cyclic code.

Section 4.2.1

1. Prove that the mapping $a \to an + k \pmod{m}$ from $\mathbf{Z}_m$ into itself is injective if and only if $\gcd(m, n) = 1$.

2. Use "computer" as the key in a Vigenere cipher, and encipher the word "cryptography."

3. Find a random sequence of 0/1 by flipping a coin and use this sequence in a one-time pad to encipher the message "secret code."

4. Using the Playfair square given in the text, encipher the word "university."

5. Show that if $\det(\mathbf{M}) \equiv -1 \pmod{26}$, then **M** is an involutory matrix if and only if $m_{11} + m_{22} \equiv 0 \pmod{26}$. Construct an involutory matrix with $m_{11} = 2$.

6. Let $\mathbf{Z}_{26}$ be the plaintext and ciphertext alphabet of a Hill cipher. Let

$$\mathbf{K} = \begin{bmatrix} 2 & 1 \\ -3 & -2 \end{bmatrix};$$

be the enciphering matrix. Encipher the word "department." What is the inverse of **K**?

Section 4.2.2

1. Using the knapsack cryptosystem, encipher the word "university." Use the same letter assignments and the same parameters as those in the corresponding example in the text.

2. Complete the example of the text on the RSA cryptosystem.

Programming Exercises

Section 4.1

1. A simple code for transmission of numeric data $x_1 x_2 \cdots x_n$, $0 \leq x_i \leq 9$ for all i, is to add a single check digit d_c consisting of the units digit of the sum

$$x_1 + 2x_2 + x_3 + 2x_4 + \cdots + X_n, \text{ where } X_n = \begin{cases} x_n, & \text{if } n \text{ is odd} \\ 2x_n, & \text{if } n \text{ is even} \end{cases}$$

The codeword then is $x_1 x_2 \cdots x_n d_c$. This code can detect many digit transposition errors, a common error introduced by human copying of data.

(a) If the received word is 12347 determine whether an error has occurred.

(b) What will the code be for the digits 13579? What will the received word be if the digits 7 and 9 are transposed?

(c) Using the above scheme, write a program to detect errors in receiving strings of up to 10 digits.

Section 4.1.1

1. Given an $r \times n$ parity check matrix **H** for a single-error-correcting (n, k) code with $r \leq 3$ and $n \leq 2^r - 1$, write a procedure to:

(a) Print out the set of binary k-tuples **H** codes and the codeword for each.

(b) Decode received binary n-tuples.

Section 4.2.2

1. Write a procedure to encipher/decipher text using the knapsack public key cryptosystem.

2. Write a procedure to encipher/decipher text using the RSA public key cryptosystem.

[Note that in certain instances this scheme may leave messages unaltered. To see this, let $n = 15$, $E = 3$; how may x are there $0 \le x \le n - 1$, such that $x^E \equiv x \pmod{n}$? What can you say about the general case?]

References

Afrati, F.: *Introduction to the theory of information* (in Greek). National Technical University of Athens, Greece, 1985.

Berlekamp, E.R.: *Algebraic coding theory*. McGraw-Hill, New York, 1968.

Blake, I. F.: Codes and designs. *Mathematics Magazine* **52**, 81–95, 1979.

Blakley, G. R., and I. Borosh: Rivest–Shamir–Adleman public key cryptosystems do not always conceal messages. *Computers and Mathematics with Applications* **5**, 169–178, 1979.

Bose, R. C., and D. K. Ray-Chaudhuri: On a class of error correcting binary group codes. *Information and Control* **3**, 68–79, 279–290, 1960.

Childs, L.: *A concrete introduction to higher algebra*. Springer Verlag, New York, 1979.

Diffie, W., and M. E. Hellman: New directions in cryptograph. *IEEE Transactions on Information Theory* **IT-22**, 644–654, 1976.

Diffie, W., and M. E. Hellman: Exhaustive cryptanalysis of the NBS data encryption standard. *Computer* **10**, 74–84, June 1977.

Ecker, A.: Ueber die mathematischen Grundlagen einiger Chiffrierverfahren. *Computing* **29**, 277–287, 1982.

Feistel, H.: Cryptography and computer privacy. *Scientific American* **228**, 15–23, May 1973.

Gass, F.: Solving a Jules Verne cryptogram. *Mathematics Magazine* **59**, 3–11, 1986.

Hamming, R.W.: Error detecting and error correcting codes. *Bell System Technical Journal* **29**, 147–150, 1950.

Hellman, M. E.: The mathematics of public-key cryptography. *Scientific American* **241**, 130–139, August 1979.

Herlestam, T.: Critical remarks on some public-key cryptosystems. *BIT* **18**, 493–496, 1978.

Hill, L. S.: Cryptography in an algebraic alphabet. *American Mathematical Monthly* **36**, 306–312, 1929.

Hill, L. S.: Concerning certain linear transformation apparatus of cryptography. *American Mathematical Monthly* **38**, 135–154, 1931.

Hocquenghem, A.: Codes correcteurs d'erreurs. *Chiffres* **2**, 147–156, 1959.

Kahn, D.: *The Codebreakers*. Macmillan, New York, 1967.

Krishnamurthy, E. V., and V. Ramachandran: A cryptographic system based on finite field transforms. *Proceedings of the Indian Academy of Sciences (Math. Sci.)* **89**, 75–93, 1980.

Lempel, A.: Cryptology in transition. *ACM Computing Surveys* **11**, 280–303, 1979.

Levinston, N.: Coding theory: A counterexample to G. H. Hardy's conception of applied mathematics. *American Mathematical Monthly* **77**, 249–258, 1970.

Lewis, H. R., and C. H. Papadimitriou: The efficiency of algorithms. *Scientific American*, 96–109, January 1978.

Lidl, R., and G. Pilz: *Applied abstract algebra*. Springer Verlag, New York, 1984.

Mackiw, G.: *Applications of abstract algebra*. Wiley, New York, 1985..

MacWilliams, F. J., and N. J. A. Sloane: *The theory of error-correcting codes*, Part I and II. North-Holland, New York, 1977.

Merkle, R. C.: Secure communications over insecure channels. *Communications of the ACM* **21**, 294–299, April 1978.

Merkle, R. C., and M. E. Hellman: Hiding information and signatures in trapdoor knapsacks. *IEEE Transactions on Information Theory* **IT-24,** 525–530, 1978.

Merkle, R. C., and M. E. Hellman: On the security of multiple encryption. *Communications of the ACM* **24**, 465–467, July 1981.

Peterson, W. W., and E. J. Weldon: *Error-correcting codes*, 2 ed. MIT Press, Cambridge, MA, 1972.

Pohlig, S. C., and M. E. Hellman: An improved algorithm for computing logarithms over $GF(p)$ and its cryptographic significance. IEEE Transactions on Information Theory **IT-24,** 106–110, 1978.

Rivest, R. L., A. Shamir, and L. Adleman: A method for obtaining digital signatures and public key cryptosystems. *Communications of the ACM* **21**, 120–126, February 1978.

Shannon, C. E.: A mathematical theory of communication. *Bell System Technical Journal* **27**, 379–423, 623–656, 1948.

Shannon, C. E.: Communication theory of secrecy systems. *Bell System Technical Journal* **28**, 656–715, 1949.

Simmons, G. J.: Cryptology: The mathematics of secure communication. *Mathematical Intelligencer* **1**, 233–246, 1979.

Sinkov, A.: *Elementary cryptanalysis—a mathematical approach*. The New Mathematical Library no. 22, Mathematical Association of America, Washington, DC, 1968.

Wakerly, J.: *Error detecting codes, self-checking circuits and applications*. North-Holland, New York, 1978.

Williams, H. C., and B. Schmid: Some remarks concerning the M.I.T. public-key cryptosystem. *BIT* **19**, 525–538, 1979.

5

Greatest Common Divisors of Polynomials over the Integers and Polynomial Remainder Sequences

5.1 Introduction and Motivation

5.2 The Sylvester–Habicht Pseudodivisions Subresultant PRS Method

5.3 The Matrix-Triangularization Subresultant PRS Method

5.4 Empirical Comparisons Between the Two Subresultant PRS Methods

Exercises

Programming Exercises

Historical Notes and References

216 GREATEST COMMON DIVISORS OF POLYNOMIALS

In this chapter we continue the discussion on computing greatest common divisors *gcd* of polynomials and polynomial remainder sequences (PRS). (The reader should review at this point Sections 3.2.1 and 3.2.2.) However, as the title indicates, we restrict our study to polynomials over the ring of integers where the basic problem is that of restricting coefficient growth. We examine in detail the two classical methods that exist in the literature: the Sylvester–Habicht pseudodivisions subresultant PRS method (consisting of two algorithms), which was initiated by Sylvester in 1853 and was completed by Habicht in 1948, and the matrix-triangularization subresultant PRS method, which was developed by the author in 1986 (Akritas, 1986, 1988) (see also historical note 1 and Figure 5.1).

As we will see below, with use of the matrix-triangularization method, the coefficients of the polynomial remainders are, in certain cases, smaller than those obtained with the Sylvester–Habicht method; hence, the former method is the best.

5.1
Introduction and Motivation

As we have already seen, $\mathbf{Z}[x]$ is neither a field nor a Euclidean domain, however, by Theorem 3.2.10 it is a unique factorization domain, that is, every non-unit element can be (essentially) uniquely expressed as a product of irreducible polynomials. (Recall that any field is a unique factorization domain in which every nonzero element is a unit and there are no primes. The integers form a unique factorization domain where the units are ± 1 and the primes are $\pm 2, \pm 3, \pm 5, \ldots$.)

In Section 3.2.3 we have said that a polynomial in $\mathbf{Q}[x]$ is called "primitive" if its coefficients are relatively prime; this definition, as well as the statement of Theorem 3.2.12, carry over to polynomials over a unique factorization domain. We also have the following.

Theorem 5.1.1. Let $J[x]$ be a unique factorization domain and nonzero $p(x) \in J[x]$. Then $p(x)$ can be uniquely factored in the form $p(x) = c \cdot p'(x)$ where $c \in J$ and $p'(x)$ is primitive. This factorization is unique with units of J; that is, if $p(x) = c_1 \cdot p_1'(x) = c_2 \cdot p_2'(x)$, then $c_1 = bc_2$ and $p_2'(x) = bp_1'(x)$ where b is a unit in J.

Proof. The proof is obvious from the results of Chapter 3. □

The constant c of Theorem 5.1.1 is the greatest common divisor of the coefficients of $p(x)$ and is called the *content* of $p(x)$, written as $cont[p(x)]$; then, obviously, $p'(x)$ is called the *primitive part* of $p(x)$, written as $pp[p(x)]$, a primitive polynomial in $J[x]$. [It is conventional to define

INTRODUCTION AND MOTIVATION 217

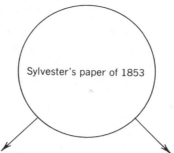

The pseudodivisions subresultant PRS method initiated by Sylvester in 1853, and completed by Habicht in 1948.

The matrix-triangularization subresultant PRS method developed by the author in 1986.

Figure 5.1. Overview of the historical development of the two classical subresultant PRS methods. The method developed by Sylvester should be used only when the PRS is complete, whereas the one by Habicht should be used, when the PRS is incomplete. The matrix-triangularization method can be used for both kinds of PRSs.

$pp[p(x)]$ so that its leading coefficient is positive.] Therefore, each nonzero polynomial $p(x)$ in $J[x]$ has a unique representation of the form

$$p(x) = cont[p(x)] \cdot pp[p(x)]$$

For example, consider the two polynomials $p_1(x) = 5x + 10$ and $p_2(x) = -3x^2 + 9$ in $\mathbf{Z}[x]$. Then

$cont[p_1(x)] = 5$ $pp[p_1(x)] = x + 2$

$cont[p_2(x)] = -3$ $pp[p_2(x)] = x^2 - 3$

$cont[p_1(x) \cdot p_2(x)] = -15$ $pp[p_1(x) \cdot p_2(x)] = x^3 + 2x^2 - 3x - 6$

Note how we made use of Theorem 3.2.12.

Let us now turn our attention to the problem of finding the greatest common divisor of two polynomials $p_1(x)$, $p_2(x)$ in $\mathbf{Z}[x]$ that is not a field, but only a unique factorization domain. Clearly, from the above discussion we can deduce the following important relations

$$cont\{gcd[p_1(x), p_2(x)]\} = gcd\{cont[p_1(x)], cont[p_2(x)]\}$$

$$pp\{gcd[p_1(x), p_2(x)]\} = gcd\{pp[p_1(x)], pp[p_2(x)]\}$$

Therefore, the problem of finding greatest common divisors of arbitrary polynomials is reduced to the problem of finding greatest common divisors of primitive polynomials.

5.1.1 A General Overview of the Classical PRS Algorithms

Consider $p_1(x)$ and $p_2(x)$, two primitive nonzero polynomials in $\mathbf{Z}[x]$, with $\deg[p_1(x)] = m$ and $\deg[p_2(x)] = n$, $m > n$. Since algorithm **PDF** (from Section 3.1) for polynomial division with remainder requires exact divisibility by the leading coefficient of $p_2(x)$, $lc[p_2(x)]$, it is usually impossible to carry out this process on $p_1(x)$ and $p_2(x)$ over the integers without relaxing the division requirement. We thus introduce the process of *pseudodivision*, which always yields a *pseudoquotient* and a *pseudoremainder* (*prem*), that will have integer coefficients.

Pseudodivision means premultiplying $p_1(x)$ times $\{lc[p_2(x)]\}^{m-n+1}$ and then applying algorithm **PDF**, knowing that all the quotients will exist; that is, we have

$$\{lc[p_2(x)]\}^{m-n+1} p_1(x) = p_2(x) q(x) + r(x), \qquad \deg[r(x)] < \deg[p_2(x)]$$

where $q(x)$ and $r(x)$ are the pseudoquotient and pseudoremainder, respectively.

Example. Using pseudodivision in $\mathbf{Z}[x]$, let us divide $p_1(x) = x^4 - 7x + 7$ by $p_2(x) = 3x^2 - 7$. In order to compute $q(x)$ and $r(x)$ we premultiply $p_1(x)$ by $3^{4-2+1} = 27$, and then, applying **PDF**, we obtain $q(x) = 9x^2 + 21$ and $r(x) = -189x + 336$. The reader should verify that **PDF** would not have worked, had we premultiplied $p_1(x)$ by just 3.

Therefore, in trying to compute a greatest common divisor of $p_1(x)$ and $p_2(x)$ we need to make sure that all the polynomial divisions involved in the process are feasible; that is, we need to form the polynomial remainder sequence using pseudodivisions. We are thus led to the following generalized Euclidean algorithm for polynomials.

GEA-P (Generalized Euclidean Algorithm for Polynomials over the Integers)

Input: $p_1(x)$, $p_2(x)$ nonzero polynomials in $\mathbf{Z}[x]$; $\deg[p_1(x)] = n_1$, $\deg[p_2(x)] = n_2$, $n_1 \geq n_2$.

Output: $gcd[p_1(x), p_2(x)]$, a greatest common divisor of $p_1(x)$ and $p_2(x)$.

1. [Compute *gcd* of the contents.] $c := gcd\{cont[p_1(x)], cont[p_2(x)]\}$. (Here we use the known Euclidean algorithm for computing the greatest common divisor of two integers.)
2. [Compute primitive parts.] $p_1'(x) := p_1(x)/cont[p_1(x)]$; $p_2'(x) := p_2(x)/cont[p_2(x)]$.
3. [Construct a PRS.] Compute $p_1'(x), p_2'(x), p_3(x), \ldots, p_h(x)$.
4. [Exit.] If $\deg[p_h(x)] = 0$, the return $gcd[p_1(x), p_2(x)] := c$; otherwise, return $gcd[p_1(x), p_2(x)] := c \cdot pp[p_h(x)]$.

INTRODUCTION AND MOTIVATION

Clearly, the computing time of this algorithm depends on how efficiently we can compute the polynomial remainder sequence $p'_1(x), p'_2(x), p_3(x), \ldots, p_h(x)$. Note that, if $n_i = \deg[p_i(x)]$, then, in general, we can say that the members of this sequence satisfy

$$\{lc[p_{i+1}(x)]\}^{n_i - n_{i+1} + 1} p_i(x) = p_{i+1}(x) q_i(x) + \beta_i p_{i+2}(x),$$
$$\deg[p_{i+2}(x)] < \deg[p_{i+1}(x)]$$

where $i = 1, 2, \ldots, h - 1$, for some h. [Of course, we have $p_i(x) := p'_i(x)$, $i = 1, 2$, where $p'_i(x)$, $i = 1, 2$ were defined in step 2 of **GEA-P**.] When a method for choosing β_i is given, the above equation provides an algorithm for constructing a PRS; of course, the terminating condition of this family of algorithms is to have the pseudoremainder equal to zero.

Below we examine the various algorithms obtained for the different values of β_i. In addition, we briefly present the matrix-triangularization subresultant PRS method, which is quite different from the other methods in the sense that it does not depend on the choice of β_i; instead of explicit polynomial pseudodivisions, the matrix-triangularization method brings a matrix into its upper triangular form and all the members of the PRS are obtained from the rows of the final matrix.

The Euclidean PRS Algorithm Here $\beta_i = 1$ for all $i = 1, 2, \ldots, h - 1$; that is, each pseudoremainder is used as it is obtained. This is by far the worst method for constructing a PRS, as a result of the exponential growth of the coefficients.

Example. Consider the polynomials $p_1(x) = x^3 - 7x + 7$, $p_2(x) = 3x^2 - 7$ in $\mathbf{Z}[x]$. Obviously, here $cont[p_1(x)] = cont[p_2(x)] = 1$ and $p_i(x) = p'_i(x)$, $i = 1, 2$. We have the following sequence:

$$p_1(x) = x^3 - 7x + 7$$
$$p_2(x) = 3x^2 - 7 \cdots\cdots\cdot \quad q_1(x) = 3x$$
$$p_3(x) = -42x + 63 \cdots\cdots\cdot q_2(x) = -126x - 189$$
$$p_4(x) = -441 \cdots\cdots\cdot \quad q_3(x) = 18522x - 27783$$

which was obtained by performing the following pseudodivisions:

$$(3)^2 p_1(x) = p_2(x) \cdot (3x) + (-42x + 63)$$
$$(-42)^2 p_2(x) = p_3(x) \cdot (-126x - 189) + (-441)$$
$$(-441)^2 p_3(x) = p_4(x) \cdot (18522x - 27783) + 0$$

Clearly, from step 4 of **GEA-P** we have that $gcd[p_1(x), p_2(x)] = 1$. Also note that in the last pseudodivision we had coefficients eight decimal digits long, since $(-441)^2 p_3(x) = -8168202x + 12252303$.

The polynomial remainder sequence of this example is called "complete" because the degree of each member is one less than the degree of the preceding member; the first two members could, of course, have the same degree. Otherwise, a sequence is called "incomplete." Observe that there is no way to tell a priori whether a PRS is going to be complete or incomplete. See also Barnett, 1970, 1974; Brown et al., 1971; and Collins, 1966, 1971.

The exponential growth of the coefficients of the terms in the PRS of the above example is due to the fact that the polynomials of the sequence are not primitive; that is, not removing their contents has detrimental effects. This situation is remedied below.

The Primitive PRS Algorithm In this case $\beta_i = cont\{\text{prem}[p_i(x), p_{i+1}(x)]\}$, $i = 1, 2, \ldots, h-1$, where "prem" denotes the pseudoremainder; that is, we now remove the content from the $(i+2)$th PRS member before we use it. [Recall that for given $p(x)$ it is conventional to define $pp[p(x)]$ so that its leading coefficent is positive.]

Example. Consider the same polynomials as in the previous example: $p_1(x) = x^3 - 7x + 7$, $p_2(x) = 3x^2 - 7$ in $\mathbf{Z}[x]$, where again $p_i(x) = p'_i(x)$, $i = 1, 2$. Now we obtain

$$p_1(x) = x^3 - 7x + 7$$
$$p_2(x) = 3x^2 - 7 \cdots\cdots\cdot q_1(x) = 3x$$
$$p_3(x) = 2x - 3 \cdots\cdots\cdot q_2(x) = 6x - 9 \cdots \beta_1 = -21$$
$$p_4(x) = 1 \cdots\cdots\cdot q_3(x) = 2x - 3 \cdots \beta_2 = -1$$

which was obtained by performing the following pseudodivisions:

$$(3)^2 p_1(x) = p_2(x) \cdot (3x) + (-21)(2x - 3)$$
$$(2)^2 p_2(x) = p_3(x) \cdot (6x + 9) + (-1)$$
$$(1)^2 p_3(x) = p_4(x) \cdot (2x - 3) + 0$$

From this example we clearly see that this algorithm is as good as can be done with respect to the coefficient growth of the terms in the PRS. However, at each step it is necessary to calculate one or more greatest common divisors of coefficients, and these computations become progressively harder as the coefficients grow. Therefore, we would like to find a way to avoid most of these computations and still reduce the coefficient growth sharply from that which occurs in the Euclidean PRS algorithm. This can be achieved using either Sylvester–Habicht's pseudodivisions or the matrix-triangularization subresultant PRS method; detailed descriptions of these methods are found in Sections 5.2 and 5.3, respectively.

The Sylvester–Habicht Pseudodivisions Subresultant PRS Method

Two algorithms are associated with this method: Sylvester's *reduced (subresultant)* PRS algorithm, if the sequence is complete; and Habicht's *subresultant* PRS algorithm, if the sequence is incomplete. (See historical note 1.)

According to this method, in case of a complete PRS we choose

$$\beta_1 = 1$$
$$\beta_i = \{lc[p_i(x)]\}^2, \quad i = 2, 3, \ldots, h-1 \qquad \text{(S)}$$

[Sylvester's reduced (subresultant) PRS algorithm], whereas, in case of an incomplete PRS we set

$$\beta_1 = (-1)^{n_1 - n_2 + 1}$$
$$\beta_i = (-1)^{n_i - n_{i+1} + 1} lc[p_i(x)] \cdot H_i^{n_i - n_{i+1}}, \quad i = 2, 3, \ldots, h-1 \qquad \text{(H)}$$

(Habicht's subresultant PRS algorithm) where

$$H_2 = \{lc[p_2'(x)]\}^{n_1 - n_2}, \text{ and}$$
$$H_i = \{lc[p_i(x)]\}^{n_{i-1} - n_i} H_{i-1}^{1 - (n_{i-1} - n_i)}, \quad i = 3, \ldots, h-1$$

The derivation of the quantities in (S) and (H) is made clear in Sections 5.2.1 and 5.2.3, respectively. Habicht's method reduces to Sylvester's in case of complete PRSs. In general, (S) can be also expressed as

$$\beta_1 = 1$$
$$\beta_i = \{lc[p_i(x)]\}^{n_i - n_{i+1} + 1}, \quad i = 2, 3, \ldots, h-1 \qquad \text{(S')}$$

which is also known as *Sylvester's reduced* (subresultant) PRS algorithm.

Observe that in both algorithms above, the polynomial remainder sequence is computed with the help of **PDF**, which is the classical division algorithm for polynomials; moreover, instead of computing the content of the $(i+2)$th member of the PRS, we simply divide its coefficients by β_i, knowing that the division can be *exactly* performed. (In Sections 5.2.1 and 5.2.3 we will see why this division is exact.) Also, in both algorithms above we have $\beta_1 = 1$, which means that $p_3(x)$ cannot be reduced, and the reader should keep this in mind.

As we mentioned before, Habicht generalized a result by Sylvester; therefore, his PRS algorithm can be also used for complete PRSs (at additional computational cost, of course). However, Sylvester's PRS algorithm [(S) or (S')] does not work very well when we have an incomplete PRS; that is, the growth can be exponential, although not as severe as in the case of the Euclidean PRS algorithm. (Recall that there is no a priori way of knowing whether a PRS is going to be complete or incomplete.)

Example. Consider again the polynomials $p_1(x) = x^3 - 7x + 7$, $p_2(x) = 3x^2 - 7$ in $\mathbf{Z}[x]$, where $cont[p_1(x)] = cont[p_2(x)] = 1$ and $p_i(x) = p'_i(x)$, $i = 1, 2$. We now have

$$p_1(x) = x^3 - 7x + 7$$
$$p_2(x) = 3x^2 - 7 \cdots\cdots\cdots\quad q_1(x) = 3x$$
$$p_3(x) = -42x + 63 \cdots\cdots\quad q_2(x) = -126x - 189$$
$$p_4(x) = -49 \cdots\cdots\quad\quad\quad q_3(x) = 18522x - 27783$$

Since we are dealing with a complete PRS, both algorithms of the Sylvester–Habicht pseudodivisions subresultant PRS method generate the same sequence, which is obtained as in the case of the Euclidean PRS algorithm. However, note that we now have $p_4(x) = -49$ instead of -441 because $9 = \{lc[p_2(x)]\}^2$ was divided out of $p_4(x)$. [Here $p_4(x)$ has just one term, but this is irrelevant.]

The Matrix-Triangularization Subresultant PRS Method In contrast with the above PRS algorithms, the matrix-triangularization method avoids explicit polynomial divisions. To see how it works, consider two polynomials in $\mathbf{Z}[x]$, $p_1(x) = c_n x^n + c_{n-1} x^{n-1} + \cdots + c_0$ and $p_2(x) = d_m x^m + d_{m-1} x^{m-1} + \cdots + d_0$, $c_n \neq 0$, $d_m \neq 0$, $n \geq m$, for which we want to compute the polynomial remainder sequence. First, we construct the matrix corresponding to Sylvester's form of the resultant (see Sections 5.2.2 and 5.3.3 for the definitions), which, in our case, is the square matrix of order $2n$ shown below [$p_2(x)$ has been transformed into a polynomial of degree n by introducing zero coefficients]:

$$\begin{bmatrix} c_n & c_{n-1} & \cdots c_0 & 0 & 0 \cdots 0 \\ d_n & d_{n-1} & \cdots d_0 & 0 & 0 \cdots 0 \\ 0 & c_n & \cdots & c_0 & 0 \cdots 0 \\ 0 & d_n & \cdots & d_0 & 0 \cdots 0 \\ & & \vdots & & \\ 0 \cdots 0 & c_n & c_{n-1} & & \cdots c_0 \\ 0 \cdots 0 & d_n & d_{n-1} & & \cdots d_0 \end{bmatrix}$$

Then, this matrix is brought into its upper triangular form using integer-preserving transformations, and all the members of the (complete or incomplete) PRS are obtained from the rows of the final matrix with the help of a theorem by the author (Theorem 5.3.6); the computing time of the matrix-triangularization method is the *same* as that of the Sylvester–Habicht pseudodivisions method.

Notice that Sylvester's form of the resultant is used for the first time in the present century; it was buried in Sylvester's paper of 1853 and was only used once by Van Vleck in 1899.

INTRODUCTION AND MOTIVATION 223

In the following example the coefficients of the polynomials obtained with the matrix-triangularization subresultant PRS method are smaller than those obtained with the Sylvester–Habicht pseudodivisions method.

Example. We consider again the polynomials $p_1(x) = x^3 - 7x + 7$ and $p_2(x) = 3x^2 - 7$. As we have seen, both algorithms of the Sylvester–Habicht pseudodivisions subresultant PRS method generate $p_3(x) = -42x + 63$, $p_4(x) = -49$, whereas the matrix-triangularization method generates $p_3(x) = -14x + 21$, $p_4(x) = -49$ (implementation details are discussed in Section 5.3.3). The corresponding final matrix is

$$\begin{bmatrix} 1 & 0 & -7 & 7 & 0 & 0 \\ 0 & 3 & 0 & -7 & 0 & 0 \\ 0 & 0 & 9 & 0 & -21 & 0 \\ 0 & 0 & 0 & -42 & 63 & 0 \\ 0 & 0 & 0 & 0 & 196 & -294 \\ 0 & 0 & 0 & 0 & 0 & -49 \end{bmatrix} *$$

The asterisked ($*$-ed) row indicates that a pivot took place between the third and fourth rows. The coefficients of $p_2(x) = -14x + 21$ are obtained from the third row *before* the pivot, and we see that they are smaller by a factor of 3 from those obtained by the Sylvester–Habicht pseudodivisions method. This occurs because the quotient on pseudodividing $p_1(x)$ by $p_2(x)$ is $3x$ with *no constant term involved*; this means that we did not have to multiply $p_1(x)$ by 3^2 but only by 3. (However, there is no way of knowing this before the actual division.)

We next consider an example of an incomplete PRS.

Example. If we consider the polynomials $p_1(x) = x^5 + 5x^4 + 10x^3 + 5x^2 + 5x + 2$ and $p_2(x) = x^4 + 4x^3 + 6x^2 + 2x + 1$, the upper triangular form of the matrix corresponding to Sylvester's form of the resultant is

$$\begin{bmatrix} 1 & 5 & 10 & 5 & 5 & 2 & 0 & 0 & 0 \\ 0 & 1 & 4 & 6 & 2 & 1 & 0 & 0 & 0 \\ 0 & 0 & 1 & 4 & 3 & 4 & 2 & 0 & 0 \\ 0 & 0 & 0 & 1 & 7 & 1 & 3 & 2 & 0 \\ 0 & 0 & 0 & 0 & 3 & -2 & -1 & 0 & 0 \\ 0 & 0 & 0 & 0 & 0 & 9 & -6 & -3 & 0 \\ 0 & 0 & 0 & 0 & 0 & 58 & 50 & 18 & 0 \\ 0 & 0 & 0 & 0 & 0 & 0 & -266 & -112 & 0 \\ 0 & 0 & 0 & 0 & 0 & 0 & 0 & -756 & -532 \\ 0 & 0 & 0 & 0 & 0 & 0 & 0 & 0 & -980 \end{bmatrix} *$$

The $*$-ed row indicates that a pivot took place. By Theorem 5.3.6, the members of the incomplete PRS generated by $p_1(x)$ and $p_2(x)$ are $3x^2 - 2x - 1$, $-266x - 112$, and -980.

5.1.2 A Nonclassical Approach: The Modular Greatest Common Divisor Algorithm

As we indicated above, in subsequent sections we examine in detail the two subresultant PRS methods. However, immediately below we present an algorithm that the reader must have anticipated.

The key observation for this algorithm is that computing a greatest common divisor of two polynomials over a finite field leads to no growth, since the coefficients cannot exceed the size of the modulus. Therefore, we can convert a greatest common divisor problem of two polynomials over the integers to a number of greatest common divisor problems of two polynomials over some finite fields. By Theorem 5.1.2, stated below, the answers to these problems are then interpolated (using the Greek–Chinese remainder algorithm) to produce the true greatest common divisor over the integers. This, basically, constitutes the *modular gcd algorithm*. In the discussion below $p^{(p)}(x)$ indicates the polynomial $p(x)$ in $\mathbf{Z}_p[x]$.

Theorem 5.1.2. Let $p_1(x)$ and $p_2(x)$ be two polynomials in $\mathbf{Z}[x]$, and $p_h(x) = gcd[p_1(x), p_2(x)]$. Then, for all primes p such that p does not divide $lc[p_1(x)] \cdot lc[p_2(x)]$ we have $\deg[p_h^{(p)}(x)] \geq \deg[p_h(x)]$, where $p_h^{(p)}(x) = gcd[p_1(x), p_2(x)]$ over $\mathbf{Z}_p[x]$, and there are only finite many primes p such that $\deg[p_h^{(p)}(x)] > \deg[p_h(x)]$.

Proof. Since p does not divide $lc[p_1(x)] \cdot lc[p_2(x)]$, it follows that p does not divide the leading coefficient of $p_h(x)$, and hence, the first assertion is trivial. The second assertion follows from the observation that the degree of $p_h^{(p)}(x)$ can happen to be greater than that of $p_h(x)$ only if p divides one of the leading coefficients of the preceding members of the PRS, and this can occur for only a finite number of primes. $\square$

The primes p for which $\deg[p_h^{(p)}(x)] > \deg[p_h(x)]$ are called "unlucky" and are discarded in the algorithm described below; moreover, for efficiency, each prime p is chosen to be less than a machine word. The following facts are also used in the modular *gcd* algorithm:

1. If $p_1(x)$ is irreducible over the integers modulo m, where m does not divide $lc[p_1(x)]$, then $p_1(x)$ is also irreducible over the integers; basically, this is Theorem 3.2.16.
2. If $p_1(x)$ and $p_2(x)$ are relatively prime over the integers modulo m, where again m does not divide $lc[p_1(x)]$ and does not divide $lc[p_2(x)]$, then $p_1(x)$ and $p_2(x)$ are relatively prime over the integers; this follows directly from Theorem 5.1.2.

M (Modular Algorithm)

Input: Two primitive univariate polynomials $p_1(x)$, $p_2(x)$, with integer coefficients.

Output: The polynomial $p_h(x) = gcd[p_1(x), p_2(x)]$ over the integers.

1. [Initialize.] Set $c := gcd\{lc[p_1(x)], lc[p_2(x)]\}$. Pick a prime p such that it does not divide $lc[p_1(x)] \cdot lc[p_2(x)]$ and compute the *monic* polynomial $p_h^{(p)}(x) = gcd[p_1(x), p_2(x)]$ in $GF(p)$. Set $d := \deg \cdot (p_h^{(p)}(x))$; if $d = 0$, then return $p_h(x) := 1$.
2. [Done?] set $g(x) := c \cdot p_h^{(p)}(x) \pmod{p}$. If $pp[g(x)] \| p_1(x)$ and $pp[g(x)] \| p_2(x)$ (where both divisions are performed over the integers), then return $p_h(x) := pp[g(x)]$.
3. [New prime.] Pick a new prime p_1 such that it does not divide $lc[p_1(x)] \cdot lc[p_2(x)]$, and compute the *monic* polynomial $p_h^{(p_1)}(x) = gcd[p_1(x), p_2(x)]$ in $GF(p_1)$.
4. [Unlucky prime?] If $\deg[p_h^{(p_1)}(x)] > d$, then p_1 was an unlucky prime; go to step 3. If $\deg[p_h^{(p_1)}(x)] < d$, then p was an unlucky prime; set $p_h^{(p)}(x) := p_h^{(p_1)}(x)$, $p := p_1$, and go to step 2.
5. [Loop over coefficients.] At this point $\deg[p_h^{(p_1)}(x)] = \deg[p_h^{(p)}(x)]$. Set $p_h'(x) := 0$. For $i := 0, 1, \ldots, d$, do the following: use **GCRA2** on the coefficients of x^i in $p_h^{(p)}(x)$ and $p_h^{(p_1)}(x)$ with moduli p and p_1; let the interpolated result be q and set $p_h'(x) := p_h'(x) + qx^i$. [Using the symmetric residue system, we have that the coefficients of $p_h'(x)$ are in absolute value $\leq \lfloor p \cdot p_1/2 \rfloor$; also note that $p_h'(x)$ is *monic*.]
6. [Update.] Set $p_h(x) := p_h'(x)$, $p := p \cdot p_1$, and go to step [2].

Computing-Time Analysis of **M**. By Theorem 5.1.2 it follows that the above algorithm terminates, since the number of unlucky primes is finite. [See also Brown (1971), Theorem 1, where a bound is given on the number of unlucky primes that might occur.] It has been indicated by Brown that the superiority of the modular algorithm over the classical ones is established only when the given polynomials are sufficiently large and sufficiently dense. In the sparse case (many missing terms) the gain is difficult to analyze.

An example is presented below. The previous algorithm did not use an a priori bound on the number of different primes employed; an interesting variation of **M** where such a bound is used is the following:

M1 (Modular Algorithm, Version 1)

Input: Two primitive univariate polynomials $p_1(x)$, $p_2(x)$, with integer coefficients.

Output: The polynomial $p_h(x) = gcd[p_1(x), p_2(x)]$ over the integers.

1. [Initialize.] Set $c := gcd\{lc[p_1(x)], lc[p_2(x)]\}$. Pick a prime p such that it does not divide $lc[p_1(x)] \cdot lc[p_2(x)]$, and compute $p_h^{(p)}(x) = gcd[p_1(x), p_2(x)]$ in $GF(p)$. Set $d := \deg[p_h^{(p)}(x)]$; if $d = 0$, return $p_h(x) := 1$. Set $B := \max[|p_1(x)|_\infty, |p_2(x)|_\infty]$.
2. [New prime.] Pick a new prime p_1 such that it does not divide $lc \cdot [p_1(x)] \cdot lc[p_2(x)]$. Compute $p_h^{(p_1)}(x) = gcd[p_1(x), p_2(x)]$ in $GF(p_1)$.
3. [Unlucky prime?] If $\deg[p_h^{(p_1)}(x)] > d$, then p_1 was an unlucky prime; go to step 2. If $\deg[p_h^{(p_1)}(x)] < d$, then p was an unlucky prime; set $p_h^{(p)}(x) := p_h^{(p_1)}(x)$, $p := p_1$, and go to step 2.
4. [Loop over coefficients.] At this point $\deg[p_h^{(p_1)}(x)] = \deg[p_h^{(p)}(x)]$. Set $p_h'(x) := 0$. For $i := 0, 1, \ldots, d$, do the following. Use **GCRA2** on the coefficients of x^i in $p_h^{(p)}(x)$ and $p_h^{(p_1)}(x)$ with moduli p and p_1, respectively; let the interpolated result be q and set $p_h'(x) := p_h'(x) + qx^i$. (Use the symmetric residue system.)
5. [Done?] Set $p_h(x) := p_h'(x)$, $p := p \cdot p_1$. If $p > 2B$, then return $p_h(x) := pp[p_h(x)]$, otherwise, go to step 2.

Actually, the quantity B used in this algorithm is a bound on the maximum of the absolute values of the coefficients of $p_h(x)$. Such a bound may be computed from a theorem by A. O. Gelfond (*Transcendental and algebraic numbers*, Dover, New York, 1960), but in implementations the maximum of the absolute values of the coefficients of $p_1(x)$ and $p_2(x)$ is used instead. The quantity $2B$ is chosen because it works *most* of the time (Brown, 1971). Clearly, there are exemptions just as when $p(x) = x^8 + 2x^7 + 3x^6 + 4x^5 + 5x^4 + 4x^3 + 3x^2 + 2x + 1$ and $q(x) = x - 1$; then $p(x)q(x) = x^9 + x^8 + x^7 + x^6 + x^5 - x^4 - x^3 - x^2 - x - 1$, and we have the case where the coefficients of the divisor are greater than those of the dividend.

Observe that either version of the modular *gcd* algorithm can be easily extended to handle multivariate polynomials. It is also possible to use the lift algorithms, described in Chapter 6, to find the greatest common divisor of two polynomials over **Z**. This is the so-called Extended Zassenhaus GCD (**EZGCD**) algorithm, which is not described here, but details can be found in the literature (Miola et al., 1974; Moses et al., 1973; and Yun, 1974).

Example. Let us find a greatest common divisor of $p_1(x) = 9x^5 + 6x^4 + 3x^3 + 3x^2 + 2x + 1$ and $p_2(x) = 3x^4 + 5x^3 + 6x^2 + 3x + 1$.

Using **M**, we first observe that $c = 3$, and then, for $p = 5$ we obtain $p_1^{(5)}(x) = 4x^5 + x^4 + 3x^3 + 3x^2 + 2x + 1$, $p_2^{(5)}(x) = 3x^4 + x^2 + 3x + 1$ and $p_h^{(5)}(x) = gcd[p_1(x), p_2(x)] = 2x^2 + 3x + 4 = x^2 + 4x + 2$ in $GF(5)$. Then, in step 2, we compute $g(x) = 3x^2 + 2x + 1$, which divides both $p_1(x)$ and $p_2(x)$. Therefore, $p_h(x) = 3x^2 + 2x + 1$.

Using **M1**, we first compute $p_h^{(5)}(x) = gcd[p_1(x), p_2(x)] = 2x^2 + 3x + 4$ in $GF(5)$, and then $p_h^{(7)}(x) = gcd[p_1(x), p_2(x)] = 5x^2 + x + 4$ in $GF(7)$. Since

$5 \cdot 7 = 35 > 2B = 18$, these two primes are enough. With the help of the Greek–Chinese remainder theorem we obtain, from $p_h^{(5)}(x)$ and $p_h^{(7)}(x)$, the polynomial $g(x) = 12x^2 + 8x + 4$ and $p_h(x) = pp[g(x)] = 3x^2 + 2x + 1$.

5.2
The Sylvester–Habicht Pseudodivisions Subresultant PRS Method

In this section we study in detail the Sylvester–Habicht pseudodivisions subresultant PRS method. Basically, we are going to see why the quantities β_i, mentioned in Section 5.1.1 with regard to Sylvester's and Habicht's PRS algorithms, divide $p_{i+2}(x)$, the $(i+2)$th member of the PRS. At this point the reader is asked to review the properties of the determinants in the Appendix at the end of the book.

5.2.1 Sylvester's Reduced (Subresultant) PRS Algorithm

In his 1853 paper Sylvester developed a method for computing a complete polynomial remainder sequence over the integers; moreover, he kept the coefficients of the members of the PRS *as small as possible* by removing their *allotrious factors* (to use Sylvester's terminology), and *not* their contents (an operation that would involve integer greatest common divisor calculations). Also, it should be noted that Sylvester was interested in computing Sturm's sequences (explained in Chapter 7), where the negative of each polynomial remainder in the PRS is calculated; that is, $p_i(x) = p_{i+1}(x)q_i(x) - p_{i+2}(x)$. Therefore, the determinants in his proof are exactly like the ones used below except for the fact that they have the second and third rows interchanged [Akritas, 1987]. (Resultants and subresultants are defined in the next section.)

Theorem 5.2.1 (Sylvester, 1853). *Let $p_1(x)$, $p_2(x)$, $p_3(x)$, ..., $p_h(x)$ be a* complete *polynomial remainder sequence, $p_i(x) \in \mathbf{Z}[x]$, for $i = 1, 2, \ldots, h$. Then $\{lc[p_i(x)]\}^2 | p_{i+2}(x)$, $1 < i \le h - 2$; that is, the square of the leading coefficient of $p_i(x)$ is a divisor of $p_{i+2}(x)$.*

Proof. Let $a, b, c, d, \ldots$ be the coefficients of any dividend $p_{i-1}(x)$ and $\alpha, \beta, \gamma, \delta, \ldots$ of the divisor $p_i(x)$, where $a = lc[p_{i-1}(x)]$ and $\alpha = lc[p_i(x)]$. Then, it is easily seen that the coefficients of the remainder $p_{i+1}(x)$, forming the second divisor, are:

$$\left(\frac{1}{a^2}\right)\det\begin{bmatrix} a & b & c \\ \alpha & \beta & \gamma \\ 0 & \alpha & \beta \end{bmatrix}, \quad \left(\frac{1}{a^2}\right)\det\begin{bmatrix} a & b & d \\ \alpha & \beta & \delta \\ 0 & \alpha & \gamma \end{bmatrix},$$

$$\left(\frac{1}{a^2}\right)\det\begin{bmatrix} a & b & e \\ \alpha & \beta & \varepsilon \\ 0 & \alpha & \delta \end{bmatrix}, \ldots$$

228 GREATEST COMMON DIVISORS OF POLYNOMIALS

[The reader should verify that the determinants are the coefficients of the remainder $p_{i+1}(x)$.] In the same way, and setting

$$m := \det \begin{bmatrix} a & b & c \\ \alpha & \beta & \gamma \\ 0 & \alpha & \beta \end{bmatrix}$$

the coefficients of the second remainder will each be of the form of the compound determinant [observe that, in assigning the value to m, we do not need to divide the determinant by a^2 (Why?)]:

$$\left(\frac{1}{m^2}\right) \det \begin{bmatrix} \alpha & \beta & \gamma \\ \begin{matrix} a & b & c \\ \alpha & \beta & \gamma \\ 0 & \alpha & \beta \end{matrix} & \begin{matrix} a & b & d \\ \alpha & \beta & \delta \\ 0 & \alpha & \gamma \end{matrix} & \begin{matrix} a & b & e \\ \alpha & \beta & \varepsilon \\ 0 & \alpha & \delta \end{matrix} \\ 0 & \begin{matrix} a & b & c \\ \alpha & \beta & \gamma \\ 0 & \alpha & \beta \end{matrix} & \begin{matrix} a & b & d \\ \alpha & \beta & \delta \\ 0 & \alpha & \gamma \end{matrix} \end{bmatrix}$$

where the above expression is the $lc[p_{i+2}(x)]$. (The 3×3 entries, inside the expression of the above determinant, are themselves determinants.) Omitting the common multiplier $(1/m^2)$ and expanding along the first column, we see that the determinant above is equal to

$$\alpha \left\{ \det \begin{bmatrix} a & b & d \\ \alpha & \beta & \delta \\ 0 & \alpha & \gamma \end{bmatrix} \cdot \det \begin{bmatrix} a & b & d \\ \alpha & \beta & \delta \\ 0 & \alpha & \gamma \end{bmatrix} - \det \begin{bmatrix} a & b & c \\ \alpha & \beta & \gamma \\ 0 & \alpha & \beta \end{bmatrix} \cdot \det \begin{bmatrix} a & b & e \\ \alpha & \beta & \varepsilon \\ 0 & \alpha & \delta \end{bmatrix} \right\}$$

$$- \det \begin{bmatrix} a & b & c \\ \alpha & \beta & \gamma \\ 0 & \alpha & \beta \end{bmatrix} \cdot \left\{ \beta \det \begin{bmatrix} a & b & d \\ \alpha & \beta & \delta \\ 0 & \alpha & \gamma \end{bmatrix} - \gamma \det \begin{bmatrix} a & b & c \\ \alpha & \beta & \gamma \\ 0 & \alpha & \beta \end{bmatrix} \right\}$$

Expanding again, we see that, for some expressions A, B, the first term is of the form

$$\alpha^2 A + \alpha(\alpha\beta\gamma \cdot \alpha\beta\gamma - a\beta^2 \cdot a\beta\delta)$$

whereas the second term is of the form (expand and simplify the expression in the brackets first)

$$\alpha^2 B - \alpha(\gamma^2 - \beta\delta)a^2\beta^2$$

Hence, after cancellations, the entire determinant is of the form $\alpha^2(A + B)$

and we see that α^2 will enter as a factor into this and every coefficient of $p_{i+2}(x)$. □

In Theorem 5.2.1, note that we cannot reduce the coefficients of $p_3(x)$; the proposition applies to the coefficients of $p_4(x)$, $p_5(x)$, and so on.

Sylvester indicates that "the same explicit method might be applied to show that if the first divisor were e degrees instead of being only one degree lower than the first dividend, α^{e+1} would be contained in every term of the second residue; the difficulty, however, of the proof by this method augments with the value of e" (Sylvester, 1853, p. 419).

Example. Let us consider again the polynomials $p_1(x) = x^3 - 7x + 7$, and $p_2(x) = 3x^2 - 7$ in $\mathbf{Z}[x]$. As you recall, the polynomial remainder sequence obtained by the Euclidean PRS algorithm is $p_1(x) = x^3 - 7x + 7$, $p_2(x) = 3x^2 - 7$, $p_3(x) = -42x + 63$, $p_4(x) = -441$.

Using Theorem 5.2.1, we compute the coefficients of $p_3(x)$ as determinants obtained from the matrix

$$\begin{bmatrix} 1 & 0 & -7 & 7 \\ 3 & 0 & -7 & 0 \\ 0 & 3 & 0 & -7 \end{bmatrix}$$

These determinants are

$$\det \begin{bmatrix} 1 & 0 & -7 \\ 3 & 0 & -7 \\ 0 & 3 & 0 \end{bmatrix} = -42 \text{ and } \det \begin{bmatrix} 1 & 0 & 7 \\ 3 & 0 & 0 \\ 0 & 3 & -7 \end{bmatrix} = 63$$

(They are actually the subresultants and are defined in the next section.) Note that at this point we cannot divide out anything, since $p_3(x)$ is the first remainder.

Likewise, the single coefficient of $p_4(x)$ is obtained from the determinant

$$(1/3^2)\det \begin{bmatrix} 3 & 0 & -7 \\ -42 & 63 & 0 \\ 0 & -42 & 63 \end{bmatrix} = (1/9) \cdot (-441) = -49$$

As we see, in this case we can exactly divide 441 by 3^2, where $3 = lc[p_2(x)]$, and we obtain a much smaller coefficient.

Using (S'), from Section 5.1.1, we have the following algorithm:

SRSPRS [Sylvester's reduced (subresultant) **PRS**]

Input: Two nonzero polynomials $p_1(x)$, $p_2(x)$ in $\mathbf{Z}[x]$, $\deg[p_1(x)] \geq \deg[p_2(x)]$.

Output: The reduced (subresultant) PRS of $p_1(x)$ and $p_2(x)$.

1. [Initialize.] $i := 1$; $c := 1$.
2. [Done?] If $p_{i+1}(x) = 0$, then exit.
3. [Compute the pseudoremainder.] Output $p_{i+2}(x) := \text{prem}[p_i(x), p_{i+1}(x)]/c^{n_i - n_{i+1} + 1}$, set $i := i+1$, $c := lc[p_i(x)]$ and go to 2.

Computing-Time Analysis of SRSPRS. The computing time of this algorithm is

$$t_{\text{SRSPRS}}[p_1(x), p_2(x)] = 0\{n^5 L^2 [|p(x)|_\infty]\}$$

where $|p(x)|_\infty = \max(|p_1(x)|_\infty, |p_2(x)|_\infty)$. The proof of this fact makes use of resultants (discussed in the next section), and it is omitted since it is the same as the proof for the computing time analysis of algorithm **HSPRS** described in Section 5.2.3. (By the way, this is also the computing time of all the classical PRS algorithms.)

Sylvester's theorem indicates a divisibility property that exists between *certain* members of *complete* PRSs, as a result of which we can easily see the existing divisibility properties. However, in order to see why Habicht's subresultant PRS algorithm works (which, as you recall, should be used in the incomplete cases), we have to have a more general divisibility property; that is, we need a property between *any* members of the PRS. This new divisibility property was hinted at by Sylvester, but was first presented by Habicht in 1948. At first, though, we need some additional concepts, discussed in the next section.

5.2.2 The Resultant of Two Polynomials

Resultants and subresultants are essential for a thorough understanding of Habicht's PRS algorithm, and in this section we study their properties. (Sylvester uses them implicitly.) Netto, 1896 is an excellent reference; see also Bocher, 1907 and Lipson, 1981.

Let us consider two polynomials in $\mathbf{Z}[x]$:

$$p_1(x) = c_n x^n + c_{n-1} x^{n-1} + \cdots + c_0, \qquad c_n \neq 0, \quad n > 0$$
$$p_2(x) = d_m x^m + d_{m-1} x^{m-1} + \cdots + d_0, \qquad d_m \neq 0, \quad m > 0$$

and let us find the conditions under which they have a common divisor, or one (at least) common root.

Denote the roots of $p_1(x) = 0$ by $\alpha_1, \alpha_2, \ldots, \alpha_n$, and of $p_2(x) = 0$ by $\beta_1, \beta_2, \ldots, \beta_m$. Then in order for $p_1(x)$ and $p_2(x)$ to have one common root, some α_i must be equal to some β_j and the following product, denoted by $\text{res}[p_1(x), p_2(x)]$:

$$\text{res}[p_1(x), p_2(x)] = c_n^m d_m^n (\alpha_1 - \beta_1)(\alpha_1 - \beta_2) \cdots (\alpha_1 - \beta_m)$$
$$(\alpha_2 - \beta_1)(\alpha_2 - \beta_2) \cdots (\alpha_2 - \beta_m)$$
$$\vdots$$
$$(\alpha_n - \beta_1)(\alpha_n - \beta_2) \cdots (\alpha_n - \beta_m)$$

must be equal to zero; this is the condition we were looking for. [The reader should note that there is no relation between $\text{res}[p_1(x), p_2(x)]$ and the variable res used in Sections 2.4 and 3.1.4; the latter was an abbreviation for the word result.] Now, grouping the n rows or the m columns and taking into consideration the fact that

$$p_1(x) = (-1)^n c_n (\alpha_1 - x)(\alpha_2 - x) \cdots (\alpha - x)$$
$$p_2(x) = d_m (x - \beta_1)(x - \beta_2) \cdots (x - \beta_m)$$

we have

$$\text{res}[p_1(x), p_2(x)] = c_n^m p_2(\alpha_1) p_2(\alpha_2) \cdots p_2(\alpha_n)$$
$$= (-1)^{mn} d_m^n p_1(\beta_1) p_1(\beta_2) \cdots p_1(\beta_m) \quad (R)$$

So we see that the factor $c_n^m d_m^n$ was necessary in order to make $\text{res}[p_1(x), p_2(x)]$ an integer function of the coefficients c_i, d_j. We clearly have

$$\text{res}[p_1(x), p_2(x)] = (-1)^{mn} \text{res}[p_2(x), p_1(x)]$$

and we call this expression the *resultant* of $p_1(x)$ and $p_2(x)$. We therefore have proved the following

Theorem 5.2.2. The vanishing of the resultant $\text{res}[p_1(x), p_2(x)]$ is the necessary and sufficient condition for the equations $p_1(x) = 0$ and $p_2(x) = 0$ to have a common root.

Some Properties of the Resultants If $p_2(x) = p_3(x) p_4(x)$, where $\deg[p_3(x)] = m_1$ and $\deg[p_4(x)] = m_2$, relation (R) yields

$$\text{res}[p_1(x), p_3(x)p_4(x)] = c_n^{m_1+m_2}[p_3(\alpha_1)p_4(\alpha_1)] \cdots [p_3(\alpha_n)p_4(\alpha_n)]$$
$$= c_n^{m_1} p_3(\alpha_1) \cdots p_3(\alpha_n) \cdot c_n^{m_2} p_4(\alpha_1) \cdots p_4(\alpha_n)$$

That is, we have

$$\text{res}[p_1(x), p_3(x)p_4(x)] = \text{res}[p_1(x), p_3(x)] \cdot \text{res}[p_1(x), p_4(x)] \quad (R1)$$

Moreover, if c is a constant, then from (R) we have

$$\text{res}[p_1(x), p_2(x) + cp_1(x)] = c_n^m [p_2(\alpha_1) + cp_1(\alpha_1)] \cdots [p_2(\alpha_n) + cp_1(\alpha_n)]$$

and since $p_1(\alpha_i) = 0$, $i = 1, 2, \ldots, n$, we have

$$\text{res}[\,p_1(x),\, p_2(x) + cp_1(x)\,] = \text{res}[\,p_1(x),\, p_2(x)\,] \tag{R2}$$

Finally, if $p_2(x) = x^m$, then all the β_i values are zero and we have the special form

$$\text{res}[\,p_1(x),\, x^m\,] = (-1)^{mn}[p_1(0)]^m = (-1)^{mn} c_0^m \tag{R3}$$

Resultant Representation We do not evaluate resultants using (R) because we would first have to compute the roots of $p_1(x)$ or $p_2(x)$. Instead, we represent the resultant as a determinant, which can be easily evaluated. In order to do this, let us consider the following system of $(m + n)$ equations [we are still dealing with the two polynomials $p_1(x)$ and $p_2(x)$ given above of degrees n and m, respectively]:

$$
\begin{aligned}
{[c_0 - p_1(x)]} + c_1 x + \cdots + c_n x^n &= 0 \\
{[c_0 - p_1(x)]} x + c_1 x^2 + \cdots + c_n x^{n+1} &= 0 \\
&\vdots \\
{[c_0 - p_1(x)]} x^{m-1} + c_1 x^m + \cdots + c_n x^{m+n-1} &= 0 \\
d_0 + d_1 x + \cdots + d_m x^m &= 0 \\
d_0 x + d_1 x^2 + \cdots + d_m x^{m+1} &= 0 \\
&\vdots \\
d_0 x^{n-1} + d_1 x^n + \cdots + d_m x^{m+n-1} &= 0
\end{aligned}
$$

We consider this to be a linear system of equations in the $(m + n)$ quantities $1, x, x^2, \ldots, x^{m+n-1}$. The first m equations are actually identities due to the definition of $p_1(x)$; the last n are valid only for $x = \beta_1, \beta_2, \ldots, \beta_m$. Rearranging columns and rows, we obtain an equation of degree m in $v = p_1(\beta_1), p_1(\beta_2), \ldots, p_1(\beta_m)$, which is represented by the determinant

$$
\det \begin{bmatrix}
c_n & c_{n-1} & \cdots & c_0 - v & 0 & \cdots & 0 \\
0 & c_n & c_{n-1} & \cdots & c_0 - v & \cdots & 0 \\
& & & \vdots & & & \\
0 & 0 & \cdots & c_n & c_{n-1} & \cdots & c_0 - v \\
d_m & d_{m-1} & \cdots & d_0 & 0 & \cdots & 0 \\
0 & d_m & d_{m-1} & \cdots & d_0 & \cdots & 0 \\
& & & \vdots & & & \\
0 & 0 & \cdots & d_m & d_{m-1} & \cdots & d_0
\end{bmatrix} = 0
\quad
\begin{matrix} \left.\vphantom{\begin{matrix}1\\1\\1\\1\end{matrix}}\right\} (m \text{ rows}) \\ \\ \left.\vphantom{\begin{matrix}1\\1\\1\\1\end{matrix}}\right\} (n \text{ rows}) \end{matrix}
$$

and the roots of this equation determine the quantities $p_1(\beta_1)$, $p_1(\beta_2), \ldots, p_1(\beta_m)$. Making use of the fact that if we have the equation

$$a_m v^m + a_{m-1} v^{m-1} + a_{m-2} v^{m-2} + \cdots + a_0 = 0$$

then, the product of all its roots is equal to $(-1)^m a_0/a_m$, we see that if we divide our equation in v, of degree m, by the coefficient of v^m, then the constant term will be equal to $(-1)^m p_1(\beta_1) p_1(\beta_2) \cdots p_1(\beta_m)$. Since in our case the coefficient of $(-v)^m$ is $(-1)^{mn} d_m^n$ (justify this), we have the following relation (observe that we have set $v = 0$, and hence, the determinant is the constant term of the mth-degree equation)

$$\det \begin{bmatrix} c_n & c_{n-1} & \cdots & & c_0 & 0 & \cdots & 0 \\ 0 & c_n & c_{n-1} & \cdots & & c_0 & \cdots & 0 \\ & & & & & \vdots & & \\ 0 & 0 & \cdots & & c_n & c_{n-1} & \cdots & c_0 \\ d_m & d_{m-1} & \cdots & d_0 & 0 & 0 & \cdots & 0 \\ 0 & d_m & d_{m-1} & \cdots & d_0 & 0 & \cdots & 0 \\ & & & & & \vdots & & \\ 0 & 0 & \cdots & d_m & d_{m-1} & & \cdots & d_0 \end{bmatrix} \begin{matrix} \\ \\ \rbrace (m \text{ rows}) \\ \\ \\ \\ \rbrace (n \text{ rows}) \\ \end{matrix}$$

$$= (-1)^{mn} d_m^n p_1(\beta_1) p_1(\beta_2) \cdots p_1(\beta_m) = c_n^m p_2(\alpha_1) p_2(\alpha_2) \cdots p_2(\alpha_n)$$
$$= \text{res}_B[p_1(x), p_2(x)]$$

Therefore, we have expressed the resultant in the form of a determinant, which can be easily evaluated; from now on we shall represent this form of the resultant as $\text{res}_B[p_1(x), p_2(x)]$, in honor of the nineteenth-century French mathematician di Bruno, who made extensive use of it.

Subresultants—Obtaining the Coefficients of Members of the PRS

Let us now define the subresultants, which play an important role in our development of the PRS algorithms. Consider the *complete* polynomial remainder sequence $p_1(x), p_2(x), p_3(x), \ldots, p_h(x)$, where now

$$p_1(x) = c_n x^n + c_{n-1} x^{n-1} + \cdots + c_0, \quad c_n \neq 0$$
$$p_2(x) = d_{n-1} x^{n-1} + d_{n-2} x^{n-2} + \cdots + d_0, \quad d_{n-1} \neq 0$$
(BT)

As we recall from the proof of Theorem 5.2.1, the coefficients of $p_3(x)$, which is a polynomial of degree $n - 2$, are given by the determinants

$$\det \begin{bmatrix} c_n & c_{n-1} & c_{n-2} \\ d_{n-1} & d_{n-2} & d_{n-3} \\ 0 & d_{n-1} & d_{n-2} \end{bmatrix}, \quad \det \begin{bmatrix} c_n & c_{n-1} & c_{n-3} \\ d_{n-1} & d_{n-2} & d_{n-4} \\ 0 & d_{n-1} & d_{n-3} \end{bmatrix}, \text{ etc.}$$

where the first determinant is the leading coefficient of $p_3(x)$, the second determinant is the coefficient of x^{n-3}, and so on; for reasons given below, these determinants are called *subresultants*. [For subsequent members of the PRS, $p_i(x), i > 3$, we know that each determinant is divisible by $\{lc[p_{i-2}(x)]\}^2$.] Comparing these determinants with the resultant of $p_1(x)$ and $p_2(x)$, expressed in the form of a determinant of order $2n - 1$, we easily see that the coefficients of $p_3(x)$ are obtained in the following way. Since the degree of $p_3(x)$ is $n - 2$, remove from the matrix $\mathbf{M}_B$ corresponding to $\text{res}_B(p_1(x), p_2(x))$ the last $n - 2$ (of the $n - 1$) rows of c values, the last $n - 2$ (of the n) rows of d values, and the last $n - 2$ columns. So, we are left with a $3 \times (n + 1)$ matrix $\mathbf{M}'_B$; that is, $\mathbf{M}'_B$ has 3 rows and $n + 1$ columns $\mathbf{M}'_B(j)$, $j = 0, 1, 2, \ldots, n$. The $n - 1$ coefficients of $p_3(x)$ are successively computed from a determinant (subresultant) of order 3, whose first two columns are always $\mathbf{M}'_B(0), \mathbf{M}'_B(1)$, and the third one is successively $\mathbf{M}'_B(j)$, $j = 2, 3, \ldots, n$; $j = 2$ yields the leading coefficient of $p_3(x)$. [This is *precisely* what Sylvester did in Theorem 5.2.1 to compute the coefficients *only* of $p_3(x)$.]

This process can be *generalized* to compute the coefficients of any member of the PRS, $p_i(x)$, of degree $n - i + 1$. We simply have to remove from the matrix $\mathbf{M}_B$ corresponding to $\text{res}_B[p_1(x), p_2(x)]$, the last $n + i + 1$ rows of c values, the last $n - i + 1$ rows of d values, and the last $n - i + 1$ columns, leaving us with a $(2i - 3) \times (n + i - 2)$ matrix $\mathbf{M}'_B$ [i.e., $\mathbf{M}'_B$ has $(i - 2)$ rows of c and $(i - 1)$ rows of d]. Then the $n - i + 2$ coefficients of $p_i(x)$ are successively computed from a determinant (subresultant) of order $(2i - 3)$ whose first $2(i - 2)$ columns are always $\mathbf{M}'_B(0)$, $\mathbf{M}'_B(1), \ldots, \mathbf{M}'_B[2(i - 2) - 1]$, and the last one is successively $\mathbf{M}'_B(j)$, $j = 2(i - 2), (2i - 3), \ldots, n + i - 3$. This general way of computing the coefficients of any member of a complete PRS was discovered in 1859 by di Bruno [see Muir (1920), pp. 327–328]; that is, di Bruno discovered that for the polynomials (BT) $p_i(x)$, the ith member of the PRS, equals

$$\lambda_i \Sigma_{0 \le m \le n - i + 1} x^{n - i + 1 - m} \cdot$$

$$\det \begin{bmatrix} c_n & c_{n-1} & c_{n-2} & \cdots & c_{n-2(i-2)+1} & c_{n-2(i-2)-m} \\ 0 & c_n & c_{n-1} & \cdots & c_{n-2(i-2)+2} & c_{n-2(i-2)-m+1} \\ \cdots & & & & & \\ 0 & 0 & \cdots & c_n & \cdots & c_{n-i+2} & c_{n-i+1-m} \\ d_{n-1} & d_{n-2} & d_{n-3} & \cdots & d_{n-2(i-2)} & d_{n-2(i-2)-1-m} \\ 0 & d_{n-1} & d_{n-2} & \cdots & d_{n-2(i-2)+1} & d_{n-2(i-2)-1-m+1} \\ \cdots & & & & & \\ 0 & 0 & \cdots & d_{n-1} & \cdots & d_{n-i+2} & d_{n-i+1-m} \end{bmatrix}$$

where λ_i is the "allotrious" factor and $c_{-k} = 0$ for any value of k.

If the polynomial remainder sequence is incomplete, or if the difference of the degrees of the polynomials $p_1(x)$ and $p_2(x)$ is greater than one, then we can basically use the approach mentioned above to compute the coefficients of any member of the PRS; however, as we have already mentioned, the allotrious factor is not so easily determined, and it will be discussed later. Moreover, if on evaluation the degree of $p_i(x)$ turns out to be smaller than $n - i + 1$—say, it is $n - i + 1 - k$, and we now compute the coefficients of $p_i(x)$ by removing from the matrix $\mathbf{M_B}$, corresponding to $\text{res}_B[p_1(x), p_2(x)]$, the last $n - i + 1 - k$ rows of c and d values, and the last $n - i + 1 - k$ columns, we obtain a polynomial that is *proportional* to the one that we would obtain had we removed the last $n - i + 1$ rows of c and d, and the last $n - i + 1$ columns. However, removing from $\text{res}_B[p_1(x), p_2(x)]$ successively the last $n - i, \ldots, n - 1 + 1 - (k - 1)$ rows of c and d and the last $n - i, \ldots, n - i + 1 - (k - 1)$ columns, *we obtain zero polynomials*. These facts are proved in Section 5.2.3.

In our discussion of Theorem 5.2.1 we have already mentioned that Sylvester was interested in obtaining the negative of the polynomial remainders (Sturm's sequence). So, for the polynomials (BT) he computed the coefficients of $p_3(x)$ as

$$\det\begin{bmatrix} c_n & c_{n-1} & c_{n-2} \\ 0 & d_{n-1} & d_{n-2} \\ d_{n-1} & d_{n-2} & d_{n-3} \end{bmatrix}, \quad \det\begin{bmatrix} c_n & c_{n-1} & c_{n-3} \\ 0 & d_{n-1} & d_{n-3} \\ d_{n-1} & d_{n-2} & d_{n-4} \end{bmatrix}, \text{ etc.}$$

In order to be able to obtain the negated coefficients of any member of the PRS [of the polynomials (BT)] in a way similar to the one discussed above, we have to use another form of the resultant, namely,

$$\text{res}_T[p_1(x), p_2(x)] = \det\begin{bmatrix} c_n & c_{n-1} & \cdots & c_0 & 0 & \cdots & 0 \\ 0 & c_n & c_{n-1} & \cdots & c_0 & \cdots & 0 \\ & & & \vdots & & & \\ 0 & 0 & \cdots & c_n & c_{n-1} & \cdots & c_0 \\ 0 & 0 & \cdots & & d_{n-1} & \cdots & d_0 \\ & & & \vdots & & & \\ 0 & d_{n-1} & \cdots & d_0 & 0 & \cdots & 0 \\ d_{n-1} & & \cdots & d_0 & 0 & 0 & \cdots & 0 \end{bmatrix}.$$

The subscript T is in honor of the Italian mathematician of the last century, N. Trudi, who gave the first methodical exposition of *bigradients* [see Muir (1920), p. 329]; the term *bigradient* refers to the form of the resultant that we denoted by $\text{res}_T[p_1(x), p_2(x)]$.

Now, to compute the coefficients of any member $p_i(x)$ of a complete PRS, where $\deg[p_i(x)] = n - i + 1$, we remove from the matrix $\mathbf{M_T}$ corresponding to $\text{res}_T[p_1(x), p_2(x)]$, the *top* $n - i + 1$ rows of c, the last $n - i + 1$ rows of d, and the *first* $n - i + 1$ columns. We are then left with a $(2i - 3) \times (n + i - 2)$ matrix $\mathbf{M'_T}$, and we work with it as in the case of $\text{res}_B[p_1(x), p_2(x)]$.

Note that

$$\text{res}_B[p_1(x), p_2(x)] = (-1)^{n(n-1)/2} \text{res}_T[p_1(x), p_2(x)]$$

We next present Trudi's theorem in its original form and notation.

Theorem 5.2.3 (Trudi, 1862). The coefficients of the rth remainder R_r arising in the course of the performance of Sturm's division process on

$$a_0 x^m + \cdots + a_m, \qquad b_0 x^n + \cdots + b_n$$

(i.e., we are computing the negative of the remainders obtained by the Euclidean PRS algorithm) are equal to the successive determinants of the (bigradient) array

$$\left[\begin{bmatrix} (a_0, a_1, \ldots, a_m)_r \\ (b_0, b_1, \ldots, b_n) \end{bmatrix} \right]$$

divided by the product of the squares of the first coefficients of all the preceding remainders and by b_0^{m-n+1} and the sign factor $(-1)^{(m-n)(m-n-1)/2}$.

Proof. The proof can be found in Householder (1970). □

Example. For the polynomials $p_1(x) = x^3 - 7x + 7$ and $p_2(x) = 3x^2 - 7$ we have the following two matrices corresponding to the respective resultants:

$$\mathbf{M}_B = \begin{bmatrix} 1 & 0 & -7 & 7 & 0 \\ 0 & 1 & 0 & -7 & 7 \\ 3 & 0 & -7 & 0 & 0 \\ 0 & 3 & 0 & -7 & 0 \\ 0 & 0 & 3 & 0 & -7 \end{bmatrix} \qquad \mathbf{M}_T = \begin{bmatrix} 1 & 0 & -7 & 7 & 0 \\ 0 & 1 & 0 & -7 & 7 \\ 0 & 0 & 3 & 0 & -7 \\ 0 & 3 & 0 & -7 & 0 \\ 3 & 0 & -7 & 0 & 0 \end{bmatrix}$$

The coefficients of $p_3(x)$, which is of degree 1, are obtained as follows:

Using $\mathbf{M}_B$:

$$\det \begin{bmatrix} 1 & 0 & -7 \\ 3 & 0 & -7 \\ 0 & 3 & 0 \end{bmatrix} = -42, \qquad \det \begin{bmatrix} 1 & 0 & 7 \\ 3 & 0 & 0 \\ 0 & 3 & -7 \end{bmatrix} = 63$$

and so, $p_3(x) = -42x + 63$.

Using $\mathbf{M}_T$:

$$\det \begin{bmatrix} 1 & 0 & -7 \\ 0 & 3 & 0 \\ 3 & 0 & -7 \end{bmatrix} = 42, \qquad \det \begin{bmatrix} 1 & 0 & 7 \\ 0 & 3 & -7 \\ 3 & 0 & 0 \end{bmatrix} = -63$$

and $p_3(x) = 42x - 63$.

As for the polynomial $p_4(x)$ that is a constant, we simply evaluate the resultants and obtain $\text{res}_B[p_1(x), p_2(x)] = -49$ and $\text{res}_T[p_1(x), p_2(x)] = 49$. The reader should compare these results with the ones obtained in the example of Section 5.2.1. (Observe that, in this case, to obtain the value ± 49, the determinant did not have to be divided by anything.)

An Additional Property of the Resultants We next present a theorem that tells us that the resultant of two polynomials can be expressed in terms of these polynomials.

Theorem 5.2.4. Let J be a commutative ring with identity, and let $p(x)$ and $q(x)$ be polynomials in $J[x]$ of positive degree. Then there exist polynomials $r(x)$ and $s(x)$ in $J[x]$ such that $p(x)r(x) + q(x)s(x) = \text{res}_B[p(x), q(x)]$, with $\deg[r(x)] < \deg[q(x)]$ and $\deg[s(x)] < \deg[p(x)]$—obviously, a similar result holds for $\text{res}_T[p(x), q(x)]$.

Proof. Let $m = \deg[p(x)]$ and $n = \deg[q(x)]$. For $1 \leq i < m + n$ multiply the ith column of the matrix $\mathbf{M}_B$ by x^{m+n-i} and add it to the last column. The result is a new matrix $\mathbf{M}'_B$ whose determinant is equal to the determinant of $\mathbf{M}_B$. The last column of $\mathbf{M}'_B$ consists of the polynomials $x^{n-1}p(x)$, $x^{n-2}p(x), \ldots, p(x), x^{m-1}q(x), x^{m-2}q(x), \ldots, q(x)$. Expanding the determinant with respect to the last column, we obtain the identity $p(x)r(x) + q(x)s(x) = \det(\mathbf{M}'_B) = \text{res}_B[p(x), q(x)]$, where the coefficients of $r(x)$ and $s(x)$ are cofactors of the last column of $\mathbf{M}'_B$, and hence of $\mathbf{M}$, and therefore belong to J. □

This theorem reminds us of the relation that exists between two polynomials and their greatest common divisor. To investigate this further, consider again two polynomials in $\mathbf{Z}[x]$:

$$p(x) = c_n x^n + c_{n-1} x^{n-1} + \cdots + c_0, \quad c_n \neq 0, \quad n > 0$$
$$q(x) = d_m x^m + d_{m-1} x^{m-1} + \cdots + d_0, \quad d_m \neq 0, \quad m > 0$$

and *let us assume that they have a common root α*. Then

$$p(x) = (x - \alpha) p_1(x), \quad q(x) = (x - \alpha) q_1(x)$$

where $p_1(x), q_1(x)$ in $\mathbf{Z}[x]$, $\deg[p_1(x)] = n - 1$, $\deg[q_1(x)] = m - 1$.
From Theorems 5.2.2 and 5.2.4, it easily follows that

$$p(x) q_1(x) - q(x) p_1(x) = 0 \qquad \text{(CR)}$$

Conversely, the existence of an equation of the above form, with $p_1(x), q_1(x)$ in $\mathbf{Z}[x]$, $\deg[p_1(x)] = n - 1$, $\deg[q_1(x)] = m - 1$ is a proof of the existence of a common root of $p(x)$ and $q(x)$. [This is so because $q_1(x)$

GREATEST COMMON DIVISORS OF POLYNOMIALS

cannot have all the linear factors $(x - \beta_\lambda)$ of $q(x)$, and, therefore, there exists at least one linear factor of $q(x)$ appearing in $p(x)$ as well.]

Now, let

$$p_1(x) = u_{n-1}x^{n-1} + u_{n-2}x^{n-2} + \cdots + u_0, \qquad u_{n-1} \neq 0$$

$$q_1(x) = v_{m-1}x^{m-1} + v_{m-2}x^{m-2} + \cdots + v_0, \qquad v_{m-1} \neq 0$$

and replace, in (CR), all the polynomials with their respective expressions. Then, grouping all the coefficients of equal powers of x together, and setting them all equal to 0, we obtain the following system of equations (where we assume $n \geq m$):

Degree

Degree					
$n+m-1$	$c_n v_{m-1}$		$-d_m u_{n-1}$		$=0$
$n+m-2$	$c_{n-1} v_{m-1}$	$+ c_n v_{m-2}$	$-d_{m-1}u_{n-1} - d_m u_{n-2}$		$=0$
$\vdots$					
$n-1$	$c_{n-m}v_{m-1} + c_{n-m+1}v_{m-2} + \cdots + c_{n-1}v_0$		$-d_0 u_{n-1} - d_1 u_{n-2} - \cdots - d_m u_{n-m-1}$		$=0$
$n-2$	$c_{n-m-1}v_{m-1} + c_{n-m}v_{m-2} + \cdots + c_{n-2}v_0$		$-d_0 u_{n-2} - \cdots - d_m u_{n-m-2}$		$=0$
$\vdots$					
$m-1$	$c_0 v_{m-1}$	$+ c_1 v_{m-2} + \cdots + c_{m-1} v_0$	$-d_0 u_{m-1} - \cdots - d_{m-1} u_0$		$=0$
$m-2$		$c_0 v_{m-2} + \cdots + c_{m-2} v_0$	$-d_0 u_{m-2} - \cdots - d_{m-2} u_0$		$=0$
$\vdots$					
0		$c_0 v_0$	$-d_0 u_0 = 0$		

This is a system of $m + n$ linear equations in the $m + n$ unknowns $u_{n-1}, \ldots, u_0, v_{m-1}, \ldots, v_0$. The existence of a common root requires that the system of equations be satisfied for nonzero values of the unknowns; hence, the determinant of the system must be zero. Interchanging rows and columns, we see that the above determinant is the resultant of $p(x)$ and $q(x)$, $\text{res}_B[p(x), q(x)]$.

This process can be continued. If the nonzero polynomials $p_1(x)$ and $q_1(x)$ in (CR) are of degrees $(n-2)$ and $(m-2)$, respectively, then, the polynomials $p(x)$ and $q(x)$ have a common factor of degree 2. So, now, (CR) becomes the necessary and sufficient condition for the existence of two common roots of $p(x)$ and $q(x)$. The proof proceeds exactly as above; we only need to set $u_{n-1} = v_{m-1} = 0$, both in the expressions of $p_1(x)$, $q_1(x)$ above and in the corresponding system of equations. Now the system will have $(m + n - 1)$ equations in $(m + n - 2)$ unknowns, and in order to have nonzero solutions, all the determinants of order $(m + n - 2)$ derived from the matrix

$$\begin{bmatrix} c_n & c_{n-1} & \cdots & c_0 & 0 & \cdots \\ 0 & c_n & \cdots & c_1 & c_0 & \\ \multicolumn{6}{c}{\cdots\cdots} \\ d_m & d_{m-1} & \cdots & d_0 & 0 & \cdots \\ 0 & d_m & \cdots & d_1 & d_0 & \cdots \\ \multicolumn{6}{c}{\cdots\cdots} \end{bmatrix} \begin{matrix} \left.\vphantom{\begin{matrix}a\\a\\a\end{matrix}}\right\} (m-1) \text{ rows} \\ \\ \left.\vphantom{\begin{matrix}a\\a\\a\end{matrix}}\right\} (n-1) \text{ rows} \end{matrix}$$

must be zero; this matrix has $(m + n - 1)$ columns. Obviously, we can continue with this process.

Existence of a Greatest Common Divisor of Specified Degree
Let us now use the above results in order to find the necessary and sufficient conditions for the existence of a greatest common divisor of $p(x)$ and $q(x)$ of specified degree. The reader should carefully study the proof of the following theorem.

Theorem 5.2.5. The degree of a greatest common divisor of two univariate polynomials $\mathbf{Z}[x]$, $p(x)$ and $q(x)$, $\deg[p(x)] = n$, $\deg[q(x)] = m$ is equal to the superscript of the first nonzero determinant

$$\operatorname{res}_B[p(x), q(x)] = \operatorname{res}_B^{(0)}[p(x), q(x)], \operatorname{res}_B^{(1)}[p(x), q(x)], \operatorname{res}_B^{(2)}[p(x), q(x)], \ldots,$$

where $\operatorname{res}_B^{(i)}[p(x), q(x)]$ denotes the subresultant obtained from the resultant $\operatorname{res}_B[p(x), q(x)]$, if we eliminate the last i rows of coefficients of both $p(x)$ and $q(x)$, and the last $2i$ columns; that is, $\operatorname{res}_B^{(i)}[p(x), q(x)]$ is a determinant of order $(m + n - 2i)$ and is called (for obvious reasons as we shall see) the *ith principal subresultant coefficient*.

Proof. We will first derive the conditions needed for $p(x)$ and $q(x)$ to have a greatest common divisor of degree 1. Let

$$p(x) = c_n x^n + c_{n-1} x^{n-1} + \cdots + c_0, \qquad c_n \neq 0, \quad n > 0$$
$$q(x) = d_m x^m + d_{m-1} x^{m-1} + \cdots + d_0, \qquad d_m \neq 0, \quad m > 0$$

From our previous discussion, we know that if $\operatorname{res}_B[p(x), q(x)] = 0$, then $p(x)$ and $q(x)$ have a divisor of (at least) first degree. Assume that this is, indeed, the case, and consider the polynomials

$$p^{(1)}(x) = v_{m-2} x^{m-2} + \cdots + v_0, \qquad v_{m-2} \neq 0$$
$$q^{(1)}(x) = u_{n-2} x^{n-2} + \cdots + u_0, \qquad u_{n-2} \neq 0$$

with unknown coefficients $u_{n-2}, \ldots, u_0, v_{m-2}, \ldots, v_0$. Then, from the equation

$$p(x) p^{(1)}(x) - q(x) q^{(1)}(x) = g_{m+n-2} x^{m+n-2} + g_{m+n-3} x^{m+n-3} + \cdots + g_0$$

form the following system of equations

$$c_n v_{m-2} - d_m u_{n-2} = g_{m+n-2}$$
$$c_{n-1} v_{m-2} + c_n v_{m-3} - d_{m-1} u_{n-2} - d_m u_{n-3} = g_{m+n-3}$$
$$\vdots$$
$$c_0 v_1 + c_1 v_0 - d_0 u_1 - d_1 u_0 = g_1$$
$$c_0 v_0 - d_0 u_0 = g_0$$

Except for the last equation, we have a system of $(m+n-2)$ linear equations in $(m+n-2)$ unknowns. Therefore, if the corresponding determinant $\operatorname{res}_B^{(1)}[p(x), q(x)]$, of order $m+n-2$, is nonzero, we can determine the unknowns, so that

$$g_{m+n-2} = g_{m+n-3} = \cdots = g_2 = 0, \text{ and } g_1 = \operatorname{res}_B^{(1)}[p(x), q(x)]$$

in which case

$$p(x)p^{(1)}(x) - q(x)q^{(1)}(x) = \operatorname{res}_B^{(1)}[p(x), q(x)] \cdot x + g_0$$

The last relation indicates that $p(x)$ and $q(x)$ have a divisor only of first degree (otherwise, the right-hand side of the above equation would also be divisible by that other factor of higher degree). So,

$$\operatorname{res}_B^{(0)}[p(x), q(x)] = 0, \qquad \operatorname{res}_B^{(1)}[p(x), q(x)] \neq 0$$

are the necessary and sufficient conditions for the greatest common divisor of $p(x)$ and $q(x)$ to be of degree 1.

If, however, $\operatorname{res}_B^{(1)}[p(x), q(x)] = 0$, then we have

$$p(x)p^{(1)}(x) - q(x)q^{(1)}(x) = g_0$$

but now, since $p(x)$ and $q(x)$ have a common factor, there is a value of x for which the left-hand side of the last equation becomes zero, and hence, it must be the case that $g_0 = 0$. So, we have

$$p(x)p^{(1)}(x) - q(x)q^{(1)}(x) = 0$$

from which we conclude that $p(x)$ and $q(x)$ have a common factor of at least second degree.

Therefore, if $\operatorname{res}_B^{(0)}[p(x), q(x)] = \operatorname{res}_B^{(1)}[p(x), q(x)] = 0$, then we set

$$p^{(2)}(x) = v_{m-3}x^{m-3} + \cdots + v_0, \qquad v_{m-3} \neq 0$$
$$q^{(2)}(x) = u_{n-3}x^{n-3} + \cdots + u_0, \qquad u_{n-3} \neq 0$$

with unknown coefficients $u_{n-3}, \ldots, u_0, v_{m-3}, \ldots, v_0$, and again, from the equation

$$p(x)p^{(2)}(x) - q(x)q^{(2)}(x) = g_{m+n-3}x^{m+n-3} + g_{m+n-4}x^{m+n-4} + \cdots + g_0$$

we build the following system of equations

$$c_n v_{m-3} - d_m u_{n-3} = g_{m+n-3}$$
$$c_{n-1} v_{m-3} + c_n v_{m-4} - d_{m-1} u_{n-3} - d_m u_{n-4} = g_{m+n-4}$$
$$\vdots$$

Except for the last *two* rows, we have a system of $(m+n-4)$ linear equations in $(m+n-4)$ unknowns. If the corresponding determinant

$$\operatorname{res}_B^{(2)}[p(x), q(x)] = \det \begin{bmatrix} c_n & c_{n-1} & \cdots & c_0 & 0 & \cdots \\ 0 & c_n & \cdots & c_1 & c_0 & \cdots \\ & & \cdots & & & \\ d_m & d_{m-1} & \cdots & d_0 & 0 & \cdots \\ 0 & d_m & \cdots & d_1 & d_0 & \cdots \\ & & \cdots & & & \end{bmatrix}$$

is not zero, then we can determine the unknowns so that

$$g_{m+n-3} = g_{m+n-4} = \cdots = 0, \qquad g_2 = \operatorname{res}_B^{(2)}[p(x), q(x)]$$

in which case

$$p(x)p^{(2)}(x) - q(x)q^{(2)}(x) = \operatorname{res}_B^{(2)}[p(x), q(x)] \cdot x^2 + g_1 x + g_0.$$

This shows that $p(x)$ and $q(x)$ have one divisor of degree 2 only; (if there were a divisor of higher degree, it would have to divide the right-hand side of the above equation).

Therefore,

$$\operatorname{res}_B^{(0)}[p(x), q(x)] = \operatorname{res}_B^{(1)}(p(x), q(x)) = 0, \qquad \operatorname{res}_B^{(2)}[p(x), q(x)] \neq 0$$

are the necessary and sufficient conditions for the greatest common divisor of $p(x)$ and $q(x)$ to be of degree 2. Proceeding in this way, we complete the proof of the theorem. □

As a corollary to Theorem 5.2.5 we have the following.

Corollary 5.2.6. A necessary and sufficient condition for two univariate polynomials to be relatively prime is that their resultant does not vanish.

The ith principal subresultant coefficient $\operatorname{res}_B^{(i)}[p(x), q(x)]$ is of extreme importance for the material covered in the next section, where it is abbreviated as R_i; see also the example below.

Results similar to the ones obtained above can be obtained for $\operatorname{res}_T[p(x), q(x)]$ where, ignoring sign differences, the same abbreviation is used; namely, we have the following.

Definition 5.2.7. The ith *principal subresultant coefficient*, $\operatorname{res}_T^{(i)}[p_1(x), p_2(x)]$, of two univariate polynomials $p_1(x), p_2(x)$ is the determinant obtained from the resultant $\operatorname{res}_T[p_1(x), p_2(x)]$ of these polynomials if we omit the first i rows and columns and the last i rows and columns.

Example. For the polynomials $p_1(x) = x^3 - 7x + 7$ and $p_2(x) = 3x^2 - 7$ we have the following subresultants $R_i = \operatorname{res}_T^{(i)}[p_1(x), p_2(x)]$, $i = 0, 1, 2$.

$$R_0 = \begin{vmatrix} 1 & 0 & -7 & 7 & 0 \\ 0 & 1 & 0 & -7 & 7 \\ 0 & 0 & 3 & 0 & -7 \\ 0 & 3 & 0 & -7 & 0 \\ 3 & 0 & -7 & 0 & 0 \end{vmatrix} \quad R_1 = \quad R_2 =$$

Theorem 5.2.8. The degree of the greatest common divisor of two univariate polynomials is equal to the superscript of the first nonzero principal subresultant coefficient $R_0 = \operatorname{res}_T^{(0)}[p_1(x), p_2(x)]$, $R_1 = \operatorname{res}_T^{(1)}[p_1(x), p_2(x)]$, $R_2 = \operatorname{res}_T^{(2)}[p_1(x), p_2(x)]$, ...

Proof. Analogous to the proof of Theorem 5.2.5. □

In the last example we find that $R_0 = \operatorname{res}_T^{(0)}[p_1(x), p_2(x)] \neq 0$ and, therefore, we conclude that the degree of the greatest common divisor of $p_1(x)$ and $p_2(x)$ is 0; that is, the $gcd[p_1(x), p_2(x)] = 1$.

We also have the following.

Theorem 5.2.9. If the hth principal subresultant coefficient of $p_1(x)$ and $p_2(x)$ is the first nonzero one [i.e., $\operatorname{res}_T^{(0)}[p_1(x), p_2(x)] = \operatorname{res}_T^{(1)}[p_1(x), p_2(x)] = \cdots = \operatorname{res}_T^{(h-1)}[p_1(x), p_2(x)] = 0$, and $\operatorname{res}_T^{(h)}[p_1(x), p_2(x)] \neq 0$], then the greatest common divisor of $p_1(x)$ and $p_2(x)$ can be obtained from $\operatorname{res}_T^{(h)}[p_1(x), p_2(x)]$ if we replace by (a) $p_1(x)$ the last element in the last row of coefficients of $p_1(x)$, by $xp_1(x)$ the element just above this, by $x^2p_1(x)$ the element above this, and so on, and (b) $p_2(x)$ the last element in the first row of coefficients of $p_2(x)$, by $xp_2(x)$ the element below this, by $x^2p_2(x)$ the element below this, and so forth.

Proof. The proof is based on Theorem 5.2.4 and is left as an exercise for the reader. □

Generalizing Theorem 5.2.9, we see that we can also obtain the ith member of the polynomial remainder sequence from $\operatorname{res}_T^{(n-i+1)}[p_1(x), p_2(x)]$ in exactly the same way as described in Theorem 5.2.9; see the example below. The same results hold for $\operatorname{res}_B[p_1(x), p_2(x)]$.

Also note that $|R_i| = |\operatorname{res}_T^{(i)}[p_1(x), p_2(x)]| = |\operatorname{res}_B^{(i)}[p_1(x), p_2(x)]|$, and therefore, ignoring sign differences, we will use R_i for both cases.

Example. For the polynomials $p_1(x) = x^3 - 7x + 7$ and $p_2(x) = 3x^2 - 7$ we already know that $R_0 = \operatorname{res}_T^{(0)}[p_1(x), p_2(x)] \neq 0$ and, therefore,

$\deg\{\gcd[p_1(x), p_2(x)]\} = 0$; the greatest common divisor is obtained from $\operatorname{res}_T^{(0)}[p_1(x), p_2(x)]$ by evaluating the determinant

$$\det \begin{bmatrix} 1 & 0 & -7 & 7 & xp_1(x) \\ 0 & 1 & 0 & -7 & p_1(x) \\ 0 & 0 & 3 & 0 & p_2(x) \\ 0 & 3 & 0 & -7 & xp_2(x) \\ 3 & 0 & -7 & 0 & x^2 p_2(x) \end{bmatrix} = 49 = p_4(x)$$

Likewise, we see that $p_3(x)$, the third member of the PRS, is obtained from $\operatorname{res}_T^{(1)}[p_1(x), p_2(x)]$:

$$\det \begin{bmatrix} 1 & 0 & p_1(x) \\ 0 & 3 & p_2(x) \\ 3 & 0 & xp_2(x) \end{bmatrix} = -9p_1(x) + 3xp_2(x)$$

$$= -9x^3 + 63x - 63 + 9x^3 - 21x$$

$$= 42x - 63 = p_3(x)$$

(Compare this with the results obtained in the example preceding Theorem 5.2.4.)

We conclude this section with the following.

Definition 5.2.10. The discriminant $\operatorname{discr}[p(x)]$ of a polynomial $p(x) = a\Pi_{1 \leq i \leq n}(x - \alpha_i)$ is defined by

$$\operatorname{discr}[p(x)] = a^{2n-2} \Pi_{1 \leq i \leq n} \Pi_{i < j \leq n} (\alpha_i - \alpha_j)^2$$

Note that the discriminant gives us a measure of how close the roots of a polynomial are. This fact will be used in the root isolation methods. Moreover, one can now prove that for a monic polynomial $p(x)$ of degree n

$$\operatorname{discr}[p(x)] = (-1)^{n(n-1)/2} \operatorname{res}_B[p(x), p'(x)],$$

where $p'(x)$ is the derivitive of $p(x)$. Additional properties of the discriminant can be found in the literature (Berlekamp, 1968).

5.2.3 Habicht's Subresultant PRS Algorithm

As we saw before, Sylvester indicated a divisibility property that exists between *certain* members of a complete polynomial remainder sequence, and this was the foundation of the reduced (subresultant) PRS algorithm. However, in order to deal with *incomplete* PRSs, we have to have a divisibility property that exists between *any* members of the PRS. This new divisibility property was presented by Habicht in 1948 and is the subject of this section.

In order to prove Habicht's theorem we consider the two polynomials in $\mathbb{Z}[x]$:

$$p_{n+1}(x) = c_0 x^{n+1} + c_1 x^n + \cdots + c_{n+1}$$
$$p_n(x) = d_0 x^n + d_1 x^{n-1} + \cdots + d_n,$$

of degrees $n+1$ and n, respectively, and the polynomial remainder sequence $p_{n-1}(x), \ldots, p_1(x), p_0(x)$, *where now the subscript indicates the degree of the corresponding polynomial.* Observe that Habicht was interested mainly in Sturm's sequences, and this explains the choice of the degrees of the first two polynomials. Many of the results presented below are valid for arbitrary degrees of the first two polynomials (the reader can easily tell when this is the case), but we follow Habicht's approach.

As we recall, for complete PRS, Sylvester proved that the ith principal subresultant coefficient squared can be divided out of the $(i-2)$-th member of the PRS:

$$\{\operatorname{res}_B^{(i)}[p_{n+1}(x), p_n(x)]\}^2 \mid p_{i-2}(x) \tag{S0}$$

in other words, the square of the leading coefficient of $p_i(x)$ exactly divides $p_{i-2}(x)$ [$\operatorname{res}_B^{(i)}[p_{n+1}(x), p_n(x)]$ was defined in Theorem 5.2.5]. For the case of incomplete PRS, where some of the members of the PRS are missing, Habicht proved that

$$\{\operatorname{res}_B^{(i+1)}[p_{n+1}(x), p_n(x)]\}^{2(i-j)} \cdot p_j(x) = p_{i+1}(x) \cdot p^{(ij)}(x) + p_i(x) \cdot q^{(ij)}(x),$$
$$0 < i < n, \quad 0 \le j < i \tag{H0}$$

where the polynomials $p^{(ij)}(x)$, $q^{(ij)}(x)$ are obtained from the coefficients of $p_{i+1}(x)$ and $p_i(x)$ in the same way (explained below) that the polynomials $p^{(i)}(x)$, $q^{(i)}(x)$ (defined in Theorem 5.2.11) are obtained from the coefficients of $p_{n+1}(x)$, and $p(x)$.

Note: The reader should carefully observe the distinction between the two pairs of polynomials with superscripts, that is, $\{p^{(ij)}(x), q^{(ij)}(x)\}$ and $\{p^{(i)}(x), q^{(i)}(x)\}$; the role of the index i is completely different. For consistency, the latter pair could be also written as $\{p^{(ni)}(x), q^{(ni)}(x)\}$, but we use the simpler notation.

The full meaning and power of (H0) will be understood below. Observe that for incomplete PRS we are lead, with the help of $p^{(ij)}(x)$ and $q^{(ij)}(x)$, to new divisibility properties that are stated at the end of this section and that lead to Habicht's subresultant PRS algorithm mentioned in Section 5.1.1.

We first present the following.

Theorem 5.2.11. Consider the polynomial remainder sequence $p_{n+1}(x)$, $p_n(x)$, $p_{n-1}(x), \ldots, p_1(x), p_0(x)$ over the integers, where the subscript indicates the degree of the corresponding polynomial; then for every member $p_i(x)$ of the PRS, $0 \le i < n$, we can find two polynomials in $\mathbf{Z}[x]$, $p^{(i)}(x)$ and $q^{(i)}(x)$:

$$p^{(i)}(x) = a_0 x^{n-i-1} + a_1 x^{n-i-2} + \cdots + a_{n-i-1}$$
$$q^{(i)}(x) = b_0 x^{n-1} + b_1 x^{n-i-1} + \cdots + b_{n-i}$$

of degrees $n - i - 1$ and $n - i$, respectively, such that

$$p_i(x) = p_{n+1}(x) p^{(i)}(x) + p_n(x) q^{(i)}(x)$$

Proof. The proof is constructive and analogous to the extended Euclidean algorithm for integers or polynomials over a field; that is, we force the condition $p_i(x) = p_{n+1}(x) p^{(i)}(x) + p_n(x) q^{(i)}(x)$ for all i.

For $i = n + 1$, obviously $p^{(i)}(x) = 1$ and $q^{(i)}(x) = 0$, whereas for $i = n$, $p^{(i)}(x) = 0$ and $q^{(i)}(x) = 1$. As we have seen before, to compute $p^{(i)}(x)$ and $q^{(i)}(x)$ for $0 \le i < n$, we update them in the same way that we update the $p_i(x)$ values; namely, if

$$p_{i-1}(x) = \{lc[p_i(x)]\}^2 p_{i+1}(x) - p_i(x) q(x)$$

where $q(x)$ is the quotient on dividing $\{lc[p_i(x)]\}^2 p_{i+1}(x)$ by $p_i(x)$, then

$$p^{(i-1)}(x) = \{lc[p_i(x)]\}^2 p^{(i+1)}(x) - p^{(i)}(x) q(x)$$
$$q^{(i-1)}(x) = \{lc[p_i(x)]\}^2 q^{(i+1)}(x) - q^{(i)}(x) q(x) . \quad \square$$

Theorem 5.2.11 indicates that the extended Euclidean algorithm works for polynomials over the integers as well. [If we compare the expressions $p_i(x) = p_{n+1}(x) p^{(i)}(x) + p_n(x) q^{(i)}(x)$, of Theorem 5.2.11, with the expressions of the form $p(x) p^{(i)}(x) - q(x) q^{(i)}(x) = \text{res}_B^{(i)}[p(x), q(x)] x^i + \cdots + g_1 x + g_0$, obtained in the proof of Theorem 5.2.5, we see that the leading coefficient of $p_i(x)$, $0 \le i < n$ (where $p_i(x)$ is any member of the PRS) is the subresultant $\text{res}_B^{(i)}[p_{n+1}(x), p_n(x)]$, and that is why the latter was called the "principal subresultant coefficient."]

We have seen that each member of the PRS, $p_i(x)$, can be computed from the corresponding principal subresultant coefficient as described in Theorem 5.2.9; in a similar way we see that the polynomials $p^{(i)}(x)$ and $q^{(i)}(x)$, of the expression $p_i(x) = p_{n+1}(x) p^{(i)}(x) + p_n(x) q^{(i)}(x)$, can also be constructed from the same principal subresultant coefficient as follows:

246 GREATEST COMMON DIVISORS OF POLYNOMIALS

The determinant for $p^{(i)}(x)$ is exactly like the one for $p_i(x)$ except for the last column, which is $x^{n-i-1}, \ldots, 1, 0, \ldots, 0$ (starting from the top).
The determinant for $q^{(i)}(x)$ is exactly like the one for $p_i(x)$ except for the last column, which is $0, \ldots, 0, x^{n-i}, \ldots, 1$ (starting from the top).

[Also, if we want Sturm's sequences and we are using Bruno's form of the determinant, there is a factor of $(-1)^{n(n+1)/2}$ involved with all the determinants; this accounts for the difference between Bruno's and Trudi's forms of the resultants. Below we will ignore this factor.]

Therefore, the polynomials $p^{(i)}(x)$ and $q^{(i)}(x)$ can be directly obtained from the following determinants of order $2(n-i)+1$ where we have interchanged the rows and columns

$$p^{(i)}(x) = \det \begin{bmatrix} x^{n-i-1} & x^{n-i-2} & \cdots & 1 & 0 & 0 & \cdots & 0 \\ c_0 & 0 & \cdots & 0 & d_0 & 0 & \cdots & 0 \\ c_1 & c_0 & \cdots & 0 & d_1 & d_0 & \cdots & 0 \\ \cdot & \cdot & & \cdot & \cdot & \cdot & & \cdot \\ \cdot & \cdot & & c_0 & \cdot & \cdot & & d_0 \\ \cdot & \cdot & & c_1 & \cdot & \cdot & & d_1 \\ \cdot & \cdot & & \cdot & \cdot & \cdot & & \cdot \\ \cdot & \cdot & \cdots & c_{n-i} & \cdot & \cdot & & d_{n-i-1} \end{bmatrix}$$

$$q^{(i)}(x) = \det \begin{bmatrix} 0 & 0 & \cdots & 0 & x^{n-i} & x^{n-i-1} & \cdots & 1 \\ c_0 & 0 & \cdots & 0 & d_0 & 0 & \cdots & 0 \\ c_1 & c_0 & \cdots & 0 & d_1 & d_0 & \cdots & 0 \\ \cdot & \cdot & & \cdot & \cdot & \cdot & & \cdot \\ \cdot & \cdot & & c_0 & \cdot & \cdot & & d_0 \\ \cdot & \cdot & & c_1 & \cdot & \cdot & & d_1 \\ \cdot & \cdot & & \cdot & \cdot & \cdot & & \cdot \\ \cdot & \cdot & \cdots & c_{n-i} & \cdot & \cdot & \cdots & d_{n-i-1} \end{bmatrix}$$

(H1)

We can also express the relations (H1) as

$$p^{(i)}(x) = \varphi_{n,i}[p_{n+1}(x), p_n(x)]$$
$$q^{(i)}(x) = \psi_{n,i}[p_{n+1}(x), p_n(x)]$$

where $\varphi_{n,i}$ and $\psi_{n,i}$ are considered to be operators, operating on any pair of polynomials of degrees $n+1$ and n.

Example. Consider $p_{2+1}(x) = x^3 - 7x + 7$, and $p_2(x) = 3x^2 - 7$, that is, $n = 2$, and let us compute $p^{(1)}(x)$, and $q^{(1)}(x)$ so that

$$p_1(x) = p_{2+1}(x)p^{(1)}(x) + p_2(x)q^{(1)}(x)$$

Observe that in this case $i = 1$ and, hence, the determinants are of order $2(n - i) + 1 = 3$; so we have

$$p^{(1)}(x) = \det\begin{bmatrix} 1 & 0 & 0 \\ 1 & 3 & 0 \\ 0 & 0 & 3 \end{bmatrix} = 9, \quad q^{(1)}(x) = \det\begin{bmatrix} 0 & x & 1 \\ 1 & 3 & 0 \\ 0 & 0 & 3 \end{bmatrix} = -3x$$

and

$$p_1(x) = p_{2+1}(x) \cdot 9 + p_2(x) \cdot (-3x) = -42x + 63$$

which is our known result. [The reader should verify that we obtain the same polynomials, $p^{(1)}(x)$ and $q^{(1)}(x)$, using the extended Euclidean algorithm for polynomials, described in Theorem 5.2.11.]

Next, to compute $p^{(0)}(x)$ and $q^{(0)}(x)$ such that $p_0 = p_{2+1}(x)p^{(0)}(x) + p_2(x)q^{(0)}(x)$, we must evaluate two determinants of order 5:

$$p^{(0)}(x) = \det\begin{bmatrix} x & 1 & 0 & 0 & 0 \\ 1 & 0 & 3 & 0 & 0 \\ 0 & 1 & 0 & 3 & 0 \\ -7 & 0 & -7 & 0 & 3 \\ 7 & -7 & 0 & -7 & 0 \end{bmatrix} = 126x + 189$$

and

$$q^{(0)} = \det\begin{bmatrix} 0 & 0 & x^2 & x & 1 \\ 1 & 0 & 3 & 0 & 0 \\ 0 & 1 & 0 & 3 & 0 \\ -7 & 0 & -7 & 0 & 3 \\ 7 & -7 & 0 & -7 & 0 \end{bmatrix} = -42x^2 - 63x + 196$$

and hence

$$p_0(x) = p_{2+1}(x)(126x + 189) + p_2(x)(-42x^2 - 63x + 196) = -49$$

In this example, note that the following astonishing equations hold:

$$lc[p_2(x)]lc[p_1(x)] = -lc[p^{(0)}(x)], \quad lc[p_{2+1}(x)]lc[p_1(x)] = lc[q^{(0)}(x)]$$

As we have already indicated, the leading coefficient of the ith member of the PRS is equal to $\text{res}_B^{(i)}[p_{n+1}(x), p_n(x)]$, which differs in sign from $\text{res}_T^{(i)}[p_{n+1}(x), p_n(x)]$; ignoring this sign difference, we denote them both as R_i. Therefore, in general, we have

$$lc[p_n(x)]R_i = -lc[p^{(i-1)}(x)], \quad lc[p_{n+1}(x)]R_i = lc[q^{(i-1)}(x)] \quad \text{(H2)}$$

These formulas can be verified, with easy computations, if one considers the determinant forms of R_i, $p^{(i-1)}(x)$ and $q^{(i-1)}(x)$ along with the fact that the former is of order $2(n-i)+1$ whereas the latter two are of order $2(n-i)+3$; see also the example above.

In this way we have created for every index i, $0 \le i < n$, a set of three polynomials, $p_i(x)$, $p^{(i)}(x)$, and $q^{(i)}(x)$; we next relate them pairwise in the following way.

Theorem 5.2.12 (Habicht, 1948). Given the assumptions of Theorem 5.2.11, we have, for $1 \le i \le n-1$,

$$\det\begin{bmatrix} p^{(i)}(x) & p^{(i-1)}(x) \\ q^{(i)}(x) & q^{(i-1)}(x) \end{bmatrix} = R_i^2, \quad \det\begin{bmatrix} p^{(i)}(x) & p^{(i-1)}(x) \\ p_i(x) & p_{i-1}(x) \end{bmatrix} = R_i^2 p_n(x)$$

$$\det\begin{bmatrix} q^{(i)}(x) & q^{(i-1)}(x) \\ p_i(x) & p_{i-1}(x) \end{bmatrix} = -R_i^2 p_{n+1}(x) \tag{H3}$$

Proof. We have

$$p_i(x) = p_{n+1}(x)p^{(i)}(x) + p_n(x)q^{(i)}(x)$$
$$p_{i-1}(x) = p_{n+1}(x)p^{(i-1)}(x) + p_n(x)q^{(i-1)}(x) \tag{H4}$$

Multiplying the first equation by $p^{(i-1)}(x)$ and the second one by $p^{(i)}(x)$, we obtain

$$\det\begin{bmatrix} p^{(i)}(x) & p^{(i-1)}(x) \\ p_i(x) & p_{i-1}(x) \end{bmatrix} = \det\begin{bmatrix} p^{(i)}(x) & p^{(i-1)}(x) \\ q^{(i)}(x) & q^{(i-1)}(x) \end{bmatrix} p_n(x) \tag{H5}$$

The left side of the equation is a polynomial of degree n, since $p_i(x)p^{(i-1)}(x)$ is of degree n whereas the polynomial $p_{i-1}(x)p^{(i)}(x)$ is only of degree $n-2$. However, since $p_n(x)$ is of degree n, the determinant on the right-hand side of (H5) is an integer constant; in fact, this constant is equal to $-lc[p^{(i-1)}(x)]R_i/lc[p_n(x)]$, the quotient on dividing the leading coefficients of $-p_i(x)p^{(i-1)}(x)$ and $p_n(x)$, which by the first of the formulas (H2) equals R_i^2 (verify this in the last example). Thus we have proved the first two relations of (H3). The last one is easily obtained if we multiply the relations in (H4) by $q^{(i-1)}(x)$ and $q^{(i)}(x)$, respectively. $\square$

[Formulas (H3) can be extended to include $p_{n+1}(x)$ and $p_n(x)$ by setting $p^{(n+1)}(x) = 1$, $p^{(n)}(x) = 0$, $q^{(n+1)}(x) = 0$, $q^{(n)}(x) = 1$. In this way (H4) and (H5) are still valid; moreover, if we make use of the sign difference that we have ignored so far and set $p^{(n-1)}(x) = -d_0^2 = -R_n^2$ (note how this is obtained in the last example), (H3) holds for $i = n$, whereas for $i = n+1$, set $R_{n+1} = 1$.]

We will now derive a relation between $p_{i+1}(x)$, $p_i(x)$ and $p_{i-1}(x)$ and a more general one between $p_{i+1}(x)$, $p_i(x)$ and $p_j(x)$, $j < i$, for $1 \le i \le n-1$.

From the last two relations of (H3) we have for $i+1$:

$$p_i(x)p^{(i+1)}(x) - p_{i+1}(x)p^{(i)}(x) = R_{i+1}^2 p_n(x)$$

$$p_i(x)q^{(i+1)}(x) - p_{i+1}(x)q^{(i)}(x) = -R_{i+1}^2 p_{n+1}(x)$$

Multiplying the first of the above equations by $q^{(i-1)}(x)$, and the second by $p^{(i-1)}(x)$, we obtain:

$$\det\begin{bmatrix} p^{(i+1)} & p^{(i-1)}(x) \\ q^{(i+1)}(x) & q^{(i-1)}(x) \end{bmatrix} p_i(x) - \det\begin{bmatrix} p^{(i)}(x) & p^{(i-1)}(x) \\ q^{(i)}(x) & q^{(i-1)}(x) \end{bmatrix} p_{i+1}(x)$$
$$= R_{i+1}^2 [p_{n+1}(x)p^{(i-1)}(x) + p_n(x)q^{(i-1)}(x)]$$

which due to (H3) is

$$R_{i+1}^2 p_{i-1}(x) = -R_i^2 p_{i+1}(x) + Q_i(x) p_i(x) \tag{H6}$$

where $Q_i(x)$ is a polynomial of degree 1 that can be expressed through the coefficients of $p_{i+1}(x)$ and $p_i(x)$ (e.g., the coefficient of x is $R_i R_{i+1}$). With the help of the operators introduced earlier, (H6) can be also written as

$$R_{i+1}^2 p_{i-1}(x) = \varphi_{i,i-1}[p_{i+1}(x), p_i(x)] p_{i+1}(x) + \psi_{i,i-1}[p_{i+1}(x), p_i(x)] p_i(x) \tag{H6'}$$

Based on the above equation, we now have the more general case.

Theorem 5.2.13 (Habicht, 1948). Under the assumptions of Theorem 5.2.11, if we set $p^{(ij)}(x) = \varphi_{ij}[p_{i+1}(x), p_i(x)]$ and $q^{(ij)}(x) = \psi_{ij}[p_{i+1}(x), p_i(x)]$, $(0 < i < n$ and $0 \leq j < i)$, then

$$R_{i+1}^{2(i-j)} p_j(x) = p^{(ij)}(x) p_{i+1}(x) + q^{(ij)}(x) p_i(x) \tag{H7}$$

Proof. Using the notation of this theorem, the polynomials $p^{(i)}(x)$ and $q^{(i)}(x)$ defined in Theorem 5.2.11 can be also written as $p^{(ni)}(x)$ and $q^{(ni)}(x)$, respectively. Therefore, the polynomials $p^{(ij)}(x)$ and $q^{(ij)}(x)$ can be computed from a determinant whose entries are the coefficients of $p_{i+1}(x)$ and $p_i(x)$, in the same way that the polynomials $p^{(ni)}(x)$ and $q^{(ni)}(x)$ can be computed from a determinant whose entries are the coefficients of $p_{n+1}(x)$ and $p_n(x)$.

Keeping i fixed, we form the polynomials

$$p'_i(x) = p_i(x) \quad \text{and} \quad p'_j(x) = p_{i+1}(x)p^{(ij)}(x) + p_i(x)q^{(ij)}(x),$$
$$j = i-1, \ldots, 0$$

Observe that the polynomials $p'_j(x)$, $j < i$, are expressed in terms of $p_{i+1}(x)$ and $p_i(x)$, whereas the polynomials $p_j(x)$ are expressed in terms of $p_{n+1}(x)$ and $p_n(x)$ (Theorem 5.2.11); therefore, the coefficients of the former are bigger than those of the latter, and we want to prove that

$$p'_j(x) = R_{i+1}^{2(i-j)} p_j(x) \tag{H8}$$

For ease of notation we also set $R_{i+1}^2 = r$ {recall that $R_{i+1} = lc[p_{i+1}(x)]$} and we form the polynomials over the rationals,

$$p''_j(x) = \left[\frac{1}{r^{(i-j)}}\right] p'_j(x), \qquad j = i, \ldots, 0$$

Denote the leading coefficients of the polynomials $p'_j(x)$ and $p''_j(x)$ by R'_j and R''_j, respectively. We will now show that (H6) holds for any $p''_{j+1}(x)$, $p''_j(x)$ and $p''_{j-1}(x)$, $1 \le j \le i-1$, and thus we will have proved the theorem.

Obviously, we have $p''_i(x) = p_i(x)$, and because of (H6') we have $p''_{i-1}(x) = p_{i-1}(x)$; therefore, we can successively conclude that we also have $p''_j(x) = p_j(x)$, $j = i-2, \ldots, 0$. However, by the definition of the operators φ_{ij} and ψ_{ij}, and by repeating the argument we used before, we see that (H6) holds for the $p'_j(x)$ values; that is

$$(R'_{j+1})^2 p'_{j-1}(x) = -(R'_j)^2 p'_{j+1}(x) + Q'_j(x) p'_j(x), \qquad j = i-1, \ldots, 1$$

Hence, dividing for every index j the corresponding relation by

$$r^{2(i-j-1)} r^{i+j+1} = r^{2(i-j)} r^{i-j-1}$$

we obtain

$$(R''_{j+1})^2 p''_{j-1}(x) = -(R''_j)^2 p''_{j+1}(x) + Q''_j(x) p''_j(x), \qquad j = i-1, \ldots, 1. \quad \square$$

This theorem indicates that (without greatest common divisor computations) the smallest coefficients for the members of the PRS are obtained as subresultants of $\text{res}_B[p_{n+1}(x), p_n(x)]$; that is, we obtain the smallest coefficients for the members of the PRS only when the coefficients of $p_{n+1}(x)$ and $p_n(x)$ are involved. However, evaluating determinants is a tedious process and the members of the PRS are computed with pseudodivisions; the coefficients of the members of the PRS are then larger, but we already know what must be divided out.

Example. Let us consider again the polynomials of the last example, $p_{2+1}(x) = x^3 - 7x + 7$ and $p_2(x) = 3x^2 - 7$, for which we already know that $p_1(x) = -42x + 63$ and $p_0(x) = -49$. Obviously, in this case, $p_1(x) = p'_1(x)$, and so we will compute $p'_0(x) = p_2(x) p^{(10)}(x) + p_1(x) q^{(10)}$. We have $i = 1$ and $j = 0$, and

$$p^{(10)}(x) = \det\begin{bmatrix} 1 & 0 & 0 \\ 3 & -42 & 0 \\ 0 & 63 & -42 \end{bmatrix} = 1764,$$

$$q^{(10)}(x) = \det\begin{bmatrix} 0 & x & 1 \\ 3 & -42 & 0 \\ 0 & 63 & -42 \end{bmatrix} = 126x + 189$$

Therefore, $p'_0(x) = 1764(3x^2 - 7) + (126x + 189)(-42x + 63) = -441$, and according to Theorem 5.2.13, 3^2, the leading coefficient of $p_2(x)$ squared, divides -441; that is, $p_0(x) = p'_0(x)/9 = -441/3^2 = -49$.

In what follows we set $R_{n+1} = c_0$, $R_n = d_0$, and we examine what happens if one or more principal subresultant coefficients are zero; that is, we will derive the formulas that constitute the foundations of Habicht's subresultant PRS algorithm [formula (H) of Section 5.1.1]. We base our investigation on (H7) and (H8).

Consider a polynomial remainder sequence $p_{n+1}(x), p_n(x), \ldots, p_0(x)$ where for all i, $0 \leq i \leq n - 1$, we have

$$p_i(x) = c_{i,0} x^i + \cdots + c_{i,i}$$

and let $p_i(x)$, $0 < i \leq n$ be a member of this sequence with

$$R_{i+1} \neq 0, \quad R_i = 0, \quad c_{i1} = 0, \ldots, c_{i,i-j-1} = 0, \quad c_{i,i-j} \neq 0$$

that is the degree of $p_i(x)$ is $j \leq i$. Then the following relations easily follow from (H7), (H8), and the form of the $p_i(x)$ values [recall that the determinants of $p^{(ij)}(x)$ and $q^{(ij)}(x)$ are constructed in the same way as those for $p^{(i)}(x)$ and $q^{(i)}(x)$, i.e., from (H1), where now n is replaced by i, i by j and the coefficients of $p_{n+1}(x)$ and $p_n(x)$ by those of $p_{i+1}(x)$ and $p_i(x)$].

H-1. $p_{i-1}(x), \ldots, p_{j+1}(x)$ are zero.
H-2. $p^{(ij)}(x) = 0$, $q^{(ij)}(x) = R_{i+1}^{i-j} c_{i,i-j}^{i-j}$
$$R_{i+1}^{i-j} p_j(x) = c_{i,i-j}^{i-j} p_i(x) \qquad (i < n)$$
$$p_j(x) = d_{n-j}^{n-j} c_0^{n-j} p_n(x) \qquad (i = n).$$
H-3. $p^{(i,j-1)}(x) = (-1)^{i-j} R_{i+1}^{i-j} c_{i,i-j}^{i-j+2}$,
$$R_{i+1}^{2(i-j+1)} p_{j-1}(x) = (-1)^{i-j} R_{i+1}^{i-j} c_{i,i-j}^{i-j+2} p_{i+1}(x) + q^{(i,j-1)}(x) p_i(x)$$
$$\hspace{10em} (i < n)$$
$$p_{j-1}(x) = d_{n-j}^{n-j+2} c_0^{n-j} p_{n+1}(x) + q^{(j-1)}(x) p_n(x) \qquad (i = n).$$

The formulas in **H-1**, **H-2**, and **H-3** are the foundations of Habicht's subresultant PRS algorithm.

Example. Let us examine the incomplete polynomial remainder sequence

252 GREATEST COMMON DIVISORS OF POLYNOMIALS

$p_4(x) = 2x^4 + 5x^3 + 5x^2 - 2x + 1$, $p_3(x) = 3x^3 + 3x^2 + 3x - 4$, $p_2(x) = 0x^2 - 21x + 45$, $p_1(x) = -?-$, $p_0(x) = -47049$. Here we see that $R_{2+1} \neq 0$, whereas $R_2 = 0$; that is, with the above notation $i = 2$ and $j = 1$. Then, to compute $p_1(x)$, we apply case **H-2** above and have

$$p^{(21)}(x) = \det \begin{bmatrix} 1 & 0 & 0 \\ 3 & 0 & 0 \\ 3 & -21 & 0 \end{bmatrix} = 0, \qquad q^{(21)}(x) = \det \begin{bmatrix} 0 & x & 1 \\ 3 & 0 & 0 \\ 3 & -21 & 0 \end{bmatrix} = 3(-21)$$

where $n := i$, $i := j$ and the coefficients of $p_4(x)$ and $p_3(x)$ have been replaced by the coefficients of $p_3(x)$ and $p_2(x)$. Therefore, from $p'_1(x) = p_3(x)p^{(21)}(x) + p_2(x)q^{(21)}$, we obtain

$$p'_1(x) = 3(-21)p_2(x), \qquad \text{or equivalently,} \qquad 3^2 p_1(x) = 3(-21)p_2(x)$$

and

$$p_1(x) = -147x + 315$$

that is, $p_1(x)$ is a polynomial proportional to $p_2(x)$. We leave it as an exercise for the reader to compute $p_0(x)$ using case **H-3**.

Figure 5.2.1 is of interest.

Using (H), from Section 5.1.1, we have the following algorithm.

HSPRS (Habicht's Subresultant **PRS**)

> *Input*: Two nonzero polynomials $p_1(x)$, $p_2(x)$ in **Z**[x], deg[$p_1(x)$] $\geq$ deg[$p_2(x)$].
> *Output*: The subresultant PRS of $p_1(x)$ and $p_2(x)$.

Figure 5.2.1. The gap structure of the PRS; a line of length $k + 1$ represents a polynomial of degree k (Habicht).

1. [Initialize.] $i := 1$; $\beta_1 := (-1)^{\deg[p_1(x)] - \deg[p_2(x)] + 1}$.
2. [Done?] If $p_{i+1}(x) = 0$, then return.
3. [Compute the pseudoremainder.] $p_{i+2}(x) := \text{prem}[p_i(x), p_{i+1}(x)]/\beta_i$, where β_i is computed from (H) of Section 5.1.1 (see also Exercise 3 for this section); set $i := i + 1$ and go to 2.

Computing-Time Analysis of **HSPRS.** Let $n = \deg[p_1(x)]$. Then, in the worst case, there will be n polynomial pseudodivisions over the integers, each of which is performed in time $0[n^2 L^2(a)]$, where a is the largest coefficient that appears during computations [analogous to the time it takes to compute the product $p_1(x) \cdot p_2(x)$]. The factor $L^2(a)$ has to do with the integer multiplications (of the coefficients) that take place. The largest coefficient that appears during computations is $\leq \text{res}_B[p_1(x), p_2(x)]$, which, in turn, is bounded by Hadamard's inequality

$$|\text{res}_B[p_1(x), p_2(x)]| \leq \Pi_{1 \leq i \leq 2n} [\Sigma_{1 \leq j \leq 2n} r_{ij}^2]^{1/2} \leq (2n)^n (|p(x)|_\infty)^{2n}$$

where r_{ij} denotes the entries of the resultant and $|p(x)|_\infty = \max(|p_1(x)|_\infty, |p_2(x)|_\infty)$. If we set $a = (2n)^n (|p(x)|_\infty)^{2n}$, then $L^2(a) = [nL(2n) + 2nL(|p(x)|_\infty)]^2 = 0(n^2 L^2[|p(x)|_\infty]$, and the time for each polynomial pseudodivision is

$$0\{n^4 L^2[|p(x)|_\infty]\}$$

Since there are at most n pseudodivisions we have that

$$t_{\text{HSPRS}}[p_1(x), p_2(x)] = 0\{n^5 L^2[|p(x)|_\infty]\}$$

5.3

The Matrix-Triangularization Subresultant PRS Method

So far we have seen that the Sylvester–Habicht pseudodivisions subresultant PRS method includes two algorithms: Sylvester's reduced (subresultant) PRS algorithm (based on Sylvester's paper of 1853), which works very well for complete PRSs, and Habicht's subresultant PRS algorithm (based on Habicht's paper of 1948), which is to be preferred for incomplete PRSs. Note that Habicht's subresultant PRS algorithm can be used for complete PRSs, but at additional computational cost; moreover, we cannot tell a priori whether the PRS is going to be complete or incomplete.

In this section we present the matrix-triangularization subresultant PRS method that was developed by the author in 1986. This method is based on the author's generalization of a theorem by Van Vleck and uniformly treats both complete and incomplete PRSs; moreover, the coefficients of the members of the PRS are the smallest that can be obtained without coefficient greatest common divisor computations and the divisibility properties are clearly demonstrated. (In Section 5.1.1 we saw an example where the coefficients obtained with this new method are smaller than those obtained with the Sylvester-Habicht PRS method.)

The astonishing fact about the matrix-triangularization subresultant PRS method is that no explicit polynomial divisions take place; instead, the gaussian elimination process (see the Appendix at the end of the book) is applied to a single matrix, corresponding to Sylvester's form of the resultant (explained below), and all the members of the PRS are obtained from the rows of the resulting upper-triangular matrix with the help of Theorem 5.3.6. It should be observed that Sylvester's form of the resultant is used for the first time in the modern literature. (See also historical note 2.)

5.3.1 Gaussian Elimination and Polynomial Division

Consider the polynomials $p_1(x) = c_n x^n + c_{n-1} x^{n-1} + \cdots + c_0$, $c_n \neq 0$, $n > 0$ and $p_2(x) = d_m x^m + d_{m-1} x^{m-1} + \cdots + d_0$, $d_m \neq 0$, $m > 0$, over the integers, $n \geq m$. From the Euclidean property we know that there exist unique polynomials over the integers, $q(x)$ and $r(x)$, such that $p_1(x) = p_2(x) q(x) + r(x)$, $\deg[r(x)] < \deg[p_2(x)]$ (at this point, the reader should review algorithm **PDF**); let $r(x) = r_{m-1} x^{m-1} + r_{m-2} x^{m-2} + \cdots + r_0$.

Our claim is that polynomial division over a field, or pseudodivision over the integers, is equivalent to bringing a matrix **M** into its upper triangular form $T(\mathbf{M})$; that is, all elements m_{ij}, $j < i$, of the matrix $T(\mathbf{M})$ are zero. Since gaussian elimination is the process by which a given matrix is transformed, by elementary row operations, into its upper triangular form, our claim is that *polynomial division is equivalent to gaussian elimination*.

To see this, consider the polynomials mentioned above, and form the $(n - m + 2)$ by $(n + 1)$ matrix **M**:

$$\mathbf{M} = \begin{bmatrix} d_m & d_{m-1} & \cdots & d_0 & \cdots & & 0 \\ & d_m & d_{m-1} & \cdots & d_0 & \cdots & 0 \\ & & & \vdots & & & \\ 0 & 0 & \cdots & d_m & d_{m-1} & \cdots & d_0 \\ c_n & c_{n-1} & & \cdots & & & c_0 \end{bmatrix} \left.\begin{matrix} \\ \\ \\ \\ \end{matrix}\right\} n - m + 1 \text{ rows}$$

where we consider the coefficients of $x^{n-m} p_2(x)$, $x^{n-m-1} p_2(x)$, ..., $p_2(x)$ as the first, second, ..., and $(n-m+1)$th row, of the $(n-m+2) \times (n+1)$ matrix; the coefficients of $p_1(x)$ are considered as the last row of the matrix. If we now transform $\mathbf{M}$ into its upper triangular form $T(\mathbf{M})$, we see that the elements of the last row are transformed into the coefficients of the remainder $r(x)$; that is, we will have

$$T(\mathbf{M}) = \begin{bmatrix} d_m & d_{m-1} & \cdots & d_0 & \cdots & 0 \\ 0 & d_m & d_{m-1} & d_0 & \cdots & 0 \\ & & & \vdots & & \\ 0 & \cdots & d_m & d_{m-1} & \cdots & d_0 \\ 0 & \cdots & 0 & r_{m-1} & \cdots & r_0 \end{bmatrix}$$

What happened here is that each of the first $n-m+1$ rows of $\mathbf{M}$ is subtracted from the last row, premultiplied accordingly, and the difference becomes the new last row; in this way the powers of x are successively eliminated until x^{m-1}.

Example. Consider $p_1(x) = x^3 - 7x + 7$ and $p_2(x) = 3x^2 - 7$. Then

$$\mathbf{M} = \begin{bmatrix} 3 & 0 & -7 & 0 \\ 0 & 3 & 0 & -7 \\ 1 & 0 & -7 & 7 \end{bmatrix}$$

and bringing it into its upper triangular form, we have

$$T(\mathbf{M}) = \begin{bmatrix} 3 & 0 & -7 & 0 \\ 0 & 3 & 0 & -7 \\ 0 & 0 & -14 & 21 \end{bmatrix}$$

from which we deduce that the pseudoremainder on dividing $p_1(x)$ by $p_2(x)$ is $r(x) = -14x + 21 = p_3(x)$. (The reader should note that the coefficients are now smaller than those obtained with direct pseudodivision.)

Below we are going to deal only with square matrices, and we are going to transform them into their upper triangular form using Bareiss's integer-preserving transformation algorithm. For an $n \times n$ matrix $\mathbf{M}$, Bareiss's algorithm works as follows. Set $m_{00}^{(-1)} := 1$, and $m_{ij}^{(0)} := m_{ij}$ (the elements of the matrix) $i, j = 1, \ldots, n$; then, for $k < i, j \leq n$ set

$$m_{ij}^{(k)} := [m_{k-1,k-1}^{(k-2)}]^{-1} \det \begin{bmatrix} m_{kk}^{(k-1)} & m_{kj}^{(k-1)} \\ m_{ik}^{(k-1)} & m_{ij}^{(k-1)} \end{bmatrix} \quad \text{(B)}$$

That is, the rows with $i \leq k$ are not changed, whereas for rows $k < i$ we obtain 0 in the lower triangle.

Note that we need pivoting (row interchange) to ensure $m_{kk}^{(k-1)} \neq 0$ for all

k, and, as we will see later, this pivoting needs to be done in such a way that it will not disturb the existing symmetry of the matrix corresponding to Sylvester's form of the resultant. Therefore, we perform *bubble pivoting*, which is defined as follows: if we need to interchange row i with row j, where $j > i + 1$, we let row j *rise* to the level of row i by successively interchanging it (pairwise) with each row which lies above it. So, after the bubble pivoting is performed, ex-row i is immediately below ex-row j.

Of great importance, in Bareiss's algorithm, is the fact that the determinant of order 2 is *exactly* divided by $m_{k-1,k-1}^{(k-2)}$ and that the resulting entries in the matrix are the smallest that can be expected without coefficient *gcd* computations and without introducing rationals. To see these, consider the matrix

$$\begin{bmatrix} a & b & c & d & e & \cdots \\ f & g & h & i & j & \cdots \\ k & l & m & n & o & \cdots \\ \cdots & & & & & \end{bmatrix}$$

After the first transformation, $k = 1$, we obtain the matrix

$$\begin{bmatrix} a & b & c & d & e & \cdots \\ 0 & ag-bf & ah-cf & ai-df & aj-ef & \cdots \\ 0 & al-bk & am-ck & an-dk & ao-ek & \cdots \\ \cdots & & & & & \end{bmatrix}$$

that is, no division takes place, since $m_{0,0}^{(-1)} = 1$. After the second transformation, $k = 2$, we obtain the following matrix (*before division*):

$$\begin{aligned} m_{3,3}^{(2)} &= (ag - bf)(am - ck) - (al - bk)(ah - cf) \\ &= aagm - acgk - abfm + bcfk - aahl + acfl + abhk - bcfk \\ &= aagm - acgk - abfm - aahl + acfl + abhk \\ m_{3,4}^{(2)} &= aagm - adgk - abfn - aail + adfl + abik, \ldots \end{aligned}$$

Therefore, we see that each element of the submatrix can be exactly divided by a. Likewise, we will be able to divide $(ag-bf)$ out of the elements of the new submatrix in the next transformation, and so on.

5.3.2 Gaussian Elimination and the Bruno, Trudi Forms of the Resultant

As we recall, given the two polynomials $p(x) = c_n x^n + c_{n-1} x^{n-1} + \cdots + c_0$, and $q(x) = d_m x^m + d_{m-1} x^{m-1} + \cdots + d_0$, $n \geq m$, the matrix $\mathbf{M}_B$ corresponding to Bruno's form of the resultant is

$$\mathbf{M}_B = \left.\begin{bmatrix} c_n & c_{n-1} & \cdots & c_0 & 0 & \cdots & 0 \\ 0 & c_n & c_{n-1} & \cdots & c_0 & \cdots & 0 \\ & & & \vdots & & & \\ 0 & 0 & \cdots & c_n & c_{n-1} & \cdots & c_0 \\ d_m & d_{m-1} & \cdots & d_0 & 0 & 0 & \cdots & 0 \\ 0 & d_m & d_{m-1} & \cdots & d_0 & 0 & \cdots & 0 \\ & & & \vdots & & & \\ 0 & 0 & \cdots & d_m & d_{m-1} & \cdots & d_0 \end{bmatrix}\right\} \begin{matrix} (m \text{ rows}) \\ \\ \\ (n \text{ rows}) \end{matrix}$$

We are going to prove that if we bring $\mathbf{M}_B$ into its upper triangular form, $T(\mathbf{M}_B)$, using row transformations only, then the last nonzero row of $T(\mathbf{M}_B)$ gives the coefficients of a $gcd[p(x), q(x)]$. Note that, *in general*, by transforming $\mathbf{M}_B$ into $T(\mathbf{M}_B)$, we can obtain *only* a gcd but *none* of the other members of the PRS generated by $p(x)$ and $q(x)$.

To prove our major result, we first need the following.

Lemma 5.3.1 (Laidacker, 1969). Let $p_i(x)$, $p_i^+(x)$ be the polynomials corresponding to the ith row of the matrices $\mathbf{M}_B$ (defined above) and $T(\mathbf{M}_B)$, respectively, where $T(\mathbf{M}_B)$ is the upper triangular form of $\mathbf{M}_B$ obtained from the latter by row transformations only. If k is the degree of $p_r^+(x)$, corresponding to the last nonzero row of $T(\mathbf{M}_B)$, then there are no polynomials of the form

$$\alpha_1 p_1(x) + \alpha_2 p_2(x) + \cdots + \alpha_{m+n} p_{m+n}(x) \qquad (L)$$

of degree less than k, where α_i is a constant, $i = 1, 2, \ldots, m+n$.

Proof. In both matrices, $\mathbf{M}_B$ and $T(\mathbf{M}_B)$, the entries of the first column are the coefficients of the term x^{m+n-1}, the entries of the second column are the coefficients of the term x^{m+n-2} and so on. Now, from linear algebra we know that the set of polynomials generated as linear combinations of the form (L) by $\{p_i(x)\}$ is the same as that generated by $\{p_i^+(x)\}$, $i = 1, 2, \ldots, m+n$, and therefore, we will prove the theorem for the latter set.

At first observe that for $k = 0$ the theorem is obviously true. So now, suppose that it is not true for $k > 0$. Then there are constants β_i, $i = 1, 2, \ldots, r$, such that $\beta_1 p_1^+(x) + \beta_2 p_2^+(x) + \cdots + \beta_r p_r^+(x) = p^+(x)$, where $\deg[p^+(x)] < k$. Since $\deg[p_1^+(x)] = m+n-1 > k$ and also $\deg[p_1^+(x)] > \deg[p_i^+(x)]$ $i = 2, \ldots, r$, β_1 must be zero. Consider the reduced expression $\beta_2 p_2^+(x) + \cdots + \beta_r p_r^+(x) = p^+(x)$; then, for the same reason as above, we conclude that $\beta_i = 0$ for $i = 2, 3, \ldots, r$, and this implies that $p^+(x)$ is a zero polynomial, which is the contradiction. $\square$

Theorem 5.3.2 (Laidacker, 1969). Under the assumptions of Lemma 5.3.1, the last nonzero row of $T(\mathbf{M}_B)$, obtained from $\mathbf{M}_B$ by row transformations only, gives the coefficients of $gcd[p(x), q(x)]$.

Proof. Let $g_k, g_{k-1}, \ldots, g_0$ be the entries of the last nonzero row of $T(\mathbf{M}_B)$, in which case $p_r^+(x) = g_k x^k + g_{k-1} x^{k-1} + \cdots + g_0$. From the construction of $\mathbf{M}_B$ we see that any polynomial of the form $\alpha_1 p_1(x) + \cdots + \alpha_{m+n} p_{m+n}(x)$ can be written in the form $p(x) g_r(x) + q(x) g_r''(x)$, where $\deg[g_r(x)] \leq m - 1$ and $\deg[g_r''(x)] \leq n - 1$. Clearly, $p_r^+(x)$ can be written in the form $p(x) g_r(x) + q(x) g_r''(x)$; then, obviously, $\gcd[p(x), q(x)] | p_r^+(x)$, from which we conclude that $\deg\{\gcd[p(x), q(x)]\} \leq \deg[p_r^+(x)]$. However, according to Lemma 5.3.1, there are no polynomials of degree $< k$ that can be of the form $p(x) g_r(x) + q(x) g_r''(x)$, and hence, our theorem is proved. $\square$

Example. Consider the pair of polynomials $p(x) = x^3 - 7x + 7$ and $q(x) = 3x^2 - 7$ which generate a complete PRS. Then

$$\mathbf{M}_B = \begin{bmatrix} 1 & 0 & -7 & 7 & 0 \\ 0 & 1 & 0 & -7 & 7 \\ 3 & 0 & -7 & 0 & 0 \\ 0 & 3 & 0 & -7 & 0 \\ 0 & 0 & 3 & 0 & -7 \end{bmatrix}, \quad T(\mathbf{M}_B) = \begin{bmatrix} 1 & 0 & -7 & 7 & 0 \\ 0 & 1 & 0 & -7 & 7 \\ 0 & 0 & 14 & -21 & 0 \\ 0 & 0 & 0 & 196 & -294 \\ 0 & 0 & 0 & 0 & -49 \end{bmatrix}$$

From the last row of $T(\mathbf{M}_B)$ we conclude that the two polynomials are relatively prime. Also note that, in $T(\mathbf{M}_B)$, we lost the polynomial $q(x)$, and it turned out that we obtained a polynomial remainder, the third row corresponding to $14x - 21$.

However, if we consider the pair of polynomials $p(x) = x^5 + 5x^4 + 10x^3 + 5x^2 + 5x + 2$ and $q(x) = x^4 + 4x^3 + 6x^2 + 2x + 1$, which generate an incomplete PRS, we have

$$\mathbf{M}_B = \begin{bmatrix} 1 & 5 & 10 & 5 & 5 & 2 & 0 & 0 & 0 \\ 0 & 1 & 5 & 10 & 5 & 5 & 2 & 0 & 0 \\ 0 & 0 & 1 & 5 & 10 & 5 & 5 & 2 & 0 \\ 0 & 0 & 0 & 1 & 5 & 10 & 5 & 5 & 2 \\ 1 & 4 & 6 & 2 & 1 & 0 & 0 & 0 & 0 \\ 0 & 1 & 4 & 6 & 2 & 1 & 0 & 0 & 0 \\ 0 & 0 & 1 & 4 & 6 & 2 & 1 & 0 & 0 \\ 0 & 0 & 0 & 1 & 4 & 6 & 2 & 1 & 0 \\ 0 & 0 & 0 & 0 & 1 & 4 & 6 & 2 & 1 \end{bmatrix}$$

and

$$T(\mathbf{M}_B) = \begin{bmatrix} 1 & 5 & 10 & 5 & 5 & 2 & 0 & 0 & 0 \\ 0 & 1 & 5 & 10 & 5 & 5 & 2 & 0 & 0 \\ 0 & 0 & 1 & 5 & 10 & 5 & 5 & 2 & 0 \\ 0 & 0 & 0 & 1 & 5 & 10 & 5 & 5 & 2 \\ 0 & 0 & 0 & 0 & -19 & -22 & -13 & -12 & -4 \\ 0 & 0 & 0 & 0 & 0 & 215 & 64 & 81 & 46 \\ 0 & 0 & 0 & 0 & 0 & 0 & -306 & 36 & 116 \\ 0 & 0 & 0 & 0 & 0 & 0 & 0 & 756 & 532 \\ 0 & 0 & 0 & 0 & 0 & 0 & 0 & 0 & 980 \end{bmatrix}$$

Considering the fact that the members of the PRS, generated by the two polynomials above, are

$$-3x^2 + 2x + 1, \ -266x - 112,980,$$

we see that from $T(\mathbf{M_B})$ we can draw general conclusions only about the $gcd[p(x), q(x)]$.

Clearly, Trudi's form of the resultant is basically the same as that of Bruno's, with the exception that the lower set of rows is pairwise-interchanged. So Theorem 5.3.2 holds in this case as well, and we can expect to safely draw conclusions only about the greatest common divisor of the two polynomials (which will now be negated).

Example. Let us consider the two pairs of polynomials examined in the previous example. First for $p(x) = x^3 - 7x + 7$ and $q(x) = 3x^2 - 7$ we have

$$\mathbf{M_T} = \begin{bmatrix} 1 & 0 & -7 & 7 & 0 \\ 0 & 1 & 0 & -7 & 7 \\ 0 & 0 & 3 & 0 & -7 \\ 0 & 3 & 0 & -7 & 0 \\ 3 & 0 & -7 & 0 & 0 \end{bmatrix}, \quad T(\mathbf{M_T}) = \begin{bmatrix} 1 & 0 & -7 & 7 & 0 \\ 0 & 1 & 0 & -7 & 7 \\ 0 & 0 & 3 & 0 & -7 \\ 0 & 0 & 0 & 42 & -63 \\ 0 & 0 & 0 & 0 & 49 \end{bmatrix}$$

Note that in $T(\mathbf{M_T})$, the row before the last, corresponds to the polynomial $42x - 63 = (-3)(-14x + 21)$; that is, it is a multiple of the negative of the remainder. Likewise, the last row is a negated greatest common divisor.

Now, for $p(x) = x^5 + 5x^4 + 10x^3 + 5x^2 + 5x + 2$ and $q(x) = x^4 + 4x^3 + 6x^2 + 2x + 1$ we have

$$\mathbf{M_T} = \begin{bmatrix} 1 & 5 & 10 & 5 & 5 & 2 & 0 & 0 & 0 \\ 0 & 1 & 5 & 10 & 5 & 5 & 2 & 0 & 0 \\ 0 & 0 & 1 & 5 & 10 & 5 & 5 & 2 & 0 \\ 0 & 0 & 0 & 1 & 5 & 10 & 5 & 5 & 2 \\ 0 & 0 & 0 & 0 & 1 & 4 & 6 & 2 & 1 \\ 0 & 0 & 0 & 1 & 4 & 6 & 2 & 1 & 0 \\ 0 & 0 & 1 & 4 & 6 & 2 & 1 & 0 & 0 \\ 0 & 1 & 4 & 6 & 2 & 1 & 0 & 0 & 0 \\ 1 & 4 & 6 & 2 & 1 & 0 & 0 & 0 & 0 \end{bmatrix}$$

and

$$T(\mathbf{M_T}) = \begin{bmatrix} 1 & 5 & 10 & 5 & 5 & 2 & 0 & 0 & 0 \\ 0 & 1 & 5 & 10 & 5 & 5 & 2 & 0 & 0 \\ 0 & 0 & 1 & 5 & 10 & 5 & 5 & 2 & 0 \\ 0 & 0 & 0 & 1 & 5 & 10 & 5 & 5 & 2 \\ 0 & 0 & 0 & 0 & 1 & 4 & 6 & 2 & 1 \\ 0 & 0 & 0 & 0 & 0 & 3 & -5 & 1 & 1 \\ 0 & 0 & 0 & 0 & 0 & 0 & 9 & -6 & -3 \\ 0 & 0 & 0 & 0 & 0 & 0 & 0 & -266 & -112 \\ 0 & 0 & 0 & 0 & 0 & 0 & 0 & 0 & -980 \end{bmatrix}$$

Observe that the last four rows of $T(\mathbf{M}_T)$ correspond to four polynomials, of which *only the last three* belong to the PRS generated by $p(x)$. So, $9x^2 - 6x - 3 = (-3)(-3x^2 + 2x + 1)$, where $-3x^2 + 2x + 1$ is the remainder on dividing $p(x)$ by $q(x)$; $-266x - 112 = (-1)(266x + 112)$, where $266x + 112$ is the second polynomial remainder and -980 is a negated *gcd*.

From this example we see that from $T(\mathbf{M}_T)$, just as from $T(\mathbf{M}_B)$, we can safely draw conclusions *only* about a greatest common divisor of two polynomials. However, in case of complete PRSs, the negated polynomial remainders can be obtained from $T(\mathbf{M}_T)$.

5.3.3 Gaussian Elimination + Sylvester's Form of the Resultant = the Matrix-Triangularization Subresultant PRS Method

So far we have seen Bruno's and Trudi's forms of the resultant; actually, in the literature, these two forms are referred to as Sylvester's. However, we choose to call Sylvester's form of the resultant the one described below; this form was "buried" in Sylvester's 1853 paper and is only mentioned in a paper by Van Vleck (1899). Sylvester indicates (p. 426 of Sylvester's 1853 paper) that he had produced this form in 1839 or 1840, and some years later Cayley unconsciously reproduced it as well. It is Sylvester's form of the resultant that forms the foundation of the matrix-triangularization subresultant PRS method.

Consider the two polynomials in $\mathbf{Z}[x]$, $p(x) = c_n x^n + c_{n-1} x^{n-1} + \cdots + c_0$ and $q(x) = d_m x^m + d_{m-1} x^{m-1} + \cdots + d_0$, $c_n \neq 0$, $d_m \neq 0$, $n \geq m$. Then, Sylvester's form of the resultant, $\text{res}_S[p(x), q(x)]$, is of order $2n$ and is as follows [$q(x)$ has been transformed into a polynomial of degree n by introducing zero coefficients]:

$$\text{res}_S[p(x), q(x)] = \det \begin{bmatrix} c_n & c_{n-1} & \cdots & c_0 & 0 & 0 & \cdots & 0 \\ d_n & d_{n-1} & \cdots & d_0 & 0 & 0 & \cdots & 0 \\ 0 & c_n & \cdots & & c_0 & 0 & \cdots & 0 \\ 0 & d_n & \cdots & & d_0 & 0 & \cdots & 0 \\ & & & & & \vdots & & \\ 0 & \cdots & 0 & c_n & c_{n-1} & & \cdots & c_0 \\ 0 & \cdots & 0 & d_n & d_{n-1} & & \cdots & d_0 \end{bmatrix} \quad (\text{S1})$$

Sylvester obtains this form of the resultant (also known as *eliminant*) from the system of equations

$$p(x) = 0$$
$$q(x) = 0$$
$$x \cdot p(x) = 0$$
$$x \cdot q(x) = 0$$
$$x^2 \cdot p(x) = 0$$
$$x^2 \cdot q(x) = 0$$
$$\vdots$$
$$x^{n-1} \cdot p(x) = 0$$
$$x^{n-1} \cdot q(x) = 0$$

and points out that if we take k pairs of the above equations, the highest power of x appearing in any of them will be x^{n+k-1}. Therefore, we shall be able to eliminate so many powers of x, that x^{n-k} will be the highest power uneliminated and $n - k$ will be the degree of a member of the negated (Sturmian) polynomial remainder sequence generated by $p(x)$ and $q(x)$; this is proved in Theorem 5.3.4. Moreover, Sylvester showed that the polynomial remainders, obtained in this way, are what he terms *simplified residues*; that is, the coefficients are the smallest possible obtained without integer greatest common divisor computations and without introducing rationals. Stated in other words, the polynomial remainders have been freed from their corresponding *allotrious factors* (to use Sylvester's terminology). (This fact is immediately obvious after our discussion of Habicht's theorem, since the only coefficients that enter in the resultant are those of the original polynomials and not of the remainders.)

Example. Consider $p_1(x) := p(x) = x^3 - 7x + 7$ and $p_2(x) := q(x) = 3x^2 - 7$. Then we have

$$\text{res}_S[p(x), q(x)] = \det \begin{bmatrix} 1 & 0 & -7 & 7 & 0 & 0 \\ 0 & 3 & 0 & -7 & 0 & 0 \\ 0 & 1 & 0 & -7 & 7 & 0 \\ 0 & 0 & 3 & 0 & -7 & 0 \\ 0 & 0 & 1 & 0 & -7 & 7 \\ 0 & 0 & 0 & 3 & 0 & -7 \end{bmatrix}$$

and we can compute the negated coefficients of the first polynomial remainder (which is of degree $n - k = 1$) if we take $k = n - 1 = 2$ pairs of rows. So, the leading coefficient is

$$\det \begin{bmatrix} 1 & 0 & -7 & 7 \\ 0 & 3 & 0 & -7 \\ 0 & 1 & 0 & -7 \\ 0 & 0 & 3 & 0 \end{bmatrix} = 3 \cdot (21) - 1 \cdot 21 = 42$$

and the second coefficient is

$$\det \begin{bmatrix} 1 & 0 & -7 & 0 \\ 0 & 3 & 0 & 0 \\ 0 & 1 & 0 & 7 \\ 0 & 0 & 3 & -7 \end{bmatrix} = 3 \cdot (-21) = -63$$

that is, the first polynomial remainder is $p_3(x) = 42x - 63$; note the sign difference between the $p_3(x)$ computed here and the $p_3(x)$ computed using the Euclidean PRS algorithm.

We leave it as an exercise for the reader to compute $p_4(x)$.

In general, if we have the polynomial remainder sequence $p_1(x), p_2(x)$,

$p_3(x), \ldots, p_h(x)$, $\deg[p_1(x)] = n$, $\deg[p_2(x)] = m$, $n \geq m$, we can obtain the (negated) coefficients of the $(i+1)$th member of the PRS, $i = 0, 1, 2, \ldots, h-1$, as minors formed from the first $2i$ rows of (S1) by successively associating with the first $2i - 1$ columns [of the $(2i) \times (2n)$ matrix] each succeeding column in turn. The polynomials formed in this way are proportional to the ones obtained using the corresponding procedure with Bruno's or Trudi's form of the resultants; moreover, in case of incomplete PRSs, the polynomials follow the pattern of Figure 5.2.1 (the proof is as before).

So far we have called a polynomial remainder sequence Sturmian if, starting with the third member, the coefficients of each polynomial remainder are the opposite of those obtained with the Euclidean PRS algorithm; this is their main characteristic. However, as we will see in Chapter 7 on root isolation, Sturmian sequences also have the following property: if the leading coefficient of any member (vanishes) is zero, the leading coefficients of the preceding and following members (polynomial remainders) have opposite signs.

To prove that the members of a PRS obtained in the way described above constitute a Sturmian sequence, we need the following.

Lemma 5.3.3 (Multiplication Rule for Computing the Product of a Determinant by One of Its Minors). (The reader is referred to Muir (1882), p. 118, Art. 90.) The product of a determinant and any one of its minors M is expressible as a sum of products of pairs of minors; these pairs are formed in the following way. The first factors of the products are obtained by taking q rows in which the rows of M are included and forming from them every minor of the qth order that contains M; the second factor of any product is that minor that includes M and the complementary of the first factor. Finally, the sign of any product is fixed by transforming the second factor so as to have its principal diagonal coincident with those of the two minors that it was formed to include, and then taking $+$ or $-$ according to the sum of the numbers indicating the rows and columns from which the first factor was formed is even or odd.

Example. Consider the determinant

$$\det \begin{bmatrix} a_1 & a_2 & a_3 & a_4 & a_5 \\ b_1 & b_2 & b_3 & b_4 & b_5 \\ c_1 & c_2 & c_3 & c_4 & c_5 \\ d_1 & d_2 & d_3 & d_4 & d_5 \\ e_1 & e_2 & e_3 & e_4 & e_5 \end{bmatrix}$$

which we represent as $|a_1 \, b_2 \, c_3 \, d_4 \, e_5|$, and its minor

$$\det \begin{bmatrix} b_3 & b_4 \\ c_3 & c_4 \end{bmatrix}$$

which we represent as $|b_3\ c_4|$. Then for two different sets of $q = 4$ rows we have

$$|b_3\ c_4| \cdot |a_1\ b_2\ c_3\ d_4\ e_5| = |a_1\ b_2\ c_3\ d_4| \cdot |b_3\ c_4\ e_5| - |a_1\ b_3\ c_4\ d_5| \cdot |b_3\ c_4\ e_2|$$
$$+ |a_2\ b_3\ c_4\ d_5| \cdot |b_3\ c_4\ e_1|$$
$$= -|a_1\ b_2\ c_3\ e_4| \cdot |b_3\ c_4\ d_5| + |a_1\ b_3\ c_4\ e_5| \cdot |b_3\ c_4\ d_2|$$
$$- |a_2\ b_3\ c_4\ e_5| \cdot |b_3\ c_4\ d_1|$$
$$= \cdots\cdots$$

Theorem 5.3.4 (Van Vleck, 1899). Consider the polynomials in $\mathbf{Z}[x]$, $p_1(x) = c_n x^n + c_{n-1} x^{n-1} + \cdots + c_0$ and $p_2(x) = d_m x^m + d_{m-1} x^{m-1} + \cdots + d_0$, $c_n \neq 0$, $d_m \neq 0$, $n \geq m$. Then, the successive polynomials that are formed from the first $2i$ rows of (S1), $i = 1, 2, \ldots, n$, constitute a Sturm sequence.

Proof. At first we observe that the signs of the leading coefficients of the polynomials formed from (S1) are in agreement with the signs of the corresponding coefficients of Sturm's remainders. Therefore, we need only prove the fact that if the leading coefficient of any member is zero, then the leading coefficients of the proceding and following polynomial remainders have opposite signs. To this end consider the leading coefficients of any three consecutive polynomial remainders $D_{2n-2}, D_{2k}, D_{2k+2}$, where D_k denotes the determinant of a $k \times k$ matrix formed from the first k rows and columns of (S1), and multiply the first by the last

$$D_{2k-2} \cdot D_{2k+2}$$

Since D_{2k-2} is a minor of D_{2k+2}, we are going to use Lemma 5.3.3. We easily see that the first determinant D_{2k-2} can be taken out as a minor of D_{2k+2} in several ways. One way to do this is to omit the first two and the last two rows, and the first column along with the last three. Next, we choose as our q rows to be all the rows from the third to the last one inclusive. Of the determinants of order q that can be formed to contain D_{2k-2}, all except three include the first column, which contains only zero elements, and hence they vanish. The remaining three determinants are obtained by omitting in addition to the first column, either the last column, the next to the last, or the second preceding the last. These will be the first factors, of the aggregate of products of pairs of minors; note that the first one is D_{2k}, while the third one, in accordance to Lemma 5.3.3, is to be multiplied by D_{2k}. Hence, when $D_{2k} = 0$, the only partial product to be considered is the remaining determinant multiplied by a factor that is easily seen to be its negative. Therefore, $D_{2k-2} \cdot D_{2k+2} < 0$ when $D_{2k} = 0$. □

From the above we see that Sylvester's form of the resultant can give us Sturm's sequence of two polynomials (the negated sequence of their polynomial remainders), which, in general, is a complete polynomial remainder sequence. Moreover, and this is the most important fact, one *does not have to compute determinants* in order to find the coefficients. This is due to the fact that in the original determinant (S1) the members of any two consecutive rows are the same as those of the two preceding rows. So, if in any row the values of the members are changed by adding a multiple of the preceding row, exactly the same change can be made in the members of each alternate row thereafter, without altering the value of any minor that appears as a coefficient in one of the polynomials of the Sturm sequence. Therefore, we can bring the corresponding matrix into its upper triangular form, using Bareiss's algorithm (B) (see Section 5.3.1), without disturbing the repetitive character of the determinant. Hence, we have the following.

Theorem 5.3.5 (Van Vleck, 1899). If we bring $\mathbf{M}_S$, the matrix corresponding to Sylvester's form of the resultant, into its upper triangular form $T(\mathbf{M}_S)$, then the *even* rows of $T(\mathbf{M}_S)$ furnish the coefficients of the successive (Sturmian) polynomial remainders. The coefficients taken from a given row are multiplied by $(-1)^k$, where k is the number of negative "constituents" in the principal diagonal above the row under consideration.

Van Vleck used this theorem to compute complete polynomial remainder sequences, updating *only* three rows at a time, and removing from each remainder the greatest common divisor of the coefficients, a computation we want to avoid; however, he did not deal with the pivoting problem. In this case, care should be exercised in determining the sign of the coefficients when a pivot is performed, since the later results in a change of signs. Observe the following example.

Example. Let $p_1(x) = x^3 - 7x + 7$, and $p_2(x) = 3x^2 - 7$; then, we have

$$\mathbf{M}_S = \begin{bmatrix} 1 & 0 & -7 & 7 & 0 & 0 \\ 0 & 3 & 0 & -7 & 0 & 0 \\ 0 & 1 & 0 & -7 & 7 & 0 \\ 0 & 0 & 3 & 0 & -7 & 0 \\ 0 & 0 & 1 & 0 & -7 & 7 \\ 0 & 0 & 0 & 3 & 0 & -7 \end{bmatrix} \Rightarrow T(\mathbf{M}_S) = \begin{bmatrix} 1 & 0 & -7 & 7 & 0 & 0 \\ 0 & 3 & 0 & -7 & 0 & 0 \\ 0 & 0 & 9 & 0 & -21 & 0 \\ 0 & 0 & 0 & -42 & 63 & 0 \\ 0 & 0 & 0 & 0 & 196 & -294 \\ 0 & 0 & 0 & 0 & 0 & -49 \end{bmatrix} *$$

The *-ed row indicates that a pivot was performed between the third and fourth rows, and this means that the Sturmian remainders are $42x - 63$ and 49. The reader should perform, *by hand*, the sequence of operations involved, and should note that in the third row, before we pivot, we obtain the terms -14 and 21. This is a crucial observation for the algorithmic implementation of the matrix-triangularization subresultant PRS method.

Note: If we consider the rows of both matrices of this example, we see that each corresponds to a given polynomial whose degree is one less than the number of elements *enclosed* between the leading and/or trailing zeros.

A slight difficulty that might arise with such an implementation is that one or more coefficients of the lower powers of x may be zero; in that case the degree of the polynomial can be easily determined by examining the degrees of all the rows of the upper triangular matrix. However, this problem does not appear if we change representation, using linked lists, for example.

What we have said so far is valid for complete polynomial remainder sequences. In order to be able to deal with incomplete PRSs we need the following theorem, which is our generalization of Theorem 5.3.5.

Theorem 5.3.6 (Akritas, 1986 and 1988). Let $p_1(x)$ and $p_2(x)$ be two polynomials of degrees n and m respectively, $n \geq m$. Using Bareiss's algorithm, transform the matrix $\mathbf{M}_S$, corresponding to $\text{res}_S[p_1(x), p_2(x)]$, into its upper triangular form $T(\mathbf{M}_S)$; let n_i be the degree of the polynomial corresponding to the ith row of $T(\mathbf{M}_S)$, $i = 1, 2, \ldots, 2n$, and let $p_k(x)$, $k \geq 2$ be the kth member of the (*complete or incomplete*) polynomial remainder sequence of $p_1(x)$ and $p_2(x)$. Then, if $p_k(x)$ is in row i of $T(\mathbf{M}_S)$, the coefficients of $p_{k+1}(x)$ are obtained from row $i + j$ of $T(\mathbf{M}_S)$, where j is the smallest integer such that $n_{i+j} < n_i$. [If $n = m$, associate both $p_1(x)$ and $p_2(x)$ with the first row of $T(\mathbf{M}_S)$.]

Proof. The proof is based on the fact that in $\mathbf{M}_S$ the members of any two consecutive rows are the same as those of the preceding rows, and that bubble pivots are performed. Suppose that, in the triangularization process, we obtain the kth member of the PRS, $p_k(x)$, from row i, so that $\deg[p_k(x)] = n_i$, and that we have triangularized up to that row. In the next step of the process, row i does not change, but suppose that row $i + 1$ changes in such a way that we have $n_{i+1} = n_i - d$, $d > 1$, and the diagonal element is 0; let $p_{k+1}(x)$ be the polynomial of degree $n_i - d$ corresponding to this row. This polynomial is the $(k + 1)$th member of the PRS, but its position in $T(\mathbf{M}_S)$ will change because next a bubble pivot takes place after which $n_{i+1} = n_i$ (justify this). This same process will be repeated until the leading coefficient of the polynomial $p_{k+1}(x)$, of degree $n_i - d$, appears exactly on the diagonal and we no longer need to pivot. So in $T(\mathbf{M}_S)$ there will be a number of rows corresponding to polynomials of degree n_i, followed by a number of rows corresponding to polynomials of degree $n_i - d$. The first one of the latter polynomials is $p_{k+1}(x)$ the coefficients of which (by now) have grown substantially as a result of the many transformations. Therefore, if we want the smallest coefficients for $p_{k+1}(x)$ we have to *store* it when we first encounter it in row $i + 1$ (where the difference of degrees first appeared) and use it in the stored version instead of the one appearing in $T(\mathbf{M}_S)$. We leave it as an exercise for the reader to complete the proof, that is, determine from what row will the coefficients of $p_{k+2}(x)$ be obtained. □

Note that as a special case of Theorem 5.3.6 we obtain Theorem 5.3.5 for complete PRSs. We see, therefore, that based on Theorem 5.3.6, we have another subresultant PRS method to compute the polynomial remainder sequence and a greatest common divisor of two polynomials. This method, where explicit divisions are *not* performed, uniformly treats both complete and incomplete PRSs and provides the smallest coefficients that can be expected without coefficient greatest common divisor computations; as we have already indicated in Section 5.1.1, in certain cases, the coefficients are *smaller* than those obtained using the Sylvester-Habicht pseudodivisions subresultant PRS method.

We have the following algorithm (examples were presented in Section 5.1.1).

ASPRS (Matrix-Triangularization Subresultant **PRS**)

Input: Two polynomials in $\mathbf{Z}[x]$, $p_1(x) = c_n x^n + c_{n-1} x^{n-1} + \cdots + c_0$ and $p_2(x) = d_m x^m + d_{m+1} x^{m-1} + \cdots + d_0$, $c_n \neq 0$, $d_m \neq 0$, $n \geq m$.

Output: The subresultant PRS of $p_1(x)$ and $p_2(x)$.

1. [Initialize.] Form the matrix $\mathbf{M}_S$, corresponding to $\text{res}_S[p_1(x), p_2(x)]$.
2. [Main step.] Using Bareiss's algorithm (B), described earlier, transform $\mathbf{M}_S$ into its upper triangular form $T(\mathbf{M}_S)$; then, the members of the PRS are obtained from the rows of $T(\mathbf{M}_S)$ according to Theorem 5.3.6.

Computing-Time Analysis of ASPRS. We know that if n is the highest degree of the two polynomials $p_1(x)$ and $p_2(x)$, then Sylvester's $\mathbf{M}_S$ matrix will be of order $2n$ by $2n$. Without taking advantage of the band form of this matrix, we see that in order to bring $\mathbf{M}_S$ into its upper triangular form (using Bareiss's method), we need to perform the following operations. At the ith stage, for *each* of the remaining $2n - i$ rows we perform $2(2n - i)$ integer multiplications and $(2n - i)$ subtractions and divisions to determine the resulting elements that are not necessarily equal to zero. So, we have that the total number of operations for the triangularization process is

$$0[\Sigma_{1 \leq i \leq 2n-1} (2n - i)(2n - i + 1)]$$

Using the formulas

$$\Sigma_{1 \leq i \leq k} i = \frac{k(k+1)}{2}, \qquad \Sigma_{1 \leq i \leq k} i^2 = \frac{k(k+1)(k+2)}{6}$$

we find that the total number of integer multiplications is $0(n^3)$. (We only count multiplications, since they are the most costly operations.)

Since we are doing exact integer arithmetic we assume that, in the worst case, each integer multiplication is performed among the largest integers that will appear. These largest integers can be as large as $\text{res}_S[p_1(x), p_2(x)]$.

Using Hadamard's inequality, we have

$$|\text{res}_S[p_1(x), p_2(x)]| \le \Pi_{1 \le i \le 2n}(\Sigma_{1 \le i \le 2n} r_{ij}^2)^{1/2} \le (2n)^n [|p(x)|_\infty]^{2n}$$

where $|p(x)|_\infty = \max(|p_1(x)|_\infty, |p_2(x)|_\infty)$. Thus, the time to multiply $(2n)^n \cdot (|p(x)|_\infty)^{2n}$ by itself is

$$L^2\{(2n)^n[|p(x)|_\infty]^{2n}\} = \{nL(2n) + 2nL[|p(x)|_\infty]\}^2 \sim n^2 L^2[|p(x)|_\infty]$$

and since there are $0(n^3)$ such multiplications, we conclude that

$$t_{\text{ASPRS}}[p_1(x), p_2(x)] = 0\{n^5 L^2[|p(x)|_\infty]\}$$

We see, therefore, that the theoretical computing time of the matrix-triangularization subresultant PRS method is equal to that of the Sylvester–Habicht pseudodivisions subresultant PRS method. In the next section we present some empirical results and comparisons.

5.4
Empirical Comparisons Between the Two Subresultant PRS Methods

In this section we present 10 examples (Anderson, 1986). Each begins with a brief statement explaining its important aspects and is followed by the polynomial remainder sequence generated by each subresultant PRS method. For the Sylvester–Habicht pseudodivisions method, we show two (sometimes different) sequences, obtained from the two different algorithms, as discussed in Section 5.2. The number in parentheses to the right of each line is the degree of a polynomial; for the matrix-triangularization method, this polynomial corresponds to a given row. Only the coefficients of the different powers of x are written, with the one corresponding to the highest power of x appearing first.

Below each method the largest predivision integer (in absolute value) is listed with the number of its digits in parentheses. This is the largest value generated by the respective method for the respective example. In Sylvester's (S') and Habicht's (H) PRS algorithms this value must be stored in memory before it is divided. In the matrix-triangularization method this value will be divided before it is stored in memory. This largest integer is included because the coefficients of the members of the PRS may seem fairly small, when, in fact, the size of the integers required to generate them is much larger.

The results obtained by Sylvester's and Habicht's PRS algorithms simply reflect the generated polynomial remainder sequence. The matrix-triangularization method, however, requires the transformation of $\mathbf{M}_S$, the matrix corresponding to Sylvester's form of the resultant, into $T(\mathbf{M}_S)$, its upper triangular form. Presented is only $T(\mathbf{M}_S)$, where the leftmost column indicates the row number. As we have stated before, to each row there corresponds a polynomial whose degree is one less than the number of elements *enclosed* between the leading and/or trailing zeros. The "#" symbol preceding a row number indicates that a pivot occurred for that row. To determine the members of the PRS, we scan top–down the rightmost column (degrees) and, from a given set of rows "of degree d," we pick the polynomial corresponding to the first one as the next member of the PRS; a ">" after the row number (leftmost column) means that row could be a member of the PRS, *provided no pivot took place*. If there was a pivot, the pivoted row is stored and printed under $T(\mathbf{M}_S)$, and it also has a ">" after its row number. In this case, we compare the coefficients of the stored row with the ones of the first row in $T(\mathbf{M}_S)$ having the same degree and pick the one with the smaller coefficients as the new member of the PRS. (See also the proof of Theorem 5.3.6 and historical note 3.)

Example 1. This is the classic incomplete PRS example used in the literature by Knuth (1981) and Brown (1971).

Sylvester–Habicht's pseudodivisions subresultant PRS method

   a. Sylvester's reduced (subresultant) PRS algorithm

$p(1) = 1 \quad 0 \quad 1 \quad 0 \quad -3 \quad -3 \quad 8 \quad 2 \quad -5 \qquad (8)$
$p(2) = 3 \quad 0 \quad 5 \quad 0 \quad -4 \quad -9 \quad 21 \qquad (6)$
$p(3) = -15 \quad 0 \quad 3 \quad 0 \quad -9 \qquad (4)$
$p(4) = 585 \quad 1125 \quad -2205 \qquad (2)$
$p(5) = -18885150 \quad 14907500 \qquad (1)$
$p(6) = 527933700 \qquad (0)$

Largest integer generated is 7864108803602112500 (18 digits).

   b. Habicht's subresultant PRS algorithm

$p(1) = 1 \quad 0 \quad 1 \quad 0 \quad -3 \quad -3 \quad 8 \quad 2 \quad -5 \qquad (8)$
$p(2) = 3 \quad 0 \quad 5 \quad 0 \quad -4 \quad -9 \quad 21 \qquad (6)$
$p(3) = -15 \quad 0 \quad 3 \quad 0 \quad -9 \qquad (4)$
$p(4) = -65 \quad -125 \quad 245 \qquad (2)$
$p(5) = -9326 \quad 12300 \qquad (1)$
$p(6) = 260708 \qquad (0)$

Largest integer generated is 21308697620 (11 digits).

The matrix-triangularization subresultant PRS method

```
 1>   1  0  1  0 -3 -3  8  2 -5   0    0    0    0  0         (8)
#2)   0  1  0  1  0 -3 -3  8  2  -5    0    0    0  0         (8)
 3>   0  0  3  0  5  0 -4 -9 21   0    0    0    0  0         (6)
 4)   0  0  0  9  0 15  0 -12 -27 63    0    0    0  0        (6)
 5)   0  0  0  0 -6  0 -15  0  9  18  -45    0    0  0        (6)
#6)   0  0  0  0  0  4  0 10  0  -6  -12   30    0  0  0      (6)
 7>   0  0  0  0  0  0 -10  0  2  0   -6    0    0  0         (4)
 8)   0  0  0  0  0  0  0 25  0 -5    0   15    0  0  0       (4)
 9)   0  0  0  0  0  0  0  0 -45 0   35   50 -125  0  0       (4)
#10)  0  0  0  0  0  0  0  0  0 81    0  -63  -90 225  0  0   (4)
11>   0  0  0  0  0  0  0  0  0  0 -117 -225  441  0  0       (2)
12)   0  0  0  0  0  0  0  0  0  0  169  325 -637  0  0       (2)
13)   0  0  0  0  0  0  0  0  0  0    0 -2035 2543 -845 0     (2)
14>   0  0  0  0  0  0  0  0  0  0    0    0 9326 -12300 0    (1)
15)   0  0  0  0  0  0  0  0  0  0    0    0    0 -7778 -46630 (1)
16>   0  0  0  0  0  0  0  0  0  0    0    0    0    0 -260708 (0)
```

Largest integer generated is 2431362808 (10 digits).

Pivoted row 2 during transformation 2. Stored row is:

```
2>  0  0  3  0  5  0 -4 -9 21  0  0  0  0  0  0   (6)
```

Pivoted row 6 during transformation 6. Stored row is:

```
6>  0  0  0  0  0  0 15  0 -3  0  9  0  0  0  0   (4)
```

Pivoted row 10 during transformation 10. Stored row is:

```
10> 0  0  0  0  0  0  0  0  0  0 65 125 -225 0 0  (2)
```

The reader should note that, in the above example, the second degree member of the PRS is not obtained from $T(M_S)$ but from the stored row printed under it. This fact is not pointed out again in the following examples.

Example 2. This is a small incomplete PRS example.

Sylvester–Habicht's pseudodivisions subresultant PRS method

   a. Sylvester's reduced (subresultant) PRS algorithm

$$p(1) = 2 \quad 5 \quad 5 \quad -2 \quad 1 \qquad (4)$$
$$p(2) = 3 \quad 3 \quad 3 \quad -4 \qquad (3)$$
$$p(3) = -21 \quad 45 \qquad (1)$$
$$p(4) = -47049 \qquad (0)$$

Largest integer generated is 37044 (5 digits).

b. Habicht's subresultant PRS algorithm

$p(1) = 2 \ \ 5 \ \ 5 \ \ -2 \ \ 1$ (4)
$p(2) = 3 \ \ 3 \ \ 3 \ \ -4$ (3)
$p(3) = -21 \ \ 45$ (1)
$p(4) = 15683$ (0)

Largest integer generated is 37044 (5 digits).

The matrix-triangularization subresultant PRS method

1>	2	5	5	−2	1	0	0	0	(4)
2>	0	3	3	3	−4	0	0	0	(3)
3)	0	0	9	9	2	3	0	0	(3)
#4)	0	0	0	27	41	−24	9	0	(3)
5>	0	0	0	0	63	−135	0	0	(1)
6)	0	0	0	0	0	147	−315	0	(1)
7)	0	0	0	0	0	0	3068	147	(1)
8>	0	0	0	0	0	0	0	−15683	(0)

Largest integer generated is 2305401 (7 digits)

Pivoted row 4 during transformation 4. Stored row is:

$\quad$ 4> $\ \ 0 \ \ 0 \ \ 0 \ \ 0 \ \ 21 \ \ -45 \ \ 0 \ \ 0$ $\quad$ (1)

Example 3. A complete PRS example where $\deg[p_1(x)] = \deg[p_2(x)]$; note that a zero appears as the constant term of $p_3(x)$.

Sylvester–Habicht's pseudodivisions subresultant PRS method

a. Sylvester's reduced (subresultant) PRS algorithm

$p(1) = 2 \ \ 1 \ \ 4 \ \ 3 \ \ 5 \ \ -2 \ \ 2$ (6)
$p(2) = 1 \ \ -1 \ \ -1 \ \ 4 \ \ -2 \ \ 1 \ \ 1$ (6)
$p(3) = 3 \ \ 6 \ \ -5 \ \ 9 \ \ -4 \ \ 0$ (5)
$p(4) = 60 \ \ -36 \ \ 75 \ \ -27 \ \ 9$ (4)
$p(5) = -1628 \ \ 240 \ \ -376 \ \ -468$ (3)
$p(6) = 42067 \ \ -27959 \ \ 12373$ (2)
$p(7) = -304996 \ \ -147119$ (1)
$p(8) = 1873839$ (0)

Largest integer generated is 3913180032193072 (16 digits).

b. Habicht's subresultant PRS algorithm

$p(1) = 2 \quad 1 \quad 4 \quad 3 \quad 5 \quad -2 \quad 2 \qquad (6)$
$p(2) = 1 \quad -1 \quad -1 \quad 4 \quad -2 \quad 1 \quad 1 \qquad (6)$
$p(3) = 3 \quad 6 \quad -5 \quad 9 \quad -4 \quad 0 \qquad (5)$
$p(4) = 60 \quad -36 \quad 75 \quad -27 \quad 9 \qquad (4)$
$p(5) = -1628 \quad 240 \quad -376 \quad -468 \qquad (3)$
$p(6) = 42067 \quad -27959 \quad 12373 \qquad (2)$
$p(7) = -304996 \quad -147119 \qquad (1)$
$p(8) = 1873839 \qquad (0)$

Largest integer generated is 3913180032193072 (16 digits).

The matrix-triangularization subresultant PRS method

1>	2	1	4	3	5	−2	2	0	0	0	0	(6)
2>	0	−3	−6	5	−9	4	0	0	0	0	0	(5)
3)	0	0	9	−22	9	−23	6	−6	0	0	0	(5)
4)	0	0	0	−60	36	−75	27	−9	0	0	0	(4)
5)	0	0	0	0	−332	−45	−379	93	−120	0	0	(4)
6>	0	0	0	0	0	−1628	240	−376	−468	0	0	(3)
7)	0	0	0	0	0	0	−2549	−8203	5113	−3256	0	(3)
8>	0	0	0	0	0	0	0	42067	−27959	12373	0	(2)
9)	0	0	0	0	0	0	0	0	255739	−151491	84134	(2)
10>	0	0	0	0	0	0	0	0	0	304996	147119	(1)
11)	0	0	0	0	0	0	0	0	0	−1992731	609992	(1)
12>	0	0	0	0	0	0	0	0	0	0	−1873839	(0)

Largest integer generated is 571513399644 (12 digits). Observe that $p_2(x)$ is not shown in $T(\mathbf{M}_S)$.

Example 4. The complete PRS example used by Van Vleck.

Sylvester–Habicht's pseudodivisions subresultant PRS method

a. Sylvester's reduced (subresultant) PRS algorithm

$p(1) = 1 \quad 1 \quad -1 \quad -1 \quad 1 \quad -1 \quad 1 \qquad (6)$
$p(2) = 6 \quad 5 \quad -4 \quad -3 \quad 2 \quad -1 \qquad (5)$
$p(3) = -17 \quad -14 \quad 27 \quad -32 \quad 37 \qquad (4)$
$p(4) = 44 \quad -114 \quad 120 \quad -7 \qquad (3)$
$p(5) = -516 \quad 828 \quad 186 \qquad (2)$
$p(6) = 9108 \quad -3114 \qquad (1)$
$p(7) = 127359 \qquad (0)$

Largest integer generated is 68687289792 (11 digits)

b. Habicht's subresultant PRS algorithm

$p(1) = 1 \quad 1 \quad -1 \quad -1 \quad 1 \quad -1 \quad 1$ (6)
$p(2) = 6 \quad 5 \quad -4 \quad -3 \quad 2 \quad -1$ (5)
$p(3) = -17 \quad -14 \quad 27 \quad -32 \quad 37$ (4)
$p(4) = 44 \quad -114 \quad 120 \quad -7$ (3)
$p(5) = -516 \quad 828 \quad 186$ (2)
$p(6) = 9108 \quad -3114$ (1)
$p(7) = 127359$ (0)

Largest integer generated is 68687289792 (11 digits).

The matrix-triangularization subresultant PRS method

1>	1	1	−1	−1	1	−1	1	0	0	0	0	(6)
2>	0	6	5	−4	−3	2	−1	0	0	0	0	(5)
3)	0	0	1	−2	−3	4	−5	6	0	0	0	(5)
4>	0	0	0	17	14	−27	32	−37	0	0	0	(4)
5)	0	0	0	0	−8	−4	6	−8	17	0	0	(4)
6>	0	0	0	0	0	−44	114	−120	7	0	0	(3)
7)	0	0	0	0	0	0	64	−72	24	−44	0	(3)
8>	0	0	0	0	0	0	0	−516	828	186	0	(2)
9)	0	0	0	0	0	0	0	0	360	552	−516	(2)
10>	0	0	0	0	0	0	0	0	0	9108	−3114	(1)
11)	0	0	0	0	0	0	0	0	0	−11916	9108	(1)
12>	0	0	0	0	0	0	0	0	0	0	−127359	(0)

Largest integer generated is 1159985772 (10 digits).

Example 5. This incomplete PRS example presents a variation of 3 in the degrees of its members (from 7 to 4) and it is the only one requiring a bubble pivot in the matrix-triangularization method; that is, a pivot will take place between rows that are not adjacent.

Sylvester–Habicht's pseudodivisions subresult PRS method

a. Sylvester's reduced (subresultant) PRS algorithm

$p(1) = 3 \quad 5 \quad 7 \quad -3 \quad -5 \quad -7 \quad 3 \quad 5 \quad 7 \quad -2$ (9)
$p(2) = 1 \quad 0 \quad 0 \quad -1 \quad 0 \quad 0 \quad -1 \quad -1 \quad -1$ (8)
$p(3) = 7 \quad 0 \quad 0 \quad -7 \quad 6 \quad 13 \quad 15 \quad 3$ (7)
$p(4) = -42 \quad -91 \quad -154 \quad -70 \quad -49$ (4)
$p(5) = -3146045 \quad -3785054 \quad -1840832 \quad -1391747$ (3)
$p(6) = -256801797 \quad -88236211 \quad -98913311$ (2)
$p(7) = 1999255468 \quad -2333396170$ (1)
$p(8) = -33438950636$ (0)

Largest integer generated is 1026442541730304580597746128 (28 digits).

b. Habicht's subresultant PRS algorithm

$p(1) = 3 \quad 5 \quad 7 \quad -3 \quad -5 \quad -7 \quad 3 \quad 5 \quad 7 \quad -2$ (9)
$p(2) = 1 \quad 0 \quad 0 \quad -1 \quad 0 \quad 0 \quad -1 \quad -1 \quad -1$ (8)
$p(3) = 7 \quad 0 \quad 0 \quad -7 \quad 6 \quad 13 \quad 15 \quad 3$ (7)
$p(4) = -42 \quad -91 \quad -154 \quad -70 \quad -49$ (4)
$p(5) = -64205 \quad -77246 \quad -37568 \quad -28403$ (3)
$p(6) = -5240853 \quad -1800739 \quad -2018639$ (2)
$p(7) = 40801132 \quad -47620330$ (1)
$p(8) = -682427564$ (0)

Largest integer generated is 8724617648516388414672 (22 digits).

The matrix-triangularization subresultant PRS method

```
1>   3  5  7 -3 -5 -7  3  5  7 -2    0     0     0     0     0    0   (9)
2>   0  1  0  0 -1  0  0 -1 -1 -1    0     0     0     0     0    0   (8)
3)   0  0  5  7  0 -5 -7  6  8 10   -2     0     0     0     0    0   (8)
4)   0  0  0 -7  0  0  7 -6 -13 -15 -3     0     0     0     0    0   (7)
5)   0  0  0  0 -49  0  0 79 23 19  -55   14     0     0     0    0   (7)
#6)  0  0  0  0  0 -343  0 -24 501 73  93  -413  98    0     0    0   (7)
#7)  0  0  0  0  0  0 -2401 -510 -1273 1637 -339 56 -2891 686  0  0  0 (7)
8>   0  0  0  0  0  0  0 2058 4459 7546 3430 2401 0   0     0    0    (4)
9)   0  0  0  0  0  0  0  0 -1764 -3822 -6468 -2940 -2058 0 0 0 0     (4)
10)  0  0  0  0  0  0  0  0 1512 3276 5544 2520 1764  0    0    0     (4)
11)  0  0  0  0  0  0  0  0 25811 -18982 4520 -811 -3024 0 0 0        (4)
12>  0  0  0  0  0  0  0  0  0 -64205 -77246 -37568 -28403 0 0 0      (3)
13)  0  0  0  0  0  0  0  0  0  0 2124693 449379 519299 128410 0 0    (3)
14>  0  0  0  0  0  0  0  0  0  0  0 -5240853 -1800739 -2018639 0 0   (2)
15)  0  0  0  0  0  0  0  0  0  0  0  0 -22909248 -24412716 10481706 0 (2)
16>  0  0  0  0  0  0  0  0  0  0  0  0  0 -40801132 47620330 0       (1)
17)  0  0  0  0  0  0  0  0  0  0  0  0  0  0 -398219984 81602264     (1)
18>  0  0  0  0  0  0  0  0  0  0  0  0  0  0  0 682427564            (0)
```

Largest integer generated is 27843817119202448 (17 digits)

Pivoted row 6 during transformation 6. Stored row is:

6> 0 0 0 0 0 0 0 42 91 154 70 49 0 0 0 0 0   (4)

Pivoted row 7 during transformation 7. Stored row is:

7) 0 0 0 0 0 0 0 294 637 1078 490 343 0 0 0 0 0

(4)

274 GREATEST COMMON DIVISORS OF POLYNOMIALS

Note that since the degree of stored row 7 is the same as that of stored row 6, the latter is not a member of the PRS.

Example 6. This example shows the PRS when $\deg[p_1(x)]$ and $\deg[p_2(x)]$ differ by more than one.

Sylvester–Habicht's pseudodivisions subresultant PRS method

   a. Sylvester's reduced (subresultant) PRS algorithm

$p(1) = 3 \quad 4 \quad 0 \quad -7 \quad 0 \quad -9 \quad 1 \quad -1 \quad 2$ (8)
$p(2) = 3 \quad 12 \quad 0 \quad -7 \quad 0 \quad 1$ (5)
$p(3) = 39960 \quad 5238 \quad -23895 \quad -945 \quad 3618$ (4)
$p(4) = 5371812 \quad 226170 \quad 56430 \quad -1003068$ (3)
$p(5) = -442106115 \quad 117089757 \quad 77380134$ (2)
$p(6) = 9708643482 \quad -4839461901$ (1)
$p(7) = 12487585839$ (0)

Largest integer generated is 41671931313194660832971041260 (29 digits].

   b. Habicht's subresultant PRS algorithm

$p(1) = 3 \quad 4 \quad 0 \quad -7 \quad 0 \quad -9 \quad -1 \quad 2$ (8)
$p(2) = 3 \quad 12 \quad 0 \quad -7 \quad 0 \quad 1$ (5)
$p(3) = 39960 \quad 5238 \quad -23895 \quad -945 \quad 3618$ (4)
$p(4) = 5371812 \quad 226170 \quad 56430 \quad -1003068$ (3)
$p(5) = -442106115 \quad 117089757 \quad 77380134$ (2)
$p(6) = 9708643482 \quad -4839461901$ (1)
$p(7) = 12487585839$ (0)

Largest integer generated is 41671931313194660832971041260 (29 digits).

The matrix-triangularization subresultant PRS method

```
 1>   3  4  0 -7   0 -9   1  -1   2      0      0     0     0    0  0  (8)
#2)   0  3  4  0  -7  0  -9   1  -1      2      0     0     0    0  0  (8)
#3)   0  0  3  4  -7  0  -9   1  -1      2      0     0     0    0  0  (8)
 4>   0  0  0  3  12  0  -7   0   1      0      0     0     0    0  0  (5)
 5)   0  0  0  0   9 36   0 -21   0      3      0     0     0    0  0  (5)
 6)   0  0  0  0   0 27 108   0 -63      0      9     0     0    0  0  (5)
 7)   0  0  0  0   0  0 -3456 -504 1746  99   -315    54    0    0  0  (5)
 8>   0  0  0  0   0  0   0 -39960 -5238 23895  945  -3618  0    0  0  (4)
 9)   0  0  0  0   0  0   0   0  75456 474480 -25560 3096 -79920 0  0  (4)
10>   0  0  0  0   0  0   0   0   0 -5371812 -226170 -56430 1003068 0 0 (3)
11)   0  0  0  0   0  0   0   0   0  63357144 -3542580 2310276 -10743624 0 0 (3)
12>   0  0  0  0   0  0   0   0   0   0 -442106115 117089757 77380134 0 0 (2)
13)   0  0  0  0   0  0   0   0   0   0  1089441759 1102788303 -884212230 0 (2)
14>   0  0  0  0   0  0   0   0   0   0   0  9708643482 -4839461901 0 (1)
15)   0  0  0  0   0  0   0   0   0   0   0 -36142658547 19417286964 (1)
16>   0  0  0  0   0  0   0   0   0   0   0   0  -12487585839 (0)
```

Largest integer generated is 121237518861722851398 (21 digits).

Pivoted row 2 during transformation 2. Stored row is:

2> 0 0 0 3 12 0 −7 0 1 0 0 0 0 0 0   (5)

Pivoted row 3 during transformation 3. Stored row is:

3) 0 0 0 3 12 0 −7 0 1 0 0 0 0 0 0   (5)

Example 7. Another small incomplete PRS example.

Sylvester–Habicht's pseudodivisions subresultant PRS method

  a. Sylvester's reduced (subresultant) PRS algorithm

    $p(1) = 1\ \ 5\ \ 10\ \ 5\ \ 5\ \ 2$ (5)
    $p(2) = 1\ \ 4\ \ 6\ \ 2\ \ 1$ (4)
    $p(3) = -3\ \ 2\ \ 1$ (2)
    $p(4) = -266\ \ -112$ (1)
    $p(5) = 980$ (0)

Largest integer generated is 212268 (6 digits).

  b. Habicht's subresultant PRS algorithm

    $p(1) = 1\ \ 5\ \ 10\ \ 5\ \ 5\ \ 2$ (5)
    $p(2) = 1\ \ 4\ \ 6\ \ 2\ \ 1$ (4)
    $p(3) = -3\ \ 2\ \ 1$ (2)
    $p(4) = -266\ \ -112$ (1)
    $p(5) = 980$ (0)

Largest integer generated is 212268 (6 digits).

The matrix-triangularization subresultant PRS method

  1> 1 5 10 5 5 2 0 0 0   (5)
  2> 0 1 4 6 2 1 0 0 0   (4)
  3) 0 0 1 4 3 4 2 0 0   (4)
 #4) 0 0 0 1 7 1 3 2 0   (4)
  5> 0 0 0 0 3 −2 −1 0 0   (2)
  6) 0 0 0 0 0 9 −6 −3 0 0   (2)
  7) 0 0 0 0 0 0 58 50 18 0   (2)
  8> 0 0 0 0 0 0 0 −266 −112 0   (1)
  9) 0 0 0 0 0 0 0 0 −756 −532   (1)
 10> 0 0 0 0 0 0 0 0 −980   (0)

Largest integer generated is 260680 (6 digits).

Pivoted row 4 during transformation 4. Stored row is:

$$4> \quad 0 \quad 0 \quad 0 \quad 0 \quad 3 \quad -2 \quad -1 \quad 0 \quad 0 \quad 0 \qquad (2)$$

Example 8. In this incomplete PRS example zero appears as the constant term of $p_3(x)$.

Sylvester–Habicht's pseudodivisions subresultant PRS method

a. Sylvester's reduced (subresultant) PRS algorithm

$$\begin{aligned}
p(1) &= 2 \quad 3 \quad 5 \quad -2 \quad -1 & (4) \\
p(2) &= 1 \quad 1 \quad 2 \quad -1 & (3) \\
p(3) &= -2 \quad 0 & (1) \\
p(4) &= 8 & (0)
\end{aligned}$$

Largest integer generated is 16 (2 digits).

b. Habicht's subresultant PRS algorithm

$$\begin{aligned}
p(1) &= 2 \quad 3 \quad 5 \quad -2 \quad -1 & (4) \\
p(2) &= 1 \quad 1 \quad 2 \quad -1 & (3) \\
p(3) &= -2 \quad 0 & (1) \\
p(4) &= 8 & (0)
\end{aligned}$$

Largest integer generated is 16 (2 digits).

The matrix-triangularization subresultant PRS method

$$\begin{array}{rrrrrrrrrr}
1> & 2 & 3 & 5 & -2 & -1 & 0 & 0 & 0 & (4) \\
2> & 0 & 1 & 1 & 2 & -1 & 0 & 0 & 0 & (3) \\
3) & 0 & 0 & 1 & 1 & 0 & -1 & 0 & 0 & (3) \\
\#4) & 0 & 0 & 0 & 1 & 5 & 0 & -1 & 0 & (3) \\
5> & 0 & 0 & 0 & 0 & 2 & 0 & 0 & 0 & (1) \\
6) & 0 & 0 & 0 & 0 & 0 & 4 & 0 & 0 & (1) \\
\#7) & 0 & 0 & 0 & 0 & 0 & 0 & 8 & -4 & (1) \\
8> & 0 & 0 & 0 & 0 & 0 & 0 & 0 & -8 & (0)
\end{array}$$

Largest integer generated is 32 (2 digits).

Pivoted row 4 during transformation 4. Stored row is:

$$4> \quad 0 \quad 0 \quad 0 \quad 0 \quad 2 \quad 0 \quad 0 \quad 0 \qquad (1)$$

Pivoted row 7 during transformation 7. Stored row is:

$$7> \quad 0 \quad 0 \quad 0 \quad 0 \quad 0 \quad 0 \quad -4 \qquad (0)$$

Example 9. Here the polynomials are not relatively prime.

Sylvester–Habicht's pseudodivisions subresultant PRS method

a. Sylvester's reduced (subresultant) PRS algorithm

$p(1) =$	6	7	−3	16	20	8	17	6	18	(8)
$p(2) =$	2	1	3	1	−2	9	−7	3		(7)
$p(3) =$	−56	28	96	−60	80	44	48			(6)
$p(4) =$	5824	1792	−1008	10528	−2912	3696				(5)
$p(5) =$	782976	530624	−321920	1104896	208704					(4)
$p(6) =$	51455488	25727744	−25727744	77183232						(3)
$p(7) =$	0									

Largest integer generated is 2925396948683036033024 (22 digits).

b. Habicht's subresultant PRS algorithm

$p(1) = 6 \quad 7 \quad -3 \quad 16 \quad 20 \quad 8 \quad 17 \quad 6 \quad 18$ (8)
$p(2) = 2 \quad 1 \quad 3 \quad 1 \quad -2 \quad 9 \quad -7 \quad 3$ (7)
$p(3) = -56 \quad 28 \quad 96 \quad -60 \quad 80 \quad 44 \quad 48$ (6)
$p(4) = 5824 \quad 1792 \quad -1008 \quad 10528 \quad -2912 \quad 3696$ (5)
$p(5) = 782976 \quad 530624 \quad -321920 \quad 1104896 \quad 208704$ (4)
$p(6) = 51455488 \quad 25727744 \quad -25727744 \quad 77183232$ (3)
$p(7) = 0$

Largest integer generated is 2925396948683036033024 (22 digits).

The matrix-triangularization subresultant PRS method

```
1>   6  7  -3  16   20    8    17    6    18    0       0        0       0      0   0    (8)
2>   0  2   1   3    1   -2     9   -7     3    0       0        0       0      0   0    (7)
3)   0  0   8 -24   26   52   -38   76    -6   36       0        0       0      0   0    (7)
4>   0  0   0  56  -28  -96    60  -80   -44  -48       0        0       0      0   0    (6)
5)   0  0   0   0 -560 1112  1216 -744  2304   24    1008        0       0      0   0    (6)
6>   0  0   0   0    0 -5824 -1792 1008 -10528 2912  -3696       0       0      0   0    (5)
7)   0  0   0   0    0    0 -133568 -116384 -27907 -210496 -39456 -104832       0      0   0    (5)
8>   0  0   0   0    0    0     0 782976  530624 -321920 1104896  208704        0      0   0    (4)
9)   0  0   0   0    0    0     0      0 3477248 11134336 2959232  518016  14093568    0   0    (4)
10>  0  0   0   0    0    0     0      0 51455488 25727744 -25727744 77183232         0   0   0    (3)
11)  0  0   0   0    0    0     0      0      0 617465856 308732928 -308732928 926198784  0   0    (3)
#12) 0  0   0   0    0    0     0      0      0      0       0        0       0      0   0    (0)
13)  0  0   0   0    0    0     0      0      0      0       0        0       0      0   0    (0)
14)  0  0   0   0    0    0     0      0      0      0       0        0       0      0   0    (0)
15)  0  0   0   0    0    0     0      0      0      0       0        0       0      0   0    (0)
16)  0  0   0   0    0    0     0      0      0      0       0        0       0      0   0    (0)
```

278 GREATEST COMMON DIVISORS OF POLYNOMIALS

Largest integer generated is 571896124988719104 (18 digits).

Example 10. A smaller example where the polynomials are not relatively prime.

Sylvester–Habicht's pseudodivisions subresultant PRS method

  a. Sylvester's reduced (subresultant) PRS algorithm

$$\begin{aligned}
p(1) &= 2 \quad -1 \quad -14 \quad 17 \quad -5 & (4)\\
p(2) &= 6 \quad -7 \quad 4 \quad -1 & (3)\\
p(3) &= -496 \quad 592 \quad -172 & (2)\\
p(4) &= 14432 \quad -7216 & (1)\\
p(5) &= 0
\end{aligned}$$

Largest integer generated is 123303313408 (12 digits).

  b. Habicht's subresultant PRS algorithm

$$\begin{aligned}
p(1) &= 2 \quad -1 \quad -14 \quad 17 \quad -5 & (4)\\
p(2) &= 6 \quad -7 \quad 4 \quad -1 & (3)\\
p(3) &= -496 \quad 592 \quad -172 & (2)\\
p(4) &= 14432 \quad -7216 & (1)\\
p(5) &= 0
\end{aligned}$$

Largest integer generated is 123303313408 (12 digits).

The matrix-triangularization subresultant PRS method

1>	2	−1	−14	17	−5	0	0	0	(4)
2>	0	6	−7	4	−1	0	0	0	(3)
3)	0	0	8	−92	104	−30	0	0	(3)
4>	0	0	0	496	−592	172	0	0	(2)
5)	0	0	0	0	−6816	8368	−2480	0	(2)
6>	0	0	0	0	0	−14432	7216	0-	(1)
7)	0	0	0	0	0	0	−144320	72160	(1)
8>	0	0	0	0	0	0	0	0	(0)

Largest integer generated is 983685120 (9 digits).

We next present Table 5.4.1, which summarizes the information on the largest predivision integers generated during computations by each example for both methods. Note that the matrix-triangularization method generates the smallest "largest integers."

TABLE 5.4.1 Number of Digits in the Largest Predivision Integer

Example	Sylvester–Habicht PRS method		Matrix-Triangularization PRS Method
	Sylvester's PRS Algorithm	Habicht's PRS Algorithm	
1	18	11	10
2	5	5	7
3	16	16	12
4	11	11	10
5	28	22	17
6	29	29	22
7	6	6	6
8	2	2	2
9	22	22	18
10	12	12	9

Exercises

Section 5.1.1

1. Compute the pseudoquotient $q(x)$ and the pseudoremainder $r(x)$, when $p_1(x) = x^3 - 3x + 1$ and $p_2(x) = 3x^2 - 3$. What is their greatest common divisor? Do the same for the polynomials $p_1(x) = x^3 - 6x^2 + 8x + 40$ and $p_2(x) = 3x^2 - 12x + 8$.

2. Let $r(x)$ be the pseudoremainder on pseudodividing $p_1(x)$ by $p_2(x)$. If $\deg[p_1(x)] \geq \deg[p_2(x)] + 2$ and $\deg[p_2(x)] \geq \deg[r(x)] + 2$, prove that $r(x)$ is a multiple of $lc[p_2(x)]$.

Section 5.1.2

1. Describe in your own words the difference between **M** and **M1**.

2. Using either one of the modular *gcd* algorithms described in the text, compute the greatest common divisor of:
   (a) $p_1(x) = 6x^4 - 24x^3 + 42x^2 - 32x + 11$ and $p_2(x) = 24x^3 - 72x^2 + 84x - 32$.
   (b) $p_1(x) = 16x^4 - 32x^3 + 88x^2 - 8x + 17$ and $p_2(x) = 64x^3 - 96x^2 + 176x - 8$.
   (c) $p_1(x) = x^4 + 2x^3 - 4x + 10$ and $p_2(x) = 4x^3 + 6x^2 - 4$.

Section 5.2.1

1. Using Theorem 5.2.1, and following the example of this section, compute the polynomial remainder sequence of the polynomials:
   (a) $p_1(x) = 2x^3 - 3x^2 + 4x - 1$ and $p_2(x) = 6x^2 - 6x + 4$.
   (b) $p_1(x) = x^4 - x^3 + 3x^2 - 2x + 1$ and $p_2(x) = 4x^3 - 3x^2 + 6x - 2$.
   (c) $p_1(x) = x^5 + x^4 - 2x^3 - 2x^2 + 4$ and $p_2(x) = 5x^4 + 4x^3 - 6x^2 - 4x$.
   (d) $p_1(x) = x^7 - x^5 - 8x^2 + 3$ and $p_2(x) = 7x^6 - 5x^4 - 16x$.

Section 5.2.2

Exercise 1 is *extremely* important for a thorough understanding of the next section; however, it is very long and tedious (use a computer to check your arithmetic).

1. Using $\mathbf{M_B}$, the matrix corresponding to $\text{res}_B[p_1(x), p_2(x)]$, where $p_1(x) = 3x^9 + 5x^8 + 7x^7 - 3x^6 - 5x^5 - 7x^4 + 3x^3 + 5x^2 + 7x - 2$ and $p_2(x) = x^8 - x^5 - x^2 - x - 1$, and the method outlined in the text, compute the members of the (incomplete) polynomial remainder sequence. What can you conclude? (*Hint*: Follow the example preceding Theorem 5.2.4.)

2. From Theorem 5.2.4 we see that the resultant can be computed by the same process used to find the greatest common divisor of two polynomials. Find the resultant of $p_1(x) = x^4 - x^3 + 2x^2 - 3x + 1$ and $p_2(x) = x^3 - 2x^2 + 3x - 1$.

3. Prove Theorem 5.2.9.

4. Compute the discriminant of $p(x) = x^2 + ax + b$.

Section 5.2.3

1. Prove the formulas (H2) of this section.

2. Apply the extended Euclidean algorithm for polynomials over the integers to the polynomials $x^3 - 7x + 7$ and $3x^2 - 7$ and compare your answer with the one obtained in the example following Theorem 5.2.11. What can you say about the last member of the PRS?

3. Complete the computations in the last example of this section.

4. Use the formulas in **H-1**, **H-2** and **H-3** to derive the expression (H) of Section 5.1.1.

Section 5.3.1

1. Determine the greatest common divisor of $x^3 - 7x + 7$ and $3x^2 - 7$ using the gaussian elimination process instead of pseudodivisions.
2. Do the same for the polynomials $x^4 + 6x^3 + 5x^2 - 4x - 2$ and $4x^3 + 18x^2 + 10x - 4$.

Section 5.3.3

1. Using the method outlined in the text, complete the first example of this section; that is, what is $p_4(x)$?
2. Following the first example of this section, compute the polynomial remainder sequence of the following pairs of polynomials:
   (a) $p_1(x) = 2x^4 + 5x^3 + 5x^2 - 2x + 1$, and $p_2(x) = 3x^3 + 3x^2 + 3x - 4$.
   (b) $p_1(x) = x^5 + 5x^4 + 10x^3 + 5x^2 + 5x + 2$, and $p_2(x) = x^4 + 4x^3 + 6x^2 + 2x + 1$.
   (c) $p_1(x) = 2x^4 + 3x^3 + 5x^2 - 2x - 1$, and $p_2(x) = x^3 + x^2 + 2x - 1$.
   (d) $p_1(x) = 2x^4 - x^3 - 14x^2 + 17x - 5$, and $p_2(x) = 6x^3 - 7x^2 + 4x - 1$.

Programming Exercises

Section 5.1.1

1. Write a procedure to compute the pseudoquotient and pseudoremainder on pseudodividing the polynomial $p_1(x)$ by $p_2(x)$.
2. Write a procedure to form the matrix corresponding to Sylvester's form of the resultant of the polynomials $p_1(x) = c_n x^n + c_{n-1} x^{n-1} + \cdots + c_0$ and $p_2(x) = d_m x^m + d_{m-1} x^{m-1} + \cdots + d_0$, $c_n \neq 0$, $d_m \neq 0$, $n \geq m$.

Section 5.2.1

1. Write a procedure to implement algorithm **SRSPRS**, and run it on the polynomials of Exercise 1 for this section.

Section 5.2.3

1. Write a procedure to implement algorithm **HSPRS** and run it on the polynomials:
   (a) $p_1(x) = 2x^4 + 5x^3 + 5x^2 - 2x + 1$, and $p_2(x) = 3x^3 + 3x^2 + 3x - 4$.
   (b) $p_1(x) = x^5 + 5x^4 + 10x^3 + 5x^2 + 5x + 2$, and $p_2(x) = x^4 + 4x^3 + 6x^2 + 2x + 1$

(c) $p_1(x) = 2x^4 + 3x^3 + 5x^2 - 2x - 1$, and $p_2(x) = x^3 + x^2 + 2x - 1$.
(d) $p_1(x) = 2x^4 - x^3 - 14x^2 + 17x - 5$, and $p_2(x) = 6x^3 - 7x^2 + 4x - 1$.

Section 5.3.2

1. Write a procedure that, given two polynomials with integer coefficients, will print out the corresponding matrices M_B and M_T. Run it on your favorite polynomials.

2. Write a procedure to implement Bareiss's algorithm (B), discussed in Section 5.3.1, and, given the matrices M_B and M_T, use it to obtain the upper triangular matrices $T(M_B)$ and $T(M_T)$.

Section 5.3.3

1. Write a procedure to implement algorithm **ASPRS** and run it on the polynomials of Exercise 2 for this section; write out the corresponding polynomial remainder sequences, and compare the answers.

Historical Notes and References

1. Brown, in both of his papers (1971, p. 485, 1978, p. 238) and Knuth (1981, p. 410) attribute to Collins (1967) the discovery of the Sylvester–Habicht subresultant PRS method. As we indicated, this discovery was initiated by Sylvester in 1853 and was completed by Habicht in 1948. In this book, for the first time, credit is given where it belongs. [In this context see also Fryer, 1959.]

2. The papers by Sylvester (1853) and Van Vleck (1899) are not mentioned in the survey article of Loos (1982).

3. We are still in the process of further investigating and developing the matrix-triangularization subresultant PRS method. Additional developments will be included in a future edition of this book.

Akritas, A. G.: *A new method for computing polynomial greatest common divisors.* University of Kansas, TR-86-9, Lawrence, KS, 1986.

Akritas, A. G.: A simple proof of the validity of the reduced prs algorithm. *Computing* **38**, 369–372, 1987.

Akritas, A. G.: A new subresultant prs method. Proceedings of the 12th IMACS World Congress on Scientific Computation (July 1988, Paris, France), **4**, 654–655, 1988.

Akritas, A. G.: A new method for computing polynomial greatest common divisors and polynomial remainder sequences. *Numerische Mathematik* **52**, 119–127, 1988.

Anderson, G.: An examination of polynomial remainder sequences. M.S. Thesis, University of Kansas, Department of Computer Science, Lawrence, KS, 1986.

Bareiss, E. H.: Sylvester's identity and multistep integer-preserving Gaussian elimination. *Mathematics of Computation* **22**, 565–578, 1968.

Barnett, S.: Greatest common divisor of two polynomials. *Linear Algebra and Applications* **3**, 7–9, 1970.

Barnett, S.: A new look at classical algorithms for polynomial resultant and gcd calculation. *SIAM Review* **16**, 193–206, 1974.

Berlekamp, E. R.: *Algebraic coding theory*. McGraw-Hill, New York, 1968.

Bocher, M.: *Introduction to higher algebra*. Macmillan, New York, 1907.

Brown, W. S.: On Euclid's algorithm and the computation for polynomial greatest common divisors. *Journal of Association for Computing Machinery* **18**, 476–504, 1971.

Brown, W. S.: The subresultant prs algorithm. *ACM Transactions on Mathematical Software* **4**, 237–249, 1978.

Brown, W. S., and J. F. Traub: On Euclid's algorithm and the theory of subresultants. *Journal of Association for Computing for Machinery* **18**, 505–514, 1971.

Collins, G. E.: Polynomial remainder sequences and determinants. *American Mathematical Monthly* **73**, 708–712, 1966.

Collins, G. E.: Subresultants and reduced polynomial remainder sequences. *Journal of Association for Computing Machinery* **14**, 128–142, 1967.

Collins, G. E.: The calculation of multivariate polynomial resultants. *Journal of Association for Computing Machinery* **19**, 515–532, 1971.

Fryer, W. D.: Applications of Routh's algorithm to network theory problems. *IEEE Transactions on Circuit Theory* **CT-6**, 144–149, 1959.

Habicht, W.: Eine Verallgemeinerung des Sturmschen Wurzelzaehlverfahrens. *Commentarii Mathematici Helvetici* **21**, 99–116, 1948.

Householder, A. S.: *The numerical treatment of a single nonlinear equation*. McGraw-Hill, New York, 1970.

Knuth, D.: *The art of computer programming*. Vol. II, 2nd ed.: *Seminumerical algorithms*. Addison-Wesley, Reading, MA, 1981.

Laidacker, M. A.: Another theorem relating Sylvester's matrix and the greatest common divisor. *Mathematics Magazine* **42**, 126–128, 1969.

Lipson, J. D.: *Elements of algebra and algebraic computing*. Addison-Wesley, Reading, MA, 1981.

Loos, R.: Generalized polynomial remainder sequences. In *Computer Algebra Symbolic and Algebraic Computations*, B. Buchberger, G. E. Collins, and R. Loos, eds. Springer Verlag, New York, 1982, *Computing Supplement* **4**, 115–137.

Miola, A., and D. Y. Y. Yun: Computational aspects of Hensel-type univariate polynomial greatest common divisor algorithms. *Proceedings of EUROSAM 1974*, pp. 46–54 (also *ACM SIGSAM Bulletin 31*).

Moses, J. and D. Y. Y. Yun: The EZGCD algorithm. *Proceedings of the ACM Annual Conference* (August 1973, Atlanta), pp. 159–166.

Muir, T.: *A treatise on the theory of determinants*. Macmillan, London, 1882.

Muir, T.: *The theory of determinants*. Vol III. Macmillan, London, 1920.

Netto, E.: *Vorlesungen ueber Algebra. Erster Band.* Teubner, Leipzig, 1896.

Sylvester, J. J.: On a theory of the syzygetic relations of two rational integral functions, comprising an application to the theory of Sturm's functions, and that of the greatest algebraical common measure. *Philosophical Transactions* **143**, 407–548, 1853.

Yun, D. Y. Y.: *The Hensel lemma in algebraic manipulation.* Ph.D. Thesis, MIT, 1974.

Van Vleck, E. B.: On the determination of a series of Sturm's functions by the calculation of a single determinant. *Annals of Mathematics, Second Series* **1**, 1–13, 1899–1900.

6

Factorization of Polynomials over the Integers

6.1 Introduction and Motivation

6.2 Factorization of Polynomials over a Finite Field

6.3 Lifting a (mod p)-Factorization to a Factorization over the Integers

Exercises

Historical Notes and References

6.1
Introduction and Motivation

Clearly, the easiest way to determine the factors of a given polynomial $p(x)$, $\deg[p(x)] = k$, is to algebraically find its roots ρ_i, in which case $p(x) = (x - \rho_1)(x - \rho_2) \cdots (x - \rho_k)$. (The reader should review Section 3.2.3 at this point.) However, as we will see in Chapter 7, we can algebraically find the roots of a given polynomial *only* if its degree is ≤ 4. For polynomials of degree greater than four we have to take a different approach.

Besides being of pure theoretical interest, the factorization of polynomials is of great practical importance (in coding theory, for example). Newton, in his *Arithmetica Universalis* (1707), gave a method for finding linear and quadratic factors of polynomials over the integers. In 1793 Newton's method was extended by the astronomer Friedrich von Schubert, who showed how to find all factors of degree n in a finite number of steps (Cantor, 1908; pp. 136–137). Schubert's method was independently rediscovered by L. Kronecker about 90 years later. However, as we will see, the Schubert–Kronecker method has an exponential computing time and, hence, it is of very little practical importance.

Much better results are obtained with the help of "modulo p" factorization methods, along with construction techniques for *lifting* a "modulo p" factorization over the integers. This approach is illustrated in Figure 6.1.1.

However, despite the fact that the approach shown in Figure 6.1.1 performs very well in almost all practical cases, in the worst case, it too has an exponential computing time bound. Quite recently (1982), A. K. Lenstra, H. W. Lenstra, Jr., and L. Lovasz (or, simply L^3) applied the theory of lattices to the factorization problem and came up with a polynomial–time factorization method; the presentation of the L^3 method, though, is beyond the scope of this book; see also Lenstra, 1981 and 1982a).

6.1.1 The Schubert–Kronecker Factorization Method over the Integers

This method works as follows. Let $p(x)$ be the given polynomial with integer coefficients, of degree n, which we want to factor. If $p(x)$ can indeed be factored, then one of the factors has to be of degree $\leq n/2$. Let $m = \lfloor n/2 \rfloor$,

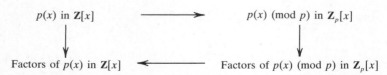

Figure 6.1.1. The roundabout way to factor any polynomial $p(x)$ in $\mathbf{Z}[x]$.

the greatest integer $\leq n/2$; then we have to see whether $p(x)$ has a factor $q(x)$ of degree $\leq m$. For $m+1$ distinct integral values $a_0, a_1, \ldots, a_m$ $(0, \pm 1, \pm 2, \ldots)$, compute $p(a_0), p(a_1), \ldots, p(a_m)$. If $q(x)$ is a factor of $p(x)$, then $q(a_0)$ must be a factor of $p(a_0)$, $q(a_1)$ of $p(a_1)$, and so on. Call f_i the finite set of all distinct integral factors of $p(a_i)$. Then, for $k = 2, \ldots, m+1$ do the following: choose a set of k elements b_i from distinct f_i and Lagrange-interpolate to find a polynomial $q(x)$ of degree $k-1$, such that $q(a_i) = b_i$ for all i. If $q(x)$ divides $p(x)$, then we have found a factor of $p(x)$ and we can recursively apply this method on $p(x)/q(x)$. Otherwise, choose another set of k elements b_i from the f_i (different from all previously chosen sets), interpolate, and test-divide again. When all possible combinations off $m+1$ or fewer integral values from the f_i have been tried, we conclude that $p(x)$ is irreducible.

Example. Let us try to factor the polynomial $p(x) = x^4 + 4$. We observe that $n = 4$ and $m = \lfloor n/2 \rfloor = 2$. Then we pick $m + 1 = 3$ points $a_0 = 0$, $a_1 = 1$, $a_2 = -1$ and form the sets f_i:

$$p(a_0) = p(0) = 4, \qquad f_0 = \{1, -1, 2, -2, 4, -4\}$$
$$p(a_1) = p(1) = 5, \qquad f_1 = \{1, -1, 5, -5\}$$
$$p(a_2) = p(-1) = 5, \qquad f_2 = \{1, -1, 5, -5\}$$

One clearly sees the tremendous number of possible combinations; we skip the case $k = 2$ because it yields no linear factors of $p(x)$ and try $k = 3$. We first choose $b_0 = 1$, $b_1 = -1$, $b_2 = 5$, and interpolating the pairs

$$\{a_0 = 0, b_0 = 1\}, \{a_1 = 1, b_1 = -1\}, \{a_2 = -1, b_2 = 5\}$$

we obtain

$$q(x) = 1\left\{\frac{(x-1)(x+1)}{(0-1)(0+1)}\right\} + (-1)\left\{\frac{(x-0)(x+1)}{(1-0)(1+1)}\right\} + 5\left\{\frac{(x-0)(x-1)}{(-1-0)(-1-1)}\right\}$$
$$= x^2 - 3x + 1$$

We next divide $p(x)$ by $q(x)$ and see that $q(x)$ is not a factor of $p(x)$ because

$$x^4 + 4 = (x^2 - 3x + 1)(x^2 + 3x + 8) + (21x - 4)$$

We then choose $b_0 = 2$, $b_1 = 5$, $b_2 = 1$, and interpolating the pairs

$$\{a_0 = 0, b_0 = 2\}, \{a_1 = 1, b_1 = 5\}, \{a_2 = -1, b_2 = 1\}$$

we obtain

$$q(x) = 2\left\{\frac{(x-1)(x+1)}{(0-1)(0+1)}\right\} + 5\left\{\frac{(x-0)(x+1)}{(1-0)(1+1)}\right\} + 1\left\{\frac{(x-0)(x-1)}{(-1-0)(-1-1)}\right\}$$
$$= x^2 + 2x + 2$$

This time, though, we see that $q(x)$ is a factor of $p(x)$ because

$$x^4 + 4 = (x^2 + 2x + 2)(x^2 - 2x + 2)$$

and this is the complete factorization of $p(x)$, since by Eisenstein's criterion (Theorem 3.2.15) the two factors are irreducible.

Obviously, the computing time of the above method is exponential, and the method works very inefficiently for $n \geq 5$.

6.1.2 A General Overview of the Roundabout Factorization Method over the Integers

In this section we give a brief description of the various steps involved in the roundabout method shown in Figure 6.1.1; detailed explanations for some of the steps are presented separately in the following sections; see also Lazard, 1982 and Moenck, 1977. (For the multivariate case see Wang, 1978).

Assume that we are given the polynomial $p(x) \in \mathbf{Z}[x]$ to be factored; that is, we are given

$$p(x) = c_n x^n + c_{n-1} x^{n-1} + \cdots + c_0, \quad c_i \in \mathbf{Z}$$

and we want to determine its irreducible factors. At first observe that we can safely conclude that $c_n = 1$, because otherwise we can set $x := x/c_n$ and multiply the resulting polynomial times c_n^{n-1}. In this way we obtain the monic polynomial $p_m(x) = c_n^{n-1} p(x/c_n)$ with integer coefficients, whose factorization problem is the same as that of $p(x)$; that is, if $p_m(x) = p_1(x) \cdots p_r(x)$, then the factorization of $p(x)$ is obtained from $p(x) = (1/c_n^{n-1})p_m(c_n x) = (1/c_n^{n-1})p_1(c_n x) \cdots p_r(c_n x)$. The rational number $1/c_n^{n-1}$ disappears after we divide out c_n from as many factors as possible and/or remove the content of the factors.

As a first example of the above process, consider the polynomial $p(x) = 2x^2 + 3x + 1$. Then

$$p_m(x) = 2p\left(\frac{x}{2}\right) = 2\left[2\left(\frac{x}{2}\right)^2 + 3\left(\frac{x}{2}\right) + 1\right] = x^2 + 3x + 2 = (x+1)(x+2)$$

The factorization of $p(x)$ is obtained from

$$p(x) = \left(\frac{1}{2}\right)p_m(2x) = \left(\frac{1}{2}\right)(2x+1)(2x+2) = (2x+1)(x+1)$$

where in this case the rational number 1/2 has been multiplied times 2, the content of the second factor.

As a second example of the same process, consider the polynomial $p(x) = 112x^4 + 58x^3 - 31x^2 + 107x - 66$. Then

$$p_m(x) = 112^3 p\left(\frac{x}{112}\right) = x^4 + 58x^3 - 3472x^2 + 1342208x - 92725248$$
$$= (x^2 + 98x - 9408)(x^2 - 40x + 9856)$$

The factorization of $p(x)$ is obtained from

$$p(x) = \left(\frac{1}{112^3}\right) p_m(112x)$$
$$= \left(\frac{1}{112^3}\right)[(112x)^2 + 98 \cdot 112x - 9408][(112x)^2 - 40 \cdot 112x + 9856]$$
$$= (8x^2 + 7x - 6)(14x^2 - 5x + 11)$$

where in this case the rational number $1/112^3$ has been multiplied times 112^3, corresponding to dividing 112 out of both factors *and* to removing the contents of both of them.

Hence, below we assume that the given polynomial $p(x)$ is monic. (This restriction is relaxed in Section 6.3.) The following are the essential steps of the roundabout factorization method over the integers (Musser, 1971, 1975 and 1978):

1. First we eliminate from $p(x)$ proper factors of degree zero and repeated factors by means of greatest common divisor calculations in **Z** and **Z**$[x]$. (In other words, we compute a primitive, squarefree polynomial; the reader should review Section 3.2.4 at this point.) As a result we are left with a monic, squarefree polynomial $p_{sf}(x) = \Pi_{1 \le i \le q} p_i(x)$, where all the $p_i(x)$ are distinct, irreducible polynomials. Our objective is to determine these q factors $p_i(x)$.

2. We choose a prime number p such that factorization in $\mathbf{Z}_p[x]$ is possible. The first consideration in the choice of a prime p is that the polynomial $p_{sf}(x)$, obtained in step 1, has the *same* degree and *remains* squarefree modulo p. Since $p_{sf}(x)$ is monic p does not divide its leading coefficient, and hence the degree will stay the same modulo p; see also Theorem 3.2.16. Moreover, we can efficiently test whether $p_{sf}(x)$ (mod p) is squarefree by testing whether $gcd[p_{sf}(x), p'_{sf}(x)] = 1$ in $\mathbf{Z}_p[x]$ [$p'_{sf}(x)$ being the derivative of $p_{sf}(x)$]. This is so because if any given polynomial $p(x) = [p_1(x)]^2 p_2(x)$, then $p'(x)$ is a multiple of $p_1(x)$. Therefore, if $gcd[p(x), p'(x)] = 1$, we know that $p(x)$ is squarefree, whereas if $gcd[p(x), p'(x)] \ne 1$ and $gcd[p(x), p'(x)] \ne p(x)$, then we have to factor both $gcd[p(x), p'(x)]$ and $p(x)/gcd[p(x), p'(x)]$; finally, if $gcd[p(x), p'(x)] = p(x)$, then $p(x) = q(xp) = [q(x)]^p$ and we have to factor $q(x)$. For more details, see Section 6.2.1. The size of the primes used is also an important factor in the choice of an algorithm for factoring in $\mathbf{Z}_p[x]$. Trying large primes would reduce the number of Hensel-type constructions (explained in Section 6.3) required to lift the factorization of $p_{sf}(x)$ (mod p) from $\mathbf{Z}_p[x]$ to $\mathbf{Z}[x]$. However, efficient factorization algorithms in $\mathbf{Z}_p[x]$ exist only for small p (Berlekamp, 1970). The third major consideration in the choice of a prime p is the number r of factors in the complete modulo p factorization. If $p_{sf}(x)$ is irreducible but splits into $r > 1$ irreducible factors modulo p (see step 3 for an explanation of this phenomenon), then r factors are obtained at the end of the lifting and a total of 2^{r-1} subsets of factors are considered in step 5, in order to determine the true factors of $p_{sf}(x)$ over the integers. (It is precisely this point that makes the maximum computing time of the algorithm exponential.) A small remedy to the problem is to factor $p_{sf}(x)$ (step 3) modulo several primes p_i for which it remains squarefree and choose as prime p the one that yields the smallest number of irreducible factors.

3. For the prime number p chosen above we obtain the factorization in $\mathbf{Z}_p[x]$

$$p_{sf}(x) \equiv \Pi_{1 \le k \le r} g_k (\text{mod } p), \qquad g_k \in \mathbf{Z}_p[x]$$

In Section 6.2 we examine ways for obtaining this factorization. Observe that the number q of true factors of $p_{sf}(x)$ over the integers (step 1) *is not necessarily equal* to the number r of factors of $p_{sf}(x)$ modulo p; for example, $x^2 + 2 \equiv (x - 2)(x - 1)$ (mod 3). It is interesting that there are polynomials irreducible over the integers, which can be factored modulo p for any prime p. For example, for any integers a, b and any prime p, $p(x) = x^4 + ax^2 + b^2$ factors in $\mathbf{Z}_p[x]$. To see that $p(x) = x^4 + ax^2 + b^2$ factors in $\mathbf{Z}_p[x]$ for any prime p, consider first the case $p = 2$. Then, for any integers a, b we have the following four possibilities: $p(x) \equiv x^4$ (mod 2), or $p(x) \equiv x^4 + x^2 = x^2(x^2 + 1)$ (mod 2), or $p(x) \equiv x^4 + 1 = (x + 1)^4$ (mod 2), or $p(x) \equiv x^4 + x^2 + 1 = (x^2 + x + 1)^2$ (mod 2). Obviously, each of these is reducible modulo 2. For the general case, p odd prime, choose c so that $a \equiv 2c$ (mod p). Then, $p(x) \equiv x^4 + 2cx^2 + b^2$ (mod p), and we have the following three possibilities:

$$p(x) \equiv (x^2 + c)^2 - (c^2 - b^2) \text{ (mod } p)$$
$$p(x) \equiv (x^2 + b)^2 - (2b - 2c)x^2 \text{ (mod } p)$$
$$p(x) \equiv (x^2 - b)^2 - (-2b - 2c)x^2 \text{ (mod } p)$$

Clearly, to show that $p(x)$ factors modulo p, it suffices to show that one of $(c^2 - b^2)$, $(2b - 2c)$, $(-2b - 2c)$ is a square mod p [$p(x)$ will then be the difference of two squares]. Let d be a primitive root modulo p (see Theorem 2.3.20). If $2b - 2c$ and $-2b - 2c$ are not squares mod p, then $2b - 2c \equiv d^e$, and $-2b - 2c \equiv d^f \pmod{p}$, where e and f are both odd. Taking the product $(2b - 2c)(-2b - 2c) = 4(b^2 - c^2) \equiv d^{e+f}$, where now $e + f = 2g$ is even, we see that $4(c^2 - b^2)$ is a square mod p. Let $2h \equiv 1 \pmod{p}$; then, $c^2 - b^2 \equiv 4(c^2 - b^2)h^2 \equiv d^{2g}h^2 = (d^g h)^2$, and we see that $c^2 - b^2$ is a square mod p. Therefore, $p(x) = x^4 + ax^2 + b^2$ factors in $\mathbf{Z}_p[x]$ modulo any prime p. Using Eisenstein's criterion (Theorem 3.2.15), we can now easily construct a polynomial of the form $p(x) = x^4 + ax^2 + b^2$, which is irreducible over the integers, and which factors mod p for every prime p. As an example, consider $x^4 + 1$. Also we leave it as an exercise for the reader to show that with $0 < b < a$, $p(x) = x^4 + ax^2 + b^2$ is irreducible over the integers.

4. Using Hensel-type constructions (discussed in Section 6.3), we lift the polynomials $g_k(x)$, obtained in step 3, to the corresponding polynomials $h_k(x) \pmod{p^j}$ such that

$$p_{\mathrm{sf}}(x) \equiv \Pi_{1 \le k \le r} h_k(x) \pmod{p^j}$$

for sufficiently large positive integer j (to be explained in Section 6.3). Now, each true factor $p_i(x)$ of $p_{\mathrm{sf}}(x)$ corresponds either to a single polynomial $h_i(x)$, or to a product (mod p^j), of several of them. This correspondence is discovered in step 5.

5. Here we partition the set of polynomials $h_k(x)$ into subsets h_i, $i = 1, 2, \ldots, f$ such that

$$p_i(x) \equiv \Pi_{h(x) \in h_i} h(x) \pmod{p^j}$$

The true factors of $p_{\mathrm{sf}}(x)$ over the integers are then determined by trial divisions. [That is, having obtained the mod p^j factors $h_1(x), h_2(x), \ldots, h_r(x)$ of $p_{\mathrm{sf}}(x)$ in the previous step, we must now consider each combination of these factors, testing by trial division, whether its mod p^j product is a true factor; if a true factor has been found, set $p_{\mathrm{sf}}(x) := p_{\mathrm{sf}}(x)/p_i(x)$ and remove the corresponding $h_i(x)$ values from the list. We need consider only those combinations with degree $\le \lfloor \deg[p(x)]/2 \rfloor$.]

*Computing-Time Analysis of the **Roundabout Factorization Method***. We easily see from the above that, in the worst case, the roundabout factorization method has an exponential computing time, since r can be as large as n, and a large number of trial divisions will have to take place (namely, 2^n). However, as we show immediately below, on average $r = \ln(n)$, and, since the average value of 2^r is about n, the *average computing time* of the roundabout algorithm is polynomial. (The average value of 2^r is obtained with generating functions and is not discussed here.)

To see that $r = \ln(n)$, we consider the equivalent problem of the average number of cycles in an n-permutation (the same result can be also obtained through generating functions). This equivalence was established in 1880 by G. Frobenius (Knuth, 1981, p. 632).

Therefore, we have the following.

Theorem 6.1.1. The average number of j-cycles in an n-permutation is $1/j$ (*independent* of n).

Proof. Let c be any fixed j-cycle. Let b be any n-permutation in which the elements of c are left fixed by b. Let $f_c(b) = b \cdot c$; obviously, c is a j-cycle of $b \cdot c$.

The permutation $f_c(b)$ is a one-to-one mapping from the set of all $(n-j)$-permutations onto the set of all n-permutations having c as a cycle. Therefore, there are $(n-j)!$ n-permutations having c as a j-cycle. We can choose the j elements of the j-cycle c in $\binom{n}{j}$ ways; then we can make these j elements into a j-cycle in $(j-1)!$ ways. Hence, the number of j-cycles c is $\binom{n}{j}(j-1)!$.

Altogether, the number of occurances of j-cycles in all n-permutations is

$$(n-j)!\binom{n}{j}(j-1)! = \frac{n!}{j}$$

and dividing by $n!$, the total number of n-permutations, we obtain the result. □

Corollary 6.1.2. The average number of cycles in an n-permutation is

$$\Sigma_{1 \le j \le n} \frac{1}{j} = H_n$$

where H_n is the nth harmonic number, approximated by $\ln(n)$.

In Section 6.2 we examine the factorization of polynomials over a finite field and in Section 6.3, the lifting process. [Of interest are Butler, 1954; Calmet et al., 1980 and 1982; Camion, 1980; Lauer, 1982; Petr, 1936; and Rabin, 1980.]

6.2
Factorization of Polynomials over a Finite Field

As we saw in the last section, factorization of polynomials in $Z_p[x]$, p prime, is interesting not only for its own sake but also because it is useful for factoring in $Z[x]$. (The reader should review Section 3.3 at this point.) Any polynomial of degree n in $Z_p[x]$ can be factored in a finite number of steps, because there are p^n possible polynomials of degree $<n$, and we can simply check them all using **PDF**. However, this trial and error approach is very inefficient.

In this section we discuss in detail Berlekamp's efficient factorization algorithm in $Z_p[x]$, for *small* primes p. This method was discovered in 1967 (Berlekamp, 1967 and 1968) and transforms the factorization problem into that of solving a system of linear equations with coefficients in $Z_p[x]$, and finding greatest common divisors; the last two problems can be solved very efficiently, and so, this transformation is a very desirable one. (Berlekamp, 1970, also developed another algorithm for factoring polynomials in $Z_p[x]$, for *large* primes p; this topic is briefly discussed in Section 6.2.4.)

We begin with the presentation of related material.

6.2.1 Squarefree Factorization over Finite Fields

In this section we modify **PSQFF**, the algorithm that we developed in Section 3.2.4, so that we can decompose polynomials over finite fields.

As you recall, the proof of Theorem 3.2.17 depended on the fact that the derivative of a nonconstant polynomial cannot vanish identically when the characteristic of the coefficient domain J is zero. Now, let J have prime characteristic p, and consider the polynomial $p(x) = c_n x^n + \cdots + c_0$. Then, the derivative $p'(x) = n c_n x^{n-1} + \cdots + c_1$ is zero if and only if $i c_i = 0$ for each i, which is true if and only if $p | i$ or $c_i = 0$ for each i. Therefore, for some polynomial $p(x) \neq 0$, $p'(x) = 0$ *if and only if $p(x)$ is a polynomial in x^p*. We have the following generalization of Theorem 3.2.17.

Theorem 6.2.1. Let J be a unique factorization domain of arbitrary characteristic and $p(x)$ be a nonconstant primitive polynomial over J. Let $p(x) = [p_1(x)]^{e_1} [p_2(x)]^{e_2} \cdots [p_n(x)]^{e_n}$ be the unique factorization of $p(x)$ into irreducible factors, and let $\delta_i = 0$ if $e_i p_i'(x) = 0$, whereas otherwise $\delta_i = 1$. [The condition $e_i p_i'(x) = 0$ occurs if and only if $p_i'(x) = 0$ or the characteristic of J divides e_i.] Then

$$gcd[p(x), p'(x)] = [p_1(x)]^{(e_1 - \delta_1)} \cdots [p_n(x)]^{(e_n - \delta_n)}$$

Proof. The proof is a slight modification of the one of Theorem 3.2.17. □

Now, suppose that the characteristic of J is p, and let $p(x) = [p_1(x)]^{e_1} [p_2(x)]^{e_2} \cdots [p_n(x)]^{e_n}$ in $J[x]$. Let $e = \max(e_1, \ldots, e_n)$ and for $1 \leq i \leq e$ define

$$s_i(x) = \begin{cases} \Pi\{p_j(x): e_j = i \text{ and } p_j'(x) \neq 0\}, & \text{if } p \text{ does not divide } i \\ 1, & \text{otherwise} \end{cases}$$

and

$$s(x) = \Pi\{[p_j(x)]^{e_j}: p | e_j \text{ or } p_j'(x) = 0\}$$

Then, we have

$$p(x) = \{s_1(x)[s_2(x)]^2 \cdots [s_e(x)]^e\}s(x)$$

and by Theorem 6.2.1 we have $gcd[s_i(x), s_i'(x)] = 1$, where it should be observed that $s'(x) = 0$; moreover, it follows from Theorem 6.2.1 that

$$gcd[p(x), p'(x)] = \{s_2(x)[s_3(x)]^2 \cdots [s_e(x)]^{e-1}\}s(x)$$

and, therefore

$$\frac{p(x)}{gcd[p(x), p'(x)]} = s_1(x)s_2(x) \cdots s_e(x)$$

It is now easy to verify that algorithm **PSQFF** applied to $p(x)$ will yield $s_1(x), \ldots, s_e(x)$, and, provided $\deg[s(x)] = 0$, it will terminate with $r(x)$; we must also append "if $t(x) = 1$, then do $\{e := 0;$ exit$\}$." to step 1, to take care of the case $p'(x) = 0$. If $\deg[r(x)] = 0$, then we have a squarefree factorization of $p(x)$. But if $\deg[r(x)] > 0$, then further calculation is required to obtain the complete squarefree factorization. Although it is not apparent how this may be done with polynomials over an arbitrary domain J of characteristic $p > 0$ (without resorting to a complete factorization algorithm), we derive an algorithm for the special case when J is a domain of polynomials in one variable over a finite field. (The reader should review at this point Theorems 3.3.15–3.3.16.)

The following two theorems are needed for squarefree factorization in $Z_p[x]$ and in the sections to follow.

Theorem 6.2.2. Let $p(x)$ be a polynomial in $Z_p[x]$. Then, we have

$$[p(x)]^p = p(x^p)$$

Proof. We can prove the theorem as follows. If $p_1(x)$ and $p_2(x)$ are two polynomials modulo p, then

$$[p_1(x) + p_2(x)]^p = [p_1(x)]^p + \binom{p}{1}[p_1(x)]^{p-1}p_2(x)$$

$$+ \cdots + \binom{p}{p-1}p_1(x)[p_2(x)]^{p-1} + [p_2(x)]^p$$

$$= [p_1(x)]^p + [p_2(x)]^p$$

since the binomial coefficients $\binom{p}{1}, \ldots, \binom{p}{p-1}$ are all multiples of p. Furthermore, by Fermat's little theorem we have $c^p \equiv c \pmod{p}$, for any integer c. Therefore, if

$$p(x) = c_n x^n + c_{n-1} x^{n-1} + \cdots + c_0$$

we find that

$$[p(x)]^p = [c_n x^n]^p + [c_{n-1} x^{n-1}]^p + \cdots + [c_0]^p$$
$$= c_n x^{np} + c_{n-1} x^{(n-1)p} + \cdots + c_0 = p(x^p). \qquad \square$$

Theorem 6.2.3. Let $p(x)$ be a polynomial in $\mathbf{Z}_p[x]$. Then $p'(x) = 0$ if and only if $p(x)$ is the pth power of some polynomial $q(x)$ in $\mathbf{Z}_p[x]$.

Proof. If $p(x) = [q(x)]^p$, then $p'(x) = p[q(x)]^{p-1} q'(x) = 0$. Conversely, if $p'(x) = 0$, then $p(x)$ can be written in the form $p(x) = c_0 + c_p x^p + c_{2p} x^{2p} + \cdots + c_{kp} x^{kp}$. Let $q(x) = (c_0)^{1/p} + (c_p)^{1/p} x + (c_{2p})^{1/p} x^2 + \cdots + (c_{kp})^{1/p} x^k$; then $q(x)$ is in $\mathbf{Z}_p[x]$, and $p(x) = [q(x)]^p$. $\qquad \square$

We now have the following algorithm.

PSQFFF (**P**olynomial **S**quare**f**ree **f**actorization over a **f**inite **f**ield)

Input: $p(x)$, a nonconstant monic polynomial in $\mathbf{Z}_p[x]$, $p > 0$ is prime.
Output: $s_1(x), \ldots, s_e(x)$ and e such that $p(x) = \prod_{1 \le i \le e}[s_i(x)]^i$ is the squarefree factorization of $p(x)$.

1. [Initialize.] $k := 0$; $m := 1$; $e := 0$.
2. [Main loop.] $j := 1$; $r(x) := \gcd[p(x), p'(x)]$; $t(x) := p(x)/r(x)$; if $t(x) = 1$, then go to 7.
3. [Update.] $e' := jm$; if $e' > e$, then do $\{s_{e+1}(x) := s_{e+2}(x) := \cdots s_{e'-1}(x) := 1; e := e'\}$.
4. [Compute $s_{e'}(x)$.] $v(x) := \gcd[r(x), t(x)]$; $s_{e'}(x) := t(x)/v(x)$.
5. [Update.] If $v(x) \ne 1$, then do $\{r(x) := r(x)/v(x); t(x) := v(x); j := j + 1;$ go to 3$\}$.
6. [Finished?] If $r(x) = 1$, then exit.
7. [$r'(x) = 0$] $p(x) := [r(x)]^{1/p}$; $k := k + 1$; $m := mp$; go to 2.

The computing time of this algorithm has been analyzed in Section 3.2.4, where an example was also presented. Throughout the execution of **PSQFFF** we have $m = p^k$, and whenever we arrive at step 6, the value of e is the largest index i such that p^k does not divide i and $p(x)$ has a nonconstant factor of order i. We also assume that the polynomials $r(x)$ and $v(x)$ calculated in steps 2 and 4, respectively, are monic.

We arrive at step 7 only if $r'(x) = 0$; therefore, by Theorem 6.2.3 we know the form of $r(x)$ and we easily compute $[r(x)]^{1/p}$ making use of the identity $a^{1/p} = a^{p^{n-1}}$.

6.2.2 Counting Irreducible Polynomials over Finite Fields

From Theorem 3.3.20 we know that in $Z_p[x]$ there is an irreducible polynomial of degree n for each n; moreover, from Theorem 3.3.21 we know that $x^{p^n} - x$ is the product of all irreducible polynomials in $Z_p[x]$ whose degrees divide n. Theorem 3.3.21 will be used below in the distinct degree (partial) factorization; in this section we present a result regarding $i_p(n)$, the number of all monic, irreducible polynomials of degree n in $Z_p[x]$. Observe that according to Theorem 3.3.21 $x^{p^n} - x$, a polynomial of degree p^n, has as factors $i_p(d)$ irreducible polynomials of degree d for each d dividing n; this implies that the degree of the product of all monic, irreducible polynomials over $Z_p[x]$ whose degree divides n is equal to $\Sigma_{d|n} i_p(d) d = p^n$.

To obtain an expression for $i_p(n)$, we need some additional results, and we begin with the following (Bender et al., 1975; Berlekamp, 1968; and Childs, 1979).

Definition 6.2.4. The *Moebius function* $\mu(n)$ is defined for $n \geq 1$ as follows:

$$\mu(n) = \begin{cases} 1, & \text{if } n = 1 \\ 0, & \text{if } p^e | n \text{ for some prime } p \text{ and some } e > 1 \\ (-1)^r, & \text{if } n \text{ is the product of } r \text{ distinct primes} \end{cases}$$

Observe that $\mu(n) = 0$ if n is divisible by the square of a prime, and that $\mu(p) = -1$, for any prime p. Also, one can easily verify that if $(m, n) = 1$, then $\mu(mn) = \mu(m)\mu(n)$; that is, μ is a *multiplicative* function. (Euler's ϕ function is another example of a multiplicative function.)

We have the following theorem concerning multiplicative functions.

Theorem 6.2.5. If f is a multiplicative function and F is the function defined by $F(n) = \Sigma_{d|n} f(d)$, then F is also multiplicative.

Proof. Let $(m, n) = 1$. Then, every divisor d of mn can be uniquely written as $d = d_1 d_2$, where $d_1 | m$, $d_2 | n$, and $(d_1, d_2) = 1$. Therefore, we have

$$F(mn) = \Sigma_{d|mn} f(d) = \Sigma_{\{d_1|m, d_2|n\}} f(d_1 d_2) = \Sigma_{\{d_1|m, d_2|n\}} f(d_1) f(d_2)$$
$$= \Sigma_{d_1|m} f(d_1) \Sigma_{d_2|n} f(d_2) = F(m) F(n). \qquad \square$$

The converse of the above proposition also holds, but we do not use it for our purposes. We next present a theorem concerning μ.

Theorem 6.2.6. $\Sigma_{d|n}\mu(d) = 0$ unless $n = 1$.

Proof. By Theorem 6.2.5 we know that the function

$$M(n) = \Sigma_{d|n}\mu(d)$$

is multiplicative, and since $M(p^e) = 1$ if $e = 0$, whereas $M(p^e) = 1 - 1 + 0 + \cdots + 0$ if $e \geq 1$, we see that $M(n) = 0$ if n is divisible by any prime; that is, $M(n) = 0$ if $n > 1$. □

Theorem 6.2.7 (Moebius Inversion Formula). For any function f defined on natural numbers (not necessarily multiplicative), if we set

$$F(n) = \Sigma_{d|n} f(d) \quad \text{for every} \quad n \geq 1$$

then

$$f(n) = \Sigma_{d|n}\mu\left(\frac{n}{d}\right)F(d) = \Sigma_{e|n}\mu(e)F\left(\frac{n}{e}\right)$$

Proof. Letting $e = n/d$, $d = n/e$, the equality of the two sums above is obvious. Then, by the definition of F we have

$$\Sigma_{e|n}\mu(e)F(n/e) = \Sigma_{e|n}\mu(e)[\Sigma_{d|(n/e)}f(d)] = \Sigma_{e|n}[\Sigma_{d|(n/e)}\mu(e)f(d)]$$

Interchanging the order of summation, and noting that if $d|(n/e)$, then $de|n$ and so $e|(n/d)$, we obtain

$$= \Sigma_{d|n}[\Sigma_{e|(n/d)}\mu(e)f(d)] = \Sigma_{d|n}[\Sigma_{e|(n/d)}\mu(e)]f(d)$$

By Theorem 6.2.6 we have that, in the last sum, the coefficient of $f(d)$ is 0 unless $n/d = 1$, or $d = n$. Therefore, the last sum reduces to a single term $f(n)$, and we have proved the theorem. □

We are now in a position to obtain the expression for $i_p(n)$.

Theorem 6.2.8. The number of all monic, irreducible polynomials of degree n over $\mathbf{Z}_p[x]$ is given by

$$i_p(n) = \left(\frac{1}{n}\right)\Sigma_{d|n}\mu\left(\frac{n}{d}\right)p^d$$

Proof. Theorem 3.3.21 implies that the degree of the product of all monic, irreducible polynomials over $\mathbf{Z}_p[x]$ whose degree divides n is equal to

$$\Sigma_{d|n}i_p(d)d = p^n$$

Applying Theorem 6.2.7 with $F(n) = p^n$ and $f(d) = i_p(d)d$, we obtain the desired result. □

It is interesting to know how rapidly $i_p(n)$ grows; the Table 6.2.1 below, for $p = 7$, is from Simmons (1970).

TABLE 6.2.1 Growth of $i_p(n)$

n	$i_p(n)$	Total Number of Monic Polynomials of Degree $n(=7^n)$
1	7	7
2	21	49
3	112	343
4	588	2401
5	3,360	16,807
6	19,544	117,649
7	117,648	823,543

Irreducibility Tests in $Z_p[x]$ Below we present two tests for determining whether a monic polynomial $p(x)$ of degree $n > 1$ is irreducible in $Z_p[x]$; a third test will be presented in Section 6.2.4. None of the tests require $p(x)$ to be squarefree, but it appears that, on average, it saves time to apply the squarefree factorization algorithm first.

Test 1. The polynomial $p(x)$ of degree $n > 1$ is irreducible in $Z_p[x]$ if and only if

$$\gcd[p(x), x^{p^i} - x] = 1, \quad \text{for} \quad i = 1, 2, \ldots, \left\lfloor \frac{n}{2} \right\rfloor.$$

This test follows directly from Theorem 3.3.21. If the polynomial $p(x)$ is reducible, then the test above is quite fast; for irreducible polynomials, though, it becomes rather inefficient because of the large number of *gcd* computations in $Z_p[x]$. We can improve on the number of *gcd* computations if we use the following test.

Test 2. The polynomial $p(x)$ of degree $n > 1$ is irreducible in $Z_p[x]$ if and only if (a) $p(x)|(x^{p^n} - x)$ and (b) $\gcd[p(x), x^{p^{n_i}} - x] = 1$, for all $n_i = n/k_i$, where k_i are the prime divisors of n.

Proof. Observe that if $p(x)$ is irreducible, then every root ξ of $p(x) = 0$ lies in $GF(p^n)$, and since $\xi^{p^n} - \xi = 0$, we have $(x - \xi)|(x^{p^n} - x)$, and condition a follows immediately. Moreover, since $p(x)$ is irreducible of degree n, it has no roots in any field $GF(p^{n_i})$, $n_i < n$, and this directly implies condition b.

To prove the converse, assume that conditions a and b hold. From condition a it follows that all the roots of $p(x)$ are in $GF(p^n)$. Suppose now that $p(x)$ has an irreducible factor $p_1(x)$ of degree $n' < n$. Then, the roots of $p_1(x)$ are all in $GF(p^{n'})$, which is generated over $\mathbf{Z}_p$ by any one of those roots. Therefore, $GF(p^{n'}) \subset GF(p^n)$ and $n'|n$. Therefore, $n'|n_i$ for some maximal divisor n_i of n, and all roots of $p_1(x)$ are in $GF(p^{n_i})$. In this case, though, $p_1(x) | \gcd[p(x), x^{p^{n_i}} - x]$, contradicting condition b. Hence, $p(x)$ is irreducible. $\square$

6.2.3 Distinct Degree (Partial) Factorization over Finite Fields

In this section we use Theorem 3.3.21 to partially factor a squarefree polynomial $p(x)$; namely, we use the fact that an irreducible polynomial $p(x)$ of degree d divides $x^{p^d} - x$ but does not divide $x^{p^c} - x$ for $c < d$. We can, therefore, cast out the irreducible factors of each degree seperately, by incorporating the relation

$$p_i(x) = \gcd\left[x^{p^i} - x, \frac{p(x)}{\prod_{1 \le j \le i-1} p_j(x)}\right], \quad \text{for} \quad i = 1, 2, \ldots, n$$

into an algorithm, making use of Theorem 6.2.2 as well.

DDF (**D**istinct **d**egree **f**actorization)

> *Input*: $p(x)$, a squarefree polynomial of degree n over $\mathbf{Z}_p[x]$.
> *Output*: Polynomials $p_d(x)$ over $\mathbf{Z}_p[x]$, $d = 1, 2, \ldots, \lfloor n/2 \rfloor$, such that $p(x) = \prod_{1 \le d \le \lfloor n/2 \rfloor} p_d(x)$; $p_d(x)$ is the product of all monic irreducible factors of degree d of $p(x)$.

1. [Initialize.] $q(x) := p(x)$; $r(x) := x$; $d := 0$.
2. [Finished?] {Note that at this point $r(x) = x^{p^d}$ [mod $q(x)$]; all the irreducible factors of $q(x)$ are distinct and have degree $> d$.} If $d + 1 > (1/2)\deg[q(x)]$, then exit [the procedure terminates since we either have $q(x) = 1$ or $q(x)$ is irreducible], or else do $\{d := d + 1;$ $r(x) := [r(x)]^p$ [mod $q(x)$]$\}$.
3. [Compute $p_d(x)$.] $p_d(x) := \gcd[r(x) - x, q(x)]$. [This is the product of all irreducible factors of $p(x)$ whose degree is d]. If $p_d(x) \ne 1$, then do $\{q(x) := q(x)/p_d(x);\ r(x) := r(x)$ [mod $q(x)$]$\}$; go to 2. [Below we discuss how to obtain the factors if the degree of $p_d(x)$ is greater than d.]

Example. Applying **DDF** to $x^{15} - 1$ in $\mathbf{Z}_2[x]$ (this polynomial was factored in Section 3.3.1) we obtain the following.

In the first pass, in step 1 we have $q(x) = x^{15} - 1$, $r(x) = x$ and $d = 0$; since $d < 7$, in step 2 we have $d = 1$ and $r(x) = x^2$. In step 3 we obtain $gcd[q(x), r(x) - x] = x - 1 = p_1(x)$, and we update $q(x)$ and $r(x)$ to $x^{14} + x^{13} + \cdots + x + 1$ and x^2, respectively.

In the second pass, in step 2 we have $d = 2$ and $r(x) = x^4$; in step 3 we then have $gcd[q(x), r(x) - x] = x^2 + x + 1 = p_2(x)$, and we update $q(x)$ and $r(x)$ to $x^{12} + x^9 + x^6 + x^3 + 1$ and x^4, respectively.

Subsequent passes through steps 2 and 3 yield no new factors and, hence, we obtain three factors altogether; namely, $x - 1$, $x^2 + x + 1$, and $x^{12} + x^9 + x^6 + x^3 + 1$, a 12-degree factor that is the product of three polynomials of degree 4. (Note that in $\mathbf{Z}_2[x]$, $+1 = -1$. See also historical note 1.)

As we see from the algorithm above, we compute x^p [mod $p(x)$], and x^{p^i} [mod $p(x)$] = $\{x^{p^{i-1}}$ [mod $p(x)$]$\}^p$ [mod $p(x)$], for $i = 2, \ldots, \lfloor n/2 \rfloor$. Therefore, the computation of the distinct degree (partial) factorization can be performed in $0[\lfloor n/2 \rfloor L(p)]$ polynomial multiplications modulo $p(x)$. We can improve on the number of polynomial multiplications modulo $p(x)$ as follows (Lenstra, 1982b).

Let the $n \times n$ matrix $\mathbf{Q}$ have as its ith row the coefficients of x^{ip} [mod $p(x)$], for $i = 0, 1, \ldots, n - 1$; below we identify polynomials of degree $< n$ with the row vectors formed by their coefficients. Then the following holds.

Theorem 6.2.9. For any polynomial $v(x) = \Sigma_{0 \le i \le n-1} v_i x^i$ in $\mathbf{Z}_p[x]$, considered as the vector $\mathbf{v}$ of its coefficients, we have

$$\mathbf{v} \cdot \mathbf{Q} = \mathbf{v}^p \ [\text{mod } p(x)]$$

Proof. Let $\mathbf{Q} = (r_{ij})$, where both $i, j = 0, 1, \ldots, n - 1$. Then

$$\{v(x)\}^p \ [\text{mod } p(x)] = v(x^p) \ [\text{mod } p(x)] = \Sigma_{0 \le i \le n-1} v_i x^{ip} \ [\text{mod } p(x)]$$
$$= \Sigma_{0 \le i \le n-1} v_i \Sigma_{0 \le k \le n-1} r_{ik} x^k \ [\text{mod } p(x)]$$
$$= \Sigma_{0 \le k \le n-1} (\Sigma_{0 \le i \le n-1} v_i r_{ik}) x^k \ [\text{mod } p(x)]$$
$$= \mathbf{v} \cdot \mathbf{Q} \qquad \square$$

From Theorem 6.2.9 we see that $\{x^{p^{i-1}}$ [mod $p(x)$]$\}\mathbf{Q} = x^{p^i}$ [mod $p(x)$], so that the $\lfloor n/2 \rfloor$ polynomials $(x^{p^i} - x)$ [mod $p(x)$], for $i = 1, \ldots, \lfloor n/2 \rfloor$ can be computed in $0(n^3)$ finite field computations, once the matrix $\mathbf{Q}$ has been calculated; the latter calculation can be performed in $0[L(p) + n - 2]$ polynomial multiplications modulo $p(x)$. Clearly, for large p the use of the matrix $\mathbf{Q}$ is preferable.

An alternative way to find the distinct degree (partial) factorization of $p(x)$ is presented in Berlekamp (1970). This method makes use of the kernel of the matrix $\mathbf{Q} - \mathbf{I}$, where $\mathbf{I}$ is the $n \times n$ identity matrix, and can be expanded to determine the complete factorization in $\mathbf{Z}_p[x]$. A discussion of this method is beyond the scope of this book.

We have seen that **DDF** determines the product of all irreducible factors of each degree d, and therefore, it tells us how many factors there are of each degree. If we want to know *only the degrees*, we can proceed as follows [$p(x)$ does not have to be squarefree now].

Define σ_i to be the number of distinct irreducible factors of $p(x)$ of degree i, and ν_i to be the the rank of the kernel of $\mathbf{Q}^i - \mathbf{I}$, for $i = 1, \ldots, n$. Let $\mathbf{A} = (a_{ij})$ be the $n \times n$ matrix where $a_{ij} = \gcd(i, j)$, $\sigma = (\sigma_1, \ldots, \sigma_n)$, $\nu = (\nu_1, \ldots, \nu_n)$. An interesting result by Smith, 1876 states that

$$\det(\mathbf{A}) = \Pi_{0 \leq i \leq n} \phi(i)$$

where ϕ is Euler's totient function, whereas in 1956 it was shown by Schwarz [see also Schwarz, 1939]

$$\mathbf{A} \cdot \sigma^T = \nu^T$$

Therefore, $\mathbf{A}$ is invertible and σ is uniquely determined by ν. So, for example, if for a given polynomial of degree 8 we obtain $\sigma = (2, 0, 2, 0, 0, 0, 0, 0)$, we know that it has two factors of degree 1 and two factors of degree 3. Additional details can be found in Gunji and Arnon (1981).

Splitting Distinct Degree Factors So far we have reduced the factoring problem to that of separating products of the same degree. To obtain a complete factorization method in $\mathbf{Z}_p[x]$, we need a way to split $p_d(x)$ into its irreducible factors when $\deg[p_d(x)] > d$ (Sims, 1984). There are several ways for doing this; we present a simple probabilistic procedure suggested by Cantor and Zassenhaus (1981), which is based on the following identity. If p is any *odd* prime, we have

$$Pd(x) = \gcd[p_d(x), q(x)] \cdot \gcd\{p_d(x), [q(x)]^{(p^d-1)/2} + 1\} \\ \cdot \gcd\{p_d(x), [q(x)]^{(p^d-1)/2} - 1\}$$

for all polynomials $q(x)$, since $[q(x)]^{p^d} - q(x)$ is a multiple of all irreducible polynomials of degree d (see Lemma 6.2.10). The above identity is based on the polynomial factorization that holds for *odd* p

$$x^{p^d} - x = x[x^{(p^d-1)/2} + 1][x^{(p^d-1)/2} - 1]$$

Lemma 6.2.10. Let $p(x) = p_1(x)p_2(x) \cdots p_r(x)$, $r > 1$, be the product of two or more distinct irreducible polynomials of the same degree d, modulo an *odd* prime p. Then, for any polynomial $q(x)$, randomly selected from among the p^{rd} polynomials of degree $<rd$ modulo p, we have that $\gcd\{p(x), [q(x)]^{(p^d-1)/2} - 1\}$ will be a proper factor of $p(x)$ with probability

$$1 - \left(\frac{1}{2^r}\right)\left[\left(1 + \frac{1}{p^d}\right)^r + \left(1 - \frac{1}{p^d}\right)^r\right]$$

which is at least $4/9$.

Proof. Every element of $GF(p^d)$ is a root of

$$x^{p^d} - x = x[x^{p^d-1} - 1] = x[x^{(p^d-1)/2} + 1]\{x^{(p^d-1)/2} - 1\}$$

Therefore, for $(p^d - 1)/2$ elements of $GF(p^d)$ we have $x^{(p^d-1)/2} = 1$, for $(p^d - 1)/2$ elements we have $x^{(p^d-1)/2} = -1$, and for one element, $x = 0$, we have $x^{(p^d-1)/2} = 0$.

Now, from the Greek–Chinese remainder theorem for polynomials (see Theorem 6.2.12), we know that $\mathbf{Z}_p[x]/p(x)$ is isomorphic to the direct sum X of r copies of $GF(p^d)$. If $x = (x_1, \ldots, x_r)$ is an element of X, then

$$x^{(p^d-1)/2} = [x_1^{(p^d-1)/2}, \ldots, x_r^{(p^d-1)/2}]$$

has only $0, 1, -1$ as components. The number of x such that $x^{(p^d-1)/2}$ has no component equal to 1 is

$$\left(\frac{p^d - 1}{2} + 1\right)^r$$

and so the number of x such that $x^{(p^d-1)/2}$ has at least one component equal to 1 is

$$p^{rd} - \left(\frac{p^d - 1}{2} + 1\right)^r$$

Of these, the number x such that $x^{(p^d-1)/2}$ has all components 1 is

$$\left(\frac{p^d - 1}{2}\right)^r$$

Hence, the number of x such that $x^{(p^d-1)/2}$ has at least one component equal to 1 and at least one component different from 1 is

$$p^{rd} - \left(\frac{p^d - 1}{2} + 1\right)^r - \left(\frac{p^d - 1}{2}\right)^r$$

For any such x, the element $x - 1$ is a nonzero invertible element of X. If $q(x)$ in $\mathbf{Z}_p[x]$ corresponds to one of these x values, then $[q(x)]^{(p^d-1)/2} - 1$ is neither relatively prime to $p(x)$ nor divisible by $p(x)$, and hence

$$\gcd\{p(x), [q(x)]^{(p^d-1)/2} - 1\}$$

is a proper factor of $p(x)$. This occurs with probability

$$\frac{p^{rd} - [(p^d - 1)/2 + 1]^r - [(p^d - 1)/2]^r}{p^{rd}}$$

$$= 1 - \left[\frac{(p^d - 1)/2 + 1)}{p^d}\right]^r - \left(\frac{p^d - 1}{2p^d}\right)^r$$

$$= 1 - \left(\frac{p^d + 1}{2p^d}\right)^r - \left(\frac{p^d - 1}{2p^d}\right)^r = 1 - \left(\frac{1}{2^r}\right)\left[\left(1 + \frac{1}{p^d}\right)^r + \left(1 - \frac{1}{p^d}\right)^r\right].$$

Since $1 - 1/p^d \le 1$ and $1 + 1/p^d \le 4/3$ (since $p^d \ge 3$), we have that this probability is at least

$$1 - \left(\frac{1}{2^r}\right)\left[\left(\frac{4}{3}\right)^r + 1\right] = 1 - \left(\frac{2}{3}\right)^r - \frac{1}{2^r}$$

Clearly, for $r = 2$, we obtain

$$1 - \left(\frac{1}{4}\right)\left[2 + \frac{2}{p^{2d}}\right] \ge 1 - \left(\frac{1}{4}\right)\left[2 + \frac{2}{9}\right] = 4/9. \qquad \square$$

To compute $gcd\{p(x), [q(x)]^{(p^d-1)/2} - 1\}$, we first raise $q(x)$ to the power $(p^d - 1)/2$ modulo $p(x)$, using the binary power algorithm **E** discussed in Section 2.3.2, and then subtract 1.

Lemma 6.2.10 is based on the factorization $x^{p^d} - x = x[x^{(p^d-1)/2} + 1][x^{(p^d-1)/2} - 1]$ which is valid for odd primes. For $p = 2$, there is another factorization of $x^{p^d} - x$ that we can use to obtain a result similar to that of Lemma 6.2.10.

In $GF(p^d)$ the *trace* polynomial, $tr(x)$, is defined by $tr(x) = x + x^p + \cdots + x^{p^{d-1}}$. Then, in $GF(2^d)$ we have the factorization $x^{2^d} - x = tr(x)[tr(x) + 1]$. This is easily seen if we use the fact that $1 = -1$ in $\mathbf{Z}_2$ and Theorem 3.3.16, according to which squaring is a linear transformation; namely, we have $tr(x)[tr(x) + 1] = tr(x) + [tr(x)]^2 = x + x^2 + \cdots + x^{2^{d-1}} + x^2 + x^4 + \cdots + x^{2^d} = x + x^{2^d} = x^{2^d} - x$.

Also observe that if $p(x)$ is an irreducible polynomial of degree d, in $\mathbf{Z}_p[x]$, and $q(x)$ is any polynomial, then the value of $\{q(x) + [q(x)]^p + [q(x)]^{p^2} + \cdots + [q(x)]^{p^{d-1}}\}$ [mod $p(x)$] is an integer (i.e., a polynomial of degree ≤ 0). To see this, define $t(x) = tr[q(x)]$ [mod $p(x)$]. Since $[q(x)]^{p^d} = q(x)$ in the field of polynomial remainders modulo $p(x)$, we have $[t(x)]^p = t(x)$ in that field. Therefore, $t(x)$ is one of the p roots of the equation $x^p - x = 0$, and hence, $t(x)$ is an integer.

Lemma 6.2.11. Let $p(x) = p_1(x)p_2(x)\cdots p_r(x)$, $r > 1$, be the product of two or more distinct irreducible polynomials of the same degree d, modulo 2. Then, for any polynomial $q(x)$, randomly selected from among the 2rd

polynomials of degree $<rd$ modulo 2, we have that $gcd\{p(x), tr[q(x)]\}$ will be a proper factor of $p(x)$ with probability $1 - 1/2^{r-1}$.

Proof. The proof is analogous to the one of Lemma 6.2.10 and is left as an exercise for the reader. □

Example. Let us split $p(x) = x^{12} + x^9 + x^6 + x^3 + 1$, the 12-degree polynomial obtained in the example following **DDF** and that is the product of three polynomials of degree 4. Following Lemma 6.2.11, we randomly choose a polynomial of degree 4, $q(x) = x^4 + x^3 + x^2 + x + 1$, compute its trace, $tr[q(x)] = q(x) + [q(x)]^2 + [q(x)]^4 + [q(x)]^8$, and then $gcd\{p(x), tr[q(x)]\} = x^4 + x^3 + x^2 + x + 1$, which is a factor of $p(x)$. (Luck!) This now leaves us with the polynomial $p(x)/gcd\{p(x), tr[q(x)]\} = x^8 - x^7 + x^5 - x^4 + x^3 - x + 1$ to be split into two factors of degree 4; we leave it to the reader to complete the example (and in so doing to try their luck).

Observe that, for *any p*, we have

$$x^{p^d} - x = \Pi_{0 \le s < p}[tr(x) - s]$$

and Lemma 6.2.11 can be used to split a distinct degree factor of $p(x)$, just as Lemma 6.2.10; that is, we now use the relation $\Pi_{0 \le s < p} gcd\{p_d(x), tr[q(x)] - s\} = p_d(x)$.

From Lemmas 6.2.10 and 6.2.11 we conclude that it will not take long to discover all the factors of $p(x)$. However, in the next section we present a deterministic method for doing this.

6.2.4 Berlekamp's Factorization Algorithm over Finite Fields

In this section we consider the factorization of a monic, squarefree polynomial $p(x)$ in $\mathbf{Z}_p[x]$; that is, given $p(x)$ with coefficients in the set $\{0, 1, 2, \ldots, p - 1\}$, we want to discover its irreducible factors $p_i(x)$, $i = 1, 2, \ldots, r$, such that

$$p(x) = p_1(x) p_2(x) \cdots p_r(x)$$

Berlekamp's idea is to make use of the Greek–Chinese remainder theorem, which, as we have already seen in Section 3.1.3, is valid for polynomials just as it is valid for integers.

Theorem 6.2.12 (Greek–Chinese Remainder Theorem for Polynomials). Let $p_1(x), \ldots, p_r(x)$ be polynomials in $\mathbf{Z}_p[x]$, with $p_j(x)$ relatively prime to $p_k(x)$ for all $j \ne k$ [equivalently, we can take the $p_i(x)$ to be distinct, irreducible polynomials over this field] and let $s_1(x), \ldots, s_r(x)$ be *arbitrary* polynomials in $\mathbf{Z}_p[x]$. Then, there is a unique polynomial $t(x)$ in $\mathbf{Z}_p[x]$ such that

$$\deg[t(x)] < \deg[p_1(x)] + \cdots + \deg[p_r(x)]$$

[i.e., $t(x)$ is defined modulo $p_1(x)p_2(x)\cdots p_r(x)$] and

$$t(x) \equiv s_i(x) \ [\mathrm{mod}\ p_i(x)], \qquad 1 \leq i \leq r.$$

Stated in other words, the mapping that associates with each polynomial $t(x)$ in $\mathbf{Z}_p[x]/p_1(x)\cdots p_r(x)$, the r-tuple $[s_1(x),\ldots,s_r(x)]$, where $t(x) \equiv s_i(x)$ [mod $p_i(x)$], $1 \leq i \leq r$, is a bijection between $\mathbf{Z}_p[x]/p_1(x)\cdots p_r(x)$ and $\mathbf{Z}_p[x]/p_1(x) + \mathbf{Z}_p[x]/p_2(x) + \cdots + \mathbf{Z}_p[x]/p_r(x)$.

Proof. We use the extended Euclidean algorithm to determine polynomials $m_i(x)$ such that $m_i(x)\Pi_{j \neq i}p_j(x) \equiv 1$ [mod $p_i(x)$]. Setting $t(x) := \Sigma_i[s_i(x)m_i(x) \Pi_{j \neq i}p_j(x)]$, we have $t(x) \equiv s_i(x)$ [mod $p_i(x)$], $1 \leq i \leq r$. If $t_1(x)$ were also a solution of the system of congruences, then $t_1(x) - t(x)$ would be divisible by each $p_i(x)$ and, therefore, $t_1(x) \equiv t(x)$ [mod $\Pi_i p_i(x)$]. □

The notation $t(x) \equiv s_i(x)$ [mod $p_i(x)$], used in Theorem 6.2.12, is equivalent to "$t(x) \equiv s_i(x)$ [mod $p_i(x)$ and p]," since we are considering polynomial arithmetic modulo p.

Theorem 6.2.12 tells us that if we consider $(s_1,\ldots,s_r)$, an arbitrary r-tuple of integers modulo p, then there is a unique polynomial $t(x)$ such that

$$t(x) \equiv s_1 \ [\mathrm{mod}\ p_1(x)], \ldots, t(x) \equiv s_r \ [\mathrm{mod}\ p_r(x)],$$
$$\deg[t(x)] < \deg[p_1(x)] + \cdots + \deg[p_r(x)] = \deg[p(x)] \qquad (B1)$$

where, as we indicated $p(x) = p_1(x)p_2(x)\cdots p_r(x)$. Note that the polynomial $t(x)$ defined in (B1) gives a way to obtain information about the factors of $p(x)$, because if $r \geq 2$ and $s_1 \neq s_2$, we will have $\gcd[p(x), t(x) - s_1]$ divisible by $p_1(x)$ but not by $p_2(x)$. Let us analyze (B1) more closely. Observe that, by Theorem 6.2.2, the polynomial $t(x)$ satisfies the condition $[t(x)]^p \equiv s_j^p = s_j \equiv t(x)$ [mod $p_j(x)$], for $1 \leq j \leq r$, and therefore

$$[t(x)]^p \equiv t(x) \ [\mathrm{mod}\ p(x)], \quad \deg[t(x)] < \deg[p(x)] \qquad (B2)$$

Moreover, we have the following results.

Theorem 6.2.13. In a finite field $GF(p)$ we have

$$x^p - x = \Pi_{s \in GF(p)}(x - s) \qquad (B3)$$

Proof. For any $s \in GF(p)$ we have by Fermat's little theorem $s^p = s$, [equality since we are in $GF(p)$], which means that s is a root of $x^p - x$, or equivalently, that $(x - s)$ is a factor of $x^p - x$. Since this is true for all $s \in GF(p)$, we have that $\Pi_{s \in GF(p)}(x - s)$ is a factor of $x^p - x$. However, $\deg[\Pi_{s \in GF(p)}(x - s)] = p = \deg(x^p - x)$, and since they are both monic, we have

$$x^p - x = \Pi_{s \in GF(p)}(x - s) . \qquad \square$$

Corollary 6.2.14. For any polynomial $t(x)$ over $GF(p)$ we have

$$[t(x)]^p - t(x) = \Pi_{s \in GF(p)}[t(x) - s] \qquad (B4)$$

If $t(x)$ satisfies (B2), it follows that $p(x)$ divides the left-hand side of (B4), so that every irreducible factor of $p(x)$ must divide one of the p relatively prime factors of the right-hand side of (B4). Therefore, all solutions of (B2) must have the form (B1), for some $s_1, \ldots, s_r$; since we have the choice of p^r elements s_i, there are exactly p^r solutions of (B2). [At first it might seem that it is hard to find all the solutions to (B2), but this is not the case since the set of solutions to (B2) is closed under addition.] Once we have found all the solutions to (B2) we can factor $p(x)$ in $\mathbf{Z}_p[x]$, as the following theorem indicates.

Theorem 6.2.15. Let $p(x)$ and $t(x)$ be two monic polynomials over $GF(p)$ such that

$$[t(x)]^p \equiv t(x) \,[\bmod\, p(x)], \qquad \deg[t(x)] < \deg[p(x)]$$

Then

$$p(x) = \Pi_{s \in GF(p)} \gcd[p(x), t(x) - s] \qquad (B5)$$

Proof. By assumption

$$[t(x)]^p \equiv t(x) \,[\bmod\, p(x)]$$

and hence, $p(x)$ divides $[t(x)]^p - t(x)$. Therefore, we have $p(x) = \gcd\{p(x), [t(x)]^p - t(x)\} = \gcd\{p(x), \Pi_{s \in GF(p)}[t(x) - s]\}$ by Corollary 6.2.14. Moreover, for $s \neq t$, $\gcd[t(x) - s, t(x) - t] = 1$, and $\gcd[p(x), t(x) - s]$ and $\gcd[p(x), t(x) - t]$ are also relatively prime; therefore, $\gcd\{p(x), \Pi_{s \in GF(p)} \cdot [t(x) - s]\} = \Pi_{s \in GF(p)} \gcd[p(x), t(x) - s]$ (see also Exercise 1 for this section). $\square$

Each factor on the right-hand side of (B5) has degree at most that of $t(x)$, which, in turn, is $< \deg[p(x)]$. Thus, there must be at least two nontrivial factors of $p(x)$ on the right-hand side of (B5), and (B5) constitutes a *nontrivial factorization of p(x)*. Moreover, if $t(x)$ is a scalar [i.e., $\deg[t(x)] = 0$], then

$$p(x) = gcd[p(x), 0] \cdot \Pi_{s \neq 0, s \in GF(p)} gcd[p(x), s] = p(x) \cdot \Pi_{s \neq 0} 1$$

Therefore, (B5) is a complete factorization formula, and it is the basic formula used in Berlekamp's method.

To summarize what we have said so far. Let $p(x)$ be the squarefree polynomial of degree n that we want to factor over $GF(p)$ and suppose that, somehow, we can find polynomials $t(x)$ in $\mathbf{Z}_p[x]$, $1 \leq \deg[t(x)] < n$ such that $[t(x)]^p \equiv t(x) \pmod{p(x)}$. [Note that if the degree of $t(x)$ is ≥ 1, we have $[t(x)]^p - t(x) \neq 0$, since the coefficient of the highest power of x is nonzero.] By Theorem 6.2.13 we know that $x^p - x$ has p roots in $GF(p)$, namely, $x = 0, 1, 2, \ldots, p-1$; hence, $x^p - x$ factors modulo p as $x^p - x = x(x-1) \cdot (x-2) \cdots (x-p+1)$. Replacing x by $t(x)$ we have

$$[t(x)]^p - t(x) = t(x)[t(x) - 1][t(x) - 2] \cdots [t(x) - p + 1]$$

in $\mathbf{Z}_p[x]$. Since $p(x)$ divides $[t(x)]^p - t(x)$, we have that

$$p(x) = gcd\{p(x), [t(x)]^p - t(x)\}$$

Moreover, since $t(x) - s$ and $t(x) - t$ are relatively prime for $s \neq t$ we have (by Corollary 6.2.14, Theorem 6.2.15, and Exercise 1)

$$p(x) = gcd\{p(x), [t(x)]^p - t(x)\} = \Pi_{s \in GF(p)} gcd[p(x), t(x) - s]$$

which is a nontrivial factorization of $p(x)$ over $GF(p)$.

Therefore, the main problem is to determine the polynomials $t(x)$. This is done by solving a *system of linear equations*, and the matrix $\mathbf{Q}$ that we introduced in Section 6.2.3 turns out to be a very useful tool; in particular, the null space (defined below) of the matrix $\mathbf{Q} - \mathbf{I}$, where $\mathbf{I}$ is the $n \times n$ identity matrix, will be of vital importance.

The system of linear equations, needed to determine the polynomials $t(x)$, such that $p(x)$ divides $[t(x)]^p - t(x)$, is obtained as follows. Let

$$t(x) = t_0 + t_1 x + \cdots + t_{n-1} x^{n-1}$$

where t_i, $i = 0, \ldots, n-1$ are the coefficients to be determined. To see if $p(x)$ divides $[t(x)]^p - t(x)$, we first look at $[t(x)]^p$; by Theorem 6.2.2 we have

$$[t(x)]^p = t_0 + t_1 x^p + t_2 x^{2p} + \cdots + t_{n-1} x^{(n-1)p} \tag{B6}$$

Dividing $p(x)$ into x^{ip}, we obtain

$$x^{ip} = p(x) q_i(x) + r_i(x), \quad i = 0, 1, 2, \ldots, n-1 \tag{B7}$$

where $r_i(x) = r_{i,0} + r_{i,1}x + \cdots + r_{i,n-1}x^{n-1}$. [Note that the polynomials $q_i(x)$ and $r_i(x)$ can be easily computed since we know $p(x)$.]

Now, replacing the x^{ip} in (B6) by their corresponding expressions of (B7), we obtain

$$[t(x)]^p = t_0 r_0(x) + t_1 r_1(x) + \cdots + t_{n-1} r_{n-1}(x) + [\text{multiple of } p(x)]$$

Thus, $p(x)$ divides $[t(x)]^p - t(x)$ if and only if $p(x)$ divides the polynomial

$$t_0 r_0(x) + t_1 r_1(x) + \cdots + t_{n-1} r_{n-1}(x) - (t_0 + t_1 x + \cdots + t_{n-1} x^{n-1})$$
$$= t_0[r_0(x) - 1] + t_1[r_1(x) - x] + \cdots + t_{n-1}[r_{n-1}(x) - x^{n-1}] \quad (B8)$$

which is of degree $\leq n - 1$. Therefore, the polynomial $p(x)$, of degree n, will divide (B8), if and only if the latter is equal to 0.

Setting (B8) equal to zero and collecting the coefficients of 1, x, $x^2, \ldots, x^{n-1}$, we obtain a system of n simultaneous linear equations in n unknowns $t_0, t_1, \ldots, t_{n-1}$; these unknowns are the coefficients of a polynomial $t(x)$ such that $p(x)$ divides $[t(x)]^p - t(x)$.

Let

$$\mathbf{Q} = \begin{bmatrix} r_{0,0} & r_{0,1} & \cdots & r_{0,n-1} \\ r_{1,0} & r_{1,1} & \cdots & r_{1,n-1} \\ & & \vdots & \\ r_{n-1,0} & r_{n-1,1} & \cdots & r_{n-1,n-1} \end{bmatrix} \quad (B9)$$

be the matrix whose rows are the coefficients of the remainder polynomials $r_0(x), \ldots, r_{n-1}(x)$. (*Note*: The coefficients of the smaller powers of x are written first.) Then, the following holds.

Theorem 6.2.16. A polynomial $t(x) = t_0 + t_1 x + \cdots + t_{n-1} x^{n-1}$ is a solution to $[t(x)]^p \equiv t(x) \pmod{p(x)}$ if and only if

$$(t_0, t_1, \ldots, t_{n-1})\mathbf{Q} = (t_0, t_1, \ldots, t_{n-1}) \quad \text{or}$$
$$(t_0, t_1, \ldots, t_{n-1})(\mathbf{Q} - \mathbf{I}) = (0, 0, \ldots, 0). \quad (B10)$$

Proof. The proof follows from the fact that (B10) holds if and only if

$$t(x) = \Sigma_j t_j x^j = \Sigma_j \Sigma_k t_k r_{k,j} x^j \equiv \Sigma_k t_k x^{p k} = t(x^p) \equiv [t(x)]^p \pmod{p(x)}$$

(See also Theorem 6.2.9.) □

Let N be the set of vectors $\mathbf{t} = (t_0, t_1, \ldots, t_{n-1})$ with $\mathbf{t}(\mathbf{Q} - \mathbf{I}) = \mathbf{0}$, where $\mathbf{t}$ is the coefficient vector of $t(x)$, and $\mathbf{0}$ is the n-dimensional zero vector. Then

N is called the *null space* of $\mathbf{Q} - \mathbf{I}$. Let $\{\mathbf{b}_1, \mathbf{b}_2, \ldots, \mathbf{b}_r\}$ be a set of vectors in N such that every vector $\mathbf{a}$ in N is a linear combination of $\mathbf{b}_1, \mathbf{b}_2, \ldots, \mathbf{b}_r$ [note that for each vector $\mathbf{b}_k$ we have the corresponding polynomial $b_k(x)$]; that is, for any vector $\mathbf{a}$ in N there are $x_1, \ldots, x_r$ in $\mathbf{Z}_p$ such that $\mathbf{a} = x_1 \mathbf{b}_1 + \cdots + x_r \mathbf{b}_r$. The smallest r for which such a set $\{\mathbf{b}_1, \mathbf{b}_2, \ldots, \mathbf{b}_r\}$ exists is called the *dimension* of N, and the set itself is called the *basis* of the null space. (See also the Appendix.)

Therefore, from Theorem 6.2.16, we see that finding the suitable polynomials $t(x)$ is equivalent to determining the null space of $\mathbf{Q} - \mathbf{I}$ (see also historical note 2). As we will see below, we can easily compute the null space, and hence, we can apply Theorem 6.2.15 and compute the factors of $p(x)$. However, how do we know that we have found the complete factorization of $p(x)$? The answer is given by the following theorem.

Theorem 6.2.17. The number of distinct, irreducible factors $p_i(x)$ of $p(x)$ in $\mathbf{Z}_p[x]$ is equal to the dimension r of the null space of the matrix $\mathbf{Q} - \mathbf{I}$.

Proof. The polynomial $p(x)$ divides $\prod_{s \in GF(p)} \gcd[p(x), t(x) - s]$ if and only if each $p_i(x)$ divides $t(x) - s_i$ for some $s_i \in \mathbf{Z}_p$. Given $s_1, \ldots, s_r \in \mathbf{Z}_p$, Theorem 6.2.12 implies the existence of a unique polynomial $t(x)$ [mod $p(x)$], such that $t(x) \equiv s_i$ [mod $p_i(x)$]. We have the choice of p^r elements s_i, and as we have already mentioned, we have exactly p^r solutions of $[t(x)]^p - t(x) \equiv 0$ [mod $p(x)$]. By Theorem 6.2.16 we know that $t(x)$ is a solution of (B2) if and only if

$$(t_0, t_1, \ldots, t_{n-1}) \mathbf{Q} = (t_0, t_1, \ldots, t_{n-1}) \quad \text{or}$$

$$(t_0, t_1, \ldots, t_{n-1})(\mathbf{Q} - \mathbf{I}) = (0, 0, \ldots, 0)$$

This system has p^r solutions. Thus, the dimension of the null space of the matrix $\mathbf{Q} - \mathbf{I}$ is r, which equals the number of distinct, monic irreducible factors of $p(x)$, and the rank of $\mathbf{Q} - \mathbf{I}$ is $n - r$. (See also Exercise 2 for this section.) □

Moreover, Theorem 6.2.17 provides us with the third irreducibility test (see Section 6.2.2 for the first two tests).

Test 3. The polynomial $p(x)$ is irreducible in $\mathbf{Z}_p[x]$ if and only if the null space of $\mathbf{Q} - \mathbf{I}$ has dimension 1 and $\gcd[p(x), p'(x)] = 1$.

Proof. By Theorem 6.2.17, the null space of $\mathbf{Q} - \mathbf{I}$ is one-dimensional if and only if $p(x) = [p_1(x)]^k$, a power of an irreducible polynomial. Then, $r = 1$ and $p(x)$ is irreducible if and only if $\gcd[p(x), p'(x)] = 1$. □

Theorem 6.2.18. Let $p(x) = p_1(x) \cdots p_r(x)$ in $\mathbf{Z}_p[x]$, and $\{\mathbf{b}_1, \mathbf{b}_2, \ldots, \mathbf{b}_r\}$

be a basis of the null space of $Q - I$. Then, for every $j \neq j'$, $1 \leq j < j' \leq r$, there exist k, $1 \leq k \leq r$, and $s \in Z_p$ such that $p_j(x)$ divides $gcd[p(x), b_k(x) - s]$ and $p_{j'}(x)$ does not divide it.

Proof. First note that there exists a vector in the null space of $Q - I$ such that its jth component differs from the j'th component. Hence, there exists k, $1 \leq k \leq r$, such that

$$b_k(x) \, [\text{mod } p_j(x)] \neq b_k(x) \, [\text{mod } p_{j'}(x)]$$

This can be also seen by the contradiction it would lead to in the opposite case. Namely, if they were equal for all k, then, since any solution of (B2) is a linear combination of $\{b_1, b_2, \ldots, b_r\}$ with coefficients in Z_p, there would exist for any such solution b an element $s \in Z_p$ with $b(x) \equiv s \, [\text{mod } p_j(x)]$, $b(x) \equiv s \, [\text{mod } p_{j'}(x)]$ [see also (B1)]. However, there is a solution to (B2) with $b(x) \equiv 0 \, [\text{mod } p_j(x)]$ and $b(x) \equiv 1 \, [\text{mod } p_{j'}(x)]$, and this contradiction proves the inequality above. Setting $b_k(x) \, [\text{mod } p_j(x)] = s \in Z_p$, we have $p_j(x)|[b_k(x) - s]$, and $p_{j'}(x)$ does not divide $[b_k(x) - s]$. □

From the preceding discussion it is now clear how to factor a polynomial $p(x)$ over a finite field.

BA (Berlekamp's Algorithm)

Input: A monic, squarefree polynomial $p(x)$ over $GF(p)$, $\deg[p(x)] = n$.
Output: The irreducible factors of $p(x)$ over $GF(p)$.

1. **[Build the Q matrix.]** Construct the $n \times n$ matrix Q as defined by (B9). As explained below, this can be done in one of two ways, depending on whether p is very large.
2. **[Triangularize $Q - I$.]** Triangularize $Q - I$ determining its rank $n - r$ and finding a basis for the null space of $Q - I$; that is, determine r linearly independent vectors $b_1, b_2, \ldots, b_r$ such that $b_j[Q - I] = 0$ for $1 \leq j \leq r$. [The first vector b_1 may always be taken as $(1, 0, \ldots, 0)$, representing the trivial solution $b_1(x) = 1$ to (B2). The triangularization may be done as explained in Section 5.3.3 or using the algorithm NS presented below.] *At this point, r is the number of irreducible factors of $p(x)$*, because the solutions to (B2) are the p^r polynomials corresponding to the vectors $a_1 b_1 + a_2 b_2 + \cdots + a_r b_r$ for all choices of integers $0 \leq a_1, \ldots, a_r \leq p$. Therefore, if $r = 1$, then we know that $p(x)$ is irreducible, and the algorithm terminates.
3. **[Compute the factors.]** Let $b_2(x)$ be the polynomial corresponding to the vector b_2. Compute $gcd[p(x), b_2(x) - s]$ for all $s \in GF(p)$. The

result is a nontrivial factorization of $p(x)$ by Theorem 6.2.15. If, using $b_2(x)$, fewer than r factors are obtained, compute $gcd[w(x), b_k(x) - s]$ for all $s \in GF(p)$ and all factors $w(x)$ found thus far, for $k = 3, 4, \ldots, r$, until r factors are obtained. Theorem 6.2.18 guarantees that we will find all the factors of $p(x)$ in this way. If p is small, the calculations of this step are quite efficient; however, for large p (say, $p > 25$) there is a better way to proceed, as we explain below.

Computing-Time Analysis of **BA**. The number of computations required for step 2 of Berlekamp's algorithm is $0(n^3)$. (This fact has been shown in the proof of the computing-time analysis of **ASPRS** in Section 5.3.3.) We now know from Section 3.2.2 that $0(n^3)$ dominates the time for one computation of a gcd; since there are at most p gcd computations for each vector **b** in the basis, and at most r of these gcd computations will be nontrivial, we have

$$t_{\mathbf{BA}}[p(x)] = 0(prn^3)$$

The reader should note the dependence of this algorithm on p; hence, for large p the algorithm is inefficient. Moreover, from Corollary 6.1.2 we know that the average number of factors r is approximately $\ln(n)$.

Example. Let us factor the squarefree polynomial $p(x) = 8x^4 + 6x^3 + 8x^2 + 3x + 12$ or its monic equivalent $x^4 + 4x^3 + x^2 + 2x + 8$ over $GF(13)$, obtained by multiplying $p(x)$ times $8^{-1} = 5$. This is a rather long example, but the reader should study it carefully.

We first compute the inverse of each nonzero element of $GF(13) = \{0, 1, 2, 3, 4, 5, 6, 7, 8, 9, 10, 11, 12\}$, since we will use them; these inverses are computed using one of the methods discussed in Chapter 2, and they are $\{1, 7, 9, 10, 8, 11, 2, 5, 3, 4, 6, 12\}$.

The first step of **BA** involves calculating the matrix **Q**, which in this case is 4×4. The first row of **Q** is always $(1, 0, 0, 0)$, representing the polynomial x^0 [mod $p(x)$] = 1. The second row represents x^{13} [mod $p(x)$], the third row x^{26} [mod $p(x)$], and finally, the last row x^{39} [mod $p(x)$]. Obviously, what we need is a fast way to compute x^{k+1} [mod $p(x)$] once x^k [mod $p(x)$] is known; below we present such a general procedure.

Let $p(x) = x^n + c_{n-1}x^{n-1} + \cdots + c_1 x + c_0$, and suppose that we have

$$x^k \equiv r_{k,n-1}x^{n-1} + \cdots + r_{k,1}x + r_{k,0} \ [\text{mod } p(x)]$$

then

$$\begin{aligned}
x^{k+1} &\equiv r_{k,n-1}x^n + \cdots + r_{k,1}x^2 + r_{k,0}x \\
&\equiv r_{k,n-1}(-c_{n-1}x^{n-1} - \cdots - c_1 x - c_0) + r_{k,n-2}x^{n-1} + \cdots + r_{k,1}x^2 + r_{k,0}x \\
&\equiv r_{k+1,n-1}x^{n-1} + \cdots + r_{k+1,1}x + r_{k+1,0} \ [\text{mod } p(x)]
\end{aligned}$$

where

$$r_{k+1,j} = r_{k,j-1} - r_{k,n-1}c_j, \qquad r_{k,-1} = 0 \tag{B11}$$

This simple recurrence formula makes it easy to calculate $x, x^2, x^3, \ldots, x^k$ [mod $p(x)$]. The calculation may be done by maintaining a one-dimensional array $(r_{n-1}, \ldots, r_1, r_0)$ and repeatedly setting $t := r_{n-1}$, $r_{n-1} := (r_{n-2} - tc_{n-1})$ (mod p), $\ldots, r_1 := (r_0 - tc_1)$ (mod p) and $r_0 := (-tc_0)$ (mod p). Thus, using arithmetic modulo 13 for our example, we obtain the following table:

k	$r_{k,3}$	$r_{k,2}$	$r_{k,1}$	$r_{k,0}$
0	0	0	0	1
1	0	0	1	0
2	0	1	0	0
3	1	0	0	0
4	9	12	11	5
5	2	2	0	6
6	7	11	2	10
7	9	8	9	9
8	11	0	4	6
9	8	6	10	3
10	0	2	0	1
11	2	0	1	0
12	5	12	9	10
13	5	4	0	12

Therefore, the second row of **Q** is $(12, 0, 4, 5)$. Similarly, we determine x^{26} [mod $p(x)$] and x^{39} [mod $p(x)$] and finally

$$\mathbf{Q} = \begin{bmatrix} 1 & 0 & 0 & 0 \\ 12 & 0 & 4 & 5 \\ 12 & 5 & 6 & 12 \\ 1 & 4 & 3 & 6 \end{bmatrix}, \qquad \mathbf{Q} - \mathbf{I} = \begin{bmatrix} 0 & 0 & 0 & 0 \\ 12 & 12 & 4 & 5 \\ 12 & 5 & 5 & 12 \\ 1 & 4 & 3 & 5 \end{bmatrix}$$

When p is large we can form x^k [mod $p(x)$] in a more efficient way than (B11); namely, this value can be obtained in $0[\log(p)]$ operations of squaring mod $p(x)$, that is, going from x^k [mod $p(x)$] to x^{2k} [mod $p(x)$]. The squaring operation is relatively easy to perform if we first make an auxiliary table of x^i [mod $p(x)$] for $i = n, n+1, \ldots, 2n-2$. So, if

$$x^k \text{ [mod } p(x)\text{]} = c_{n-1}x^{n-1} + \cdots + c_1 x + c_0$$

then

$$x^{2k} \,[\text{mod } p(x)] = [c_{n-1}^2 x^{2n-2} + \cdots + (c_1c_0 + c_1c_0)x + c_0^2]\,[\text{mod } p(x)]$$

where $x^{2n-2}, \ldots, x^n$ can be replaced by polynomials in the auxiliary table. Thus we compute $x^p\,[\text{mod } p(x)]$, the second row of **Q**. To obtain further rows of **Q**, we can compute $x^{2p}\,[\text{mod } p(x)]$, $x^{3p}\,[\text{mod } p(x)]$, ..., simply by multiplying repeatedly by $x^p\,[\text{mod } p(x)]$, in a fashion analogous to squaring mod $p(x)$.

This finishes step 1 of Berlekamp's algorithm.

The next step of **BA** requires finding the "null space" of $\mathbf{Q} - \mathbf{I}$. In general, let **M** be the $n \times n$ matrix over a field, whose rank $n - r$ is to be determined; moreover, suppose that we wish to determine linearly independent vectors $\{\mathbf{b}_1, \mathbf{b}_2, \ldots, \mathbf{b}_r\}$ such that $\mathbf{b}_i\mathbf{M} = \mathbf{0}$, $i = 1, 2, \ldots, r$. The algorithm for this calculation is based on the observation that any column of **M** may be multiplied by a nonzero quantity, and any multiple of one of its columns may be added to a different column, without changing the rank or the vectors $\{\mathbf{b}_1, \mathbf{b}_2, \ldots, \mathbf{b}_r\}$. {Observe that the polynomials $b_i(x)$ corresponding to the basis vectors $\mathbf{b}_i$ of the null space of $\mathbf{Q} - \mathbf{I}$ are all the solutions to $[t(x)]^p \equiv t(x)\,[\text{mod } p(x)]$.} Therefore, we have the following triangularization algorithm (Knuth, 1981 and Lidl et al., 1984).

NS (Null-Space Algorithm)

Input: An $n \times n$ matrix $\mathbf{M} = (m_{ij}), 0 \le i, j \le n - 1$, whose elements belong to a field.

Output: Linearly independent vectors $\mathbf{b}_1, \mathbf{b}_2, \ldots, \mathbf{b}_r$ such that $\mathbf{b}_i\mathbf{M} = \mathbf{0}$, $i = 1, 2, \ldots, r$, where $n - r$ is the rank of **M**.

1. [Initialize column flags.] Set $r := 0$ and $c_0 := c_1 := c_2 := \cdots := c_{n-1} := -1$.
2. [Main loop.] For $h = 0, 1, \ldots, n - 1$ do {if there is some column j such that $m_{hj} \ne 0$ and $c_j < 0$, $j = 0, 1, \ldots, n - 1$, then do the following [multiply column j of **M** times $-1/m_{hj}$ so that m_{hj} becomes equal to -1; then add m_{hi} times column j to column i for all $i \ne j$; finally set $c_j := h$.] If there is no column j such that $m_{hj} \ne 0$ and $c_j < 0$, $j = 0, 1, 2, \ldots, n - 1$, then set $r := r + 1$ and output the vector $\mathbf{b}_r = (b_{r,0}, b_{r,1}, \ldots, b_{r,n-1})$, where for $j = 0, 1, \ldots, n - 1$

$$b_{r,j} = \begin{cases} m_{hk}, & \text{if } c_k = j > 0, \text{ (if } c_k > 0 \text{ for more than one } k, \text{ pick } any \text{ one of them)} \\ 1, & \text{if } j = h \\ 0, & \text{otherwise\}} \end{cases}$$

Applying the above procedure to the matrix $\mathbf{M} := \mathbf{Q} - \mathbf{I}$ computed above, and whose elements are in $GF(13)$, we have the following (keep in mind that algorithm **NS** numbers the rows and columns of the matrix starting with 0, *not* from 1).

When $h = 0$, we output the vector $\mathbf{b}_1 = (1, 0, 0, 0)$, corresponding to the constant polynomial 1. When $h = 1$, we may take j to be either 0, 1, 2, or 3 since $c_i = -1$ for $i = 0, 1, 2, 3$; the choice is completely arbitrary, although it affects the particular vectors that are chosen to be output. We pick $j = 0$, and using the table of inverses computed at the beginning of this example, we have $-1/m_{10} = -(1/12) = -12 \equiv 1 \pmod{13}$; the column operations ($\text{col}_0 := 1 \cdot \text{col}_0$, $\text{col}_1 := \text{col}_1 + 12 \cdot \text{col}_0$, $\text{col}_2 := \text{col}_2 + 4 \cdot \text{col}_0$, $\text{col}_3 := \text{col}_3 + 5 \cdot \text{col}_0$) described in the algorithm above, change the matrix $\mathbf{M}$ to the matrix

$$\begin{bmatrix} 0 & 0 & 0 & 0 \\ \mathbf{12} & 0 & 0 & 0 \\ 12 & 6 & 1 & 7 \\ 1 & 3 & 7 & 10 \end{bmatrix}$$

The boldface element **12** in row 1, column 0, indicates that $c_0 = 1$. When $h = 2$, we may choose $j = 1$ and proceed in a similar way [i.e., first compute $-(1/6) = -11 \equiv 2 \pmod{13}$, and then set $\text{col}_1 := 2 \cdot \text{col}_1$, $\text{col}_0 := \text{col}_0 + 12 \cdot \text{col}_1$, $\text{col}_2 := \text{col}_2 + 1 \cdot \text{col}_1$, $\text{col}_3 := \text{col}_3 + 7 \cdot \text{col}_1$], obtaining the matrix

$$\begin{bmatrix} 0 & 0 & 0 & 0 \\ \mathbf{12} & 0 & 0 & 0 \\ 0 & \mathbf{12} & 0 & 0 \\ 8 & 6 & 0 & 0 \end{bmatrix}$$

As above, the boldface element **12** in row 2, column 1 indicates that $c_1 = 2$. Now every column that does not have a boldface entry is completely zero; therefore, when $h = 3$, the algorithm outputs the vector $\mathbf{b}_2 = (0, 8, 6, 1)$ corresponding to the polynomial $x^3 + 6x^2 + 8x$.

From the form of the matrix $\mathbf{M}(:= \mathbf{Q} - \mathbf{I})$ after $h = 3$, it is clear that the vectors $\mathbf{b}_1$ and $\mathbf{b}_2$ satisfy the equation $\mathbf{b}_i \mathbf{M} = \mathbf{0}$. Since the above computation has produced two linearly independent vectors, $p(x)$ must have exactly two irreducible factors over $GF(13)$, which are obtained from step 3 of Berlekamp's algorithm.

In this final step 3 of **BA** we need to perform the necessary greatest common divisor computations in order to find the two factors of $p(x)$; that is, we need to compute $gcd[p(x), b_2(x) - s]$ for *all* $s \in GF(13)$, where $p(x) = x^4 + 4x^3 + x^2 + 2x + 8$, and $b_2(x) = x^3 + 6x^2 + 8x$. Performing the computations, we see that the two factors are obtained from $s = 1$ and $s = 9$; that is, we have

$$s = 1: \gcd[p(x), b_2(x) - 1] = x^2 + 9x + 9$$
$$s = 9: \gcd[p(x), b_2(x) - 9] = x^2 + 8x + 11$$

Check: $(x^2 + 9x + 9)(x^2 + 8x + 11) = x^4 + 4x^3 + x^2 + 2x + 8 \pmod{13}$.

Clearly, when p is large and we want to calculate greatest common divisors for all $s \in GF(p)$, there is an enormous amount of work involved. Cantor and Zassenhaus (1981) suggested a better way to proceed. If $t(x)$ is any solution to (B2), and p is *odd*, we know that $p(x) | \{[t(x)]^p - t(x)\} = t(x)\{[t(x)]^{(p-1)/2} + 1\}\{[t(x)]^{(p-1)/2} - 1\}$. This suggests that we compute

$$\gcd\{p(x), [t(x)]^{(p-1)/2} - 1\} \tag{B12}$$

We then have a certain probability that (B12) will be a nontrivial factor of $p(x)$. To determine this probability, observe that $\gcd\{p(x), [t(x)]^{(p-1)/2} + / - 1\} = p(x)$ implies $\gcd\{p(x), [t(x)]^{(p-1)/2} - / + 1\} = 1$ and, therefore, the probability that $\gcd\{p(x), [t(x)]^{(p-1)/2} - 1\}$ or $\gcd\{p(x), [t(x)]^{(p-1)/2} + 1\}$ will be a nontrivial factor equals

$$1 - \text{prob}(\gcd\{p(x), [t(x)]^{(p-1)/2} - 1\} = p(x))$$
$$- \text{prob}(\gcd\{p(x), [t(x)]^{(p-1)/2} + 1\} = p(x)).$$

[Note that we cannot have both $\gcd\{p(x), [t(x)]^{(p-1)/2} - 1\} = 1$ and $\gcd\{p(x), [t(x)]^{(p-1)/2} + 1\} = 1$, since $\deg[t(x)] < n$.] Now $\gcd\{p(x), [t(x)]^{(p-1)/2} - 1\} = p(x)$ if and only if $p_j(x) | \{[t(x)]^{(p-1)/2} - 1\}$, $1 < j < r$, and this is the case if and only if $s_j^{(p-1)/2} \equiv 1 \pmod{p}$, $1 < j < r$ [consider (B1)], where we know that exactly $(p-1)/2$ of the integers s in the range $0 \le s < p$ satisfy $s^{(p-1)/2} \equiv 1 \pmod{p}$. If $t(x)$ is a random solution of (B2), where all p^r solutions are equally likely, then the probability the gcd in (B12) equals $p(x)$ is $[(p-1)/2p]^r$. The same reasoning holds for $\text{prob}(\gcd\{p(x), [t(x)]^{(p-1)/2} + 1\} = p(x))$, and therefore, the probability that a nontrivial factor of $p(x)$ will be obtained by computing (B12) is $1 - 2[(p-1)/2p]^r > 1 - 2(1/2)^r \ge 1/2$ for all $r \ge 2$ and $p \ge 3$.

Therefore, unless p is quite small, it is a good idea to replace the last step in Berlekamp's algorithm with the following procedure: Set $t(x) := a_1 b_1(x) + a_2 b_2(x) + \cdots + a_r b_r(x)$, where the coefficients a_i are randomly chosen in the range $0 \le a_i < p$. Let $p(x) = w_1(x) \cdots w_h(x)$ be the current partial factorization of $p(x)$, where h is initially 1. Compute

$$g_i(x) = \gcd\{w_i(x), [t(x)]^{(p-1)/2} - 1\}$$

for all i such that $\deg[w_i(x)] > 1$, replace $w_i(x)$ by $g_i(x)[w_i(x)/g_i(x)]$, and increase the value of h, whenever a nontrivial *gcd* has been found. This process is repeated for different choices of $t(x)$ until $h = r$. Details are left to the reader.

6.3
Lifting a (mod p)-Factorization to a Factorization over the Integers

In this section we will see how to raise a (mod p)-factorization to one over the integers, and in the process we will do away with the requirement that $p(x)$ in $\mathbf{Z}[x]$ be monic. The reader should review Section 6.1.2 at this point.

At first let us put together some thoughts on the factorization of $p(x)$, in general. Suppose that the polynomial $p(x)$ factors over the integers and we have $p(x) = p_1(x)p_2(x)$; then by Theorem 3.2.16 we have that $p(x) \equiv p_1(x)p_2(x) \pmod{p}$ for all primes p, so there is a nontrivial factorization modulo p unless $p | lc[p(x)]$. An efficient algorithm for polynomial factorization in $GF(p)$ (discussed in Section 6.2) can, therefore, be used in an attempt to reconstruct possible factorizations of $p(x)$ over the integers. If, for example, for a given polynomial $p(x)$ of degree 8 we have that $p(x) \equiv p_1(x)p_2(x) \pmod{p_1}$, $\deg[p_1(x)] = 6$ and $\deg[p_2(x)] = 2$, whereas for another prime p_2 we have $p(x) \equiv p_1(x)p_2(x)p_3(x) \pmod{p_2}$, $\deg[p_1(x)] = 4$, $\deg[p_2(x)] = 3$ and $\deg[p_3(x)] = 1$, then, since in the (mod p_2)-factorization there is no factor of degree 2, $p(x)$ must be irreducible over the integers.

The above example is perhaps too simple and straightforward, but it should be observed that irreducibility is not always so easily established. It should be kept in mind that for all $k \geq 2$, Swinnerton-Dyer (1970) has exhibited polynomials of degree 2^k that are irreducible over the integers, but they factor completely into linear and quadratic factors modulo every prime.

If we try to determine the factors of $p(x)$ by considering its behavior modulo different primes, the results will not be easy to combine, in general. Therefore, it is desirable to stick to a single prime and to see "how much mileage" we can get out of it, provided we feel that the factors modulo this prime have the right degrees. We will examine in detail the following two approaches: choosing a single *large* prime, and choosing a single small prime combined with "lifting."

We begin with the first idea, namely, to factor $p(x)$ modulo a *large* prime p, where p is some number that is larger than twice the absolute values of the coefficients of all possible factors of $p(x)$; that is, the coefficients of any true factorization $p(x) = p_1(x)p_2(x)$ over the integers must actually lie between $-p/2$ and $p/2$. [Below we compute bounds on the coefficients of the factors of $p(x)$ without actually computing these factors.] Then all possible factors over the integers can be "read off" from the (mod p)-factors that we know how to compute. The reason behind this approach is as follows. If we choose a small prime p and $p(x) \equiv p_1(x)p_2(x) \pmod{p}$, in general, there may be many ways this factorization could be combined to a factorization of $p(x)$ over the integers. For example, consider the polynomial

$$p(x) = 112x^4 + 58x^3 - 31x^2 + 107x - 66$$
$$= (14x^2 - 5x + 11)(8x^2 + 7x - 6) \text{ in } \mathbf{Z}[x]$$

Modulo 13, $p(x)$ factors as

$$p(x) \equiv 8(x^2 + 8x + 11)(x^2 + 9x + 9) \pmod{13}$$
$$\equiv (8x^2 + 12x + 10)(x^2 + 9x + 9) \pmod{13}$$
$$\equiv (x^2 + 8x + 11)(8x^2 + 7x + 7) \pmod{13}$$

(See also the example in Section 6.2.4.) Observe that there are many polynomials that are congruent to either factor (mod 13) that could possibly be factors of $p(x)$ over the integers, and it would take some effort to discover the true factors in $\mathbf{Z}[x]$. [This problem does *not* exist if we work with the monic version of $p(x)$.]

So, suppose, instead, that $p(x) \equiv p_1(x)p_2(x) \pmod{p}$, where p is *large enough* so that the coefficients of any possible factor of $p(x)$, of degree that of $p_1(x)$, lie between $-p/2$ and $p/2$; suppose, moreover, that we choose $p_1(x)$ so that each of its coefficients is in this range. Then, if $p(x) = P_1(x)P_2(x)$ in $\mathbf{Z}[x]$, where $P_1(x) \equiv p_1(x) \pmod{p}$ and $P_2(x) \equiv p_2(x) \pmod{p}$, it must be that $P_1(x) = p_1(x)$. [Otherwise, there is a coefficient of $P_1(x)$ that differs from the corresponding coefficient of $p_1(x)$ by a nonzero multiple of p. But then, that coefficient of $P_1(x)$ must be in absolute value $\geq p/2$, and is, therefore, too big in absolute value to qualify $P_1(x)$ as a possible factor of $p(x)$.] Therefore, if $p(x) \equiv p_1(x)p_2(x) \pmod{p}$, then either $p_1(x) | p(x)$ in $\mathbf{Z}[x]$ or the factorization $p(x) \equiv p_1(x)p_2(x) \pmod{p}$ corresponds to no factorization over the integers.

Let us now examine how to find a suitably large p. (At this point the reader should review the definitions of the various norms for polynomials defined in Section 1.2.)

Bounding the Coefficients of Factors of a Polynomial We begin with the following well-known theorem, which we prove in the general case of complex coefficients.

Theorem 6.3.1. Let $p(x) = x^n + c_{n-1}x^{n-1} + \cdots + c_0$ be a monic polynomial with complex coefficients, and suppose that all its complex roots are in absolute value less than some positive real number b. If $f(x)$ is a monic factor of $p(x)$, $\deg[f(x)] = r$, then all the coefficients of $f(x)$ are in absolute value $\leq \max_{1 \leq k \leq r}\{\binom{r}{k}b^k\}$.

Proof. In $\mathbf{C}[x]$ let $f(x) = \Pi_{1 \le i \le r}(x + s_i) = x^r + f_{r-1}x^{r-1} + \cdots + f_0$, where $-s_i, 1 \le i \le r$ are the roots of $f(x)$ in the complex numbers. Multiplying out the factorization $f(x) = \Pi_{1 \le i \le r}(x + s_i)$ and equating coefficients, we obtain

$$f_{r-1} = s_1 + s_2 + \cdots + s_r$$
$$f_{r-2} = s_1 s_2 + s_1 s_3 + s_1 s_4 + \cdots = \Sigma_{i<j} s_i s_j$$
$$f_{r-3} = \Sigma s_i s_j s_k$$
$$\cdots$$
$$f_0 = s_1 s_2 \cdots s_r$$

Since each $-s_i$, $1 \le i \le r$ is a root of $p(x)$, we have $|-s_i| \le b$. Therefore, using the triangle inequality, for each i, we see that $|f_i|$ is $\le$ the sum obtained by replacing all the s_i by b. For example, when $r = 4$, we obtain

$$|f_3| = |s_1 + s_2 + s_3 + s_4| \le |s_1| + |s_2| + |s_3| + |s_4| \le b + b + b + b = 4b$$
$$|f_2| = |s_1 s_2 + s_1 s_3 + s_1 s_4 + s_2 s_3 + s_2 s_4 + s_3 s_4| \le |s_1 s_2| + |s_1 s_3| + |s_1 s_4|$$
$$+ |s_2 s_3| + |s_2 s_4| + |s_3 s_4| \le 6b^2$$
$$|f_1| = |s_1 s_2 s_3| + |s_1 s_2 s_4| + |s_1 s_3 s_4| + |s_2 s_3 s_4| \le 4b^3$$
$$|f_0| = |s_1 s_2 s_3 s_4| \le b^4$$

Replacing each s_i by b, we obtain the polynomial $f_1(x) = (x + b)^r = \Sigma_{0 \le k \le r} \binom{r}{k} b^k x^{r-k}$. Therefore,

$$|f_{r-1}| \le \binom{r}{1} b = rb$$
$$|f_{r-2}| \le \binom{r}{2} b^2 = \left[\frac{r(r-1)}{2}\right] b^2$$
$$\cdots$$
$$|f_0| = \binom{r}{r} b^r = b^r$$

and this proves the theorem. □

From Theorem 6.3.1. we see that in order to find a good bound on the coefficients of the factors of $p(x)$ we must find a good bound on the roots of $p(x)$. We have the following.

Theorem 6.3.2 (Specht, 1949). Let $p(x) = x^n + c_{n-1} x^{n-1} + \cdots + c_0$ be a monic polynomial with complex coefficients; moreover, let $z_1, z_2, \ldots, z_k$ be those roots of $p(x)$ (counted with their multiplicities), such that $1 \le |z_1| \le |z_2| \le \cdots \le |z_k|$. Then

$$\Pi_{1\le i\le k}|z_i| \le |p(x)|_2$$

Proof. The proof is rather extended and is omitted. It can be found in German in Specht's paper (1949), and (a different one) in English in Mignotte's paper (1974). (See also historical note 3.) □

From the well-known expression of the coefficients of a polynomial we observe that for the polynomial $p(x)$ above we have

$$|c_i| \le \binom{n}{i}|z_1 \cdots z_k|$$

and

$$\Sigma_{0\le i\le n}|c_i| \le 2^n|z_1 \cdots z_k|$$

We are now ready to obtain a bound on the coefficients of factors of $p(x)$.

Theorem 6.3.3 (Mignotte, 1974). Let $p(x) = x^n + c_{n-1}x^{n-1} + \cdots + c_0$ be a monic polynomial in $\mathbf{Z}[x]$, and suppose that $p(x) = p_1(x)p_2(x) \cdots p_m(x)$, where $p_i(x)$, $1 \le i \le m$, are monic polynomials in $\mathbf{Z}[x]$. Then

$$\Pi_{1\le i\le m}|p_i(x)|_1 \le 2^d|p(x)|_2 \tag{M1}$$

where $d = \Sigma_{1\le i\le m}\deg[p_i(x)]$; moreover, if $p_h(x)$ is one of these factors, where $p_h(x) = x^h + \cdots + c_{h,1}x + c_{h,0}$, then

$$|c_{h,j}| \le \binom{h}{j}|p(x)|_2, \; j = 0, 1, \ldots, h-1 \tag{M2}$$

Proof. This follows immediately from Theorems 6.3.1 and 6.3.2 and the well-known formula on the coefficients of a polynomial stated above. □

From the preceding discussion we easily see that the large prime p must be chosen so that $p \ge 2^d|p(x)|_2$ so that if $p(x) \equiv p_1(x)p_2(x) \pmod{p}$, then either $p_1(x)|p(x)$ in $\mathbf{Z}[x]$ or the factorization $p(x) \equiv p_1(x)p_2(x) \pmod{p}$ corresponds to no factorization over the integers.

Let us now consider the second approach: Instead of using a large prime p, which might have to be extremely large, if $p(x)$ has large degree or large coefficients (or both), we can also make use of small primes provided $p(x)$ (mod p) is squarefree. Currently, one of the best methods available for factoring a polynomial $p(x)$ over the integers is to factor it modulo p, for a small prime p, and then extend (or "lift") this factorization in a unique way to a factorization modulo p^k for a suitable k determined from (M2).

6.3.1 Linear and Quadratic Lifting

In this section we assume that we have factored $p(x)$ in $\mathbf{Z}_p[x]$, for some small prime p, and we will lift this factorization to one over the integers.

Let $p(x)$ be a polynomial of degree n in $GF(p^k)[x]$ such that $lc[p(x)] \neq 0$ in $GF(p)$; moreover, let $g_1(x)$ and $h_1(x)$ be polynomials in $GF(p)[x]$ such that $p(x) = g_1(x)h_1(x)$ over $GF(p)$, and $gcd[g_1(x), h_1(x)] = 1$. Below we present two methods to lift this factorization over $GF(p)$ to the unique factorization over $GF(p^k) = GF(q)$, that is, to find $g_k(x)$, $h_k(x)$ in $GF(p^k)[x]$ such that $p(x) = g_k(x)h_k(x)$ over $GF(p^k)$, and such that $g_k(x) \equiv g_1(x)$, $h_k(x) \equiv h_1(x)$ modulo p. To achieve this, we need the following two algorithms; the first algorithm is the well-known extended Euclidean algorithm in $GF(p)[x]$, whereas the second algorithm solves the polynomial equation $a'(x)g_j(x) + b'(x)h_j(x) = c(x)$ for given $g_j(x), h_j(x)$ and $c(x)$ in $GF(p^j)[x]$.

Algorithm 6.3.1 {Extended Euclidean Algorithm in $GF(p)[x]$}. Given $g(x)$ and $h(x)$ in $GF(p)[x]$, this algorithm computes unique (up to units) polynomials $a(x), b(x)$, and $c(x)$ in $GF(p)[x]$ such that $g(x)a(x) + h(x)b(x) = c(x) = gcd[g(x), h(x)]$ over $GF(p)$, with $\deg[a(x)] < \deg[h(x)] - \deg[c(x)]$ and $\deg[b(x)] < \deg[g(x)] - \deg[c(x)]$. A description of this algorithm has already been presented in Section 3.2.2 (**XEA-P**).

Algorithm 6.3.2 {Solution of a Polynomial Equation in $GF(p^j)[x]$}. Given the polynomials $a(x), b(x), g_j(x), h_j(x)$ in $GF(p^j)[x]$ such that $g_j(x)a(x) + h_j(x)b(x) = 1$ over $GF(p^j)$, this algorithm computes polynomials $a'(x)$ and $b'(x)$ in $GF(p^j)[x]$ such that $g_j(x)a'(x) + h_j(x)b'(x) = c(x)$ over $GF(p^j)$, with $\deg[a'(x)] < \deg[h_j(x)]$ as follows.

1. Using **PDF**, the division algorithm for polynomials over a field, compute in $GF(p^j)[x]$ $q(x)$ and $r(x)$ such that $a(x)c(x) = h_j(x)q(x) + r(x)$, $\deg[r(x)] < \deg[h_j(x)]$
2. Set $a'(x) := r(x)$, $b'(x) := b(x)c(x) + g_j(x)q(x)$.
   Then, we have

$$g_j(x)a'(x) + h_j(x)b'(x) = g_j(x)[a(x)c(x) - h_j(x)q(x)]$$
$$+ h_j(x)[b(x)c(x) + g_j(x)q(x)]$$
$$= [g_j(x)a(x) + h_j(x)b(x)]c(x)$$
$$= c(x) \quad \text{in } GF(p^j)$$

With these facts we can now proceed to the following linear lift algorithm, which is derived from the proof of the following lemma.

Lemma 6.3.4 (Hensel, 1908). Let $p(x)$ be a polynomial of degree n in $\mathbf{Z}[x]$ such that $lc[p(x)] \neq 0$ in $GF(p)$, p prime. Let $g_1(x)$ and $h_1(x)$ be polynomials in $GF(p)[x]$ such that $p(x) = g_1(x)h_1(x)$ over $GF(p)$, and $gcd[g_1(x), h_1(x)] = 1$ over $GF(p)$. Then there exist unique polynomials $g_k(x), h_k(x)$ in $GF(p^k)[x]$ such that $p(x) = g_k(x)h_k(x)$ over $GF(p^k)$, $\deg[h_k(x)] = \deg[h_1(x)]$ and $g_k(x) \equiv g_1(x)$, $h_k(x) \equiv h_1(x)$ over $GF(p)$.

Proof. Using Algorithm 6.3.1, we determine $a(x), b(x)$ such that $g_1(x) \cdot a(x) + h_1(x)b(x) = 1$. It suffices to show how to construct $g_j(x), h_j(x)$ in $GF(p^j)[x]$ for $j = 2, \ldots, k$ such that $p(x) = g_j(x)h_j(x)$ over $GF(p^j)$, $\deg[h_j(x)] = \deg[h_1(x)]$ and $g_j(x) \equiv g_1(x)$, $h_j(x) \equiv h_1(x)$ over $GF(p)$. Suppose that such $g_j(x), h_j(x)$ exist for some $j \geq 1$, and let $c(x)$ in $GF(p)[x]$ be such that $p(x) - g_j(x)h_j(x) = p^j c(x)$ over $GF(p^{j+1})$. Using Algorithm 6.3.2, compute $a'(x), b'(x)$ in $GF(p)[x]$ such that $g_1(x)a'(x) + h_1(x)b'(x) = c(x)$ over $GF(p)$. Define

$$g_{j+1}(x) = g_j(x) + p^j \cdot b'(x)$$
$$h_{j+1}(x) = h_j(x) + p^j \cdot a'(x)$$

where it should be noted that $\deg[h_{j+1}(x)] = \deg[h_j(x)]$ and $lc[h_{j+1}(x)] = lc[h_j(x)]$. Obviously, $g_{j+1}(x), h_{j+1}(x)$ are polynomials in $GF(p^{j+1})[x]$, $g_{j+1}(x) \equiv g_1(x)$, $h_{j+1}(x) \equiv h_1(x)$ over $GF(p)$ and

$$\begin{aligned} g_{j+1}(x)h_{j+1}(x) &= [g_j(x) + p^j \cdot b'(x)][h_j(x) + p^j \cdot a'(x)] \\ &= g_j(x)h_j(x) + p^j \cdot [g_j(x)a'(x) + h_j(x)b'(x)] \\ &= g_j(x)h_j(x) + p^j \cdot c(x) \\ &= p(x) \quad \text{over } GF(p^{j+1}) \end{aligned}$$
□

It is obvious that Lemma 6.3.4 provides us with a method to lift a factorization step by step; that is, from $GF(p)$ to $GF(p^2)$, $GF(p^3)$, $\ldots, GF(p^k)$. Therefore, it is called *linear lift algorithm*. Observe that in practice it frequently occurs that there are more than two factors to be lifted. In this case the lifting is done by repeated application of Lemma 6.3.4 to pairs of factors in which one is $p_i(x)$ and the other is the product $p_{i+1}(x)p_{i+2}(x) \cdots p_r(x)$. Namely, if

$$p(x) \equiv p_1(x)p_2(x) \cdots p_r(x) \pmod{p}$$

then, we take at first $g_1(x) = p_1(x)$ and $h_1(x) = p_2(x) \cdots p_r(x)$ and lift to $GF(p^k)[x]$; then set $p(x) := p(x)/g_1(x)$ in $GF(p^k)$, $g_1(x) := p_2(x)$, $h_1(x)$

:= $p_3(x) \cdots p_r(x)$ and lift to $GF(p^k)[x]$, and so on. If $p(x)$ is monic, it is possible to modify the algorithm so that an arbitrary number of relatively prime factors can be lifted simultaneously.

The reader should carefully study the following two examples; in the first we use the nonnegative residue set of integers whereas in the second, the symmetric residue set.

Example [Linear Hensel Construction; $p(x)$ is not Monic]. Consider the polynomial $p(x) = 112x^4 + 58x^3 - 31x^2 + 107x - 66$ in $\mathbf{Z}[x]$, which is

$$p(x) \equiv (8x^2 + 12x + 10)(x^2 + 9x + 9) \pmod{13}$$

We will lift this factorization from $GF(13)$ to $GF(13^2)$ using Hensel's linear construction and nonnegative residue arithmetic. For better appreciation of the process we reveal that, over the integers, we have $p(x) = (8x^2 + 7x - 6)(14x^2 - 5x + 11)$; these two factors of $p(x)$ are discovered with one application of the lift algorithm.

Obviously, in our example we have $g_1(x) = 8x^2 + 12x + 10$ and $h_1(x) = x^2 + 9x + 9$. Using Algorithm 6.3.1, we compute polynomials $a(x) = x + 11$ and $b(x) = 5x + 11$ in $GF(13)[x]$ such that $g_1(x)a(x) + h_1(x)b(x) = 1$ over $GF(13)$. [Actually, from Algorithm 6.3.1 we obtain $a(x) = 5x + 3$, $b(x) = 12x + 3$, but then $g_1(x)a(x) + h_1(x)b(x) = 5$ in $GF(13)$, and hence we have to adjust them by multiplying by a suitable constant.]

Next, from $p(x) - g_1(x)h_1(x) = 13c(x)$—where $g_1(x)$, $h_1(x)$ are considered in $\mathbf{Z}[x]$—we obtain $c(x) = 8x^4 + 11x^3 + 9x^2 + 6x + 1$ over $GF(13)$. Note that $p \cdot c(x)$ is the difference between $p(x)$ and $g_1(x)h_1(x)$ in the next bigger space $GF(p^2)[x]$, and we must now correct our factors; that is, add to $g_1(x)$ and $h_1(x)$ what is missing in an effort to make the difference disappear. Using Algorithm 6.3.2, we compute polynomials $a'(x) = x + 9$ and $b'(x) = 8x^2 + 12x + 6$ in $GF(13)[x]$ such that $g_1(x)a'(x) + h_1(x)b'(x) = c(x)$ over $GF(13)$. We then define

$$g_2(x) = g_1(x) + p \cdot b'(x) = (8x^2 + 12x + 10) + 13(8x^2 + 12x + 6)$$
$$= 112x^2 + 168x + 88$$

$$h_2(x) = h_1(x) + p \cdot a'(x) = (x^2 + 9x + 9) + 13(x + 9) = x^2 + 22x + 126$$

At this point we have $p(x) = g_2(x)h_2(x)$ over $GF(169 = 13^2)$, as can be easily verified, but neither $g_2(x)$ nor $h_2(x)$ divide $p(x)$ in $\mathbf{Z}[x]$. However, note that $g_2(x) = 8(14x^2 + 21x + 11)$, and if we combine the factor 8 with $h_2(x)$, we have $8 \cdot h_2(x) = 8(x^2 + 22x + 126) = 8x^2 + 176x + 1008 \equiv 8x^2 + 7x + 163 \equiv 8x^2 + 7x - 6 \pmod{169}$. At this point we have that $(8x^2 + 7x - 6) | p(x)$ over the integers, and hence it is one of its factors; dividing, we obtain the other factor $14x^2 - 5x + 11$.

It should be noted that in this example, since $p(x)$ was not monic, there was some effort in discovering its true factors. Fortunately, there is an easy way out of this; namely, the factorization $p(x) = p_1(x)p_2(x)$ implies a factorization $lc[p(x)] \cdot p(x) = p_1'(x)p_2'(x)$, where $lc[p_1'(x)] = lc[p_2'(x)] = lc[p(x)]$. ("I hope you do not mind if I multiply the polynomial by its leading coefficient before factoring it.") We can then proceed as before, but using a bound on the coefficients of the factors of $lc[p(x)] \cdot p(x)$ instead of $p(x)$.

Before we proceed with the second lift example it is instructive, at this point, to go over the various considerations involved in trying to factor $p(x) = 112x^4 + 58x^3 - 31x^2 + 107x - 66$ over the integers (see also Section 6.1.2).

At first we had to make sure that $p(x)$ is squarefree over the integers; that is, we had to determine $gcd[p(x), p'(x)]$, where $p'(x) = 448x^3 + 174x^2 - 62x + 107$. Given the size of the coefficients involved, we used one of the modular gcd algorithms described in Section 5.1.2; that is, we had to compute $gcd[p(x), p'(x)]$ over various fields $GF(p_i)$, where $p_1, p_2, \ldots, p_k$ are such that $p_1 \cdot p_2 \cdots p_k > 2B$, where B is the maximum coefficient in absolute value of $p(x)$ and $p'(x)$: Namely, we tried to compute $gcd[p^{(p)}(x), p'^{(p)}(x)]$ for $p = 5, 7, 11$. [Recall that $p^{(p)}(x) = p(x) \pmod{p}$.]

$p = 5$. In this case $p^{(5)}(x) \equiv 2x^4 + 3x^3 + 4x^2 + 2x + 4 \pmod{5}$, $p'^{(5)}(x) \equiv 3x^3 + 4x^2 + 3x + 2 \pmod{5}$ and $gcd[p^{(5)}(x), p'^{(5)}(x)] \equiv 4x + 4 \pmod{5}$.

$p = 7$. This was *not* allowed since $7 | 112$.

$p = 11$. In this case $p^{(11)}(x) \equiv 2x^4 + 3x^3 + 2x^2 + 8x \pmod{11}$, $p'^{(11)}(x) \equiv 8x^3 + 9x^2 + 4x + 8 \pmod{11}$ and $gcd[p^{(11)}(x), p'^{(11)}(x)] \equiv 2 \pmod{11}$.

Since the degree of this last gcd was less than the degree of $gcd[(p^{(5)}(x), p'^{(5)}(x)]$, we conclude that 5 was an unlucky prime; moreover, we conclude that we do not have to try any more primes since $gcd[p(x), p'(x)] = 1$ and $p(x)$ is squarefree over the integers. (This is also easily verified using **PSQFF**.)

Having made sure that $p(x)$ is squarefree, we next had to choose an appropriate prime number p in order to apply Berlekamp's algorithm; the prime number p had to be such that the polynomial $p(x) \pmod{p}$ remained squarefree and had the same degree as $p(x)$. We made the following observations.

$p = 2$. This prime could not be used since $2 | 112$ and $\deg[p^{(2)}(x)] < 4$.

$p = 3$. This prime could not be used because $p^{(3)}(x)$ was not squarefree; that is, we had $p^{(3)}(x) = x^4 + x^3 + 2x^2 + 2x = x(x+2)(x+1)^2$.

$p = 5$. This prime could not be used because $p^{(5)}(x)$ was not squarefree; that is, we had $p^{(5)}(x) = 2x^4 + 3x^3 + 4x^2 + 2x + 4 = (x + 4)(x + 1)^2(2x + 1)$.

$p = 7$. This prime could not be used because $7|112$ and $\deg[p^{(7)}(x)] < 4$.

$p = 11$. This prime *could* be used for our computations since $p^{(11)}(x)$ was squarefree over $GF(11)$ (verify this) and had three factors; namely, $p^{(11)}(x) = x(x + 2)(2x^2 + 10x + 4)$.

$p = 13$. This prime *could* be used for our computations since $p^{(13)}(x)$ was squarefree over $GF(13)$ (verify this) and had two factors; namely, $p^{(13)}(x) = (8x^2 + 12x + 10)(x^2 + 9x + 9)$.

As we have already seen, we choose to work with the factorization modulo 13, since it provided the fewest factors, and we had the less amount of work in lifting them.

Below we present the second example of lifting a factorization mod p to one mod p^j, for some j, using the linear Hensel construction; the symmetric residue set is used now (Yun, 1973).

***Example* [Linear Hensel Construction; $p(x)$ is Monic].** Consider the polynomial $p(x) = x^5 + 12x^4 - 22x^3 - 163x^2 + 309x - 119$ in $\mathbf{Z}[x]$. For $p = 5$ we have

$$p(x) \equiv g_1(x)h_1(x) \pmod{5}$$

where $g_1(x) = x^3 + 2$ and $h_1(x) = x^2 + 2x - 2$. We are going to carry out two iterations of the Hensel construction; that is, we will find $g_3(x), h_3(x)$ such that $p(x) \equiv g_3(x)h_3(x) \pmod{5^3}$; in fact, in this case $p(x) = g_3(x)h_3(x)$ in $\mathbf{Z}[x]$. As before, in order to gain a greater insight as to how the coefficients of $g_3(x), h_3(x)$ are discovered, we reveal the answer; namely, we have

$$g_3(x) = x^3 - 15x + 17, \qquad h_3(x) = x^2 + 12x - 7$$

Using Algorithm 6.3.1, we obtain $a(x) = -x + 1$ and $b(x) = x^2 + 2x - 2$ in $GF(5)[x]$ such that $g_1(x)a(x) + h_1(x)b(x) = 1$ in $GF(5)$. Next, from $p(x) - g_1(x)h_1(x) = 5 \cdot c(x)$—where $g_1(x), h_1(x)$ are considered in $\mathbf{Z}[x]$—we obtain

$$p(x) - g_1(x)h_1(x) = 10x^4 - 20x^3 - 165x^2 + 305x - 115$$

from which we have

$$c(x) = \frac{p(x) - g_1(x)h_1(x)}{5} = 2x^4 + x^3 + 2x^2 + x + 2 \quad \text{in } GF(5)$$

Note that $p \cdot c(x)$ is the difference between $p(x)$ and $g_1(x)h_1(x)$ in the next bigger space $GF(p^2)[x]$ and we must now correct our factors; that is, add to

$g_1(x)$ and $h_1(x)$ what is missing in an effort to make the difference disappear. Using Algorithm 6.3.2, we obtain

$$a'(x) = 2x - 1, \qquad b'(x) = 2x - 2$$

such that $g_1(x)a'(x) + h_1(x)b'(x) = c(x)$ over $GF(5)$ and then define

$$g_2(x) = g_1(x) + p \cdot b'(x) = x^3 + 10x - 8$$
$$h_2(x) = h_1(x) + p \cdot a'(x) = x^2 + 12x - 7$$

It is now easily verified that $p(x) \equiv g_2(x)h_2(x)$ (mod p^2); moreover, note that $h_2(x) \equiv h_1(x)$ and $g_2(x) \equiv g_1(x)$ (mod p), and in addition (since we know the final answer) $h_2(x) \equiv h_3(x)$ and $g_2(x) \equiv g_3(x)$ (mod p^2) [in fact, we already have $h_2(x) = h_3(x)$]

For the second iteration, now, we have the new polynomial

$$c(x) = \frac{p(x) - g_2(x)h_2(x)}{p^2} \pmod{p} = -(x^3 + x^2 + x + 2)$$

and, as we did above, we need to find new polynomials $a'(x)$ and $b'(x)$ in $GF(p)[x]$ such that $g_2(x)a'(x) + h_2(x)b'(x) = c(x)$ over $GF(5)$. Using Algorithm 6.3.2, we obtain $a'(x) = 0$ and $b'(x) = -x + 1$ in $GF(p)[x]$, and then define

$$g_3(x) = g_2(x) + p^2 \cdot b'(x) = x^3 - 15x + 17$$
$$h_3(x) = h_2(x) = x^2 + 12x - 7$$

which are the polynomials we were looking for.

We see that if we denote the successive pairs $\{a'(x), b'(x)\}$ by $\{a_j'(x), b_j'(x)\}$, we have the following relations:

$$g_3(x) = g_1(x) + p \cdot b_1'(x) + p^2 \cdot b_2'(x)$$
$$h_3(x) = h_1(x) + p \cdot a_1'(x) + p^2 \cdot a_2'(x)$$

which clearly indicate that $g_3(x) \equiv g_1(x)$ and $h_3(x) \equiv h_1(x)$ (mod p).

We now proceed with the second lift algorithm, proposed by Zassenhaus in 1969; this is called the *quadratic-lift algorithm* because it extends a factorization "in one step" from $GF(p^j)$ to $GF(p^{2j})$; that is, a factorization is lifted successively mod p^2, p^4, p^8, and so on. In the proof below, we follow the convention that the subscripts $j, 2j$ denote a polynomial in $\mathbf{Z}[x]$ whose coefficients are uniquely defined only mod p^j, mod p^{2j}, respectively.

Lemma 6.3.5 (Zassenhaus, 1969). Let $p(x)$ be a polynomial in $GF(p^{2j})[x]$, such that $lc[p(x)] \neq 0$ in $GF(p)$, and let $p(x) \equiv g_j(x)h_j(x) \pmod{p^j}$ with $g_j(x)$ and $h_j(x)$ being relatively prime (mod p^j); that is, there exist $a_j(x)$, $b_j(x)$ in $GF(p^j)[x]$ such that $g_j(x)a_j(x) + h_j(x)b_j(x) = 1$ over $GF(p^j)$. Then we can determine polynomials $a_{2j}(x), b_{2j}(x), g_{2j}(x), h_{2j}(x)$ in $GF(p^{2j})[x]$, such that $p(x) \equiv g_{2j}(x)h_{2j}(x) \pmod{p^{2j}}$, $g_{2j}(x)a_{2j}(x) + h_{2j}(x)b_{2j}(x) = 1$ over $GF(p^{2j})$ and $g_{2j}(x) \equiv g_j(x) \pmod{p^j}$, $h_{2j}(x) \equiv h_j(x) \pmod{p^j}$.

Proof. Let $c_j(x)$ satisfy the relation $p(x) - g_j(x)h_j(x) = p^j \cdot c_j(x)$ over $GF(p^{2j})$. Using Algorithm 6.3.2, compute $a'_j(x), b'_j(x)$ in $GF(p^j)[x]$ such that $g_j(x)a'_j(x) + h_j(x)b'_j(x) = c_j(x)$ over $GF(p^j)$. Define $g_{2j}(x) = g_j(x) + p^j \cdot b'_j(x)$ and $h_{2j}(x) = h_j(x) + p^j \cdot a'_j(x)$ over $GF(p^{2j})$. It is now trivial to verify that $g_{2j}(x), h_{2j}(x)$ satisfy the conditions above.

To prove that $g_{2j}(x), h_{2j}(x)$ are relatively prime over $GF(p^{2j})$, let $c'_j(x)$ in $GF(p^j)[x]$ be such that $g_{2j}(x)a_j(x) + h_{2j}(x)b_j(x) = 1 + p^j \cdot c'_j(x)$ over $GF(p^{2j})$. Compute $a''_j(x), b''_j(x)$ in $GF(p^j)[x]$ such that $g_j(x)a''_j(x) + h_j(x)b''_j(x) = c'_j(x)$ over $GF(p^j)$, and then define $a_{2j}(x) = a_j(x) - p^j \cdot a''_j(x)$ and $b_{2j}(x) = b_j(x) - p^j \cdot b''_j(x)$ over $GF(p^{2j})$. We then have

$$g_{2j}(x)a_{2j}(x) + h_{2j}(x)b_{2j}(x) = g_{2j}(x)[a_j(x) - p^j \cdot a''_j(x)]$$
$$+ h_{2j}(x)[b_j(x) - p^j \cdot b''_j(x)]$$
$$= a_j(x)g_{2j}(x) + b_j(x)h_{2j}(x)$$
$$- p^j[g_{2j}(x)a''_j(x) + h_{2j}(x)b''_j(x)]$$
$$= 1 \quad \text{over } GF(p^{2j}) \qquad \square$$

Implementation details of the quadratic lift algorithm can be found in the literature. It has been demonstrated that the quadratic lift algorithm is substantially faster than the linear lift algorithm [see Lenstra (1982b, pp. 184–185), Miola et al. (1974), Yun (1974) and Zassenhaus (1978)].

6.3.2 Finding the True Factors over the Integers

Given a polynomial $p(x)$ in $\mathbf{Z}[x]$ to be factored, we have done the following:

1. We chose a suitable prime number p and factored $p(x)$ over $GF(p)$.
2. We took a minimal value of k such that

$$p^k \geq 2 \cdot |lc[p(x)]| \cdot \binom{\lfloor \frac{n}{2} \rfloor}{\lfloor \frac{n}{4} \rfloor} \cdot |p(x)|_2$$

and lifted the factorization of $p(x)$ over $GF(p)$ to the factorization over $GF(p^k)$, $p(x) \equiv lc[p(x)] \cdot \Pi_{1 \leq i \leq r} h_i(x) \pmod{p^k}$. The value of k determined by this relation is derived from (M2), considering the fact we are now looking for factors of $lc[p(x)] \cdot p(x)$ (due to the mechanism to restore the leading coefficient, as we saw in our discussion in

Section 6.3.1 after the first lift example), and that we can restrict ourselves to factors of degree $\leq \lfloor n/2 \rfloor$. After we obtain the factorization of $p(x)$ (mod p^k), working with polynomials that are not necessarily monic, we have to do the following (step 3).

3. Compute the polynomial $h(x) = lc[p(x)] \cdot \prod_{i \in S} h_i(x)$ over $GF(p^k)$ for all subsets S of the set $\{1, \ldots, r\}$ such that $\deg[h(x)] \leq \lfloor n/2 \rfloor$, and test whether $h(x)$ is a divisor of $lc[p(x)] \cdot p(x)$. If so, then the primitive part of $h(x)$ is a factor of $p(x)$ over the integers.

At present there is in general no way to know which subset of the irreducible factors modulo p^k form an irreducible factor of $p(x)$ over the integers, and so we have to test all possible combinations of degree $\leq \lfloor n/2 \rfloor$ to determine whether they lead to a factor of $p(x)$ over $\mathbf{Z}$. If $p(x)$ is irreducible over $\mathbf{Z}$, then none of these trial divisions will be successful, and as we have already mentioned, the tests are exponential in the number r, of the factors modulo p^k. However, it is possible to reduce the number of trial divisions in $\mathbf{Z}[x]$ considerably by (1) first trying the constant coefficient and then (2) factoring $p(x)$ modulo several primes. Use a prime with the smallest number of factors in steps 2 and 3 above, and combine the information about the degrees of the factors modulo the different primes to reduce the possibilities for the degrees of the factors over $\mathbf{Z}$. As we have already seen, if $p(x)$ factors modulo p_1 in 3 factors of degree 2, and if $p(x)$ factors modulo p_2 in 2 factors of degree 3, then $p(x)$ is irreducible over the integers.

Finally, there are two obvious ways to implement step 3 above. The first is the *cardinality procedure*; that is, take combinations of s factors at a time, for $s = 1, 2, \ldots, \lfloor r/2 \rfloor$. The second approach is the *degree procedure*; that is, take combinations of total degree s, for $s = 1, 2, \ldots, \lfloor r/2 \rfloor$. It has been found that the cardinality procedure is preferable.

Exercises

Section 6.1.1

1. Using the Schubert–Kronecker method, factor the following polynomials:
   (a) $x^4 + x^3 + x^2 - x - 2$. (*Hint*: The factors are $x^2 + x + 2, x - 1, x + 1$.)
   (b) $x^4 - x^2 + 1$.
   (c) $x^6 + 1$.
   (d) $x^4 + x^3 + x^2 + x + 1$.

Section 6.2.1

1. Consider the polynomial $p(x) = 112x^4 + 58x^3 - 31x^2 + 107x - 66$ in $\mathbf{Z}[x]$. Is this squarefree in $\mathbf{Z}_p[x]$, $p = 3, 5, 11, 13$? (*Hint*: To answer this

part of the exercise, consult step 2 of Section 6.1.2; that is, perform *gcd* computations.) Use algorithm **PSQFFF** to find the squarefree factors of $p(x)$ in $Z_p[x]$, $p = 3, 5, 11, 13$.

2. Consider the polynomial $p(x) = x^5 + x + 1$ in $Z[x]$. Is this squarefree in $Z_p[x]$, $p = 3, 5, 7, 11, 13$? Use algorithm **PSQFFF** to find the squarefree factors of $p(x)$ in $Z_p[x]$, $p = 3, 5, 7, 11, 13$.

Section 6.2.2

1. Prove Theorem 6.2.7 in the opposite direction; that is, if F is a function defined on natural numbers and f is defined by $f(n) = \Sigma_{d|n} \mu(d) F(n/d)$, prove that $F(n) = \Sigma_{d|n} f(d)$.

2. Compute $i_3(4)$ and find all the irreducible polynomials of degree 4 in $Z_3[x]$.

Section 6.2.3

1. Prove Lemma 6.2.11.
2. Using the techniques discussed in this section factor the polynomials
   (a) $x^9 - x$ in $Z_3[x]$.
   (b) $x^{25} - x$ in $Z_5[x]$.
   (c) $x^5 + x^4 + 1$ in $Z_2[x]$.
   (d) $x^6 + x^5 + x^3 + x^2 + x + 1$ in $Z_2[x]$.

Section 6.2.4

1. Show that if $b_1(x)$ and $b_2(x)$ are relatively prime polynomials in $Z_p[x]$ (or over any field, for that matter) and $p(x)$ is in $Z_p[x]$, we have

   $$gcd[p(x), b_1(x)b_2(x)] = gcd[p(x), b_1(x)] gcd[p(x), b_2(x)]$$

2. Prove that if **V** is a vector space of dimension r over the field $GF(p)$, then **V** has p^r elements.

3. Using Berlekamp's algorithm, factor the polynomials
   (a) $x^5 + x^4 + 1$ in $Z_2[x]$.
   (b) $x^{12} + x^9 + x^6 + x^3 + 1$ in $Z_2[x]$.
   (c) $x^8 - x^7 + x^5 - x^4 + x^3 - x + 1$ in $Z_2[x]$.
   (d) $x^9 - x$ in $Z_3[x]$.
   (e) $x^{25} - x$ in $Z_5[x]$.
   (f) $2x^4 + 3x^3 + 2x^2 + 8$ in $Z_{11}[x]$.

Section 6.3

1. Find the bounds on the coefficients of the factors, in $\mathbf{Z}[x]$, of the polynomials
   (a) $x^5 + x^4 + 1$.
   (b) $x^6 + x^5 + x^3 + x^2 + x + 1$.
   (c) $x^6 + x^5 + x^4 + x^3 + x^2 + x + 1$.
   (d) $x^9 - x$.
   (e) $x^{25} - x$.
   (f) $2x^4 + 3x^3 + 2x^2 + 8$.

Section 6.3.1

1. Consider the polynomials
   (a) $x^5 + x^4 + 1$.
   (b) $x^6 + x^5 + x^3 + x^2 + x + 1$.
   (c) $x^6 + x^5 + x^4 + x^3 + x^2 + x + 1$.
   (d) $x^9 - x$.
   (e) $x^{25} - x$.
   (f) $2x^4 + 3x^3 + 2x^2 + 8$.

   In the exercises for Section 6.2.4 you are asked to factor each one of them in $\mathbf{Z}_p[x]$, for a specified value of p; moreover, in the exercises for Section 6.3 you are asked to compute the respective bounds on the coefficients of their factors in $\mathbf{Z}[x]$. Pick the polynomial of your liking and linear-lift its mod p factorization to a factorization over the integers.

Section 6.3.2

1. Can the polynomial $p(x) = x^3 - 7x + 7$ be factored over the integers?
2. Factor the polynomials $p(x) = x^5 + x^4 + 1$, and $p(x) = x^5 + 1$ over the integers.

Historical Notes and References

1. There exists also *Stickelberger's* theorem (presented below) that helps us to decide whether a given polynomial in $\mathbf{Z}_p[x]$, p an odd prime, has an odd or an even number of distinct irreducible factors. *Stickelberger's theorem*: Let p be an odd prime, $p(x)$ a monic, squarefree {so that its discriminant $\text{discr}[p(x)] \neq 0$} polynomial of degree n in $\mathbf{Z}_p[x]$, and let r be the number of irreducible factors of $p(x)$ in $\mathbf{Z}_p[x]$. Then $r \equiv n \pmod{2}$ if and only if the discriminant $\text{discr}[p(x)]$ is a square in $\mathbf{Z}_p$. [Its proof can be found in Childs (1979).]

2. McEliece (1969) describes an easy to program algorithm to determine a set of polynomials $\{b_1(x), \ldots, b_r(x)\}$ that separates the factors of $p(x)$; this set of polynomials can be computed without matrix diagonalization and each polynomial of the set is of the form $b(x) = x^i + x^{ip} + x^{ip^2} + \cdots + x^{ip^s}$ [mod $p(x)$], for several values of i and s, so that $[b(x)]^p \equiv b(x)$ [mod $p(x)$]. The drawback of this method is that its computing time depends on the degrees of the factors of $p(x)$, whereas Berlekamp's algorithm depends only on n, p and the number of factors.
3. It should be noted that Theorem 6.3.2 corresponds to Theorem 1 in Mignotte's paper of 1974. Actually this theorem was first mentioned by Landau in 1905 and then by W. Specht in 1949. Combined with the well-known Theorem 6.3.1, a bound is obtained on the coefficients of the factors of a polynomial.

Bender, E. A., and J. R. Goldman: On the application of Moebius inversion in combinatorial analysis. *American Mathematical Monthly* **82**, 789–802, 1975.

Berlekamp, E. R.: Factoring polynomials over finite fields. *The Bell System Technical Journal* **46**, 1853–1859, 1967.

Berlekamp, E. R.: *Algebraic coding theory*. McGraw-Hill, New York, 1968.

Berlekamp, E. R.: Factoring polynomials over large finite fields. *Mathematics of Computation* **24**, 713–735, 1970.

Butler, M. C. R.: On the reducibility of polynomials over a finite field. *The Quarterly Journal of Mathematics Oxford Second Series* **5**, 102–107, 1954.

Calmet, J., and R. Loos: An improvement of Rabin's probabilistic algorithm for generating irreducible polynomials over $GF(p)$. *Information Processing Letters* **11**, 94–95, 1980.

Calmet, J., and R. Loos: Deterministic versus probabilistic factorization of integral polynomials. In *Computer Algebra*, EUROCAM 1982 Lecture Notes in Computer Science, Springer Verlag, New York, 1982, pp. 117–125.

Camion, P.: Un algorithm de construction des idempotents primitifs d'ideaux d'algebres sur F_q. *Comptes Rendus des Séances de l'Académie des Sciences* **291**, 479–482, 1980.

Cantor, D. G., and H. Zassenhaus: A new algorithm for factoring polynomials over finite fields. *Mathematics of Computation* **36**, 587–592, 1981.

Cantor, M.: *Geschichte der Mathematik*, Vol. 4. Teubner Verlag, Leipzig, 1908.

Childs, L.: *A concrete introduction to higher algebra*. Springer Verlag, New York, 1979.

Gunji, H., and D. Arnon: On polynomial factorization over finite fields. *Mathematics of Computation* **36**, 281–287, 1981.

Hensel, K.: *Theorie der algebraischen Zahlen*. Teubner Verlag, Leipzig, 1908.

Kaltofen, E.: Factorization of polynomials. *Computing* (Suppl. 4) 95–113, 1982.

Knuth, D. E.: The art of computer programming. Vol. 2, 2nd ed. *Seminumerical Algorithm*. Addison-Wesley, Reading, MA, 1981.

Landau, E.: Sur quelques theoremes de M. Petrovitch relatifs aux zeros des fonctions analytiques. *Bulletin de la Societe Mathematique de France* **33**, 251–261, 1905.

Lauer, M.: Computing by homomorphic images. *Computing* (*Suppl.* 4), 139–168, 1982.

Lazard, D.: On polynomial factorization. In *Computer Algebra, EUROCAM* 1982 *Lecture Notes in Computer Science*, Springer Verlag, New York, 1982, pp. 126–134.

Lenstra, A. K.: *Lattices and factorization of polynomials*. Technical Report 190/81, Department of Computer Science, Mathematisch Centrum, Amsterdam, 1981.

Lenstra, A. K.: *Factoring polynomials over algebraic number fields*. Technical Report 213/82, Department of Computer Science, Mathematisch Centrum, Amsterdam, 1982a.

Lenstra, A. K.: Factorization of polynomials. In *Computational methods in number theory*. H. W. Lenstra, Jr. and R. Tijdeman, eds. *Mathematical Centre Tracts*, Vol. 154, Amsterdam, 1982b, pp. 169–198.

Lenstra, A. K., H. W. Lenstra, Jr., and L. Lovasz: Factoring polynomials with rational coefficients. *Mathematische Annalen* **261**, 515–534, 1982.

Lidl, R., and G. Pilz: *Applied abstract algebra*. Springer Verlag, New York, 1984.

McEliece, R. J.: Factorization of polynomials over finite fields. *Mathematics of Computation* **23**, 861–867, 1969.

Mignotte, M.: An inequality about factors of polynomials. *Mathematics of Computation* **28**, 1153–1157, 1974.

Miola, A., and D. Y. Y. Yun: Computational aspects of Hensel-type univariate polynomial greatest common divisor algorithms. *Proceedings of EUROSAM '74*, 46–54, 1974.

Moenck, R. T.: On the efficiency of algorithms for polynomial factoring. *Mathematics of Computation* **31**, 235–250, 1977.

Musser, D. R.: *Algorithms for polynomial factorization*. Ph.D. Thesis, University of Wisconsin–Madison, Department of Computer Science, 1971.

Musser, D. R.: Multivariate polynomial factorization. *Journal of the Association for Computing Machinery* **22**, 291–308, 1975.

Musser, D. R.: On the efficiency of a polynomial irreducibility test. *Journal of the Association for Computing Machinery* **25**, 271–282, 1978.

Petr, K.: Ueber die Reduzibilitaet eines Polynoms mit ganzzahligen Koeffizienten nach einem Primzahlmodul. *Casopis pro Pestovani Matematiky a Fysiky* **66**, 85–94, 1936.

Rabin, M. O.: Probabilistic algorithms in finite fields. *SIAM Journal of Computing* **9**, 273–280, 1980.

Schwarz, S.: Sur le nombre des racines et des facteurs irréductibles d'une congruence donnée. *Casopis pro Pestovani Matematiky a Fysiky* **69**, 128–145, 1939.

Schwarz, S.: On the reducibility of polynomials over a finite field. *The Quarterly Journal of Mathematics Oxford, Second Series* **7**, 110–124, 1956.

Sims, C. C.: *Algebra, a computational approach*. Wiley, New York, 1984.

Simmons, G.: On the number of irreducible polynomials of degree d over $GF(p)$. *American Mathematical Monthly* **77**, 743–745, 1970.

Smith, S.: On the value of a certain arithmetical determinant. *Proceedings of the London Mathematical Society* **7**, 208–212, 1876.

Specht, W.: Abschaetzungen der Wurzeln algebraischer Gleichungen. *Mathematische Zeitschrift* **52**, 310–321, 1949.

Swinnerton-Dyer, H. P. F.: Cited in Berlekamp's article "Factoring polynomials over large finite fields." Mathematics of Computation **24**, 733–734, 1970.

Wang, P. S.: An improved multivariate polynomial factoring algorithm. *Mathematics of Computation* **32**, 1215–1231, 1978.

Yun, D. Y. Y.: *The Hensel lemma in algebraic manipulation*. Ph.D. Thesis, Department of Mathematics, M.I.T., 1973.

Zassenhaus, H.: On Hensel factorization. *Journal of Number Theory* **1**, 291–311, 1969.

Zassenhaus, H.: A remark on the Hensel factorization method. *Mathematics of Computation* **32**, 287–292, 1978.

7

Isolation and Approximation of the Real Roots of Polynomial Equations

7.1 Introduction and Motivation

7.2 Fourier's Theorem and Sturm's Bisection Method for Isolation of the Real Roots

7.3 Budan's Theorem and the Two Continued Fractions Methods for Isolation of the Real Roots

7.4 Empirical Comparisons Between the Two Real Root Isolation Methods

7.5 Approximation of the Real Roots of a Polynomial Equation

7.6 Empirical Comparisons Between the Two Real Root Approximation Methods

Exercises

Programming Exercises

Historical Notes and References

334 ISOLATION AND APPROXIMATION OF REAL ROOTS OF POLYNOMIAL EQUATIONS

Methods for finding the real roots of, or solving, polynomial equations in $Z[x]$ of any degree can be found in all texts on numerical analysis and theory of equations. Our treatment of the subject basically differs in three major points:

1. We clearly differentiate between isolation and approximation of the roots (Section 7.1), whereas in most texts, root approximation is equivalent to the solution of equations.
2. We give precise statements of the theorems by Budan and Fourier and state their implications (see also Figure 7.1.1); it should be noted that the statement of Budan's theorem can be hardly found in the literature, whereas Fourier's theorem is often attributed to Budan.
3. We present Vincent's theorem of 1836 and the two continued fractions methods, derived from it, for the isolation of the real roots of an equation; the first of these methods is due to Vincent (1836) and has exponential computing time, whereas the second is due to the author (1978) and has polynomial computing time (the best of the classical methods). Vincent's theorem was so totally forgotten that even such a major work as *Enzyclopaedie der mathematischen Wissenschaften* ignores it.

We first examine in detail Sturm's bisection method of 1829 and the continued fractions method of 1978, which is the fastest method existing; these are the two classical methods that exist in the literature for the isolation of the real roots of a polynomial equation with integer coefficients (see historical note 1 and Figure 7.1.1). We then examine two corresponding methods for the approximation of these (isolated) roots to any desired degree of accuracy; namely, a bisection method, and one using continued fractions.

Fourier's Theorem (1820) ← two equivalent theorems → Budan's theorem (1807)
↓ ↓
Sturm's theorem (1829) Vincent's theorem (1836)
↓ ↓
Sturm's bisection method (1829) The continued fractions
method of 1978 developed
by the author

Figure 7.1.1. Overview of the historical development of the two classical methods for the isolation of the real roots of a polynomial equation with integer coefficients. Vincent's continued fractions method of 1836 is not shown because it is an exponential method (see also historical note 7).

7.1
Introduction and Motivation

The earliest known equivalents of algebraic equations occur in the Rhind papyrus, evidently compiled from earlier works by the Egyptian Ahmes about 1650 or 1700 B.C. (van der Waerden, 1956). We find, for instance, the following problem: "A quantity and its seventh added together become 19. What is the quantity?" Clearly, the problem is to solve the equation $x + (1/7)x = 19$, as we would express it today. Lacking a convenient algebraic notation, the Egyptians proceeded by a cumbersome method later known as that of the "false position." Although the Greeks are sometimes credited with solving equations of the second degree, in general neither the Egyptians nor the Greeks made any progress that is significant from a modern point of view. The Arabs achieved more, but it was not until the Renaissance, when the Italian mathematicians of the fifteenth and sixteenth centuries (Tartaglia, Cardano, and Ferrari) succeeded in solving by radicals the general equations of the third and fourth degree, that any work of lasting interest was done. [The historically interested reader can find out about the priority struggle between Tartaglia and Cardano elsewhere, see, e.g., the book by Burnside and Panton (1960), pp. 271–274. Other references are Dickson, 1922; Todhunter, 1882; Turnbull, 1957; and Weisner, 1938.]

In the seventeenth and eighteenth centuries numerous attempts were made to solve the general quintic equation; deeper insight was also gained into the nature of the roots of an algebraic equation (especially with the work of Descartes), but despite all this, no one succeeded in solving by radicals the fifth-degree equation. It was only natural that mathematicians should inquire as to whether any such solution was possible at all. The answer was given by Ruffini (1804) who was the first to demonstrate the impossibility of solving algebraically the quintic equation. Later, Abel (1826) proved that it is impossible to solve, in general, algebraic equations of degree greater than four.

In the beginning of the nineteenth century the attention of the mathematicians had already been focused on numerical methods for the solution of general polynomial equations with integer coefficients. During this period, Fourier conceived of the idea to proceed in *two* steps; that is, first to isolate the real roots, and then to approximate them to any desired degree of accuracy. (According to a well-known procedure in mathematics, when a problem is too difficult to tackle by itself, it is split into simpler ones.)

Isolation of the real roots of a polynomial equation is the process of finding real, disjoint intervals such that each contains exactly one real root and every real root is contained in some interval. *Approximation*, on the other hand, is the process of making the isolation intervals as small as desired, thus approximating the roots to the required degree of accuracy.

The major problem was isolation, which attracted the attention of the mathematicians.

Once the objectives were made clear, success followed. In the beginning of the nineteenth century F. D. Budan and J. B. J. Fourier presented two different (but equivalent) theorems that enable us to determine the *maximum* possible number of real roots that an equation with real coefficients has within a given interval. Budan's theorem appeared in 1807 in the memoir "Nouvelle méthode pour la résolution des équations numériques," whereas Fourier's theorem was first published in 1820 in "Le bulletin des sciences par la société philomatique de Paris." Because of the importance of these two theorems, there was a great controversy regarding priority rights. In his book (1859) *Biographies of distinguished scientific men* (p. 383), F. Arago informs us that Fourier "deemed it necessary to have recourse to the certificates of early students of the Polytechnic School or Professors of the University" in order to prove that he had taught his theorem in 1796, 1797, and 1803.

As we will see in Section 7.2, based on Fourier's proposition, C. Sturm presented in 1829 an improved theorem whose application yields the *exact* number of real roots which a polynomial equation in $Z[x]$, without multiple zeros, has within a real interval; thus he solved the real root isolation problem (Bocher, 1911–1912). Since 1830, Sturm's method has been the only one widely known and used, and consequently Budan's theorem was pushed into oblivion. To our knowledge, Budan's theorem can be found only in a paper by A. J. H. Vincent and in the works of the author, whereas Fourier's proposition appears in almost all texts on the theory of equations. Budan's theorem merits special attention because, as we will see in Section 7.3, it constitutes the basis of Vincent's theorem of 1836, which, in turn, is the foundation of the continued fractions method of 1978 (Akritas, 1980a, 1980b), for the isolation of the real roots of an equation, a method that far surpasses Sturm's in efficiency. [See also Figure 7.1.1 and Akritas, 1982.]

7.2
Fourier's Theorem and Sturm's Bisection Method for Isolation of the Real Roots

In this section we begin first with Fourier's theorem, which enables us to determine the *maximum* possible number of real roots that an equation with integer coefficients has within a given interval and then see how Sturm modified it in order to obtain the *exact* number of real roots within this interval.

7.2.1 Fourier's Theorem

Fourier's theorem, first published in 1820, was also included in his *Analyse des Equations*, published posthumously by C. L. M. N. Navier in 1831. Found in almost all texts on the theory of equations, it is sometimes stated under the name *Budan–Fourier* or even *Budan*. [Hurwitz (1912) presents it as a special case of a more general theorem, and Obreschkoff (1963) generalizes it for complex roots.] We present it after we prove the following two lemmas; although we are interested in polynomials with integer coefficients, we prove these lemmas for the general case of real coefficients.

Lemma 7.2.1. Let $p(x) = 0$ be a polynomial equation of degree $n > 0$ with real coefficients, and let α be one of its real roots (if there exists any). For $\varepsilon > 0$ sufficiently small, the polynomials $p(x)$ and $p'(x)$ [the derivative of $p(x)$] have *opposite* signs in the interval $(\alpha - \varepsilon, \alpha)$ and *same* signs in the interval $(\alpha, \alpha + \varepsilon)$.

Proof. The lemma follows immediately if we consider Taylor's expansion formula $p(a \pm x) = \Sigma_{0 \le i \le n}[p^{(i)}(a)/i!]x^i$, where $p^{(i)}(a)$ is the ith derivative. Suppose that α has multiplicity $m \ge 1$; then, for $p(\alpha \pm \varepsilon)$ and $p'(\alpha \pm \varepsilon)$ we have

$$p(\alpha \pm \varepsilon) = \left[\frac{p^{(m)}(\alpha)}{m!}\right](\pm \varepsilon)^m + \cdots$$

and

$$p'(\alpha \pm \varepsilon) = \left[\frac{p^{(m)}(\alpha)}{(m-1)!}\right](\pm \varepsilon)^{m-1} + \cdots$$

from which we see that the signs of these expressions [which signs, for ε sufficiently small, depend on the sign of the term $p^{(m)}(\alpha)$] are opposite in $(\alpha - \varepsilon, \alpha)$ and the same in $(\alpha, \alpha + \varepsilon)$. □

Applying the previous result to successive derivatives, we have the following.

Lemma 7.2.2. Let $p(x) = 0$ be a polynomial equation with real coefficients of degree $n > 0$, and let α be one of its real roots with multiplicity m. In the sequence of the m functions $p(x), p^{(1)}(x), \ldots, p^{(m-1)}(x)$ [where $p^{(i)}(x)$ is the ith derivative], for $\varepsilon > 0$ sufficiently small, we can replace x by $\alpha - \varepsilon$, so that the signs in the resulting numerical sequence *alternate*, and we can replace x by $\alpha + \varepsilon$, so that the signs in the resulting numerical sequence are all the *same* as that of $p^{(m)}(\alpha)$, the first function for which α is not a root.

Proof. The proof is left as an exercise to the reader. □

Definition 7.2.3. We say that a sign variation (or simply variation) exists between two nonzero numbers c_p and c_q ($p < q$) of a finite or infinite sequence of real numbers $c_1, c_2, c_3, \ldots$, if the following hold:

  i. For $q = p + 1$, c_p, and c_q have opposite signs.
  ii. For $q \geq p + 2$, the numbers $c_{p+1}, \ldots, c_{q-1}$ are all zero and c_p and c_q have opposite signs.

Example. Consider the polynomial $p(x) = x^3 - 7x + 7$, whose coefficients form the finite sequence of numbers $\{1, 0, -7, 7\}$. Clearly, in this sequence there are two sign variations.

Definition 7.2.4. Let $p(x) = 0$ be a polynomial equation of degree $n > 0$ with real coefficients. Then, the sequence of the $n + 1$ functions

$$\text{fseq}(x) = \{p(x), p^{(1)}(x), p^{(2)}(x), \ldots, p^{(n)}(x)\}$$

where $p^{(i)}(x)$ is the ith derivative, is called the *Fourier sequence*.

Example. Consider the polynomial equation $p(x) = x^3 - 7x + 7 = 0$. The Fourier sequence corresponding to this polynomial is $\text{fseq}(x) = \{x^3 - 7x + 7, 3x^2 - 7, 6x, 6\}$.

We are now ready to state and prove Fourier's theorem. The reader should carefully study the proof because it provides useful insight.

An Upper Bound on the Number of Real Roots an Equation Has in an Open Interval

Theorem 7.2.5 (Fourier, 1820). Let $p(x) = 0$ be a polynomial equation with real coefficients of degree $n > 0$. If in the Fourier sequence $\text{fseq}(x) = \{p(x), p^{(1)}(x), p^{(2)}(x), \ldots, p^{(n)}(x)\}$ we replace x by any two real numbers p, q ($p < q$), then the following hold for the two resulting sequences of numbers $\text{fseq}(p)$ and $\text{fseq}(q)$:

  i. $\text{fseq}(p)$ cannot present fewer sign variations than $\text{fseq}(q)$.
  ii. The number of real roots of the equation $p(x) = 0$, located between p and q, can never be greater than the number of sign variations lost in $\text{fseq}(x)$ in passing from the substitution $x := p$ to the substitution $x := q$.
  iii. When the number of real roots of $p(x) = 0$ located between p and q is less than the number of sign variations lost in $\text{fseq}(x)$ in passing from

the substitution $x := p$ to the substitution $x := q$, then the difference is an even number.

Proof. As x varies, the number of sign variations in the Fourier sequence can change *only* when x is a zero of $p(x)$ or of any of its derivatives; we will examine these two cases separately.

> *Case 1.* Let α be a zero of $p(x)$ of multiplicity m. Using Lemma 7.2.2, we see that as x varies in $(\alpha - \varepsilon, \alpha + \varepsilon)$, for sufficiently small $\varepsilon > 0$, there are m sign variations in the Fourier sequence immediately before passage through α, and zero sign variations immediately after. Therefore, there are m sign variations lost in fseq(x).
>
> *Case 2.* Let α now be a zero of one of the derivatives, of multiplicity m; that is, for some $i \neq 0$ and $i \neq n$ we have $p^{(i-1)}(\alpha) \neq 0$ and $p^{(i)}(\alpha) = 0$. Consider the sequence
>
> $$p^{(i-1)}(x), p^{(i)}(x), p^{(i+1)}(x), \ldots, p^{(i+m)}(x)$$
>
> where $p^{(i+m)}(x)$ is the first function for which α is not a root. Observe that here, as x varies in $(\alpha - \varepsilon, \alpha + \varepsilon)$ for sufficiently small $\varepsilon > 0$, the signs of $p^{(i-1)}(x)$ and $p^{(i+m)}(x)$ remain unchanged since they are nonvanishing. Working our the different cases, that is, taking the signs of $p^{(i-1)}(x)$ and $p^{(i+m)}(x)$ to be the same or not, and the multiplicity m to be even or odd, we see that the total number of sign variations lost is an *even* number.

Therefore, we see that as x varies in a given interval, the number of sign variations lost must be either equal to the number of roots of $p(x)$ in the interval, or must exceed it by an even number. □

Example. Consider the polynomial $p(x) = x^3 - 7x + 7 = 0$, whose Fourier sequence, as we have already seen, is fseq(x) = $\{x^3 - 7x + 7, 3x^2 - 7, 6x, 6\}$; then, to find an upper bound on the number of its real roots located in the interval $(0, 2)$, we evaluate the Fourier sequence at 0 and at 2 and we obtain the following number sequences:

$$\text{fseq}(0) = \{7, -7, 0, 6\}, \quad \text{fseq}(2) = \{1, 5, 12, 6\}$$

where fseq(0) has two sign variations and fseq(2) has none. By Fourier's theorem we know that $p(x)$ either has two real roots in the interval $(0, 2)$ or it has none; this is determined by further investigation.

Obviously, Fourier's theorem gives us an upper bound on the number of real roots that the equation $p(x) = 0$ (with real coefficients and of degree n)

has inside the interval (p, q) (see historical note 2); observe that by direct substitution we can easily determine whether p, q are roots.

Theorem 7.2.5 can be used to easily prove the Cardano–Descartes rule of signs.

An Upper Bound on the Number of Positive Roots of an Equation

Theorem 7.2.6 (Cardano–Descartes). Let $p(x) = c_n x^n + c_{n-1} x^{n-1} + \cdots + c_1 x + c_0 = 0$ be a polynomial equation with real coefficients. If v is the number of sign variations in the sequence of coefficients $c_n, c_{n-1}, \ldots, c_0$ (zero coefficients are simply omitted) and p is the number of positive roots of $p(x) = 0$, then

$$v = p + 2\lambda$$

where $\lambda \geq 0$ is an integer.

Proof. What we are looking for is an upper bound on the number of the roots of $p(x) = 0$, located in the interval $(0, \infty)$. The Fourier sequence in this case is

$$p(x) = c_n x^n + \cdots + c_0$$
$$p^{(1)}(x) = nc_n x^{n-1} + \cdots + c_1$$
$$p^{(2)}(x) = n(n-1)c_n x^{n-2} + \cdots + (2!)c_2$$
$$\cdots$$
$$p^{(n)}(x) = (n!)c_n$$

Applying Fourier's theorem, we see that for $x = 0$, we obtain

$$\text{fseq}(0) = \{c_0, c_1, (2!)c_2, \ldots, (n!)c_n\}$$

which has v sign variations, whereas fseq(∞) has no sign variations (since all terms have the sign of c_n). Therefore, by Theorem 7.2.5, we have

$$v = p + 2\lambda$$

where $\lambda \geq 0$ is an integer, and this is exactly the Cardano–Descartes rule of signs. □

Example. Consider the polynomial equation $p(x) = x^3 - 7x + 7 = 0$. Since in the sequence of its coefficients there are two sign variations, we know from the Cardano–Descartes rule of signs that $p(x)$ either has two positive roots or it has none; this is determined by further investigation.

7.2.2. Sturm's Theorem and Sturm's Bisection Method for Isolation of the Real Roots

We remind the reader that the two main subjects of Fourier's life work were the theory of heat and the theory of the solution of numerical equations. Both of these subjects were carried forward by Sturm, who had personal and scientific relations with Fourier. By 1829 the manuscript of Fourier's treatise on the solution of numerical equations had been communicated to several people, including Sturm, who mentions explicitly what a great influence it had on his own work.

What Sturm did was to replace Fourier's sequence by the sequence

$$\text{sseq}(x) = \{p(x), p^{(1)}(x), r_1(x), \ldots, r_k(x)\}, \tag{S}$$

which is called *Sturm's sequence* or *chain*. (This is Sturm's classical sequence; see Definition 7.2.7 for a general definition of Sturm's sequences.) This new sequence is obtained by applying the Euclidean algorithm to the polynomials $p(x)$ and $p^{(1)}(x)$ and taking $r_i(x)$, $i = 1, \ldots, k$ as the *negative* of the remainder polynomial; that is, the sequence is defined by the following relations:

$$p(x) = p^{(1)}(x)q_1(x) - r_1(x)$$
$$p^{(1)}(x) = r_1(x)q_2(x) - r_2(x)$$
$$\ldots \tag{S1}$$
$$r_{k-2}(x) = r_{k-1}(x)q_k(x) - r_k(x)$$

The advantage of Sturm's sequence is that we can now obtain the *exact* number of real roots that the equation $p(x) = 0$ has within a given interval. Observe that if n is the degree of the polynomial $p(x)$, then, usually, there are n auxiliary functions $p^{(1)}(x), r_1(x), \ldots, r_k(x)$, because in the search for the greatest common divisor of $p(x)$ and $p^{(1)}(x)$, each remainder is usually of degree one less than the degree of the preceding remainder. Moreover, if there are no multiple roots, then $r_k(x)$ is a constant. [In Chapter 5 we have examined in detail various ways for constructing (S) over the integers.]

Example. The Sturm sequence corresponding to $p(x) = x^3 - 7x + 7$ is $\text{sseq}(x) = [x^3 - 7x + 7, 3x^2 - 7, 2x - 3, 1\}$.

We have the following.

***Definition 7.2.7** (General Definition of Sturm's Sequence).* Let $p(x) = 0$ be an equation with rational coefficients and without multiple roots. Then, starting with $p(x)$, it is possible to form a sequence of polynomials (as described below), called (*generalized*) *Sturm's sequence*,

$$\text{general-sseq}(x) = \{p(x), p_1(x), \ldots, p_{k+1}(x)\}$$

with the following properties in the interval (p, q) (i.e., when x increases from p to q):

i. In a sufficiently small neighborhood of a real root α of $p(x)$, the polynomials $p(x)$ and $p_1(x)$ have opposite signs if $x < \alpha$ and same signs if $x > \alpha$.
ii. Two consecutive members of Sturm's general-sseq(x) cannot vanish simultaneously.
iii. If one of the functions in Sturm's general-sseq(x) vanishes for some value x_0, then its neighboring functions in the sequence, evaluated at that same value, have opposite signs.
iv. The last function $p_{k+1}(x)$ does not vanish, and hence it keeps constant sign.

After Theorem 7.2.9 we present a way for constructing an infinite number of sets of functions [different from those defined by (S1)] that satisfy properties i–iv of Definition 7.2.7; moreover, as we see in the proof of Theorem 7.2.8, the sequence sseq(x) defined by (S1) does satisfy these properties.

Theorems 7.2.8 and 7.2.9, although valid for *any* general-sseq(x), are stated and proved only for the "classical" Sturm sequence sseq(x) (S).

The Exact Number of Real Roots an Equation Has in an Open Interval

Theorem 7.2.8 (Sturm, 1829). Consider the polynomial equation $p(x) = 0$ with integer coefficients that has only simple roots. Then, the number of its real roots in the interval (p, q) is equal to the difference $v(p) - v(q)$, where $v(\xi)$ denotes the number of sign variations in Sturm's sequence, sseq(x), for $x = \xi$.

Proof. The proof is based on the properties i–iv of Definition 7.2.7 that characterize the functions of a Sturm sequence sseq(x) defined by (S1).

a. In a sufficiently small neighborhood of a real root α of $p(x)$, the polynomials $p(x)$ and $p^{(1)}(x)$ have opposite signs if $x < \alpha$ and same signs if $x > \alpha$ (see Lemma 7.2.1).
b. Two consecutive members of the Sturm sequence cannot vanish simultaneously. To see why this is so, assume the contrary; that is, let $r_i(x)$ and $r_{i+1}(x)$ be both zero for some value x_0. Then from (S1) we see that for the same value x_0 we also have $r_{i+1}(x_0) = r_{i+2}(x_0) = \cdots = r_k(x) = 0$. This, however, is a contradiction since $r_k(x)$ is a nonzero constant. [Obviously, the same result holds also for $p(x)$ and $p^{(1)}(x)$.]

c. If one of the functions in the Sturm sequence vanishes for some value x_0, then, for that same value, its neighboring functions in the sequence have opposite signs. Indeed, if $r_i(x_0) = 0$, then we have

$$r_{i-1}(x_0) = r_i(x_0)q_{i+1}(x_0) - r_{i+1}(x_0)$$

from which it follows that

$$r_{i-1}(x_0) = -r_{i+1}(x_0). \text{ [This applies also for } p^{(1)}(x).]$$

Having established these characteristic properties, we now must show that, as x varies from p to q, the Sturm sequence losses one sign variation if x goes through a root α of $p(x) = 0$ and that, unlike Fourier's sequence, it does not lose a sign variation if x goes through a root of another member of the sequence. Indeed, from property a we see that as x goes through a root α of $p(x) = 0$, we lose exactly one sign variation. So let us now suppose that x goes through α_i, a root of $r_i(x) = 0$. From properties b and c it follows that $r_{i-1}(\alpha_i)$ and $r_{i+1}(\alpha_i)$ are nonzero and have opposite signs. We can then choose a small neighborhood $(\alpha_i - \varepsilon, \alpha_i + \varepsilon)$, $\varepsilon > 0$, where these two functions do not change signs, and construct the following table:

x	r_{i-1}	r_i	r_{i+1}	r_{i-1}	r_i	r_{i+1}
$\alpha_i - \varepsilon$	$+$	$\pm$	$-$	$-$	$\pm$	$+$
α_i	$+$	0	$-$	$-$	0	$+$
$a_i + \varepsilon$	$+$	$\mp(\pm)$	$-$	$-$	$\mp(\pm)$	$+$

[Note that α_i does *not* have to be a simple root of $r_i(x) = 0$.]

From this table we see that in the group of the three functions $r_{i-1}(x), r_i(x), r_{i+1}(x)$ there is no loss of sign variation as x goes through a root of $r_i(x)$, and this completes the proof. □

Example. Consider the polynomial equation $p(x) = x^3 - 7x + 7 = 0$ whose Sturm sequence is $\text{sseq}(x) = \{x^3 - 7x + 7, 3x^2 - 7, 2x - 3, 1\}$. Then, using Theorem 7.2.8, we can say with certainty that $p(x)$ has two real roots in the interval $(0, 2)$ since on evaluating the Sturm sequence at $x = 0$ we obtain $\text{sseq}(0) = \{7, -7, -3, 1\}$ with two sign variations, whereas on evaluating it at $x = 2$ we obtain $\text{sseq}(2) = \{1, 5, 1, 1\}$ with no sign variations and $v(0) - v(2) = 2 - 0 = 2$.

Sturm himself tells us that Theorem 7.2.8 was merely a by-product of his extensive investigations on the subject of linear difference equations of the second order. The requirement that $p(x) = 0$ have only simple roots is no restriction of the generality because we can first apply squarefree factorization (see Section 3.2.4) and then use Sturm's theorem. Moreover, whenever the Sturm sequence has $n + 1$ members, where $n = \deg[p(x)]$, we can easily determine the number of pairs of complex roots of the equation $p(x) = 0$ by the following rule.

Theorem 7.2.9 (Sturm, 1835). Consider the polynomial equation $p(x) = 0$ with integer coefficients, without multiple roots and of degree n. Then, $p(x)$ has as many pairs of complex roots as there are sign variations in the sequence of the first terms of the n functions $p^{(1)}(x), r_1(x), \ldots, r_k(x)$ of the Sturm sequence sseq(x).

Proof. The validity of this rule is based on the fact that one of any two consecutive functions of the Sturm sequence has degree even and the other degree odd. Therefore, if these two functions have the same sign for $x = +\infty$, they will have opposite signs for $x = -\infty$, and vice versa. So, if we evaluate the Sturm sequence sseq(x) at $x = +\infty$ and at $x = -\infty$, each sign variation in either one of the sequences will correspond to a *permance* in the other; that is, the number of permanences in the Sturm sequence evaluated at $x = -\infty$ equals the number of sign variations in the Sturm sequence evaluated at $x = +\infty$.

Let i be the number of sign variations in sseq$(+\infty)$. These variations are obtained from the signs of the coefficients of the highest powers of x in the n auxiliary functions $p^{(1)}(x), r_1(x), \ldots, r_k(x)$, where the first terms of $p(x)$ and $p^{(1)}(x)$ are taken positive.

However, we just saw that sseq$(-\infty)$ will contain i permanences, or equivalently, $n - i$ sign variations. [Here we use the facts that there are $n + 1$ functions in the Sturm sequence, and that in sseq(x) the number of variations plus the number of permanences add up to n.]

Using Theorem 7.2.8, we know that the number of the real roots of $p(x) = 0$, located between $-\infty$ and $+\infty$, equals the number of variations in sseq$(-\infty)$ minus the number of variations in sseq$(+\infty)$. Hence $p(x)$ has $n - 2i$ real roots, and therefore, $2i$ complex roots that appear in pairs. So, we have i pairs of complex roots. □

Example. Consider the polynomial of the last example $p(x) = x^3 - 7x + 7$ whose Sturm sequence is sseq$(x) = \{x^3 - 7x + 7, 3x^2 - 7, 2x - 3, 1\}$, Clearly, $\{1, 3, 2, 1\}$, the coefficients of the first terms of all the members of sseq(x) are positive and, since there is no sign variation, $p(x)$ does not have complex roots.

For completeness we now present another way of constructing a generalsseq(x); thus, we show that there is an infinite number of sets of functions that can be used to isolate the real roots of a polynomial equation.

Let $p_1(x)$ be the derivative of $p(x)$, $\deg[p(x)] = n$; multiply $p_1(x)$ by the binomial $(p_1 x + q_1)$, where p_1 and q_1 are unknowns and subtract $p(x)$ from this product. The result is a polynomial of degree n, which we divide by the polynomial $a_1 x^2 + b_1 x + c_1$, where a_1, b_1, c_1 are known numbers such that $a_1 x^2 + b_1 x + c_1$ remains positive for all real values of x [or, at least it does not vanish but for one value of x, which value is not a root of $p_1(x)$, and remains positive for all other values of x]. On dividing $p_1(x)(p_1 x + q_1) - p(x)$ by $a_1 x^2 + b_1 x + c_1$, we obtain a quotient $p_2(x)$, $\deg[p_2(x)] = n - 2$, containing p_1 and q_1 to the first power in all its terms, and a remainder of degree one of the form $Kx + L$, where the coefficients K, L also contain p_1 and q_1 to the first power. Equating these coefficients K, L to 0, we obtain numerical values for p_1 and q_1; on substitution of these values of p_1 and q_1 in $p_2(x)$, the latter becomes completely specified. Therefore, we have the relation

$$p_1(x)(p_1 x + q_1) - p(x) = p_2(x)(a_1 x^2 + b_1 x + c_1)$$

or

$$p(x) = p_1(x)(p_1 x + q_1) - p_2(x)(a_1 x^2 + b_1 x + c_1)$$

If the coefficient of x^{n-2} in $p_2(x)$ is not zero, we can continue this process and obtain a function $p_3(x)$ such that

$$p_1(x) = p_2(x)(p_2 x + q_2) - p_3(x)(a_2 x^2 + b_2 x + c_2)$$

However, if $\deg[p_2(x)] = n - 3$, then we replace $p_2 x + q_2$ by the trinomial $p_2 x^2 + q_2 x + r_2$ and have the relation

$$p_1(x) = p_2(x)(p_2 x^2 + q_2 x + r_2) - p_3(x)(a_2 x^2 + b_2 x + c_2)$$

If $p(x)$ does not contain multiple roots, then at the end we obtain a function $p_{k+1}(x)$, which is a numerical constant. The reader should verify that this new sequence satisfies the conditions of Definition 7.2.7, and hence it is a Sturm sequence.

Example. Let us apply the above procedure to the polynomial $p(x) = x^3 - 7x + 7$, where clearly we have $p_1(x) = 3x^2 - 7$. Then $p_1(x)(p_1 x + q_1) - p(x) = (3p_1 - 1)x^3 + 3q_1 x^2 + (7 - 7p_1)x - 7q_1 - 7$, and on dividing it by $x^2 + x + 2$, we obtain the quotient $p_2(x) = (3p_1 - 1)x - 3p_1 + 3q_1 + 1$ and the remainder $[-10p_1 - 3q_1 + 8]x + [6p_1 - 13q_1 - 9]$. Forcing the remainder to 0, we obtain $p_1 = 131/148$ and $q_1 = -21/74$, and substituting these values in $p_2(x)$, we obtain $p_2(x) = (245/148)x - 371/148$; that is, we have

$$x^3 - 7x + 7 = (3x^2 - 7)\left[\left(\frac{131}{148}\right)x - \frac{21}{74}\right] - \left[\left(\frac{245}{148}\right)x - \frac{371}{148}\right](x^2 + x + 2)$$

We leave it as an exercise for the reader to compute the constant term and complete the example.

Sturm's theorem can be used to isolate the real roots of a polynomial equation $p(x) = 0$ with integer coefficients and without multiple roots. The most efficient way to do this is to first isolate the positive roots and then the negative ones (Sturm's original proposal). The process is quite simple and is described below; for the analysis of the computing time of this algorithm, we need the following.

Definition 7.2.10. Let $p(x)$ be a univariate polynomial with integer coefficients, $\deg[p(x)] = n \geq 1$; moreover, assume that $p(x)$ has k, $k \leq n$, distinct roots $\rho_1, \ldots, \rho_k$. If $k \geq 2$, we define the *minimum root separation* of $p(x)$, $\Delta > 0$, by

$$\Delta = \min_{1 \leq i < j \leq k} |\rho_i - \rho_j|$$

if $k = 1$, then $\Delta = \infty$.

As we will see in Section 7.2.4, for $n \geq 2$, a lower bound on Δ is given by

$$\Delta \geq \sqrt{3} \cdot n^{-(n+2)/2} \cdot |p(x)|_1^{-(n-1)} \tag{S2}$$

STURM (Sturm's Bisection Method for Isolation of the Real Roots of an Equation)

Input: $p(x) = 0$, a polynomial equation with integer coefficients and without multiple roots.

Output: The isolating intervals of the real roots of $p(x)$ or its exact roots.

1. [Initialize.] Set $p_w(x) := p(x)$; if $p_w(0) = 0$, then output the closed interval $[0, 0]$ and set $p_w(x) := p_w(x)/x$. Compute the Sturm sequence sseq(x), defined by (S1), corresponding to $p_w(x) = 0$ and set *pn*-flag $:= 0$. (When *pn*-flag $= 0$, we isolate the positive roots, whereas when *pn*-flag $= 1$, we isolate the negative.)
2. [Compute a root bound.] Using Cauchy's rule, described in Section 7.2.3, compute an upper bound b on the values of the *positive* roots of $p_w(x)$, if *pn*-flag $= 0$, or on the values of the positive roots of $p_w(-x)$, if *pn*-flag $= 1$, so that the positive or negative roots of $p_w(x)$ lie in the interval $(0, b]$ or $[-b, 0)$, respectively. [In $p_w(-x)$ the negative roots of $p_w(x)$ have become positive; $(0, b]$ is an interval closed at the right endpoint, whereas $[-b, 0)$ is closed at the left endpoint.] If *pn*-flag $= 0$, then do {if $p_w(b) = 0$, then the closed interval $[b, b]$ is output; $l_1 := 0$; $r_1 := b$}, or else do {if $p_w(-b) = 0$, then the closed interval $[-b, -b]$ is output; $l_1 := -b$; $r_1 := 0$}. Set $I := \emptyset$, where I is a list of intervals.

3. [Update the interval.] Set $l := l_1$ and $r := r_1$.
4. [Main loop.] Using Theorem 7.2.8, compute the number of positive roots in the interval (l, r). If there is only one root, then (l, r) is its isolating interval and is output. If the interval contains more than one root, then it is subdivided into the two subintervals $(l, (l + r)/2)$ and $((l + r)/2, r)$ of equal length that are appended to the list I; if $p_w[(l + r)/2] = 0$, then the closed interval $[(l + r)/2, (l + r)/2]$ is output.
5. [Done?] If $I \neq \emptyset$, then remove from it the first interval (l_1, r_1) and go to 3; if $I = \emptyset$ and pn-flag = 1, then exit.
6. [Isolate the negative roots.] If $p_w(x) \neq p_w(-x)$, then do {set pn-flag := 1; go to 2}, or else exit. (If we exit here, the negative roots are symmetric to the positive ones and we already know their isolating intervals. Of course, in this case, the intervals that we obtain for the negative roots are in the positive half-plane and have to be mapped to the corresponding intervals in the negative half-plane, a trivial matter.)

Computing-Time Analysis of **STURM**. From the description of the algorithm above it becomes obvious that Sturm's method is basically a *bisection* method. The way we implement it, computing separate bounds for the positive and negative roots and isolating them separately, is very efficient because we minimize the number of bisections that have to be performed. Moreover, if $p(x) = p(-x)$, then we do not have to do anything to isolate the negative roots, since then are symmetric to the positive ones.

In step 1 we compute Sturm's sequence sseq(x), and we know from Chapter 5 that this is performed in time $0\{n^5 L^2[|p(x)|_\infty]\}$.

In step 2 we compute an upper bound b on the values of the positive roots of $p(x)$[or $p(-x)$], and as we will see in Section 7.2.3, this is executed in time $0\{n^2 L[|p(x)|_\infty]\}$. Moreover, we take the bound b to be a binary rational number; that is, its denominator is a power of 2 and, hence, in step 4 all the rational numbers resulting from bisecting the original interval $(0, b)$ [or $(-b, 0)$] will also be binary rationals.

Let us now compute the time it takes to execute step 4. Note that by using Sturm's theorem, we *do not* actually have to evaluate each member of Sturm's sequence at a given rational number, since we only need the sign of the polynomial evaluated at that number. It follows easily that if $p(x) = \Sigma_{0 \leq i \leq n} c_i x^i$ and a/d is a nonzero rational number $d > 0$, then the sign of $p(a/d)$ is the same as the sign of $\Sigma_{0 \leq i \leq n} c_i a^i d^{n-i}$ (see also the numerical example below). Therefore, we can determine the sign of a polynomial at a rational point using only integer arithmetic. From Section 3.1.2 it is easily seen that the time needed for this computation is

$$0\{n^2 L^2(e) L[|p(x)|_\infty]\} \tag{S3}$$

where $e = \max\{|a|, d\}$. We assume the worst possible case, namely, that *each* polynomial evaluation is calculated with the biggest value e that will appear during the computations; of course e depends on the number of bisections that have to be performed in order to isolate the roots. To compute a bound on $L(e)$, we take logarithms in (S2) and clearly see that

$$L(e) \le |L(\Delta^{-1})| = 0\{nL[|p(x)|_\infty] + nL(n)\}$$

Moreover, in most cases of interest n is small, $L(n) = 1$, and we can safely conclude that

$$|L(\Delta^{-1})| = 0[nL(|p(x)|_\infty)]$$

which also bounds the number of bisections. Therefore

$$L(e) = 0[nL(|p(x)|_\infty)]$$

and replacing this last expression in (S3), we see that in the worst case *each* polynomial evaluation is executed in time

$$0\{n^4 L^2[|p(x)|_\infty]\}$$

Given that we have $0\{nL[|p(x)|_\infty]\}$ bisections for each of the at most n intervals that contain the roots, and that we have at most $n + 1$ polynomials of the Sturm sequence, we see that

$$t_{\text{STURM}}[p(x)] = 0\{n^7 L^3[|p(x)|_\infty]\}$$

(See also historical note 3.)

Example. Let us isolate the roots of $p(x) = x^3 - 7x + 7 = 0$, where $p(0) \ne 0$; $p_w(x) = p(x)$ throughout this example. We already know that sseq$(x) = \{x^3 - 7x + 7, 3x^2 - 7, 2x - 3, 1\}$, and we will use it to isolate both the positive and the negative roots.

> *Isolation of the positive roots.* Using algorithm **BPR** of Section 7.2.3, we obtain $b = 4$ as an upper bound on the values of the positive roots of $p(x)$ (see also the corresponding example in Section 7.2.3); that is, they all are in the interval $(0, 4)$, where $p(0) \ne 0$ and $p(4) \ne 0$. Using Sturm's theorem, we find that there are two roots in the interval $(0, 4)$; namely, at $x = 0$ we obtain sseq$(0) = \{7, -7, -3, 1\}$ with two sign variations, whereas at $x = 4$ we obtain sseq$(4) = \{43, 41, 5, 1\}$ with zero sign variations. Subdividing the interval $(0, 4)$, we find that $p(2) \ne 0$ and that sseq$(2) = \{1, 5, 1, 1\}$ with zero sign variations; therefore, the two roots are in $(0, 2)$ and the interval $(2, 4)$ is disregarded. Subdividing now the interval $(0, 2)$, we find that $p(1) \ne 0$ and that sseq$(1) = \{1, -4, -1, 1\}$ also with two sign variations; therefore, the

two roots are in $(1, 2)$ and the interval $(0, 1)$ is diregarded. We next subdivide the interval $(1, 2)$ and find that $p(3/2) \neq 0$ and that sseq$(3/2) = \{-1/8, -1/4, 0, 1\}$ with one sign variation; therefore, the isolating intervals for the two positive roots are $(1, 3/2)$ and $(3/2, 2)$. Note that in computing sseq$(3/2)$ we did not have to resort to rational arithmetic. Following what we have stated in the proof of the computing time of Sturm's algorithm, it suffices to compute only the signs of the sequence, namely, sseq$(3/2) = \{-, -, 0, +\}$ where, for example, the first sign corresponds to the sign of the *numerator* of $[3^3 - 7 \cdot 3 \cdot 2^2 + 7 \cdot 2^3]/2^3 = (3/2)^3 - 7 \cdot (3/2) + 7$ obtained using only integer arithmetic.

Isolation of the negative roots. We replace x by $-x$ in $p(x) = 0$ and obtain $p(-x) = x^3 - 7x - 7 \neq p(x)$. We pretend that we do not know that there is only one negative root and we compute an upper bound on the values of the positive roots of $p(-x)$; this time again $b = 4$ and, since $p(-4) \neq 0$, all the negative roots of $p(x)$ are in the interval $(-4, 0)$. Using Sturm's theorem, we find that there is only one root in $(-4, 0)$; namely, at $x = -4$ we obtain sseq$(-4) = \{-29, 41, -11, 1\}$ with three sign variations, whereas at $x = 0$ we obtain sseq$(0) = \{7, -7, -3, 1\}$ with two sign variations. Therefore, the isolating interval for the negative root of $p(x)$ is $(-4, 0)$.

From the preceding example it becomes obvious that we have to elaborate on how to (1) compute an upper bound b on the values of the positive roots and (2) determine the number of bisections needed to isolate the roots. As we saw in the discussion of the computing time of Sturm's method, the second topic above relates how close the roots are and is of extreme importance in both guaranteeing that the process terminates in a finite number of steps and calculating its theoretical computing-time bound. Both topics are treated below.

In Section 7.2.3 we present Cauchy's rule, a very efficient rule for the computation of an upper bound b on the values of the *positive* roots of an equation, and then, in Section 7.2.4 we prove (S2), a result due to Mahler (1964) that provides us with a lower bound on the minimum root separation; see also Rump, 1979.

7.2.3 Computation of an Upper (Lower) Bound on the Values of the Positive Roots of a Polynomial Equation

In this section we state and prove Cauchy's rule for the computation of an upper bound on the values of the positive roots of a polynomial equation with integer coefficients. Subsequently, a discussion follows on how this rule is best implemented. The bound we compute is a rational number whose denominator is a power of 2 (see historical note 4).

Theorem 7.2.11 (Cauchy's Rule). Let $p(x) = x^n + c_{n-1}x^{n-1} + \cdots + c_1 x + c_0 = 0$ be a monic polynomial equation with integer coefficients of degree $n > 0$, with $c_{n-k} < 0$ for at least one k, $1 \le k \le n$, and let λ be the number of its negative coefficients. Then

$$b = \max_{\{1 \le k \le n \,:\, c_{n-k} < 0\}} \{|\lambda c_{n-k}|^{1/k}\} \qquad \text{(C)}$$

is an upper bound on the values of the positive roots of $p(x) = 0$.

Proof. From the way b is defined we conclude that

$$b^k \ge \lambda |c_{n-k}|$$

for each k such that $c_{n-k} < 0$; for these k the last inequality is also written as

$$b^n \ge \lambda |c_{n-k}| b^{n-k}$$

Summing over all the appropriate k, we obtain

$$\lambda b^n \ge \lambda \Sigma_{\{1 \le k \le n \,:\, c_{n-k} < 0\}} |c_{n-k}| b^{n-k}$$

or

$$b^n \ge \Sigma_{\{1 \le k \le n \,:\, c_{n-k} < 0\}} |c_{n-k}| b^{n-k}$$

From the last inequality we conclude that if we substitute b for x in $p(x) = 0$, the first term, namely, b^n, will be greater than or equal to the sum of the absolute values of all the negative coefficients. Therefore, $p(x) \ne 0$ for all $x > b$. □

At first glance, it might be thought that Theorem 7.2.11 requires a great amount of computation, since it seems that the calculation of kth roots is necessary. However, this is not so because, for each k with $c_{n-k} < 0$ we compute the smallest integer k' such that

$$\left| \frac{\lambda c_{n-k}}{c_n} \right|^{1/k} \le 2^{k'}$$

and then we set $b = 2^{k''+1}$, where k'' is the maximum of all k'. [Observe that we are treating the general case where $p(x)$ is not monic and $c_n > 0$; see Akritas, 1981b.] The computation of each k' is as follows.

Let

$$\frac{b}{c} = \left| \frac{\lambda c_{n-k}}{c_n} \right|$$

be a quotient for some k, $1 \le k \le n$, and suppose that

$$2^i \le b < 2^{i+1} \quad \text{and} \quad 2^j \le c < 2^{j+1}$$

Then clearly

$$2^{i-j-1} < \frac{b}{c} < 2^{i-j+1}$$

If we set $p = i - j - 1$ and $p + 2 = i - j + 1$, the above inequality becomes

$$2^p < \frac{b}{c} < 2^{p+2}$$

Taking the kth root of the last expression yields

$$2^{p/k} < (b/c)^{1/k} < 2^{(p+2)/k} \tag{C1}$$

If $p = q \cdot k + r$, $0 \le r < k$, then clearly

$$2^q \le 2^{p/k}$$

moreover, $p + 2 = q \cdot k + r + 2$ and

$$2^{(p+2)/k} = 2^q \cdot 2^{(r+2)/k} \le 2^{q+2} \tag{C2}$$

since $(r+2)/k \le 2$ and given $r \le k - 1$. Combining (C1) and (C2) results in $k' = q + 2$. Actually, we can obtain a smaller value for k' if $r \le k - 2$. In this case $(r+2)/k \le 1$ and

$$2^{(p+q)/k} \le 2^{q+1} \tag{C3}$$

so that $k' = q + 1$.

From the above discussion we conclude that the main operations in Cauchy's rule are:

1. The computation of $\lfloor \log_2 |i| \rfloor$, the greatest integer $\le \log_2 |i|$, for any integer $i \ne 0$.
2. The computation of 2^k for a nonnegative integer k.
3. The computation of $[|i|/2^k]$ for any integer i and k positive or negative, where we define

$$[x] = \lfloor x \rfloor \quad \text{if} \quad x \ge 0 \quad \text{and} \quad [x] = \lceil x \rceil \quad \text{if} \quad x < 0$$

Assuming that we have efficient algorithms for these operations (see also the programming exercise at the end of this section), we have the following algorithm.

BPR (Upper **B**ound on Values of **P**ositive **R**oots of a Polynomial Equation)

Input: The polynomial equation $p(x) = c_n x^n + c_{n-1} x^{n-1} + \cdots + c_0 = 0$ with integer coefficients.

Output: The rational number b whose denominator is a power of 2 (b is the smallest power of 2 such that $|\lambda c_{n-k}/c_n|^{1/k} \le b$, for $1 \le k \le n$, and $c_{n-k} < 0$). If $p(x)$ has no negative coefficients, then $b = 1$.

1. [Initialize.] If $lc[p(x)] < 0$ set $p(x) := (-1)p(x)$; set λ equal to the number of negative coefficients of $p(x)$, and $k'' := 0$. If $\lambda = 0$ or $\deg[p(x)] = 0$, then go to 3; otherwise, {set $j := \lfloor \log_2(c_n) \rfloor$; $t := 0$}.
2. [Process negative terms.] For each term $c_i x^{n_i}$, $c_i < 0$, do the following: set $k := n - n_i$; $c_i' := |\lambda c_i|$; $i := \lfloor \log_2(c_i') \rfloor$; $p := i - j - 1$; $q := $ **QUO** $\cdot (p, k)$; $r := p - kq$; if $r < 0$, then do $\{r := r + k; q := q - 1\}$; $k' := q + 1$; if $r = k - 1$, then do $\{c_n' := \lceil c_n/2^{-k'k} \rceil$; if $c_i' > c_n'$, then $k' := k' + 1\}$; if $t = 0$ or $k' > k''$, then $k'' := k'$; $t := 1$.
3. [Exit.] Return $b := 2^{k''}$, a rational number.

Computing-Time Analysis of **BPR**. Taking into consideration the programming exercise for this section, we see the following:

Step 1 is executed in time $0\{L[|p(x)|_\infty]\}$.

One execution of step 2, which computes k' such that $|\lambda c_{n-k}/c_n| \le 2^{k'}$, is executed in time $\sim c\{[k + L[|p(x)|_\infty]\}$. Since step 2 is performed at most n times, we easily conclude (summing over $k = 1, \ldots, n$) that its computing time is $0\{n^2 L[|p(x)|_\infty]\}$.

Step 3 is executed in time $0(|k''| + 1)$. Since k'' is a single-precision integer, we see that the execution time of step 2 dominates the execution time of the whole algorithm and, hence,

$$t_{\mathbf{BPR}}[p(x) = 0\{n^2 L[|p(x)|_\infty]\}.$$

Example. Let us use Cauchy's rule to compute upper bounds on the positive and negative roots of $p(x) = x^3 - 7x + 7 = 0$.

For the positive roots we use $p(x)$ as is. Applying **BPR**, we obtain in the first step $\lambda = 1$, $j = 0$ and $t = 0$. The second step is executed only once and we have $k = 2$, $c'_{3-2} = c'_1 = 7$, $i = 2$, $p = 1$, $q = 0$, $r = 1$, $k' = 1$, $c'_3 = 4$, and since $c'_1 > c'_3$, we update $k' := 2$; finally we set $k'' := 2$ and from the third step obtain $b = 4$.

For the negative roots we replace x by $-x$ so that they become positive. We then obtain the polynomial $-x^3 + 7x + 7$ and again apply **BPR**. This time in the first step the polynomial is changed to $x^3 - 7x - 7$ and so on. Details are left as an exercise to the reader; the answer is $b = 4$.

So far we have computed an upper bound b on the values of the positive roots of an equation $p(x) = 0$. Using the same procedure **BPR**, we can also compute a *lower* bound b_{lo} on the values of the positive roots of $p(x) = 0$. As can be easily verified, $b_{lo} = 1/b_{lo\text{-inv}}$, where $b_{lo\text{-inv}}$ is an upper bound on the values of the positive roots of $p(1/x) = 0$. In Section 7.3.2 we will discuss an efficient way for obtaining $p(1/x)$.

7.2.4 Computation of a Lower Bound on the Distance Between Any Two Roots of a Polynomial Equation

Given a polynomial equation $p(x) = 0$, in this section we will compute an expression [in terms of the degree and the norm of $p(x)$] that is a lower bound on the distance between any two roots (minimum root separation) of $p(x) = 0$ (see also Definition 7.2.10). This result is derived from a more general theorem of K. Mahler (1964) and is of extreme importance in the computing time analysis not only of Sturm's method but, in general, of any method used to isolate the real roots of an equation. (Of interest is Mignotte et al., 1979.)

Theorem 7.2.12 (Mahler). If $p(x) = \Sigma_{0 \leq i \leq n} c_{n-i} x^i$ is a univariate, squarefree polynomial of degree $n \geq 2$ with integer coefficients and Δ is its minimum root separation, then

$$\Delta \geq \sqrt{3} n^{-(n+2)/2} |p(x)|_1^{-(n-1)} \tag{M1}$$

Proof. The main tools in this proof are the inequality of Theorem 6.3.2 and Hadamard's theorem on determinants that may be stated as follows:
  If the elements d_{ij}, $i, j = 1, 2, \ldots, n$, of the determinant

$$d = \det \begin{bmatrix} d_{11} & \cdots & d_{1n} \\ & \cdots & \\ d_{n1} & \cdots & d_{nn} \end{bmatrix}$$

are arbitrary complex numbers, then

$$|d| \leq \{\Pi_{1 \leq j \leq n} (\Sigma_{1 \leq i \leq n} |d_{ij}|^2)\}^{1/2} \tag{M2}$$

and equality holds if and only if

$$\Sigma_{1 \leq i \leq n} d_{ij} \bar{d}_{ik} = 0 \quad \text{for} \quad 1 \leq j < k \leq n$$

where $\bar{d}_{ik}$ denotes the complex conjugate of d_{ik}. [A proof of Hadamard's theorem can be found in Marcus and Minc (1965), Exercise 5, p. 208.]
  To prove Theorem 7.2.12, let $\rho_1, \rho_2, \ldots, \rho_n$ be the roots of $p(x) = \Sigma_{0 \leq i \leq n} c_{n-i} x^i$ and define

$$r[p(x)] = |c_0| \Pi_{1 \leq i \leq n} \max(|\rho_i|, 1)$$

Then, from (Specht's, 1949) Theorem 6.3.2 we have the inequality

$$r[p(x)] \leq |p(x)|_2 \qquad \text{(M3)}$$

where from Definition 1.2.6 it follows that $|p(x)|_\infty \leq |p(x)|_2 \leq |p(x)|_1 \leq (n+1)|p(x)|_\infty$.

Now let the roots of $p(x)$ be so numbered that

$$|\rho_1| \geq |\rho_2| \geq \cdots \geq |\rho_N| > 1 \geq |\rho_{N+1}| \geq \cdots \geq |\rho_n| \qquad \text{(M4)}$$

and let

$$v[p(x)] = \Pi_{1 \leq i < j \leq n}(\rho_i - \rho_j) = \Pi_{1 \leq i \leq n} \Pi_{i < j \leq n}(\rho_i - \rho_j)$$

with the convection that $v[p(x)] = 1$ in the excluded case $n = 1$. It is well known that $v[p(x)] = d_v$, where d_v is the Vandermonde determinant

$$d_v = \det \begin{bmatrix} 1 & 1 & \cdots & 1 \\ \rho_1 & \rho_2 & \cdots & \rho_n \\ \rho_1^2 & \rho_2^2 & \cdots & \rho_n^2 \\ \cdots \\ \rho_1^{n-1} & \rho_2^{n-1} & \cdots & \rho_n^{n-1} \end{bmatrix}$$

Moreover, define $v''[p(x)] = (\rho_1 \rho_2 \cdots \rho_N)^{-(n-1)} v[p(x)]$. Then $v''[p(x)] = d_v''$, where

$$d_v'' = \det \begin{bmatrix} \rho_1^{-(n-1)} & \cdots & \rho_N^{-(n-1)} & 1 & \cdots & 1 \\ \rho_1^{-(n-2)} & \cdots & \rho_N^{-(n-2)} & \rho_{N+1} & \cdots & \rho_n \\ \cdots \\ \rho_1^{-1} & \cdots & \rho_N^{-1} & \rho_{N+1}^{n-2} & \cdots & \rho_n^{n-2} \\ 1 & \cdots & 1 & \rho_{N+1}^{n-1} & \cdots & \rho_n^{n-1} \end{bmatrix}$$

That is, d_v'' is obtained from d_v after multiplying the ith column of the matrix corresponding to d_v times $\rho_i^{-(n-1)}$, $1 \leq i \leq N$. Since, now, the absolute value of each element in the above determinant is ≤ 1, it follows from Hadamard's inequality that

$$|v''[p(x)]| = |d_v''| \leq n^{n/2} \qquad \text{(M5)}$$

Here equality holds if both

$$|\rho_1| = |\rho_2| = \cdots = |\rho_n| = 1$$

and

$$\Sigma_{0\leq k\leq n-1}\bar{\rho}_i^k \rho_j^k = 0 \quad \text{for} \quad 1\leq i<j\leq n \quad \text{(M6)}$$

Let $i = 1$; then, taking into consideration a well-known relation on complex numbers, we see that (M6) becomes

$$\Sigma_{0\leq k\leq n-1}\bar{\rho}_1^k \rho_j^k = \Sigma_{0\leq k\leq n-1}(\rho_1^{-1}\rho_j)^k = 0$$

and multiplying the above expression by $\rho_1^{-1}\rho_j - 1$, we obtain $(\rho_1^{-1}\rho_j)^n = 1$, from which we see that the n quotients $\rho_j/\rho_1, 1\leq j\leq n$, are the n distinct roots of $x^n - 1 = 0$. However

$$x^n - 1 = \Pi_{1\leq j\leq n}\left(x - \frac{\rho_j}{\rho_1}\right)$$

or

$$\rho_1^n x^n - \rho_1^n = \Pi_{1\leq j\leq n}(n - \rho_j)$$

So, we see that $|v''[p(x)]| = n^{n/2}$, just in case $p(x) = c_0 x^n + c_n$, where $|c_0| = |c_n| > 0$.

Now, let r and s be fixed, with $1\leq r<s\leq n$. We are going to obtain a bound on $|d_v''/(\rho_r - \rho_s)|$ by a method very similar to that just applied to $|d_v''|$.

First, in the matrix corresponding to the Vandermonde determinant d_v, substract the sth column from the rth column, so that the new rth column consists of the elements

$$0, \rho_r - \rho_s, \rho_r^2 - \rho_s^2, \ldots, \rho_r^{n-1} - \rho_s^{n-1}$$

all of which are multiples of $\rho_r - \rho_s$. Then, divide this column by $\rho_r - \rho_s$ to get as new elements of the rth column of the matrix corresponding to $|d_v/(\rho_r - \rho_s)|$ the quantities $q_0, q_1, \ldots, q_{n-1}$, where $q_0 = 0$ and $q_i = (\rho_r^i - \rho_s^i)/(\rho_r - \rho_s) = \rho_r^{i-1} + \rho_r^{i-2}\rho_s + \cdots + \rho_r \rho_s^{i-2} + \rho_s^{i-1}$, $i\geq 1$.

Therefore, the quotient $d_v/(\rho_r - \rho_s)$ can now be written as a determinant in which the rth column consists of the elements $q_0, q_1, \ldots, q_{n-1}$, while the other $n - 1$ columns are the same as in the original determinant d_v. Dividing now again the 1st, 2nd, ..., Nth columns of the matrix corresponding to the new determinant by the factors $\rho_1^{n-1}, \rho_2^{n-2}, \ldots, \rho_N^{n-1}$, respectively, we obtain a determinant with value $|d_v''/(\rho_r - \rho_s)|$. Except for the rth column, this determinant is identical with that for d_v''. Its rth column consists of the elements

$$q_0 \rho_r^{-(n-1)}, q_1 \rho_r^{-(n-1)}, \ldots, q_{n-1}\rho_r^{-(n-1)}$$

if $r \leq N$, and of the elements

$$q_0, q_1, \ldots, q_{n-1}$$

if $r > N$. Since

$$|\rho_r| \geq |\rho_s| \quad \text{and} \quad |\rho_r| \quad \text{is} \quad \begin{cases} >1 & \text{for} \quad r \leq N \\ <1 & \text{for} \quad r > N \end{cases}$$

the absolute values of the consecutive elements of the rth column of the determinant do not exceed the values $0, 1, 2, \ldots, n-2, n-1$, respectively. Therefore, by Hadamard's inequality we have

$$\left| \frac{d_v''}{\rho_r - \rho_s} \right|^2 \leq n^{n-1} \{0^2 + 1^2 + 2^2 + \cdots + (n-1)^2\}$$

Since the sum in the brackets is equal to $n(n-1)(2n-1)/6$, which is $< n^3/3$, the final result takes the form

$$\left| \frac{d_v''}{\rho_r - \rho_s} \right|^2 < (1/3) n^{(n+2)} \tag{M7}$$

Solving (M7) for $|\rho_r - \rho_s|^2$, we obtain

$$|\rho_r - \rho_s|^2 > 3n^{-(n+2)} (d_v'')^2$$

or

$$|\rho_r - \rho_s|^2 > 3n^{-(n+2)} (\rho_1 \rho_2 \cdots \rho_N)^{-2(n-1)} \{[v[\rho(x)]]\}^2$$

From Section 5.2.2 we know that the discriminant of $p(x)$, $\text{discr}[p(x)]$, is equal to

$$c_0^{2n-2} \Pi_{1 \leq i \leq n} \Pi_{i < j \leq n} (\rho_i - \rho_j)^2 = c_0^{2n-2} \{v[p(x)]\}^2$$

and, hence, we can also write

$$|\rho_r - \rho_s|^2 > 3n^{-(n+2)} (c_0 \rho_1 \rho_2 \cdots \rho_N)^{-2(n-1)} c_0^{2n-2} \{v[p(x)]\}^2$$

or

$$|\rho_r - \rho_s|^2 > 3n^{-(n+2)} \{r[p(x)]\}^{-2(n-1)} \text{discr}[p(x)]$$

Finally, since the coefficients of $p(x)$ are integers, we have that $|\text{discr}[p(x)]| \geq 1$, since $\text{discr}[p(x)]$ is not zero; moreover, choosing r and s so that $|\rho_r - \rho_s| = \Delta$, and using the inequality (M3), we complete the proof of (M1). □

Example. For the polynomial $p(x) = x^3 - 7x + 7$ we obtain

$$\Delta > \sqrt{3} \cdot 3^{-5/2} \cdot 15^{-2} = \frac{\sqrt{3}}{\sqrt{3^5} \cdot 15^2} \approx \frac{1.7}{\sqrt{243} \cdot 225} \approx \frac{1.7}{3510} \approx 0.0005$$

This means that the smallest distance between any two roots of $p(x)$ is greater than 0.0005.

7.3
Budan's Theorem and the Two Continued Fractions Methods for Isolation of the Real Roots

So far we have examined Sturm's bisection method for the isolation of the real roots of an equation; historically, this method was the first to be developed and was a major breakthrough. In this section we examine the two continued fractions methods that exist in the literature for the isolation of the real roots of an equation; the first of these methods was developed in 1836 by Vincent and is exponential, whereas the second was developed in 1978 (Akritas, 1978a, 1980a, 1980b) by the author and is much faster than Sturm's (and all others). We begin with Budan's theorem, which, as you recall, is equivalent to the one by Fourier.

7.3.1 Budan's Theorem

Although Budan's theorem appeared much earlier than Fourier's, it seems to have been ignored, and it hardly appears in the standard texts on the theory of equations. It is, however, of extreme importance because it constitutes the basis of Vincent's theorem. The following statement of the theorem is from Vincent's paper of 1836.

Another Upper Bound on the Number of Real Roots an Equation Has in an Open Interval

Theorem 7.3.1 (Budan, 1807). If in an equation in x of degree $n > 0$, $p(x) = 0$, we make two substitutions, $x := p + x'$ and $x := q + x''$, where p and q are real numbers such that $p < q$, then the following hold:

  i. The transformed equation in $x' = x - p$ cannot have fewer sign variations than the transformed equation in $x'' = x - q$.
  ii. The number of real roots of the equation $p(x) = 0$, located between p and q, can never be greater than the number of sign variations lost in the sequence of coefficients in passing from the transformed equation in $x' = x - p$ to the transformed equation in $x'' = x - q$.

iii. When the number of real roots of $p(x) = 0$ located between p and q is less than the number of sign variations lost in the sequence of coefficients in passing from the transformed equation in $x' = x - p$ to the transformed equation in $x'' = x - q$, the difference is an even number.

Proof. The proof is analogous to that of Theorem 7.2.5. □

Theorems 7.2.5 and 7.3.1 are equivalent; this fact can be easily seen if in Fourier's sequence we replace x by any real number α. Then, the $n + 1$ resulting numbers are proportional to the corresponding coefficients of the transformed polynomial equation $p(x + \alpha) = \Sigma_{0 \leq i \leq n}[p^{(i)}(\alpha)/i!]x^i$, obtained by Taylor's expansion formula.

Budan's theorem, just like Theorem 7.2.5, also gives us an upper bound on the number of real roots of the equation $p(x) = 0$ inside the interval (p, q). However, it makes use only of the substitutions $x := p + x'$ and $x := q + x''$ and does not depend on any sequence of polynomials; these substitutions are called *Moebius* or *linear fractional substitutions*, and because of their importance, they are studied separately below.

7.3.2 Moebius Substitutions and Their Effect on the Roots of an Equation

Of greatest importance to the discussion of Vincent's theorem (in Section 7.3.3) are substitutions of the form $x := a + 1/x$. They belong to the class of linear fractional, or Moebius substitutions, named after A. F. Moebius (1790–1868), who first studied the corresponding transformations in projective geometry.

Definition 7.3.2. A *general substitution* is defined by the expression

$$x := \mathbf{M}(x) = \frac{ax + b}{cx + d}$$

where a, b, c, d are complex numbers such that their determinant $ad - bc \neq 0$; it is also called a *Moebius substitution* and is abbreviated by $x := \mathbf{M}(x)$, $\det(\mathbf{M}) \neq 0$.

For any $x \in \mathbf{C}$, $x := \mathbf{M}(x)$ is given directly by the expression in Definition 7.3.2, provided $cx + d \neq 0$; otherwise, we define $\mathbf{M}(-d/c) = \infty$. If $c = 0$, then we must have $ad \neq 0$, since the determinant of the coefficients has to be nonzero, and the defining expression of the substitution becomes

$$x := \mathbf{M}(x) = \left(\frac{a}{d}\right)x + \frac{b}{d}$$

moreover, in this case we have $M(\infty) = \infty$, whereas, if $c \neq 0$, then $M(\infty) = a/c$.

It is easily seen that to each Moebius substitution there corresponds the square matrix of its coefficients, and this completely defines the substitution. Let

$$M = \left\{ M(x) \quad \text{such that} \quad x := M(x), M = \begin{bmatrix} m_{11} & m_{12} \\ m_{21} & m_{22} \end{bmatrix}, \right.$$

$$\left. \det M \neq 0, x \in \mathbf{C}' \right\}$$

be the set of all Moebius substitutions; $\mathbf{C}' = \mathbf{C} \cup \{\infty\}$, is the compact complex plane. In M we introduce an equality relation, by noting that two substitutions $A(x), B(x) \in M$ coincide [in other words, $A(x) = B(x)$ for all $x \in \mathbf{C}'$] if and only if there exists $\lambda \in \mathbf{C}$, $\lambda \neq 0$, such that $\mathbf{A} = \lambda \mathbf{B}$ (where the latter is a matrix equality).

Theorem 7.3.3. The set M of the Moebius substitutions forms a group, which is isomorphic to the group of the square matrices of rank 2.

Proof. We have already defined an equality relation in M. We next define the product of two substitutions $A(x), B(x) \in M$ to be the product of their matrices. The so-defined product is also a substitution because

$$x := AB(x) = A[B(x)] = \frac{a_{11}B(x) + a_{12}}{a_{21}B(x) + a_{22}}$$

$$= \frac{(a_{11}b_{11} + a_{12}b_{21})x + a_{11}b_{12} + a_{12}b_{22}}{(a_{21}b_{11} + a_{22}b_{21})x + a_{21}b_{12} + a_{22}b_{22}}$$

$$= \begin{bmatrix} a_{11}b_{11} + a_{12}b_{21} & a_{11}b_{12} + a_{12}b_{22} \\ a_{21}b_{11} + a_{22}b_{21} & a_{21}b_{12} + a_{22}b_{22} \end{bmatrix}(x) = (AB)(x)$$

Note that the order of application of the substitutions is from *left to right*.

As the unit element in M we take the identity substitution $x := x$, whose matrix is

$$\mathbf{I} = \begin{bmatrix} 1 & 0 \\ 0 & 1 \end{bmatrix}$$

Through the manner in which we have defined the equality, we see that $\lambda \mathbf{I}$, where $\lambda \neq 0$ and $\lambda \in \mathbf{C}$, coincides with the identity substitution. For $M(x) \in M$, the inverse substitution of $x := M(x)$ is $x := M^{-1}(x)$, where

$$\mathbf{M}^{-1} = \left[\frac{1}{\det(\mathbf{M})} \right] \begin{bmatrix} m_{22} & -m_{12} \\ -m_{21} & m_{11} \end{bmatrix}$$

is the inverse matrix of **M**; obviously, det $(\mathbf{M}^{-1}) \neq 0$. [Observe that, by the definition of equality, we can take

$$\mathbf{M}^{-1} = \begin{bmatrix} -m_{22} & m_{12} \\ m_{21} & -m_{11} \end{bmatrix}$$

which is derived immediately by solving $w = \mathbf{M}(z)$ for z.] □

The group M is not Abelian, since in general $\mathbf{A}(x)\mathbf{B}(x) \neq \mathbf{B}(x)\mathbf{A}(x)$.

Definition 7.3.4. The following three Moebius substitutions are called the *generating substitutions* of the group M:

i. *Translation*: $x := a + x = \begin{bmatrix} 1 & a \\ 0 & 1 \end{bmatrix}(x)$

ii. *Stretching*: $x := ax = \begin{bmatrix} a & 0 \\ 0 & 1 \end{bmatrix}(x)$,

iii. *Inversion*: $x := 1/x = \begin{bmatrix} 0 & 1 \\ 1 & 0 \end{bmatrix}(x)$

When a is a complex number, stretching is called *rotation*.

The following theorem makes clear the importance of the generating substitutions.

Theorem 7.3.5. Every $\mathbf{M}(x) \in M$ is derived by a suitable multiplication of the generating substitutions (*the order of multiplication being from left to right*). Therefore, every general substitution is obtained by a sequence of the generating substitutions.

Proof. Let

$$\mathbf{M} = \begin{bmatrix} a & b \\ c & d \end{bmatrix}$$

In order to prove the theorem, we distinguish the following two cases:

Case i: $c = 0$. In this case we easily see that $\mathbf{M} = \mathbf{M}_1 \mathbf{M}_2$, where

$$\mathbf{M}_1 = \begin{bmatrix} 1 & \dfrac{b}{d} \\ 0 & 1 \end{bmatrix}, \quad \mathbf{M}_2 = \begin{bmatrix} \dfrac{a}{d} & 0 \\ 0 & 1 \end{bmatrix}$$

since

$$\mathbf{M}_1 \mathbf{M}_2 = \begin{bmatrix} \dfrac{a}{d} & \dfrac{b}{d} \\ 0 & 1 \end{bmatrix} = (1/d)\begin{bmatrix} a & b \\ 0 & d \end{bmatrix}$$

In other words, when $c = 0$, $\mathbf{M}(x)$ is equivalent to a translation, $\mathbf{M}_1(x)$, *followed* by a stretching, $\mathbf{M}_2(x)$.

Case ii: $c \neq 0$. In this case, $\mathbf{M}(x)$ is equivalent to a translation, $\mathbf{M}_1(x)$, followed by an inversion, $\mathbf{M}_2(x)$, followed by another translation, $\mathbf{M}_3(x)$, which finally is followed by a stretching, $\mathbf{M}_4(x)$, where

$$\mathbf{M}_1 = \begin{bmatrix} 1 & \dfrac{a}{c} \\ 0 & 1 \end{bmatrix}, \quad \mathbf{M}_2 = \begin{bmatrix} 0 & 1 \\ 1 & 0 \end{bmatrix}$$

$$\mathbf{M}_3 = \begin{bmatrix} 1 & \dfrac{cd}{bc - ad} \\ 0 & 1 \end{bmatrix}, \quad \mathbf{M}_4 = \begin{bmatrix} \dfrac{c^2}{bc - ad} & 0 \\ 0 & 1 \end{bmatrix}$$

Clearly, their product is

$$\mathbf{M}_1 \mathbf{M}_2 \mathbf{M}_3 \mathbf{M}_4 = \begin{bmatrix} \dfrac{a}{c} & 1 \\ 1 & 0 \end{bmatrix} \mathbf{M}_3 \mathbf{M}_4$$

$$= \begin{bmatrix} \dfrac{a}{c} & \dfrac{bc}{bc - ad} \\ 1 & \dfrac{cd}{bc - ad} \end{bmatrix} \mathbf{M}_4 = \begin{bmatrix} \dfrac{ac}{bc - ad} & \dfrac{bc}{bc - ad} \\ \dfrac{c^2}{bc - ad} & \dfrac{dc}{bc - ad} \end{bmatrix}$$

$$= \dfrac{c}{bc - ad} \begin{bmatrix} a & b \\ c & d \end{bmatrix} \qquad \square$$

Example. Consider the substitution $x := 1/(1 + x)$, whose matrix is

$$\mathbf{M} = \begin{bmatrix} 0 & 1 \\ 1 & 1 \end{bmatrix}$$

Applying Theorem 7.3.5, we have

$\mathbf{M}_1(x) = \begin{bmatrix} 1 & 0 \\ 0 & 1 \end{bmatrix}(x)$, an identity, $\qquad \mathbf{M}_2(x) = \begin{bmatrix} 0 & 1 \\ 1 & 0 \end{bmatrix}(x)$, an inversion,

$\mathbf{M}_3(x) = \begin{bmatrix} 1 & 1 \\ 0 & 1 \end{bmatrix}(x)$, a translation, $\qquad \mathbf{M}_4(x) = \begin{bmatrix} 1 & 0 \\ 0 & 1 \end{bmatrix}(x)$, an identity.

Therefore, $x := 1/(1 + x)$ is equivalent to an inversion, $\mathbf{M}_2(x)$, followed by a *unit* translation, $\mathbf{M}_3(x)$.

From the previous example and Exercise 1 for this section, we easily see that the substitutions $x := 1 + x_1, x_1 := 1 + x_2, \ldots, x_{a-1} := 1 + x_a$, followed by $x_a := 1/(1 + x)$ are equivalent to $x := a + 1/x$ followed by $x := 1 + x$.

We are now prepared to examine the effect of the Moebius, or equivalently of the generating substitutions, on the roots of a univariate polynomial equation $p(x)$. Let

$$p(x) = x^n + c_{n-1}x^{n-1} + \cdots + c_1 x + c_0 = (x - \alpha_1)(x - \alpha_2) \cdots (x - \alpha_n) = 0$$

When a translation takes place, we substitute x by $x + k$ for some real k and obtain the polynomial

$$p(x + k) = p_t(x) = (x + k - \alpha_1) \cdots (x + k - \alpha_n)$$
$$= [x - (\alpha_1 - k)] \cdots [x - (\alpha_n - k)] = 0$$

Thus, with translation, the real part of the roots of the transformed polynomial equation will increase or decrease, according as k is negative or positive.

When a stretching is performed, then x is substituted by kx, for some real k, $k \neq 0$. Consequently,

$$p(kx) = p_s(x) = k^n x^n + c_{n-1}k^{n-1}x^{n-1} + \cdots + c_1 kx + c_0$$
$$= (kx - \alpha_1)(kx - \alpha_2) \cdots (kx - \alpha_n)$$
$$= k^n \left(x - \frac{\alpha_1}{k}\right)\left(x - \frac{\alpha_2}{k}\right) \cdots \left(x - \frac{\alpha_n}{k}\right) = 0$$

Dividing through by k^n, we have

$$\left(\frac{1}{k^n}\right)p(kx) = x^n + \left(\frac{1}{k}\right)c_{n-1}x^{n-1} + \cdots + \left(\frac{1}{k^{n-1}}\right)c_1 x + \left(\frac{1}{k^n}\right)c_0$$
$$= \left(x - \frac{\alpha_1}{k}\right)\left(x - \frac{\alpha_2}{k}\right) \cdots \left(x - \frac{\alpha_n}{k}\right) = 0 \qquad \text{(M1)}$$

Therefore, after a stretching of the form $x := kx$ or $x := x/k$ $(k \neq 0)$, the roots of the transformed polynomial equation will be divided or multiplied by k, respectively. [Observe that if $k = b > 0$, where b is an upper bound on the absolute values of the roots of $p(x)$, then $p(bx) = p_s(x) = 0$ will have all its roots inside the unit circle.]

Finally, if an inversion takes place, then the transformed polynomial equation is

$$p\left(\frac{1}{x}\right) = p_i(x) = \frac{1}{x^n} + \frac{c_{n-1}}{x^{n-1}} + \cdots + \frac{c_1}{x} + c_0$$
$$= \left(\frac{1}{x^n}\right)(1 - x\alpha_1)(1 - x\alpha_2) \cdots (1 - x\alpha_n)$$
$$= \left[\frac{(-1)^n}{x^n}\right]\left(x - \frac{1}{\alpha_1}\right)\left(x - \frac{1}{\alpha_2}\right) \cdots \left(x - \frac{1}{\alpha_n}\right) = 0$$

Multiplying through by x^n, we obtain

$$x^n p\left(\frac{1}{x}\right) = c_0 x^n + c_1 x^{n-1} + \cdots + c_{n-1} x + 1$$

$$= \left(x - \frac{1}{\alpha_1}\right)\left(x - \frac{1}{\alpha_2}\right) \cdots \left(x - \frac{1}{\alpha_n}\right) = 0 \quad \text{(M2)}$$

from which we see that, through an inversion, we obtain a transformed polynomial equation whose roots are the inverses of the roots of the original polynomial equation.

Having seen the effect of the generating substitutions on the roots of a univariate polynomial equation, we now briefly discuss how these substitutions are actually performed.

Translating a polynomial equation is by far of the greatest interest. Analytically, this can be achieved with the help of Taylor's expansion formula $p(\alpha + x) = \Sigma_{0 \le k \le n}[p^{(k)}(\alpha)/k!]x^k$. If $p(x) = c_n x^n + c_{n-1} x^{n-1} + \cdots + c_1 x + c_0 = 0$, and x is substituted by $\alpha + x$, then the coefficients of the transformed polynomial are

$$b_k = \frac{p^{(k)}(\alpha)}{k!}, \quad k = 0, 1, \ldots, n$$

and can be obtained from the formula

$$b_k = \Sigma_{k \le j \le n} \binom{j}{k} \alpha^{j-k} c_j, \quad k = 0, 1, \ldots, n$$

This computation is somewhat simplified when $\alpha = 1$, but even so, one could easily be discouraged by the amount of calculations involved. Fortunately, we have seen that there exists the Ruffini–Horner method (see Section 3.1.2 and Cajori, 1911).

Inversion is very easy to implement. From (M2) we clearly see that, in order to perform the substitution $x := 1/x$ it suffices to invert the order of its coefficients.

Stretching can be achieved [see also (M1)] by scaling the coefficients by powers of k—starting from the one corresponding to the second highest degree.

7.3.3 Vincent's Theorem: Extension and Application

In this section we discuss Vincent's theorem of 1836, which is the foundation of both continued fractions methods for the isolation of the real roots of an equation.

We begin with a careful examination of the Cardano–Descartes rule of signs (Theorem 7.2.6), which states that the number p of positive roots of a polynomial equation $p(x) = 0$ cannot exceed the number v of sign variations in the sequence of its coefficients, and if $n = v - p > 0$, then n is an even number.

A closer examination of Theorem 7.2.6 reveals that it is a rather weak proposition; it gives us the exact number of the positive roots of a polynomial equation, $p(x) = 0$, only in the following two special cases [below, when we lapse into sloppy terminology and say "sign variation," we mean a sign variation in the sequence of coefficients of $p(x)$]:

i. If there is no sign variation, then there is no positive root.
ii. If there is one sign variation, then there is one positive root.

As we will subsequently see, these two special cases are of great importance. Moreover, the converse of case i is also true because we have the following.

Lemma 7.3.6 (Stodola). If the polynomial equation

$$p(x) = c_0 x^d + c_1 x^{d-1} + \cdots + c_d = 0, \qquad (c_0 > 0)$$

with real coefficients c_j, $j = 0, 1, 2, \ldots, d$, has only roots with negative real parts, then all its coefficients are positive, and hence, they present no sign variation.

Proof. Let $-\alpha_n$, $n = 1, 2, \ldots, k$ be the real roots, and let $-\gamma_m \pm i\delta_m$, $m = 1, 2, \ldots, s$ be the complex roots of $p(x) = 0$, where by assumption α_n and γ_m are >0 for all n and m. The polynomial $p(x)$ can be expressed as a product of c_0, $(x + \alpha_n)$ and $[(x + \gamma_m)^2 + \delta_m^2]$, for all $n = 1, \ldots, k$ and $m = 1, \ldots, s$. However, these factors only have positive coefficients, and consequently the coefficients of the product itself will all be positive, thus presenting no sign variation. □

Regarding the second special case of Theorem 7.2.6 (case ii, mentioned above) we observe that its converse is not in general true as can be seen from the polynomial equation $x^3 - x^2 + x - 1 = (x - 1)(x - i)(x + i) = 0$, where we have one positive root but three sign variations. However, under more restrictive conditions the converse is also true; formally this is stated as follows.

Lemma 7.3.7 (Akritas and Danielopoulos, 1985). Let $p(x) = 0$ be a polynomial equation of degree $n > 1$, with real coefficients, without multiple roots, which has one positive root $\xi \neq 0$ and $n - 1$ roots $\xi_1, \xi_2, \ldots, \xi_{n-1}$ with negative real parts (the complex roots appearing in conjugate pairs) and which roots can be expressed in the form

$$\xi_j = -(1 + \alpha_j), \qquad j = 1, 2, \ldots, n - 1$$

with $|\alpha_j| < \varepsilon_n$, where

$$\varepsilon_n = \left(1 + \frac{1}{n}\right)^{1/(n-1)} - 1$$

Then $p(x)$, in its expanded form, presents exactly one sign variation in the sequence of its coefficients.

Proof. Except for a constant factor, the polynomial $p(x)$ can be written in the form

$$p(x) = (x - \xi)(x - \xi_1) \cdots (x - \xi_{n-1})$$
$$= (x - \xi)(x + 1 + \alpha_1) \cdots (x + 1 + \alpha_{n-1})$$
$$= (x - \xi)(x^{n-1} + c_1 x^{n-2} + \cdots + c_{n-1}) \quad \text{(AD1)}$$

where

$$c_k = \Sigma(1 + \alpha_1)(1 + \alpha_2) \cdots (1 + \alpha_k)$$

is a sum consisting of $\binom{n-1}{k}$ terms ($k \leq n - 1$). Clearly, (AD1) can be further written as

$$p(x) = x^n + (c_1 - \xi)x^{n-1} + (c_2 - c_1 \xi)x^{n-2} + \cdots + (c_{n-1} - c_{n-2}\xi)x - c_{n-1}\xi$$

If $c_k > 0$, $k = 1, 2, \ldots, n - 1$, and the ratio c_k/c_{k-1}, with $c_0 = 1$, diminishes with increasing k, then obviously $p(x)$ has exactly one sign variation. To show that $c_k > 0$, $k = 1, 2, \ldots, n - 1$, we note that for each of the $\binom{n-1}{k}$ terms we have

$$|(1 + \alpha_1)(1 + \alpha_2) \cdots (1 + \alpha_k) - 1| \leq (1 + |\alpha_1|)(1 + |\alpha_2|) \cdots (1 + |\alpha_k|) - 1$$

and since by hypothesis $|\alpha_j| < \varepsilon_n$ for $j = 1, 2, \ldots, n - 1$, we obtain

$$(1 + |\alpha_1|)(1 + |\alpha_2|) \cdots (1 + |\alpha_k|) - 1 \leq (1 + \varepsilon_n)^k - 1 \leq (1 + \varepsilon_n)^{n-1} - 1 = \frac{1}{n}$$

Therefore, it is possible to write

$$c_k = \binom{n-1}{k}(1 + \delta_k) \quad \text{(AD2)}$$

where $|\delta_k| \leq 1/n$, and hence, $c_k > 0$, $k = 1, 2, \ldots, n - 1$.

Next we need to show that the ratio c_k/c_{k-1} diminishes with increasing k, in other words, $c_k/c_{k-1} > c_{k+1}/c_k$, $k = 1, 2, \ldots, n - 1$. However, this is trivial since, using (AD2), we obtain

$$\frac{c_k}{c_{k-1}} = \left(\frac{n-k}{k}\right) \cdot \left(\frac{1 + \delta_k}{1 + \delta_{k-1}}\right)$$

and

$$\frac{c_{k+1}}{c_k} = \left(\frac{n-k-1}{k+1}\right) \cdot \left(\frac{1 + \delta_{k+1}}{1 + \delta_k}\right)$$

and we now have to prove that

$$\frac{k(n-k-1)}{(k+1)(n-k)} < \frac{(1+\delta_k)^2}{(1+\delta_{k-1})(1+\delta_{k+1})}$$

Indeed, this is true since on one hand

$$\frac{k(n-k-1)}{(k+1)(n-k)} = 1 - \frac{n}{(k+1)(n-k)} \le 1 - \frac{4n}{(n+1)^2} = \frac{(n-1)^2}{(n+1)^2}$$

and, on the other hand,

$$\frac{(1+\delta_k)^2}{(1+\delta_{k-1})(1+\delta_{k+1})} > \frac{(1-1/n)^2}{(1+1/n)^2} = \frac{(n-1)^2}{(n+1)^2}$$

Therefore, since $c_k > 0$, $k = 1, 2, \ldots, n-1$, and the ratio c_k/c_{k-1} diminishes with increasing k, we have proved the theorem □

Having thoroughly analyzed the two special cases of Theorem 7.2.6, we can now state Vincent's theorem, which depends heavily on them (see also historical note 5; Akritas et al., 1978; Lloyd, 1979; and Poggendorff, 1863).

Theorem 7.3.8 (Vincent, 1836). If in a polynomial equation with rational coefficients and without multiple roots one makes successive substitutions of the form

$$x := a_1 + \frac{1}{x'}, \quad x' := a_2 + \frac{1}{x''}, \quad x'' := a_3 + \frac{1}{x'''}, \ldots$$

where a_1 is an arbitrary nonnegative integer and $a_2, a_3, \ldots$ is any positive integer, then the resulting, transformed equation has either zero or one sign variation. In the latter, the equation has a single positive root represented by the continued fraction

$$a_1 + \cfrac{1}{a_2 + \cfrac{1}{a_3 + \cdots}}$$

whereas in the former case ther is no root.

Proof. The proof of this theorem can be found in Vincent's original paper and it is omitted here. Instead, immediately below we present a proof of a more general theorem. □

Obviously, this theorem treats only positive roots; the negative roots are investigated by replacing x by $-x$ in the original polynomial equation. The generality of the theorem is not restricted by the fact that there should be no multiple roots, because, as in the case of Sturm's theorem, we can first apply squarefree factorization. Vincent himself states that Theorem 7.3.8 was hinted in 1827 by Fourier, who never did give any proof of it (or if he did, it was never found); moreover, Lagrange had used the main idea of this theorem much earlier.

The dependence of Vincent's theorem on the one by Budan is easily seen if each substitution of the form $x := a_i + 1/x$ is replaced by the equivalent pair of substitutions $\{x := a_i + x, x := 1/x\}$.

Intuitively speaking, the purpose of the series of successive substitutions of the form $x := a_i + 1/x$, performed on $p(x) = 0$, is to force one of its positive roots inside the interval $(0, 1)$ and all the others inside $(1, \infty)$ or vice versa—excluding, of course, the case when 1 is a root. In the first case, the subsequent substitution $x := 1/(1 + x)$ will result in an equation with only one root in $(0, \infty)$, whereas in the second case the same is achieved with the subsequent substitution $x := 1 + x$.

In Theorem 7.3.8, the question naturally arises as to the maximum number of substitutions of the form $x := a_i + 1/x$, necessary to obtain the polynomial equation with at most one sign variation. The answer is given by the following theorem (see also historical note 6).

Theorem 7.3.9 (Vincent–Uspensky–Akritas). Let $p(x) = 0$ be a polynomial equation of degree $n > 1$, with rational coefficients and without multiple roots, and let $\Delta > 0$ be the smallest distance between any two of its roots. Let m be the smallest index such that

$$F_{m-1} \frac{\Delta}{2} > 1, \quad F_{m-1} F_m \Delta > 1 + \frac{1}{\varepsilon_n} \tag{V1}$$

where F_k is the kth member of the Fibonacci sequence 1, 1, 2, 3, 5, 8, 13, 21, ... and

$$\varepsilon_n = \left(1 + \frac{1}{n}\right)^{1/(n-1)} - 1 \tag{V2}$$

Let a_1 be an arbitrary nonnegative integer, and let $a_2, \ldots, a_m$ be arbitrary positive integers. Then the substitution

$$x := a_1 + \cfrac{1}{a_2 + \cfrac{}{\ddots + \cfrac{1}{a_m + \cfrac{1}{\xi}}}} \tag{V3}$$

(which is equivalent to the series of successive substitutions of the form $x := a_i + 1/\xi$, $i = 1, 2, \ldots, m$) transforms the equation $p(x) = 0$ into the equation $p_{ti}(\xi) = 0$, which has not more than one sign variation.

Proof. In order to prove the theorem it suffices to show that, after the successive substitutions of the form $x := a_i + 1/\xi$, the real part of all complex roots, as well as all real roots except for at most one, become negative. (It is of extreme importance to note that the roots of the transformed equation are *clustered together around* -1.)

Indeed, let p_k/q_k be the kth convergent to the continued fraction

$$a_1 + \cfrac{1}{a_2 + \cfrac{1}{a_3 + \cdots}}$$

From Section 2.2.3 we know that for $k \geq 0$, $p_0 = 1$, $p_{-1} = 0$, $q_0 = 0$ and $q_{-1} = 1$ we have

$$p_{k+1} := a_{k+1} p_k + p_{k-1}$$
$$q_{k+1} := a_{k+1} q_k + q_{k-1}$$

Since $q_1 = 1$ and $q_2 = a_2 \geq 1$, it follows that $q_k \geq F_k$. Further, (V3) can be expressed in the form

$$x = \frac{p_m \xi + p_{m-1}}{q_m \xi + q_{m-1}}$$

from which it follows that

$$\xi = -\left(\frac{p_{m-1} - q_{m-1} x}{p_m - q_m x}\right) \tag{V4}$$

Clearly, if x_0 is any root of the equation $p(x) = 0$, the quantity ξ_0, determined by (V4), is the corresponding root of the transformed equation $p_{ti}(\xi) = 0$.

a. Assume that x_0 is a complex root of $p(x) = 0$; that is, $x_0 = a \pm ib$, $b \neq 0$. In this case the real part of the corresponding root ξ_0 is

$$rp(\xi_0) = -\left[\frac{(p_{m-1} - q_{m-1} a)(p_m - q_m a) + q_{m-1} q_m b^2}{(p_m - q_m a)^2 + q_m^2 b^2}\right] \tag{V5}$$

which is certainly negative if $(p_{m-1} - q_{m-1} a)(p_m - q_m a) \geq 0$. If, on the contrary, $(p_{m-1} - q_{m-1} a)(p_m - q_m a) < 0$, then clearly the value of a is contained between the two consecutive convergents p_{m-1}/q_{m-1} and p_m/q_m, whose difference in absolute value is $1/q_{m-1}q_m$. Hence,

$$\left|\frac{p_{m-1}}{q_{m-1}} - a\right| < \frac{1}{q_{m-1}q_m} \quad \text{and} \quad \left|\frac{p_m}{q_m} - a\right| < \frac{1}{q_{m-1}q_m}$$

from which it follows that

$$|(p_{m-1} - q_{m-1}a)(p_m - q_m a)| < \frac{1}{q_{m-1}q_m} \leq 1 \quad \text{(V6)}$$

From (V5) and (V6) we conclude that $rp(\xi_0)$ will be negative if $q_{m-1}q_m b^2 > 1$. To prove that this is true in our case, first observe that, since Δ is the minimum distance between any two roots of $p(x) = 0$, we have

$$|(a + ib) - (a - ib)| = |2ib| = 2|b| \geq \Delta$$

from which we obtain $|b| \geq \Delta/2$; moreover, we know that $q_m \geq q_{m-1} \geq F_{m-1}$, and, from (V1), $F_{m-1}\Delta/2 > 1$. Then clearly $F_{m-1}|b| > 1$, which implies $q_{m-1}|b| > 1$ and $q_m|b| > 1$. From the last two inequalities we obtain $q_{m-1}q_m b^2 > 1$, thus proving that the $rp(\xi_0) < 0$; this is obviously true for all complex roots of the transformed equation $p_{ti}(\xi) = 0$.

b. Assume now that x_0 is a real root of $p(x) = 0$, and consider, at first, the case where for all real roots x_i:

$$(p_{m-1} - q_{m-1}x_i)(p_m - q_m x_i) > 0$$

From (V4) it follows that all real roots of the transformed equation $p_{ti}(\xi) = 0$ will be negative; moreover, we know from assumption a that all the complex roots of $p_{ti}(\xi) = 0$ have negative real parts. As a consequence of Lemma 7.3.6, $p_{ti}(\xi)$ presents no sign variation.

Suppose, now, that for some real root x_0

$$(p_{m-1} - q_{m-1}x_0)(p_m - q_m x_0) \leq 0 \quad \text{(V7)}$$

Then, clearly, x_0 is contained between the two consecutive convergents p_{m-1}/q_{m-1} and p_m/q_m, and hence, $|p_m/q_m - x_0| \leq 1/q_{m-1}q_m$. Let x_k, $k \neq 0$, be any other root, real or complex, of $p(x) = 0$, and ξ_k the corresponding root of the transformed equation. Then, keeping in mind that

$$p_m q_{m-1} - p_{m-1} q_m = (-1)^m$$

it follows from (V4) that

$$\xi_k + \frac{q_{m-1}}{q_m} = \frac{(-1)^m}{q_m(p_m - q_m x_k)}$$

or

$$\xi_k = -\left(\frac{q_{m-1}}{q_m}\right)\left[1 - \frac{(-1)^m}{q_{m-1}q_m(p_m/q_m - x_k)}\right] = -\left(\frac{q_{m-1}}{q_m}\right)(1 + \alpha_k)$$

where

$$\alpha_k = \frac{(-1)^{m-1}}{q_{m-1}q_m(p_m/q_m - x_k)}$$

Now

$$\left|\frac{p_m}{q_m} - x_k\right| = \left|\frac{p_m}{q_m} - x + x - x_k\right| \geq |x - x_k| - \left|\frac{p_m}{q_m} - x\right|$$

$$\geq \Delta - \frac{1}{q_{m-1}q_m} > 0$$

and consequently

$$|\alpha_k| \leq \frac{1}{q_{m-1}q_m \Delta - 1} \leq \frac{1}{F_{m-1}F_m \Delta - 1}$$

From the last expression and the second inequality of (V1) we deduce that $|\alpha_k| < \varepsilon_n$. Thus, the roots ξ_k, $k = 1, 2, \ldots, n-1$, of the transformed equation $p_{ti}(\xi) = 0$, corresponding to the roots x_k, $k = 1, 2, \ldots, n-1$, of the equation $p(x) = 0$, which are all different from x_0, are of the form

$$\xi_k = -\left(\frac{q_{m-1}}{q_m}\right)(1 + \alpha_k), \qquad |\alpha_k| < \varepsilon_n \tag{V8}$$

That is, the roots of the transformed equation have negative real parts and *are clustered together around* -1. If we make the substitutions

$$\xi := -\left(\frac{q_{m-1}}{q_m}\right)u, \qquad \xi_k := -\left(\frac{q_{m-1}}{q_m}\right)\xi'_k, \qquad k = 0, 1, \ldots, n-1$$

where

$$\xi'_0 > 0, \qquad \xi'_k = -(1 + \alpha_k), \qquad k = 1, 2, \ldots, n-1$$

then the transformed polynomial $p_{ti}(\xi)$ can be written in the form

$$p_{ti}(\xi) = \left(\frac{q_{m-1}}{q_m}\right)^n p_{ti}(u) = c\left(\frac{q_{m-1}}{q_m}\right)^n (u - \xi'_0) \cdots (u - \xi'_{n-1})$$

Since $p_{ti}(u)$ satisfies all the assumptions of Lemma 7.3.7, it presents exactly one sign variation, and, obviously, the same is true for the transformed polynomial $p_{ti}(\xi)$.

The last thing to consider now is the case when (V7) holds as an equality; that is

$$(p_{m-1} - q_{m-1}x_0)(p_m - q_m x_0) = 0$$

If $p_{m-1} - q_{m-1}x_0 = 0$, then we see from (V4) that $\xi_0 = 0$, and clearly the transformed equation $p_{ti}(\xi) = 0$ has no sign variation (Lemma 7.3.6). For $p_m - q_m x_0 = 0$ we have $\xi_0 = \infty$, and the transformed equation reduces to degree $n - 1$. Since again all the roots have negative real parts, we conclude from Lemma 7.3.6 that $p_{ti}(\xi) = 0$ presents no sign variation. Thus, we have proved the theorem completely. □

From Theorem 7.3.9 we clearly see that m is the desired bound on the number of substitutions of the form $x := a_i + 1/x$ that have to performed in order to obtain the equation with at most one sign variation in the sequence of its coefficients.

Corollary 7.3.10. Under the assumptions of Theorem 7.3.9 we have

$$m = 0\{nL[|p(x)|_\infty] + nL(n)\}$$

Proof. By definition m is the smallest index such that both inequalities (V1) hold simultaneously. Clearly, one of these inequalities (and possibly both) will not be satisfied if we reduce m by one; suppose that the first one fails, so that

$$F_{m-2}\frac{\Delta}{2} \leq 1 \qquad (V9)$$

Applying the relation $F_k = \phi^k/\sqrt{5}$ (with the right-hand side of the equation rounded to the nearest integer), where $\phi = 1.618\ldots$, (V9) yields $\phi^{m-2} \leq 2\sqrt{5} \cdot (1/\Delta)$, from which we deduce that

$$m \leq 2 + \log_\phi 2 + (1/2)\log_\phi 5 - \log_\phi \Delta \qquad (V10)$$

Moreover, recall that from Theorem 7.2.12 (Mahler, 1964) we have

$$\Delta \geq \sqrt{3} n^{-(n+2)/2} |p(x)|_1^{-(n-1)} \tag{V11}$$

The corollary is now proved if we combine (V10) and (V11), taking also into consideration the fact that $L[|p(x)|_2] \sim L[|p(x)|_\infty]$. [The same result is obtained if we assume that the second of the inequalities (V1) fails.] □

In most cases of interest $L(n) = 1$, and, so, we can safely conclude that

$$m = 0\{nL[|p(x)|_\infty]\} \tag{V12}$$

Theorem 7.3.9 can be used in the isolation of the real roots of a polynomial equation. To see how it is applied, observe the following (for clarity and better understanding, we repeat certain parts from the proof of Theorem 7.3.9, where ξ is now replaced by y):

i. The continued fraction substitution (V3) can also be written as

$$x := \frac{p_m y + p_{m-1}}{q_m y + q_{m-1}} \tag{V13}$$

where p_k/q_k is the kth convergent to the continued fraction

$$a_1 + \cfrac{1}{a_2 + \cfrac{1}{a_3 + \cdots}}$$

and, as we stated before, for $k \geq 0$, $p_0 = 1$, $p_{-1} = 0$, $q_0 = 0$, and $q_{-1} = 1$ we have

$$p_{k+1} := a_{k+1} p_k + p_{k-1}, \qquad q_{k+1} := a_{k+1} q_k + q_{k-1} \tag{V14}$$

ii. The distance between two consecutive convergents is $|p_{m-1}/q_{m-1} - p_m/q_m| = 1/q_{m-1} q_m$. Clearly, the smallest values of the q_i occur when $a_i = 1$ for all i. Then, $q_m = F_m$, the mth Fibonacci number. This explains why there is a relation between the Fibonacci numbers and the distance Δ in Theorem 7.3.9.

iii. Let $p_{ti}(y) = 0$ be the equation obtained from $p(x) = 0$ after a substitution of the form (V13), corresponding to a series of *t*ranslations and *i*nversions. Observe that (V13) maps the interval $0 < y < \infty$ onto the x-interval whose unordered endpoints are the consecutive convergents p_{m-1}/q_{m-1} and p_m/q_m. If this x-interval has length less than Δ, then it contains at most one root of $p(x) = 0$, and the corresponding equation $p_{ti}(y) = 0$ has at most one root in $(0, \infty)$.

iv. If y' were this positive root of $p_{ti}(y) = 0$, then the corresponding root x' of $p(x) = 0$ could be easily obtained from (V13). We only know, though, that y' lies in the interval $(0, \infty)$; therefore, substituting y in (V13) once by 0 and once by ∞ we obtain for the positive root x its isolating interval whose unordered endpoints are p_{m-1}/q_{m-1} and p_m/q_m. To each positive root there corresponds a different continued fraction; at most m partial quotients have to be computed for the isolation of any positive root. (As we mentioned before, negative roots can be isolated if we replace x by $-x$ in the original equation.)

7.3.4 The Two Continued Fractions Methods for Isolation of the Real Roots

From the preceding discussion it is obvious that the calculation of the partial quotients $a_1, a_2, \ldots, a_m$, for the substitutions of the form (V3), which lead to an equation with exactly one sign variation, constitutes the real root isolation procedure. [From Budan's theorem we know that the value of a particular partial quotient a_i has been computed if $p(a_i + x) = 0$ has more sign variations in the sequence of its coefficients than $p(a_i + 1 + x) = 0$.]

There are two continued fractions methods, Vincent's of 1836 and the author's of 1978, corresponding to the two different ways in which the computation of the partial quotients a_i may be performed (see historical note 7). As we see below, the difference between these two methods can be considered analogous to the difference between the integrals of Riemann and Lebesgue. That is, it is well known that the sum $1 + 1 + 1 + 1 + 1$ can be computed in the following two ways:

i. $1 + 1 = 2, 2 + 1 = 3, 3 + 1 = 4, 4 + 1 = 5$ (Riemann).
ii. $5 \cdot 1 = 5$ (Lebesgue).

Vincent's continued fractions method of 1836 basically consists of computing a particular partial quotient a_i by a series of unit incrementations, $a_i := a_i + 1$, with each one of which we have to preform the translation $p_{ti}(x) := p_{ti}(1 + x)$ [for some polynomial equation $p_{ti}(x) = 0$], and check for a change in the number of sign variations. This "brute force" approach results in a method with exponential behavior and, hence, of little practical value; no special algorithm is presented for this method.

As an example of Vincent's method, let us isolate the roots of the polynomial equation

$$p(x) = (x - \alpha)(x - \beta) = 0 \tag{A1}$$

where $\alpha = 5 \cdot 10^9 + \varepsilon$ and $\beta = \alpha + 1$. Consider $a_1^{(\alpha)}$, the first partial quotient for α, which is $5 \cdot 10^9$. Using Vincent's method, we initially set $a_1^{(\alpha)} := 1$, $p_{ti}(x) := p(x)$ and compute $p_{ti}(x) := p_{ti}(1 + x)$. Since the number of sign variations in the sequence of coefficients of the transformed polynomial

$p_{ti}(x)$ has not changed, we increment $a_1^{(\alpha)} := a_1^{(\alpha)} + 1$, and compute a new $p_{ti}(x) := p_{ti}(1 + x)$, checking again the number of sign variations. This process is repeated $5 \cdot 10^9$ times and, on the fastest computer available, it would take several years. However, Vincent's method is quite efficient when the values of the partial quotients are small.

On the contrary, the continued fractions method developed by the author in 1978 consists of immediately computing a particular partial quotient a_i as the *lower* bound b on the values of the positive roots of some polynomial $p_{ti}(x)$; that is, using Cauchy's rule (Section 7.2.3), we immediately determine b and we set $a_i := b$. [Recall that computing a lower bound b_{lo} on the values of the positive roots of some polynomial equation $p(x) = 0$ is equivalent to computing an upper bound $b_{lo\text{-inv}}$ on the values of the positive roots of $p(1/x) = 0$ and then setting $b_{lo} := 1/b_{lo\text{-inv}}$.] Once we set $a_i := b, b \geq 1$, we only need to perform on the corresponding polynomial $p_{ti}(x)$ the substitution $x := b + x$, which takes approximately the same amount of time as the substitution $x := 1 + x$; therefore, with this method, we have enormous savings of computing time, and (A1) is solved in a matter of seconds. A detailed algorithm for this method is presented below.

Notice that for all i, $a_i = \lfloor \alpha_s \rfloor$, where α_s is the samllest positive root of some polynomial equation. Hence, it is obvious that, in general, Cauchy's rule is applied more than once in order to compute $\lfloor \alpha_s \rfloor$; for example, 18 applications were needed to compute $[5 \cdot 10^9 + \varepsilon]$ for the polynomial (A1). However, since the number of applications of Cauchy's rule is very small compared to the value of a_i, and it cannot be predetermined, we can safely conclude that $b = \lfloor \alpha_s \rfloor$ in our discussion.

This interpretation of each a_i as a lower bound on the values of the positive roots of $p(x) = 0$ is clarified if we consider that our objective is to force one of its positive roots inside the interval $(0, 1)$ and all others inside the interval $(1, \infty)$ or vice versa. The following lemmas are relevant.

Lemma 7.3.11. Let $p(x) = 0$ be an univariate polynomial equation of degree $d \geq 2$, with integer coefficients and without multiple roots, which has p real roots inside the interval $(0, 1)$, $2 \leq p \leq d$, and let $\Delta_p > 0$ be the smallest distance between any two of these roots. Then the inversion $x := 1/x$, performed on $p(x) = 0$, maps these p roots in the interval $(1, \infty)$, where now the smallest distance between any two of them is $\Delta'_p > \Delta_p$.

Proof. Let $0 < \alpha_1 < \cdots < \alpha_i < \alpha_j < \cdots < \alpha_m < \alpha_n < \cdots < \alpha_p < 1$ be the p roots of $p(x) = 0$ inside the interval $(0, 1)$, and suppose that $\Delta_p = \alpha_j - \alpha_i$, whereas $\Delta'_p = 1/\alpha_m - 1/\alpha_n$. The lemma follows immediately since $\Delta'_p = (\alpha_n - \alpha_m)/\alpha_m \alpha_n > \alpha_n - \alpha_m \geq \alpha_j - \alpha_i = \Delta_p$. □

Lemma 7.3.12. Let $p(x) = 0$ be an univariate polynomial equation of degree $d \geq 2$, with integer coefficients and without multiple roots, which has two complex conjugate roots α_1 and α_2 inside the circle with center $(1/2, 0)$ and radius $1/2$; moreover, let $\delta_p = |\alpha_1 - \alpha_2|$. Then the inversion $x := 1/x$, performed on $p(x) = 0$, maps α_1 and α_2 in the half-plane with real part >1, where now their distance is $\delta'_p > \delta_p$.

Proof. Similar to that for Lemma 7.3.11. □

Below we continue to elaborate on the distinction between the two continued fractions methods for the isolation of the real roots of a polynomial equation.

Consider an infinite binary tree with each node of which we associate a triplet of the form $\{p_M(x), \mathbf{M}(x), v_M\}$, where $p_M(x)$ is obtained from an original polynomial $p(x)$ after the substitution $x := \mathbf{M}(x) = (a_1 x + a_0)/(b_1 x + b_0)$, and v_M is the number of sign variations in the sequence of coefficients of $p_M(x)$. [It is necessary to associate with each transformed polynomial the corresponding function $\mathbf{M}(x)$ so that at the end we can easily obtain the isolating intervals of the roots; see also (V13).] If $p(x) = 0$ is the original polynomial equation with v sign variations in the sequence of its coefficients, then the root of the tree corresponds to the triplet $\{p_M(x) := p(x), \mathbf{M}(x) := x, v_M := v\}$.

The path from each node to the right descendant corresponds to the substitution $x := 1 + x$, whereas the path to the left descendant corresponds to the substitution $x := 1/(1 + x)$; note that for any partial quotient a_i, a series of a_i successive substitutions of the form $x := 1 + x$, followed by $x := 1/(1 + x)$ is equivalent to $x := a_i + 1/x$ followed by $x := 1 + x$. All the nodes belonging to a specific path, finite or infinite, will be considered as members of disjoint sets that can be of three types. A set of type V_0, V_1, or V_n contains nodes corresponding to polynomials with zero, one, or more than one sign variations, respectively. Sets of type V_0 or V_1 are called *terminal sets*. In the case of sets belonging to the same path, a set X is said to *precede* a set Y if and only if for all x in X and all y in Y pathlength(x) < pathlength(y). In a terminal set, the node having the shortest path from the root of the treed will be called a *terminal node*.

With these definitions in mind, the difference between the two continued fractions methods for the isolation of the real roots of a polynomial equation is seen in Figure 7.3.1.

The following proposition is also true.

Theorem 7.3.13. Let $p(x) = c_n x^n + \cdots + c_1 x + c_0 = 0$ be a polynomial equation of degree $n > 1$, with rational coefficients and without any multiple roots, which corresponds to the root of the binary tree. Suppose, moreover, that the substitution

$$x := a_1 + \cfrac{1}{a_2 + \cfrac{\ddots}{ + \cfrac{1}{a_h + \cfrac{1}{y}}}}$$

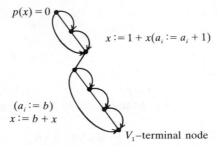

Figure 7.3.1. Geometric interpretation of the two different ways of computing the value of a particular a_i (the length of a branch).

with a_1 arbitrary nonnegative integer and $a_2, \ldots, a_h$, $1 \le h \le m$ positive, integer elements (where m is defined by (V1) in Theorem 7.3.9) transforms $p(x) = 0$ into a new equation corresponding to a V_0 or V_1 terminal node. Then for all k, $1 \le k \le h$ we have

$$a_k = 0[|p(x)|_\infty] \tag{A2}$$

Proof. For $k = 1$ consider the following:

i. An upper bound on the absolute values of the roots of $p(x) = 0$ is given by $b = 2\max_{1 \le k \le n} |c_{n-k}/c_n|^{1/k}$, and obviously, $b \le 2|p(x)|_\infty$.
ii. If $h > 1$, then a_1 is equal to the integer part of some root of $p(x)$ having positive real part.

From conditions i and ii it follows easily that $a_1 = 0[|p(x)|_\infty]$. In order to prove the general case, set $p^{(0)}(x) := p(x)$ and assume that $p_t^{(i+1)}(x) = 0$ and $p^{(i)}(x) = 0$ are two equations such that $p_t^{(i+1)}(x) = p^{(i)}(a_i + x)$. Subsequently apply the previous considerations to $p^{(i)}(x) = 0$, also taking into account the fact that $|p^{(i)}(x)|_\infty \sim |p_t^{(i+1)}(x)|_\infty$ and $|p_t^{(i+1)}(x)|_\infty = |p^{(i+1)}(x)|_\infty$, where $p^{(i+1)}(x) = p_t^{(i+1)}(1/x) = 0$. □

Putting the above together, we have the following algorithm, which is the only one with polynomial computing time using continued fractions.

ACF1978 (Continued Fractions Method of 1978 for Isolation of the Real Roots of an Equation)

Input: $p(x) = 0$, a polynomial equation with integer coefficients and without multiple roots.
Output: The isolating intervals of the real roots of $p(x)$ or its exact roots.

1. [Initialize.] Set $p_w(x) := p(x)$; if $p_w(0) = 0$, then output the closed interval $[0, 0]$ and set $p_w(x) := p_w(x)/x$; pn-flag $:= 0$; ti-flag $:= 0$; $T := \emptyset$ and compute the number v of sign variations in the sequence of coefficient of $p_w(x)$ [T is the set of triplets $\{p_M(x), \mathbf{M}(x), v_M\}$ defined above, where $\mathbf{M}(x) = (a_1 x + a_0)/(b_1 x + b_0)$; when pn-flag $= 0$ we isolate the positive roots, whereas when pn-flag $= 1$ we isolate the negative; ti-flag is the translation–inversion flag].

2. [$v = 0$ or $v = 1$?] If $v = 0$ or $v = 1$, then from the Cardano–Descartes rule of signs (Theorem 7.2.6) we know that $p_w(x)$ has no positive roots, or exactly one positive root, respectively; in the latter case $(0, \infty)$ is its isolating interval and is output. In either case above, no substitutions are necessary; if pn-flag $= 0$, then go to 10 or else exit.

2a. [$v > 1$.] At this point we know that $v > 1$, and $p_w(x)$ has to be further investigated; set $p_M(x) := p_w(x)$, $\mathbf{M}(x) := x$, $v_M := v$ and go to 4.

3. [Test for completion.] If $T \neq \emptyset$, then remove the first triplet $\{p_M(x), \mathbf{M}(x), v_M\}$ and go to 4; if $T = \emptyset$ and pn-flag $= 0$, then go to 10 or else exit.

4. [Compute b.] Using Cauchy's rule (Section 7.2.3), we compute a lower bound b on the values of the positive roots of $p_M(x)$; if $b < 1$, then go to 6.

5. [$x := b + x, b > 1$.] We set $p_M(x) := p_M(b + x)$; $\mathbf{M}(x) := \mathbf{M}(b + x)$ (v_M does not change). If $p_M(0) = 0$, we have found a (rational) root of the original equation, in which case we output the closed interval $[a_0/b_0, a_0/b_0]$ [obtained from the corresponding $\mathbf{M}(x)$] and set $p_M(x) := p_M(x)/x$.

6. [$x := 1 + x$.] Set $v' := v_M$; $p_{M1}(x) := p_M(1 + x)$; $\mathbf{M1}(x) := \mathbf{M}(1 + x)$. If $p_{M1}(0) = 0$, then we have found a (rational) root of the original equation, in which case we output the closed interval $[a_0/b_0, a_0/b_0]$ [obtained from the corresponding $\mathbf{M}(x)$] and set $p_{M1}(x) := p_{M1}(x)/x$. Go to 8.

7. [$x := 1/(1 + x)$.] If $v' = v_{M1}$, then do $\{ti$-flag $:= 0$; go to 3$\}$. At this point we know that $v' \neq v_{M1}$, which means that there are roots of the equation $p_M(x) = 0$ in the interval $(0, 1)$; set $p_i(x) := p_M(1/x)$, making sure that $lc[p_i(x)] > 0$ [multiply $p_i(x)$ by (-1) if necessary]; $\mathbf{M}_i(x) := \mathbf{M}(1/x)$; $p_{M1}(x) := p_i(1 + x)$; $\mathbf{M1}(x) := \mathbf{M}_i(1 + x)$.

8. [Change ti-flag.] ti-flag $:= ti$-flag $+ 1$; compute v_{M1}, the number of sign variations in the sequence of coefficients of $p_{M1}(x)$. As in Step 2, if $v_{M1} = 0$ or $v_{M1} = 1$, then from the Cardano–Descartes rule of signs we know that $p_{M1}(x)$ has either no positive roots, or one positive root; in the latter case we output its isolating interval, which is: (i) $(0, a_0/b_0)$, if $a_1 = 0$ and $b_1 > 0$, (ii) $(a_0/b_0, \infty)$, if $a_1 > 0$ and $b_1 = 0$, and (iii) the open interval whose unordered endpoints are $a_0/b_0, a_1/b_1$ otherwise [obtained, of course, in all cases from the corresponding

M1(x)]. If $v_{M1} > 1$, then we prefix the triplet $\{p_{M1}(x), \mathbf{M1}(x), v_{M1}\}$ to the set T.

9. [Go back to loop.] If *ti*-flag = 1, then go to 7; otherwise, do {*ti*-flag := 0; go to 3}.

10. [Isolate the negative roots.] If $p(x) \neq p(-x)$, then do {set *pn*-flag := 1; *ti*-flag := 0; $p_w(x) := p(-x)$, so that the negative roots become positive; $T := \emptyset$; compute the number v of sign variations in the sequence of the coefficients of $p_w(x)$ and go to 2} or else exit. (If we exit here, the negative roots are symmetric to the positive ones and we already know their isolating intervals. Of course, in this case, the intervals that we obtain for the negative roots are in the positive half-plane and have to be mapped to the corresponding intervals in the negative half-plane, a trivial matter.)

Computing-Time Analysis of ACF1978. Given the polynomial $p(x)$, deg[$p(x)$] = n, we know from Section 3.1.2 that the substitution $p(x) := p(a + x)$ is executed in time

$$0\{n^3 L^2(a) + n^2 L(a) L[|p(x)|_\infty]\}$$

From (V12) we know that for each real root of $p(x) = 0$ we have to perform at most m such substitutions, where

$$m = 0\{nL[|p(x)|_\infty]\}$$

Moreover, from (A2) of Theorem 7.3.13, we know that for each real root of $p(x)$ and for each substitution $p(x) := p(a + x)$ we have

$$a = 0[L(|p(x)|_\infty)]$$

Combining the above three results, we have that one real root of $p(x)$ can be isolated in time

$$0\{n^4 L^3[|p(x)|_\infty]\}$$

Since $p(x)$ has at most n real roots, we have that

$$t_{\text{ACF1978}}[p(x)] = 0\{n^5 L^3[|p(x)|_\infty]\}$$

It has been demonstrated, both theoretically and empirically, that using exact integer arithmetic the continued fractions method of 1978 is the fastest method for the isolation of the real roots of a polynomial equation; see also Section 7.4.

Example. Let us isolate the real roots of $p(x) = x^3 - 7x + 7 = 0$, where $p(0) \neq 0$. We apply the algorithm **ACF1978** and obtain the following results:

Step 1. Here we set $p_w(x) := x^3 - 7x + 7$, pn-flag $:= 0$, ti-flag $:= 0$, $T := \emptyset$ and compute v, which is 2. (We first isolate the positive roots.)

Step 2a. Since $v > 1$, we set $p_M(x) := x^3 - 7x + 7$, $\mathbf{M}(x) := x$, $v_M := 2$ and go to 4.

Step 4. To use Cauchy's rule, we first set $x := 1/x$ in $p_M(x) = 0$ and obtain $7x^3 - 7x^2 + 1 = 0$; we then apply **BPR** to the last equation and obtain $1/b = 1$. Therefore, $b = 1$.

Step 5. At this step we update $p_M(x) := p_M(1 + x) = x^3 + 3x^2 - 4x + 1$ and $\mathbf{M}(x) := \mathbf{M}(1 + x) = x + 1$; obviously, $p_M(0) \neq 0$.

Step 6. Set $v' := 2$, update $p_{M1}(x) := p_M(1 + x) = x^3 + 6x^2 + 5x + 1$ and $\mathbf{M1}(x) := \mathbf{M}(1 + x) = x + 2$, and go to 8.

Step 8. At this step we set ti-flag $:= 1$ and compute v_{M1}, which is 0.

Step 9. Since ti-flag $= 1$, we go to 7.

Step 7. $v' \neq v_{M1}$, and we compute a new $p_{M1}(x) = x^3 - x^2 - 2x + 1$ along with $\mathbf{M1}(x) = (x + 2)/(x + 1)$. [Note that $p_{M1}(x) := p_i(1 + x)$ and $\mathbf{M1}(x) := \mathbf{M}_i(1 + x)$, where $p_i(x) = x^3 - 4x^2 + 3x + 1$ and $\mathbf{M}_i(x) = (x + 1)/x$ were obtained by setting $x := 1/x$ in $p_M(x)$ and $\mathbf{M}(x)$, respectively.]

Step 8. At this step we set ti-flag $:= 2$ and compute v_{M1}, which turns out to be 2. In this case we set $T := \{[x^3 - x^2 - 2x + 1, (x + 2)/(x + 1), 2]\}$.

Step 9. Since ti-flag $= 2$, we set ti-flag $:= 0$ and go to 3.

Step 3. $T \neq \emptyset$ and so we remove from it the first triplet $p_M(x) = x^3 - x^2 - 2x + 1$, $\mathbf{M}(x) = (x + 2)/(x + 1)$, $v_M = 2$ and go to 4.

Step 4. Again we first set $x := 1/x$ in $p_M(x) = 0$ and obtain $x^3 - 2x^2 - x + 1 = 0$; we then apply **BPR** to the last equation and obtain $1/b = 4$. Therefore, $b = 1/4 < 1$ and go to 6.

Step 6. Set $v' := 2$, update $p_{M1}(x) := p_M(1 + x) = x^3 + 2x^2 - x - 1$ and $\mathbf{M1}(x) := \mathbf{M}(1 + x) = (x + 3)/(x + 2)$ and go to 8.

Step 8. Here we set ti-flag $:= 1$ and compute v_{M1}, which turns out to be 1. In this case we output the first isolating interval $(1, 3/2)$ obtained from $\mathbf{M1}(x) = (x + 3)/(x + 2)$.

Step 9. Since ti-flag $= 1$, we go to 7.

Step 7. $v' \neq v_{M1}$, and we compute a new $p_{M1}(x) = x^3 + x^2 - 2x - 1$ along with $\mathbf{M1}(x) = (2x + 3)/(x + 2)$.

Step 8. Here we set ti-flag $:= 2$ and compute v_{M1}, which turns out to be 1. In this case we output the second isolating interval $(3/2, 2)$ obtained from $\mathbf{M1}(x) = (2x + 3)/(x + 2)$.

Step 9. Since ti-flag $= 2$, we set ti-flag $:= 0$ and go to 3.

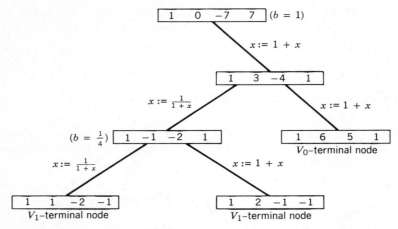

Figure 7.3.2. The binary tree obtained for the isolation of the positive roots of the equation $x^3 - 7x + 7 = 0$.

Step 3. $T = \emptyset$ and pn-flag $= 0$; therefore, we go to 10.

Step 10. We now isolate the negative roots. Since $p(x) \neq p(-x)$, we set pn-flag $:= 1$, ti-flag $:= 0$, $p_w(x) := p(-x) = x^3 - 7x - 7$, $T := \emptyset$, compute v, which turns out to be 1, and go to 2.

Step 2. Here $v = 1$, which implies that the isolating interval for the negative root is $(-\infty, 0)$, and since pn-flag $= 1$, we exit. (Using Cauchy's rule we can obtain a better bound than $-\infty$ for this root.)

This whole process is illustrated in Figure 7.3.2.

7.4
Empirical Comparisons between the two Real Root Isolation Methods

In Sections 7.2 and 7.3 we discussed at length the two classical ways, Sturm's bisection method and the continued fractions method of 1978, for the isolation of the real roots of polynomial equations with integer coefficients. Below we present two tables showing the observed computing times for two classes of polynomials for these two methods. All times are in seconds and were obtained by using the **sac-1** computer algebra system on an IBM S/370 Model 165. Each polynomial in Table 7.4.1 was formed by taking the product of the corresponding number of linear terms. All the coefficients of the polynomials in Table 7.4.2 were nonzero, each 10 decimal digits long and randomly generated.

TABLE 7.4.1 Polynomials with Randomly Generated Roots in the Interval $(0, 10^3)$

Degree	Continued Fractions Method of 1978	Sturm's Bisection Method
5	0.71	0.73
10	23.22	22.50
15	95.35	151.42
20	288.49	>600

TABLE 7.4.2 Polynomials with Randomly Generated Coefficients

Degree	Continued Fractions Method of 1978	Sturm's Bisection Method
5	0.26	2.05
10	0.46	33.28
15	0.94	156.40
20	2.36	524.42

The polynomials in Table 7.4.2 are the same ones used by Collins [in Rice (1977)] and Collins and Loos (1976) to test the Collins–Akritas and differentiation methods; see also historical notes 1 and 7. Therefore, one can easily verify the superiority of the continued fractions method of 1978 (**ACF1978**) by a simple comparison of the ratios of the times of the various methods to the corresponding times obtained with Sturm's method. For the sake of comparison we have computed these ratios in Table 7.4.3. (See also Mignotte, 1976.)

TABLE 7.4.3 Comparison of the Various Methods for the Polynomials of Table 7.4.2

Degree	ACF1978/Sturm	Collins-Akritas/Sturm	Differentiation/Strum
5	0.13	0.28	1.31
10	0.014	0.10	0.55
15	0.004	0.05	0.28
20	0.0045	0.03	0.21

7.5
Approximation of the Real Roots of a Polynomial Equation

In the previous sections we have studied, in detail, two different approaches for the isolation of the real roots of a polynomial equation with integer coefficients; that is, we can now find real, disjoint intervals such that each contains exactly one real root and every real root is contained in some interval. However, according to Fourier this is only the first step (of the two steps) involved in the computation of the real roots of a polynomial equation; the second step consists of approximating these roots to any desired degree of accuracy ε. In other words, this second step consists of making the length of the isolating intervals less than or equal to ε.

Below we examine two approximation methods, one based on bisection and another one utilizing continued fractions in conjunction with Vincent's theorem; see Akritas et al., 1983, Cajori, 1910; Ng, 1980; Nordgaard, 1922; and Verbaeten, 1975.

7.5.1 Real Root Approximation Using Bisection

In this section we approximate a real root of a polynomial equation by continuously bisecting its (given) isolation interval until the distance between the endpoints of the interval is less than or equal to some ε.

The Bisection Algorithm Given a squarefree, polynomial equation $p(x) = 0$ with integer coefficients and an isolating interval (a, b) for the root we want to approximate, where a, b are rationals, proceed as follows.

Evaluate the sign of $p(x)$ at the point $(a + b)/2$; if it is zero, replace (a, b) by $[(a + b)/2, (a + b)/2]$ and terminate. If $p[(a + b)/2]$ has the same sign as $p(b)$, then the root is in the interval $(a, (a + b)/2)$, whereas if $p[(a + b)/2]$ has the same sign as $p(a)$, then the root is in the interval $((a + b)/2, b)$. This process is repeated until the length of the current interval is less than or equal to ε.

From Sections 3.1.2 and 7.2.2 we can easily see that the bisection method will approximate a root within ε in time

$$0\left\{n^2 L\left(\frac{h}{\varepsilon}\right) L\left(\frac{eh}{\varepsilon}\right) L\left(\frac{eh|p(x)|_\infty}{\varepsilon}\right)\right\} \tag{B1}$$

where n is the degree of the squarefree polynomial $p(x)$, h is the initial length of the isolating interval, ε is the degree of accuracy (limit of approximation), and $e = \max\{|a_1|, |a_2|, b_1, b_2\}$, where $(a_1/b_1, a_2/b_2)$ is the initial interval. Empirical results showed that bisection is a very slow method; its performance was later improved when it was combined with Newton's method (see also Tables 7.6.1 and 7.6.2 of Section 7.6).

Example. Given the equation $p(x) = x^3 - 7x + 7$, let us approximate to within $\varepsilon = 0.01$ its positive root whose isolating interval is $(1, 3/2)$; note that $p(1) = 1$, whereas $p(3/2) = -1/8$. (Here we are actually interested in the sign of the resulting number rather than on the number itself.)

At first we observe that $3/2 - 1 = 0.5 > \varepsilon$, and so we evaluate the sign of

APPROXIMATION OF THE REAL ROOT OF A POLYNOMIAL EQUATION   383

$p(x)$ at the midpoint; namely, we obtain $p(5/4) = 13/64$, which implies that a tighter isolating interval for the root is $(5/4, 3/2)$. Continuing in the same way, we obtain the following table:

Current Isolation Interval	Endpoint Distance	Midpoint	Sign of $p(x)$ Evaluated at the Midpoint	Next Isolation Interval
$(5/4, 3/2)$	0.25	$11/8$	$-13/512$	$(5/4, 11/8)$
$(5/4, 11/8)$	0.125	$21/16$	$+301/4096$	$(21/16, 11/8)$
$(21/16, 11/8)$	0.0625	$43/32$	$+659/32768$	$(43/32, 11/8)$
$(43/32, 11/8)$	0.031	$87/64$	$-953/262144$	$(43/32, 87/64)$
$(43/32, 87/64)$	0.0156	$173/128$	$+16757/2097152$	$(173/128, 87/64)$
$(173/128, 87/64)$	$0.0078 < \varepsilon$			

and we stop.

Expressing in decimal representation the endpoints of the last current isolation interval above, $(173/128, 87/64)$, we obtain $173/128 = 1.3515625$ and $87/64 = 1.359375$. Therefore, to within $\varepsilon = 0.01$ the approximated root is 1.35.

7.5.2 Real Root Approximation Using Continued Fractions

The idea to approximate the real roots of a polynomial equation using continued fractions is due to Lagrange, whose objective was to develop a procedure free of the defects plaguing the well-known method by Newton. Lagrange's idea may be stated as follows.

Suppose that a root of the polynomial equation $p(x) = 0$ lies between the *consecutive integers* a_1 and $a_1 + 1$; diminish the roots of the equation by a_1; that is, perform the substitution $p(x) := p(a_1 + x)$, and take the reciprocal equation—that is, perform the substitution $p(x) := p(1/x)$. Find, by *trial*, a root of the last equation lying between a_2 and $a_2 + 1$, diminish the roots by a_2 and take the reciprocal equation. Proceed in this way. Then the continued fraction

$$a_1 + \cfrac{1}{a_2 + \cfrac{1}{a_3 + \cdots}}$$

approximates a root of the equation to the desired accuracy.

Clearly, Lagrange's approach has certain drawbacks. Note that it is straightforward if there is one, and only one, root between the consecutive integers a_i and $a_i + 1$; however, the process does not work if there are two or more roots within $(a_i, a_i + 1)$ (see historical note 8). (In this latter case Lagrange actually uses Vincent's not yet published theorem to first isolate the roots.) Moreover, just as in Vincent's continued fractions method of 1836, the integer part of the root is computed by trial; that is, to find where lies the single root > 1, one tries the values $1, 2, 3, \ldots$ in the polynomial equation, performing a series of substitutions of the form $p(x) := p(1 + x)$, until a change of sign is observed. Clearly, this is an exponential procedure.

Pursuing studies in the direction outlined in Section 7.3, it was observed that Theorem 7.3.9 can be also used to approximate the real roots of a polynomial equation to any desired degree of accuracy (see historical note 9).

Clearly, this is easily achieved by *extending* (computing more partial quotients of) the continued fraction (V3), which transforms the original polynomial equation into one with exactly one sign variation in the sequence of its coefficients (see historical note 10).

Note: Observe that the approximation method we are describing depends heavily on the isolation process; that is, it *cannot* work if it is provided simply with the isolating intervals of the roots and the polynomial $p(x) = 0$.

Suppose that the limit of approximation is ε, and that in approximating the real root ρ of a polynomial equation we have computed k partial quotients of the continued fraction expansion of this root. Then, from the discussion of Section 7.3, it becomes obvious that ρ lies between the consecutive convergents p_{k-1}/q_{k-1} and p_k/q_k, whose difference in absolute value is $1/q_{k-1}q_k$. Hence, we have

$$\left| \frac{p_k}{q_k} - \rho \right| \leq \frac{1}{q_{k-1}q_k} \leq \frac{1}{q_{k-1}^2},$$

and the method will terminate when

$$\frac{1}{q_{k-1}^2} \leq \varepsilon \qquad (CF1)$$

for some k.

Below we describe two ways for extending the continued fraction (V3) in order to approximate within ε a real root ρ of an equation. These two methods have the same theoretical computing time bound, but different empirical performance.

The first way to extend the continued fraction (V3) is to compute each additional partial quotient with the help of Cauchy's rule as in Section 7.3.4. However, mainly due to Cauchy's rule, this approach is inefficient as can be seen from Table 7.6.2 in Section 7.6. Actually, it is even slower than the bisection method, a method well-known for its slowness.

The second way to extend the continued fraction (V3) is to take advantage of the special nature of the polynomials we are now dealing with. These polynomials are special in the sense that they have one sign variation (and, hence, only one positive root) and, consequently, they cross the x-axis only once. Therefore, the integer part of the positive root (which is the next partial quotient a_i) can be computed by successively bisecting (and evaluating at midpoints) the interval $(0, b)$, where b is an easily computed upper bound on the value of the single root; see Table 7.6.2 in Section 7.6 for the improvement. This upper bound b is now computed with the help of the following.

Theorem 7.5.1. Let $p(x) = c_n x^n + \cdots + c_{r+1} x^{r+1} - c_r x^r - \cdots - c_0 = 0$ be a polynomial equation of degree n, with only one sign variation in the sequence of its integer coefficients. Then

$$b = \max_{0 \leq j \leq r} \left\{ \frac{|c_j|}{\sum_{r+1 \leq i \leq n} c_i + 1} \right\} \quad \text{(CF2)}$$

is an upper bound on the value of its (single) positive root.

Proof. For any power m of x we have

$$x^m = (x - 1)(x^{m-1} + x^{m-2} + \cdots + x + 1$$

In the expression of $p(x)$ we expand each term with *positive* coefficient and obtain

$$p(x) = c_n(x-1)x^{n-1} + c_n(x-1)x^{n-2} + c_n(x-1)x^{n-3} + \cdots + c_n$$
$$+ c_{n-1}(x-1)x^{n-2} + c_{n-1}(x-1)x^{n-3} + \cdots + c_{n-1}$$
$$\cdots$$
$$- c_r x^r - \cdots - c_0$$

Consider the successive vertical columns where there is no negative coefficient. The value of each such column is positive provided that $x > 1$. To ensure a positive value of the last columns in which a negative coefficient occurs, we must have

$$(c_n + c_{n-1} + \cdots + c_{r+1})(x - 1) \geq c_i$$

for $0 \leq i \leq r$, and this proves the theorem. □

The implementation of this theorem is much simpler than the one of Cauchy's rule, and we leave it as an exercise for the reader to show that it can be performed in time

$$O\{nL[|p(x)|_\infty] + L^2[|p(x)|_\infty]\} \tag{CF3}$$

Below we assume the existance of a procedure **A7.5.1** that, using Theorem 7.5.1 and bisection, computes the integer part a_i of the single positive root of $p(x) = 0$ in time

$$t_{A7.5.1}[p(x)] = O\{n^2 L^3[|p(x)|_\infty]\} \tag{CF4}$$

(See the Programming Exercise for this section.)

APPROX-CF (Approximate a Real Root of an Equation Using Continued Fractions)

Input: The limit of approximation ε that is a rational number, and $p_M(x) = 0$ along with $\mathbf{M}(x)$, from which we obtain the isolating interval of the root; $p_M(x) = 0$ is a polynomial equation of degree n with *one* sign variation in the sequence of its integer coefficients and is obtained from an original equation $p(x) = 0$ after the substitution $x := \mathbf{M}(x) = (a_1 x + a_0)/(b_1 x + b_0)$ for some specific values a_0, a_1, b_0, b_1. We want to approximate within ε that root of $p(x) = 0$ that is located in the open interval with unordered endpoints a_1/b_1 and a_0/b_0. Obviously, $|a_1/b_1 - a_0/b_0| \geq \varepsilon$ because otherwise we would have nothing to do.

Output: The isolating interval of that root of $p(x) = 0$ that is located in the open interval with endpoints e_1 and e_2 where now $|e_1 - e_2| \leq \varepsilon$, or the exact root of $p(x) = 0$; e_1 and e_2 are both rational numbers.

1a. [Initialize.] Set $p_w(x) := p_M(x)$, $\mathbf{M}_w(x) := \mathbf{M}(x)$.

1b. [Extend the continued fraction.] Using **A7.5.1**, compute a_i, the integer part of the positive root of $p_w(x) = 0$. If $a_i = 0$, go to 5.

2. [$x := a_i + x, a_i > 0$.] Update $p_w(x) := p_w(a_i + x)$ and $\mathbf{M}_w(x) := \mathbf{M}_w(a_i + x)$.

3. [Root?] If $p_w(0) = 0$, then we set $e_1 := e_2 := a_0/b_0$, where a_0 and b_0 are obtained from $\mathbf{M}_w(x)$, and we return the closed interval $[e_1, e_2]$ and terminate. (The positive root we were trying to approximate is a rational number.)

4. [Done?] If $|a_1/b_1 - a_0/b_0| \leq \varepsilon$, then we return the open interval with unordered endpoints $e_1 := a_1/b_1$ and $e_2 := a_0/b_0$, where a_0, a_1, b_0, and b_1 are obtained from $\mathbf{M}_w(x)$, and we terminate. [This interval approximates the root of $p(x)$ to within the specified accuracy.]

5. [$x := 1/x$.] Set $p_w(x) := p_w(1/x)$, $\mathbf{M}_w(x) := \mathbf{M}_w(1/x)$ and go to 1b.

*Computing-Time Analysis of **APPROX-CF**.* We observe that the execution time of the algorithm is dominated by the execution times of steps 1 and 2. For the first iteration of the algorithm we have the following:

Step 1 is executed in time $0\{n^2 L^3[|p_M(x)|_\infty]\}$ (see the Programming Exercise for this section).

Step 2 is executed in time $0\{n^3 L^2(a_i) + n^2 L(a_i) L[|p_M(x)|_\infty]\}$, which is the execution time of the substitution $p_w(x) := p_w(a_i + x)$ [and includes that of $\mathbf{M}_w(x) := \mathbf{M}_w(a_i + x)$]; combining this expression with the fact that $a_i = 0[|p_M(x)|_\infty]$, for all i [see (A2)], we obtain $0\{n^3 L^2[|p_M(x)|_\infty]\}$, which is bounded by

$$0\{n^3 L^3[|p_M(x)|_\infty]\} \qquad \text{(CF5)}$$

Comparing the computing time for Cauchy's rule (Section 7.2.3), (CF3), and (CF4) with (CF5), we see that (CF5) dominates the computing times of steps 1 and 2.

For subsequent iterations of the algorithm we have that (CF5) again dominates the computing times of steps 1 and 2, since for any subsequent $p_w(x)$ we have $|p_w(x)|_\infty \sim |p_M(x)|_\infty$ (Theorem 7.3.13).

To find i, the number of iterations needed to approximate the root to within ε, we use (CF1), that is, the relation $1/q_i^2 \leq \varepsilon$, and the following facts:

i. $q_i \leq F_i$, where F_i is the ith member of the Fibonacci sequence.
ii. $F_i = \phi^i/\sqrt{5}$ (rounded to the nearest integer), where $\phi = 1.618\ldots$.

Obviously (CF1) yields $5/\phi^{2i} \leq \varepsilon$, from which we obtain

$$i = 0\left(\log_\phi \frac{1}{\varepsilon}\right) = 0\left[L\left(\frac{1}{\varepsilon}\right)\right] \qquad \text{(CF6)}$$

Multiplying (CF5) and (CF6), we obtain

$$t_{\textbf{APPROX-CF}}[p_M(x), \mathbf{M}(x), \varepsilon] = 0\{n^3 L(1/\varepsilon) L^3[|p_M(x)|_\infty]\}$$

Example. Let us approximate to within $\varepsilon = 0.01$ the positive root of the equation $p(x) = x^3 - 7x + 7$ whose isolating interval is $(1, 3/2)$. We are given $p_M(x) = x^3 + 2x^2 - x - 1$ along with $\mathbf{M}(x) = (x+3)/(x+2)$ from which the isolating interval is obtained; $p_M(x)$ and $\mathbf{M}(x)$ were both computed in the example of Section 7.3.4.

Applying **APPROX-CE**, we obtain the following results:

Step 1a. We set $p_w(x) := x^3 + 2x^2 - x - 1$ and $\mathbf{M}_w(x) := (x+3)/(x+2)$.
Step 1b. Application of **A7.5.1** yields $a_i = 0$, and we go to 5.

Step 5. Here we compute a new $p_w(x) = x^2 - 2x - 1$ along with $\mathbf{M}_w(x) = (3x+1)/(2x+1)$ and go to 1b.
Step 1b. Application of **A7.5.1** yields $a_i = 1$.
Step 2. At this step we update $p_w(x) := p_w(1+x) = x^3 + 4x^2 + 3x - 1$ and $\mathbf{M}_w(x) := \mathbf{M}_w(1+x) = (3x+4)/(2x+3)$.
Step 3. Obviously, $p_w(0) \neq 0$.
Step 4. $|3/2 - 4/3| = 0.169$.
Step 5. Here we compute a new $p_w(x) = x^3 - 3x^2 - 4x - 1$ along with $\mathbf{M}_w(x) = (4x+3)/(3x+2)$ and go to 1b.
Step 1b. Application of **A7.5.1** yields $a_i = 4$.
Step 2. At this step we update $p_w(x) := p_w(4+x) = x^3 + 9x^2 + 20x - 1$ and $\mathbf{M}_w(x) := \mathbf{M}_w(4+x) = (4x+19)/(3x+14)$.
Step 3. Obviously, $p_w(0) \neq 0$.
Step 4. $|19/14 - 4/3| = 0.0238$.
Step 5. Here we compute a new $p_w(x) = x^3 - 20x^2 - 9x - 1$ along with $\mathbf{M}_w(x) = (19x+4)/(14x+3)$ and go to 1b.
Step 1b. Application of **A7.5.1** yields $a_i = 20$.
Step 2. At this step we update $p_w(x) := p_w(20+x) = x^3 + 40x^2 + 391x - 181$ and $\mathbf{M}_w(x) := \mathbf{M}_w(20+x) = (19x+384)/(14x+283)$.
Step 3. Obviously, $p_w(0) \neq 0$.
Step 4. $|19/14 - 384/283| = 0.00025 < \varepsilon$.

and we stop.

Expressing in decimal representation the endpoints of the last isolation interval above, we obtain $384/283 = 1.3568904$ and $19/14 = 1.3571428$ (see also the Programming Exercise for Section 7.5.1). Therefore, to within $\varepsilon = 0.01$ the approximated root is 1.35.

7.6
Empirical Comparisons between the Two Real Root Approximation Methods

In this section we present several tables comparing the bisection and continued fractions methods for the approximation of the real roots of a polynomial equation. We compare both theoretical aspects as well as the actual computing times for the Chebyshev polynomials.

We first determine the number of bisections and partial quotients, respectively, needed for each method under consideration, in order to approximate a root to within a specified degree of accuracy. Under the assumption that the original polynomial has only one positive root, we can take $(0, \infty)$ as the initial interval for the bisection method. Moreover, for the continued fractions method we assume the worst possible case; that is, each partial quotient is 1, in which case we have $q_m = F_m$, where F_m is the mth member of the Fibonacci sequence. Under the above conditions, we can see from Table 7.6.1 that it takes more bisections than partial quotients in order to approximate the root to the same degree of accuracy.

In Table 7.6.2 we show the computing times needed for the approximation ($\varepsilon = 10^{-15}$) of the real roots of the Chebyshev polynomials using the bisection and continued fractions methods. All the times are in seconds. We compare three versions of the continued fractions method: the versions that utilize (1) Cauchy's rule, (2) Theorem 7.5.1 with bisection, and (3) preconditioning. In version 3, we assume that a list of partial quotients is supplied as input. In this way we spend no time in computing the floor functions. (The results of version 3 reflect the optimum time for the

TABLE 7.6.1 Comparison of the Number of Partial Quotients and Bisections Needed to Obtain the Required Degree of Accuracy

m	Bisection $1/2^m$	Continued Fraction $1/F_m^2$
1	0.5	1
5	0.03125	0.04
10	9.7×10^{-4}	3.3×10^{-4}
15	3.1×10^{-5}	2.7×10^{-6}
20	9.5×10^{-7}	2.2×10^{-8}
25	1.9×10^{-8}	1.7×10^{-10}
30	9.3×10^{-10}	1.4×10^{-2}
35	1.5×10^{-11}	3.1×10^{-14}
40	2.3×10^{-13}	9.5×10^{-17}

TABLE 7.6.2 Computing Times (in Seconds) for Approximations of All the Real Roots of Chebyshev Polynomials ($\varepsilon = 10^{-15}$)

		Continued Fractions using		
Degree	Bisection	Cauchy's Rule	Theorem 7.5.1 with Bisection	Preconditioning
2	17.2	11.5	6.7	5.4
3	17.9	10.3	4.9	3.8
4	42.3	38.7	15.7	10.3
5	45.8	40.0	16.4	10.8
6	83.1	99.8	46.2	29.2
7	90.9	105.1	44.6	27.0
8	146.3	277.8	93.0	50.2
9	170.6	257.6	106.2	62.2
10	243.2	524.3	202.8	116.2

390 ISOLATION AND APPROXIMATION OF REAL ROOTS OF POLYNOMIAL EQUATIONS

approximation of a real root using the continued fractions method.) The difference between the versions using Theorem 7.5.1 with bisection and preconditioning reflects the time spent in computing the floor functions. We remind the reader that Cauchy's rule had to be applied a number of times in order to obtain the floor function of the root.

These results show that the version using Theorem 7.5.1 with bisection is substantially better than the one using Cauchy's rule. On the other hand, comparing it with the results of the version using preconditioning, we note that there is still room for improvement.

As an illustration of the approximation algorithm, in Table 7.6.3 we show the derived list of partial quotients along with the approximating intervals for each root of the Chebyshev polynomials of degrees 2–10 ($\varepsilon = 10^{-15}$). Since the Chebyshev polynomials are symmetric, we present only the positive roots. For odd-degree polynomials we omit the root $x = 0$.

TABLE 7.6.3 Approximation of the Real Roots of the Chebyshev Polynomials of Degrees 2–10 ($\varepsilon = 10^{-15}$)

Degree	List of Partial Quotients for		Approximation Intervals
	Isolation	Approximation	
2	( )	(0;1,2,2,2,2,2,2,2,2,2, 2,2,2,2,2,2,2,2,2,2,2)	0.70710678118654683338 0.70710678118654764296
3	( )	(0;1,6,2,6,2,6,2,6,2,6, 2,6,2,6,2)	0.86602540378443847925 0.86602540378443865879
4	(0;1,1)	(11,7,3,2,1,1,1,1,20,5, 3,11,1,7)	0.92387953251128634045 0.92387953251128676332
	(0;2)	(0,1,1,1,1,2,2,4,3,1, 19,6,8,3,2,9)	0.38268343236508971655 0.38268343236509064218
5	(0;1,1,1)	(1,2,1,6,1,56,1,54,1,1, 1,10,1,16)	0.58778525229247311679 0.58778525229247338805
	(0;1,2)	(17,2,3,6,5,1,1,1,3,2, 1,25,2,2,1,1)	0.95105651629515309991 0.95105651629515359558
6	(0;1,2,1)	(1,2,2,2,2,2,2,2,2,2,2, 2,2,2,2,2,2,2,2)	0.70710678118654683338 0.70710678118654764296
	(0;1,3)	(25,2,1,7,21,1,8,1,3, 10,1,2,3)	0.96592582628906811211 0.96592582628906894040
	(0;2)	(1,1,6,2,1,30,5,2,9,3, 1,10,5,5)	0.25881904510252075804 0.25881904510252080914
7	(0;1,3,1)	(0,1,2,2,24,2,3,2,2,2, 2,2,8,1,8,1)	0.78183148246802977984 0.78183148246803000835
	(0;1,4)	(34,1,7,1,2,3,1,1,2,3, 1,132,2,3)	0.97492791218182315610 0.97492791218182365484
	(0;2)	(0,3,3,1,1,3,1,29,1,3, 1,18,16,1,1,2)	0.43388373911755805748 0.43388373911755880719
8	(0;1,1,3,1)	(0,14,14,17,1,1,10,2,1, 2,18,1)	0.83146961230254518379 0.83146961230254537502

TABLE 7.6.3 Cont.

Degree	List of Partial Quotients for		Approximation Intervals
	Isolation	Approximation	
9	(0;1,1,4)	(46,23,43,8,1,2,1,3,1,15)	0.98078528040322959629 0.98078528040323045751
	(0;1,1,1)	(2,1,840,2,1,4,1,3,21,1,17)	0.55557023301960177859 0.55557023301960224274
	(0;2)	(3,7,1,17,1,13,3,2,7,1,1,8,1,1,1,40)	0.19509032201612826749 0.19509032201612828909
	(0;1,1,5,1)	(1,6,2,6,2,6,2,6,2,6,2,6,2)	0.86602540378443847925 0.86602540378443865879
	(0;1,1,6)	(57,1,4,1,1,1,6,1,1,6,10,1,1,3,1,1,2,6)	0.98480775301220799222 0.98480775301220806644
	(0;1,1,1)	(0,3,1,72,1,1,2,3,1,1,3,14,1,1,1,16,1)	0.64278760968653925445 0.64278760968653933737
10	(0;2)	(0,1,12,8,17,1,1,2,1,1,1,49,1,20,3)	0.34203014332566870745 0.34203014332566873472
	(0;1,2,6,1,1)	(3,1,14,13,1,1,10,1,72,23,1)	0.45399049973954674160 0.45399049973954687273
	(0;1,2,6,2)	(3,2,1,1,4,1,2,2,1,35,1,2,4,1,3,4,65)	0.15643446504023080945 0.15643446504023086913
	(0;1,2,6,1)	(4,1,2,1,1,3,1,12,5,19,11,1,1)	0.89100652418836719434 0.89100652418836801487
	(0;1,2,7)	(71,4,2,7,3,1,1,42,1,12,18)	0.98768834059513772582 0.98768834059513783641
	(0;1,2,1)	(1,2,2,2,2,2,2,2,2,2,2,2,2,2,2,2,2,2,2)	0.70710678118654683338 0.70710678118654764296

Observe that the last digit of the isolation-list (of partial quotients) and the first digit of the approximation list (of partial quotients) constitute one and the same partial quotient.

Exercises

Section 7.2.1

1. Prove Lemma 7.2.2.

2. Compute an upper bound on the number of real roots that the following polynomial equations have in the interval $(-2, 4)$.
   (a) $p(x) = x^3 - 3x + 1 = 0$.
   (b) $p(x) = x^4 - 4x^3 + x^2 + 6x + 2 = 0$.
   (c) $p(x) = 20x^5 + 5x^4 - 20x^2 - 10x + 2 = 0$.

Section 7.2.2

1. Complete the example preceding **STURM**'s algorithm.
2. Using Sturm's method, isolate the real roots of the polynomial equations in the interval $(-2, 4)$:
   (a) $p(x) = x^3 - 3x + 1 = 0$.
   (b) $p(x) = x^4 - 4x^3 + x^2 + 6x + 2 = 0$.
   (c) $p(x) = 20x^5 + 5x^4 - 20x^2 - 10x + 2 = 0$.

Section 7.2.3

1. Complete the example of this section.
2. Prove that an upper bound on the absolute values of the roots of $p(x) = c_n x^n + c_{n-1} x^{n-1} + \cdots + c_0 = 0$, c_i in **Z**, is

$$b = 2 \max_{1 \leq k \leq n} |c_{n-k}/c_n|^{1/k} \tag{C4}$$

3. Using (C), (C4), and the algorithm **BPR** compute, and compare, the upper bounds on the positive, negative, and absolute values of the roots of the polynomial equations:
   (a) $p(x) = x^3 - 3x + 1 = 0$.
   (b) $p(x) = x^4 - 4x^3 + x^2 + 6x + 2 = 0$.
   (c) $p(x) = 20x^5 + 5x^4 - 20x^2 - 10x + 2 = 0$.

Section 7.2.4

1. Compute the smallest distance between any two roots of the following polynomial equations:
   (a) $p(x) = x^3 - 3x + 1 = 0$.
   (b) $p(x) = x^4 - 4x^3 + x^2 + 6x + 2 = 0$.
   (c) $p(x) = 20x^5 + 5x^4 - 20x^2 - 10x + 2 = 0$.

Section 7.3.2

1. Show that the substitution $x := a + 1/x$ is equivalent to a *general* translation of length a, followed by an inversion.

Section 7.3.4

1. On the basis of Theorem 7.3.9, can you explain why we *have* to use Cauchy's rule? What would happen to the continued fractions method of 1978 had we used a lower bound on the *absolute* values of the roots?

2. What happens to the complex roots of $p(x)$ when we apply the continued fractions method of 1978? State a proposition analogous to Lemma 7.3.11.

3. Using the continued fractions method of 1978, isolate the real roots of the polynomial equations:
   (a) $p(x) = x^3 - 3x + 1 = 0$.
   (b) $p(x) = x^4 - 4x^3 + x^2 + 6x + 2 = 0$.
   (c) $p(x) = 20x^5 + 5x^4 - 20x^2 - 10x + 2 = 0$.

Section 7.5.1

1. Using the bisection method described above, approximate to within $\varepsilon = 0.01$ the second positive root of $x^3 - 7x + 7 = 0$, whose isolating interval is $(3/2, 2)$.

2. Using the bisection method described above, approximate to within $\varepsilon = 0.01$ the real roots of the following polynomial equations (their isolating intervals were obtained in the exercises for Sections 7.2.2 and/or 7.3.4):
   (a) $p(x) = x^3 - 3x + 1 = 0$.
   (b) $p(x) = x^4 - 4x^3 + x^2 + 6x + 2 = 0$.
   (c) $p(x) = 20x^5 + 5x^4 - 20x^2 - 10x + 2 = 0$.

Section 7.5.2

1. Using the continued fractions method described above, approximate to within $\varepsilon = 0.01$ the second positive root of $x^3 - 7x + 7 = 0$, whose isolating interval is $(3/2, 2)$; as we saw in the example of Section 7.3.4, this isolating interval was obtained from $p_M(x) = x^3 + x^2 - 2x - 1$ and $M(x) = (2x + 3)/(x + 2)$.

2. Using the continued fractions method described above, approximate to within $\varepsilon = 0.01$ the real roots of the following polynomial equations (their isolating intervals were obtained in the exercises for Section 7.3.4):
   (a) $p(x) = x^3 - 3x + 1 = 0$.
   (b) $p(x) = x^4 - 4x^3 + x^2 + 6x + 2 = 0$.
   (c) $p(x) = 20x^5 + 5x^4 - 20x^2 - 10x + 2 = 0$.

Programming Exercises

Section 7.2.3

1. Develop procedures to
   (a) Compute $\lfloor \log_2 |i| \rfloor$, for any integer $i \neq 0$, in time $0[L(|i|)]$.

(b) Compute 2^k for a nonnegative integer k, in time $0(|k|+1)$.
(c) Compute $[|i|/2^k]$, for any integer i and k positive or negative, in time $0[L(i|)+|k|]$.

Section 7.5.1

1. Develop a procedure to convert a rational number into its decimal representation. The number of digits to be obtained in the decimal representation should be a parameter; the digits are to be obtained *one at a time*. This procedure can then be used to express in their decimal representation the rational endpoints of the isolating interval of a root.

Section 7.5.2

1. a. Develop a procedure to implement Theorem 7.5.1 in time

$$0\{nL[|p(x)|_\infty] + L^2[|p(x)|_\infty]\}$$

b. Develop a procedure **A7.5.1** that, using Theorem 7.5.1 and bisection, computes the integer part of the single positive root of $p(x) = 0$ in time

$$t_{A7.5.1}[p(x)] = 0\{n^2 L^3[|p(x)|_\infty]\}$$

Historical Notes and References

1a. We present in detail only the methods developed by Sturm and the author, because it is only these two that can be considered classical, since they are based on two very old and related theorems. There exist two additional real root isolation methods: (i) the method by differentiation, which is based on Rolle's theorem [see the papers by Collins and Loos (1976, 1982)], and (ii) the Collins–Akritas method, which is based on a complete modification of Vincent's theorem. However, these last two methods are of little interest, since they are both *bisection* procedures (just like Sturm's), whose theoretical and empirical performance is inferior to that of the author's continued fractions method of 1978.

1b. The author's continued fractions method of 1978 received the *First Prize* in the student paper competition when it was first presented in the sixteenth annual southeastern regional ACM conference in Atlanta, Georgia (April 1978).

1c. The Collins–Akritas method was developed immediately after Vincent's theorem was discovered by Akritas in 1975/76 and is described in Collins and Akritas (1976), Collins and Loos (1982), and Collins [in Rice (1977)] under the name "*modified* Uspensky's method". Regarding this point, see Akritas', paper of 1986 and historical note 7.

2. In Obreschkoff (1963), pp. 61–87, there are other proofs of Fourier's theorem due to Hurwitz and Laguerre, as well as a generalization of it due to Obreschkoff.

3. Heindel (1971) obtained a different bound for Sturm's method; namely, he showed that if $p(x) = 0$ is an univariate polynomial equation with integer coefficients of degree $n > 0$, without multiple roots, then the computing time of Sturm's method is

$$0\{n^{13}L^3[|p(x)|_\infty]\}$$

Moreover, instead of isolating first the positive and then the negative roots, as Sturm originally suggested, Heindel computes an upper bound b on the absolute values of the roots so that they all lie in the interval $(-b, b)$; this interval is then subdivided.

4. Cauchy's rule, despite its importance, is not widely known; we found it stated only in Obreschkoff's book (1963), pp. 50–51. Moreover, van der Sluis (1940) proved a theorem (Theorem 2.7) from which we can conclude that Cauchy's rule is the best.

5. Vincent's theorem was not mentioned by any authors, with the exception of Obreschkoff (1963) and Uspensky (1948). Akritas discovered it in 1975/76 while reviewing methods for the isolation of the real roots of equations as presented by Uspensky. There also exists the following generalization of Vincent's theorem [see Chen's paper (1987)]. *Wang's theorem* (1960): Let $p(x) = 0$ be an integral polynomial equation of degree $n \geq 3$, and assume that it has at least two sign variations in the sequence of its coefficients; moreover, let $\Delta > 0$ be the smallest distance between any two of its roots. Let m' be the smallest positive index such that $F_{m'-1}^2 > 1/\Delta$, where F_k is the kth member of the Fibonacci sequence $1,1,2,3,5,8,13,21,\ldots$, and let m'' be the smallest positive integer such that $m'' > 1 + \lceil \log_\phi n \rceil /2$. If we let $m = m' + m''$, then the arbitrary continued fraction substitution

$$x := a_1 + \cfrac{1}{a_2 + \cfrac{\ddots}{ + \cfrac{1}{a_m + \cfrac{1}{y}}}}$$

with a_1 nonnegative integer and $a_2, \ldots, a_m$ positive integers, transforms $p(x) = 0$ into the equation $p_{ti}(y) = 0$, which has r sign variations in the sequence of its coefficients. If $r = 0$, then there are no roots of $p(x)$ in the interval I_m with (unordered) endpoints p_m/q_m, p_{m-1}/q_{m-1} [obtained from (V13)]. If $r > 0$, then $p(x) = 0$ has a unique positive real root of multiplicity r in I_m.

Despite its theoretical interest, empirical results indicate that the implementation of Wang's theorem behaves poorly and, hence, is of little practical interest for the isolation of multiple roots of polynomial equations; instead, Vincent's theorem should be used combined with squarefree factorization. (See also, Cantor et al., 1972.)

6. Uspensky (1948), pp. 298–304, extended Vincent's theorem in order to obtain an upper bound on the number of substitutions that have to be performed. His treatment, though, contains certain errors, in the statement and the proof, which were corrected by Akritas (1978b). In this text we give the correct version of the extension of Vincent's theorem; for completeness we also add its proof, which is much shorter than the one by Uspensky due to the fact that we use Lemma 7.3.7; see also the papers by Akritas (1981a) and Akritas and Danielopoulos (1985).

7a. In the articles by Collins and Akritas (1976), Collins and Loos (1982), and Collins [in Rice, (1977)], Vincent's (exponential) continued fractions method of 1836 has been erroneously attributed to Uspensky, as a consequence of which the Collins–Akritas method appears under the name "*modified* Uspensky's method"; this was probably due to Uspensky's claim (in the preface of his book) that he himself invented the method. However, as it was pointed out in Akritas's work {(1978a), pp. 85–86, (1981a), and (1986)}, what Uspensky did was to take Vincent's method and *double* its computing time because he was not aware of Budan's theorem; namely, after the substitution $x := 1 + x$ applied to some $p_{ti}(x) = 0$ Uspensky has to perform the substitution $x := 1/(1 + x)$ to make sure that $p_{ti}(x) = 0$ has no roots in the interval $(0, 1)$. (Vincent, of course, obtains that information using Budan's theorem.)

What can be considered a contribution on Uspensky's part is only the fact that he used the Ruffini–Horner method to perform the substitutions of the form $x := a_i + x$; Vincent, on the contrary, used Taylor's expansion theorem.

7b. Taylor's expansion theorem is somewhat tedious for hand computations, but it was the "technology" available then. Horner's paper had appeared in 1819, but Vincent, obviously, was not aware of it; nor was Horner aware of the fact that he himself had been anticipated by Ruffini in 1804 (Cajori, 1911). In any case it has been shown by Akritas and Danielopoulos (1980) that both Taylor's expansion theorem and the Ruffini–Horner method have approximately equal computing times.

8. Sturm, in his 1835 paper, presented a solution to the problem involved with Lagrange's method (Lagrange, 1798), namely, when the number of roots between two consecutive integers is greater than one. Sturm is actually using both, the Sturm sequence, and Vincent's theorem to isolate and approximate the roots to any degree of accuracy. More details can be found in pp. 292–297 and 299–305 of Sturm's 1835 paper.

9. This is in full agreement with Vincent's remark in his 1836 paper, pp. 352–353: "Pour approcher davantage de la valeur de cette racine, nous pourrions continuer le calcul en suivant toujours la meme marche; et nous serions surs de n'avoir, dans toutes les transformées subsequentes, qu'une seule variation, et par conséquent une seule racine positive, laquelle, de plus, serait toujours necessarement plus grande que l'unité." ("To better approximate the value of that root, we could continue the calculations following always the same procedure; and we would be sure of having, in all the subsequent transformed equations, only one sign variation, and consequently only one positive root, which, moreover, would always necessarily be greater than one.")

10. Vincent himself did not follow this approach because in his paper (p. 353) he states: "la réduction en fraction continue ne croissant que tres lentement,

changeons maintenant notre marche" ("since the continued fraction expansion proceeds very slowly, let us now change our procedure").

Abel, N. H.: Beweis der Unmoeglichkeit algebraische Gleichungen von hoeheren Graden als den vierten allgemein aufzuloesen. *Journal fur die reine und angewandte Mathematik* **1**, 65–84, 1826.

Akritas, A. G.: *Vincent's theorem in algebraic manipulation*. Ph.D. Thesis, Operations Research Program, North Carolina State University, Raleigh, NC, 1978a.

Akritas, A. G.: A correction on a theorem by Uspensky. *Bulletin of the Greek Mathematical Society* **19**, 278–285, 1978b.

Akritas, A. G.: The fastest exact algorithms for the isolation of the real roots of a polynomial equation. *Computing* **24**, 219–313, 1980a.

Akritas, A. G.: An implementation of Vincent's theorem. *Numerische Mathematik* **36**, 53–52, 1980b.

Akritas, A. G.: Vincent's forgotten theorem, its extension and application. *International Journal of Computers and Mathematics with Applications* **7**, 309–317, 1981a.

Akritas, A. G.: Exact algorithms for the implementation of Cauchy's rule. *International Journal of Computer Mathematics* **9**, 323–333, 1981b.

Akritas, A. G.: Reflections on a pair of theorems by Budan and Fourier. *Mathematics Magazine* **55**, 292–298, 1982.

Akritas, A. G.: There is no "Uspensky's method." Extended Abstract. *Proceedings of the 1986 Symposium on Symbolic and Algebraic Computation* (Waterloo, Ontario, Canada), 1986, pp. 88–90.

Akritas, A. G., and S. D. Danielopoulos: On the forgotten theorem of Mr. Vincent. *Historia Mathematica* **5**, 427–435, 1978.

Akritas, A. G., and S. D. Danielopoulos: On the complexity of algorithms for the translation of polynomials. *Computing* **24**, 51–60, 1980.

Akritas, A. G., and S. D. Danielopoulos: A converse rule of signs for polynomials. *Computing* **34**, 283–286, 1985.

Akritas, A. G., and K. H. Ng: Exact algorithms for polynomial real root approximation using continued fractions. *Computing* **30**, 63–76, 1983.

Bocher, M.: The published and unpublished work of Charles Sturm on algebraic and differential equations. *Bulletin of the American Mathematical Society* **18**, 1–18, 1911–1912.

Burnside, W. S., and A. W. Panton: *The theory of equations*. Vol. 1, 2nd ed., Dover, New York, 1960.

Cajori, F.: Horner's method of approximation anticipated by Ruffini. *American Mathematical Society Bulletin* **17**, 409–414, 1911.

Cajori, F.: A history of the arithmetical methods of approximation to the roots of numerical equations of one unknown quantity. *Colorado College Publications, General Series No. 51,* Science Series Vol. XII, No. 7, pp. 171–287, Colorado Springs, CO, 1910.

Cantor, D. G., P. H. Galyean, and H. G. Zimmer: A continued fraction algorithm for real algebraic numbers. *Mathematics of Computation* **26**, 785–791, 1972.

Chen, J.: A new algorithm for the isolation of the real roots of polynomial equations. *Second International Conference on Computers and Applications* (Beijing, People's Republic of China), 714–719, 1987.

Collins, G. E. and A. G. Akritas: Polynomial real root isolation using Descartes' rule of signs. *Proceedings of the 1976 ACM Symposium on Symbolic and Algebraic Computation*, Yorktown Heights, NY, 1976, pp. 272–275.

Collins, G. E., and R. Loos: Polynomial real root isolation by differentiation. *Proceedings of the 1976 ACM Symposium on Symbolic and Algebraic Computation*, Yorktown Heights, NY, 1976, pp. 15–20.

Collins, G. E., and R. Loos: Real zeros of polynomials. In *Computer algebra symbolic and algebraic computations*, B. Buchberger, G. E. Collins, and R. Loos, eds. Springer Verlag, New York, *Computing Supplement* **4**, 83–94, 1982.

Dickson, L. E.: *First course in the theory of equations*. Wiley, New York, 1922.

Heindel, L. E.: Integer arithmetic algorithms for polynomial real zero determination. *Journal of the ACM* **18**, 533–548, 1971.

Hurwitz, A.: Ueber den Satz von Budan-Fourier. *Mathematische Annalen* **71**, 584–591, 1912.

Lagrange, J. L.: *Traité de la résolution des équations numériques*. Paris (n.p.), 1798.

Lloyd, E. K.: On the forgotten Mr. Vincent. *Historia Mathematica* **6**, 448–450, 1979.

Mahler, K.: An inequality for the discriminant of a polynomial. *Michigan Mathematical Journal* **11**, 257–262, 1964.

Marcus, M., and H. Minc: *Introduction to linear algebra*. Macmillan, New York, 1965.

Mignotte, M.: Sur la complexité de certains algorithms ou intervient la séparation des racines d'un polynome. *Revue Française d' Automatique, Informatique et Recherce Opérationnelle (RAIRO)* **10**, 51–55, 1976.

Mignotte, M., and M. Payafar: Distance entre les racines d'un polynome. *RAIRO–Analyse Numérique* **13**, 181–192, 1979.

Ng, K. H.: *Polynomial real root approximation using continued fractions*. M. S. Research Report, University of Kansas, Department of Computer Science, Lawrence, KS, 1980.

Nordgaard, M. A.: *A historical survey of algebraic methods of approximating the roots of numerical higher equations up to the year 1819*. Teachers College, Columbia University, New York, 1922.

Obreschkoff, N.: *Verteilung und Berechnung der Nullstellen reeller Polynome*. VEB Deutscher Verlag der Wissenschaften, Berlin, 1963.

Poggendorff, J. C.: *Biographisch–Literarisches Handwoerterbuch zur Geschichte der exacten Wissenschaften*. J. A. Barth, Leipzig, 1863.

Rice, J.: *Mathematical software III*. Academic Press, New York, 1977, pp. 35–68.

Ruffini, P.: Sopra la determinazione delle radici nelle equazioni numeriche di qualunque grado.... *Societa Italiana delle Scienze*, 1804.

Rump, S. M.: Polynomial minimum root separation. *Mathematics of Computation* **33**, 327–336, 1979.

Specht, W.: Abschaetzungen der Wurzeln algebraischer Gleichungen. *Mathematische Zeitschrift* **52**, 310–321, 1949.

Sturm, C.: Mémoire sur la résolution des équations numériques. *Mémoires des Savants Etrangers* **6**, 271–318, 1835.

Todhunter, I.: *Theory of equations*. Macmillan, London, 1882.

Turnbull, H. W.: *Theory of equations*, 5th ed. Oliver & Boyd, Edinburgh and London, 1957.

Uspensky, J. V.: *Theory of equations*. McGraw-Hill, New York, 1948.

van der Sluis, A.: Upperbounds for roots of polynomials. *Numerische Mathematik* **15**, 250–262, 1970.

van der Waerden, B. L.: *Erwachende Wissenschaft*. Birkhaeuser Verlag, Basel and Stuttgart, 1956.

Verbaeten, P.: Computing real zeros of polynomials with SAC-1. *ACM-SIGSAM Bulletin* **9**, 8–10, 1975.

Vincent, A.J.H.: Sur la résolution des équations numériques. *Journal de Mathématiques Pures et Appliquées* **1**, 341–372, 1836.

Weisner, L.: *Introduction to the theory of equations*. Macmillan, New York, 1938.

Appendix: Linear Algebra

This chapter is intended to give a brief survey of those aspects of linear algebra that are of relevance in this book. Below we assume that R is a commutative ring like $\mathbf{Q}$, $\mathbf{R}$, $\mathbf{Z}$, or $\mathbf{Z}_m$.

Matrix Multiplication

A row vector, or just vector, is a row of elements of R; that is, $\mathbf{v} = (a_1, \ldots, a_n)$. A column vector, or the *transpose* of a row vector, is a column of elements of R; that is

$$\mathbf{v}^T = \begin{bmatrix} a_1 \\ \vdots \\ a_n \end{bmatrix}$$

An $m \times n$ matrix $\mathbf{A}$ is a rectangular array of mn elements of R, where m is the number of rows and n is the number of columns; that is, we have

$$\mathbf{A} = \begin{bmatrix} a_{11} & a_{12} & \cdots & a_{1n} \\ a_{21} & a_{22} & \cdots & a_{2n} \\ \cdot & \cdot & & \cdot \\ \cdot & \cdot & & \cdot \\ a_{m1} & a_{m2} & \cdots & a_{mn} \end{bmatrix}$$

which can be thought of as a collection of row vectors placed in a column, or as a collection of column vectors laid out in a row. The transpose of an

$m \times n$ matrix, $\mathbf{A}^T$, is the $n \times m$ matrix obtained by making the rows columns (or the columns rows) in the original matrix.

Given a row vector with n elements (placed on the left) and a column vector with the same number of elements (placed on the right), we can multiply then to obtain an element of R; that is,

$$(a_1, \ldots, a_n) \begin{bmatrix} b_1 \\ \vdots \\ b_n \end{bmatrix} = a_1 b_1 + a_2 b_2 + \cdots + a_n b_n$$

As an example, consider the case $R = \mathbf{Z}$; then

$$(1, 2, 3) \begin{bmatrix} 4 \\ 5 \\ 6 \end{bmatrix} = 1 \cdot 4 + 2 \cdot 5 + 3 \cdot 6 = 32$$

Given an $m \times n$ matrix $\mathbf{A}$ (placed on the left) and an n-element column vector $\mathbf{v}^T$ (placed on the right), we can form the product $\mathbf{A}\mathbf{v}^T$ by regarding the matrix as a collection of m n-element row vectors and doing m multiplications of the row vectors of $\mathbf{A}$ with $\mathbf{v}^T$. The result, $\mathbf{A}\mathbf{v}^T$, is a column of m elements. For example,

$$\begin{bmatrix} 1 & 2 & 3 & 4 \\ 5 & 6 & 7 & 8 \end{bmatrix} \begin{bmatrix} 1 \\ 1 \\ 1 \\ 1 \end{bmatrix} = \begin{bmatrix} 10 \\ 26 \end{bmatrix}$$

Given an $m \times n$ matrix $\mathbf{A}$ (on the left) and an $n \times p$ matrix $\mathbf{B}$ (on the right), we can multiply them by considering $\mathbf{A}$ as a collection of n-element rows and $\mathbf{B}$ as a collection of n-element columns. The result is an $m \times p$ matrix, $\mathbf{AB}$, where the element in the ith row and jth column is obtained by multiplying the ith row of $\mathbf{A}$ times the jth column of $\mathbf{B}$. For example,

$$\begin{bmatrix} 1 & 2 & 3 & 4 \\ 5 & 6 & 7 & 8 \end{bmatrix} \begin{bmatrix} 1 & 1 \\ 1 & 2 \\ 1 & 2 \\ 1 & 1 \end{bmatrix} = \begin{bmatrix} 10 & 15 \\ 26 & 39 \end{bmatrix}$$

It should be noted that the order in which the matrices are multiplied (i.e., which matrix is on the left and which is on the right) is very important. There are cases where it is impossible to multiply two matrices if we change their order (the reader should construct an example); even when it makes sense to multiply in either order, the results may be different, as, for example in

$$\begin{bmatrix}1\\3\end{bmatrix}(2,1) = \begin{bmatrix}2 & 1\\6 & 3\end{bmatrix} \quad \text{whereas} \quad (2,1)\begin{bmatrix}1\\3\end{bmatrix} = 5$$

If **A** is a matrix of any size (in particular, a column or row vector) and s is an element of R, that is, a scalar, then we define the matrix $s\mathbf{A}$ to be the matrix in which each element of **A** is multiplied by s; for example,

$$s\mathbf{A} = s\begin{bmatrix} a_{11} & a_{12} & \cdots & a_{1n} \\ a_{21} & a_{22} & \cdots & a_{2n} \\ \cdots & \cdots & & \cdots \\ \cdots & \cdots & & \cdots \\ a_{m1} & a_{m2} & \cdots & a_{mn}\end{bmatrix} = \begin{bmatrix} sa_{11} & sa_{12} & \cdots & sa_{1n} \\ sa_{21} & sa_{22} & \cdots & sa_{2n} \\ \cdots & \cdots & & \cdots \\ \cdots & \cdots & & \cdots \\ sa_{m1} & sa_{m2} & \cdots & sa_{mn}\end{bmatrix}$$

In particular, we have

$$x\begin{bmatrix}a_1\\a_2\\a_3\end{bmatrix} = \begin{bmatrix}xa_1\\xa_2\\xa_3\end{bmatrix}$$

Given two vectors $\mathbf{v}_1 = (a_1, \ldots, a_n)$ and $\mathbf{v}_2 = (b_1, \ldots, b_n)$, we define $\mathbf{v}_1 + \mathbf{v}_2 = (a_1 + b_1, \ldots, a_n + b_n)$. Likewise, we can add two matrices, **A** and **B**, of the same size by thinking of them as sequences of row vectors; that is, we have:

$$\begin{bmatrix} a_{11} & a_{12} & \cdots & a_{1n} \\ a_{21} & a_{22} & \cdots & a_{2n} \\ \cdots & \cdots & & \cdots \\ \cdots & \cdots & & \cdots \\ a_{m1} & a_{m2} & \cdots & a_{mn}\end{bmatrix} + \begin{bmatrix} b_{11} & b_{12} & \cdots & b_{1n} \\ b_{21} & b_{22} & \cdots & b_{2n} \\ \cdots & \cdots & & \cdots \\ \cdots & \cdots & & \cdots \\ b_{m1} & b_{m2} & \cdots & b_{mn}\end{bmatrix} = \begin{bmatrix} a_{11}+b_{11} & \cdots & a_{1n}+b_{1n} \\ a_{21}+b_{21} & \cdots & a_{2n}+b_{2n} \\ \cdots & & \cdots \\ \cdots & & \cdots \\ a_{m1}+b_{m1} & \cdots & a_{mn}+b_{mm}\end{bmatrix}$$

Note that $\mathbf{A} + \mathbf{B}$ makes sense only if **A** and **B** have the same "shape." In particular, $\mathbf{A} + \mathbf{B}$ is defined if **A, B** are square $n \times n$ matrices.

Let $\mathbf{A}_n(R)$ be the set of $n \times n$ (square) matrices with elements from the set R. Let **0** be the $n \times n$ matrix consisting of all zeros; then, clearly, $\mathbf{0} + \mathbf{A} = \mathbf{A} + \mathbf{0} = \mathbf{A}$ for any $n \times n$ matrix **A**. Moreover, let

$$\mathbf{I} = \begin{bmatrix} 1 & 0 & \cdots & 0 \\ 0 & 1 & \cdots & 0 \\ \cdots & \cdots & & \cdots \\ \cdots & \cdots & & \cdots \\ 0 & 0 & \cdots & 1 \end{bmatrix}$$

then, $\mathbf{AI} = \mathbf{IA} = \mathbf{A}$ for any $n \times n$ matrix **A**. We now have the following.

Theorem A.1. If R is a commutative ring with identity, then $\mathbf{A}_n(R)$ is a ring with identity.

Proof. The proof consists of checking a number of axioms and is left as an exercise for the reader. □

Observe that $\mathbf{A}_n(R)$ has zero divisors for any $n \geq 2$; for example, for $n = 2$ consider

$$\begin{bmatrix} 1 & 0 \\ 0 & 0 \end{bmatrix} \begin{bmatrix} 0 & 0 \\ 0 & 1 \end{bmatrix} = \begin{bmatrix} 0 & 0 \\ 0 & 0 \end{bmatrix}$$

Moreover, as we will see below, there exist nonzero $n \times n$ matrices without an inverse, for any $n \geq 2$.

Linear Equations

Matrices and vectors are a convenient way to describe systems of linear equations. Consider the following system of m equations in n unknowns:

$$\begin{aligned} a_{11}x_1 + a_{12}x_2 + \cdots + a_{1n}x_n &= b_1 \\ a_{21}x_1 + a_{22}x_2 + \cdots + a_{2n}x_n &= b_2 \\ &\cdots \\ a_{m1}x_1 + a_{m2}x_2 + \cdots + a_{mn}x_n &= b_m \end{aligned} \quad \text{(LE)}$$

This system of equations is called *homogeneous* if $b_1 = b_2 = \cdots = b_m = 0$ and *nonhomogeneous* otherwise.

We can also think of the above system of linear equations as an equality of column vectors, because two column vectors are equal precisely when their respective components are equal. Thus we can rewrite (LE) in either of two ways:

1. Using the definition of addition and scalar multiplication of column vectors, (LE) is written as

$$x_1 \begin{bmatrix} a_{11} \\ \vdots \\ a_{m1} \end{bmatrix} + x_2 \begin{bmatrix} a_{12} \\ \vdots \\ a_{m2} \end{bmatrix} + \cdots + x_n \begin{bmatrix} a_{1n} \\ \vdots \\ a_{mn} \end{bmatrix} = \begin{bmatrix} b_1 \\ \vdots \\ b_m \end{bmatrix}$$

This says that to solve the original system is the same as to write the vector $(b_1, \ldots, b_m)^T$ as a linear combination (i.e., a sum of scalar multiples) of the vectors $(a_{11}, \ldots, a_{m1})^T, \ldots, (a_{1n}, \ldots, a_{mn})^T$.

2. Observing that the left-hand side of (LE) is the product of the matrix $\mathbf{A}$ of coefficients of the original system, with the vector $(x_1, \ldots, x_n)^T$, (LE) is written as $\mathbf{A}\mathbf{x}^T = \mathbf{b}^T$, where $\mathbf{x} = (x_1, \ldots, x_n)$, $\mathbf{b} = (b_1, \ldots, b_m)$, and

$$\mathbf{A} = \begin{bmatrix} a_{11} & a_{12} & \cdots & a_{1n} \\ a_{21} & a_{22} & \cdots & a_{2n} \\ \cdot & \cdot & \cdot & \cdot \\ \cdot & \cdot & \cdot & \cdot \\ a_{ml} & a_{m2} & \cdots & a_{mn} \end{bmatrix}$$

Example. The set of equations

$$\begin{aligned} x_1 + x_2 + x_3 &= 6 \\ 2x_1 + 3x_2 - x_3 &= 5 \\ 4x_1 \qquad\quad - x_3 &= 1 \end{aligned}$$

may be written as

$$x_1 \begin{bmatrix} 1 \\ 2 \\ 4 \end{bmatrix} + x_2 \begin{bmatrix} 1 \\ 3 \\ 0 \end{bmatrix} + x_3 \begin{bmatrix} 1 \\ -1 \\ -1 \end{bmatrix} = \begin{bmatrix} 6 \\ 5 \\ 1 \end{bmatrix}$$

or as

$$\begin{bmatrix} 1 & 1 & 1 \\ 2 & 3 & -1 \\ 4 & 0 & -1 \end{bmatrix} \begin{bmatrix} x_1 \\ x_2 \\ x_3 \end{bmatrix} = \begin{bmatrix} 6 \\ 5 \\ 1 \end{bmatrix}$$

Suppose now that, for the above system of equations $\mathbf{A}\mathbf{x}^T = \mathbf{b}^T$, there were an $n \times m$ matrix $\mathbf{A}^{-1}$, the inverse of $\mathbf{A}$, such that $\mathbf{A}^{-1}\mathbf{A} = \mathbf{I}$. If we could find such a matrix $\mathbf{A}^{-1}$, then $\mathbf{A}^{-1}\mathbf{b}^T = \mathbf{A}^{-1}\mathbf{A}\mathbf{x}^T = \mathbf{I}\mathbf{x}^T = \mathbf{x}^T$ would be a solution of the equations. Therefore, solving equations is closely related to finding inverses of matrices. For example, the matrix

$$\begin{bmatrix} 1 & 1 & 1 \\ 2 & 3 & -1 \\ 4 & 0 & -1 \end{bmatrix}$$

turns our to have the inverse

$$\begin{bmatrix} 3/17 & -1/17 & 4/17 \\ 2/17 & 5/17 & -3/17 \\ 12/17 & -4/17 & -1/17 \end{bmatrix}$$

so that

$$\begin{bmatrix} x_1 \\ x_2 \\ x_3 \end{bmatrix} = \begin{bmatrix} 3/17 & -1/17 & 4/17 \\ 2/17 & 5/17 & -3/17 \\ 12/17 & -4/17 & -1/17 \end{bmatrix} \begin{bmatrix} 6 \\ 5 \\ 1 \end{bmatrix} = \begin{bmatrix} 1 \\ 2 \\ 3 \end{bmatrix}$$

Determinants and Inverses

Consider a square, $n \times n$ matrix

$$\mathbf{A} = \begin{bmatrix} a_{11} & a_{12} & \cdots & a_{1n} \\ a_{21} & a_{22} & \cdots & a_{2n} \\ \cdot & \cdot & \cdot & \cdot \\ \cdot & \cdot & \cdot & \cdot \\ a_{n1} & a_{n2} & \cdots & a_{nn} \end{bmatrix}$$

where $\mathbf{A}$ is merely a tabulation of numbers and has no numerical value as such. We are going to assign a numberical value to $\mathbf{A}$ in the following way:

1. Choose any element in row 1, any element in row 2 *not in the same column*, any element in row 3 in a column *different from the other two*, and so on, until we have n numbers such that no two of them belong to the same row or column; form all such possible groups of numbers. For example, given the 3×3 matrix

$$\begin{bmatrix} a_{11} & a_{12} & a_{13} \\ a_{21} & a_{22} & a_{23} \\ a_{31} & a_{32} & a_{33} \end{bmatrix}$$

we have the following possibilities:

Choices	Permutations
a_{11}, a_{22}, a_{33}	1 2 3
a_{11}, a_{23}, a_{32}	1 3 2
a_{12}, a_{21}, a_{33}	2 1 3
a_{12}, a_{23}, a_{31}	2 3 1
a_{13}, a_{21}, a_{32}	3 1 2
a_{13}, a_{22}, a_{31}	3 2 1

Permutation refers to the order of the second subscripts when the first subscripts are in natural order $1, 2, \ldots$.

2. Calculate the number of *inversions* for each permutation; that is, calculate the number of interchanges needed to bring the permutation back to the natural order. This can be easily done in the following way: for each permutation, proceed from left to right, counting how many digits to *the right* are less than the one being considered. For example, for the permutation 3 1 2 we have $2 + 0 + 0 = 2$ because 3 has two smaller numbers to its right, and 1, 2 have none. So the number of inversions for 3 1 2 is 2; likewise, that for 3 2 1 is $3 = 2 + 1 + 0$.

3. Now, the numerical value corresponding to **A** is the sum

$$\Sigma(-1)^m a_{1i_1} a_{2i_2} \cdots a_{ni_n}$$

where the summation is taken over all i from 1 to n, each term of the sum is the product of the elements chosen in step 1 above, and m is the number of inversions of the corresponding permutation. (*We have $n!$ terms.*)

Example. Given $a_{22} a_{13} a_{31}$, let us find i_1, i_2, i_3; also, is this term added or subtracted? To find i_1, i_2, i_3, we simply rearrange the elements so that the first subscripts are in natural order: $a_{13} a_{22} a_{31}$. Now, we clearly have $i_1 = 3$, $i_2 = 2$, and $i_3 = 1$. The term is subtracted because $m = 3$.

Definition A.2. Given a square, $n \times n$ matrix **A**, its *determinant* is

$$\det(\mathbf{A}) = \Sigma(-1)^m a_{1i_1} a_{2i_2} \cdots a_{ni_n}$$

where the summation is taken over all i from 1 to n, and m is the number of inversions of the corresponding permutation.

Sometimes the determinant is also written as $|\mathbf{A}|$, or $|a_{ij}|$, $i, j = 1, 2, \ldots, n$; moreover, $\det(\mathbf{A})$ can be also considered as the sum of products of the entries in one row or column of **A** with the corresponding *cofactors*. The cofactor of the element a_{ij} is the determinant of the matrix obtained from **A** if we remove the ith row and jth column (see also Definition A.8). For the 3×3 case that we considered above we have (see also Theorem A.9).

$$|\mathbf{A}| = \det \begin{bmatrix} a_{11} & a_{12} & a_{13} \\ a_{21} & a_{22} & a_{23} \\ a_{31} & a_{32} & a_{33} \end{bmatrix}$$

$$= a_{11} a_{22} a_{33} + a_{12} a_{23} a_{31} + a_{13} a_{21} a_{32} - a_{12} a_{21} a_{33} - a_{13} a_{22} a_{31} - a_{11} a_{23} a_{32}$$

For the 2×2 case we have

$$|\mathbf{A}| = \det \begin{bmatrix} a_{11} & a_{12} \\ a_{21} & a_{22} \end{bmatrix} = a_{11} a_{22} - a_{12} a_{21}$$

If **A** is a triangular matrix, that is, a matrix of the form

$$\mathbf{A} = \begin{bmatrix} a_{11} & 0 & 0 \cdots 0 \\ a_{21} & a_{22} & 0 \cdots 0 \\ \cdots \cdots \\ \cdots \cdots \\ a_{n1} & a_{n2} & \cdots a_{nn} \end{bmatrix}$$

then $\det(\mathbf{A}) = a_{11} a_{22} \cdots a_{nn}$.

If **A** is an $n \times n$ matrix with elements from R, sometimes **A** has an inverse, an $n \times n$ matrix A^{-1} such that $\mathbf{AA}^{-1} = \mathbf{A}^{-1}\mathbf{A} = \mathbf{I}$. A well known theorem in linear algebra says that $\mathbf{A}^{-1}$ exists if and only if det(**A**) is an invertible element of R. In particular, for 2×2 matrices, if the determinant of **A** is invertible, the inverse of **A** can be found as follows: If

$$\mathbf{A} = \begin{bmatrix} a & b \\ c & d \end{bmatrix}$$

with $1/(ad - bc)$ in R, then

$$\mathbf{A}^{-1} = \frac{1}{ad - bc} \begin{bmatrix} d & -b \\ -c & a \end{bmatrix}$$

as is easily checked. For matrices **A** of order 3×3 or bigger, the formula for obtaining the inverse of **A** using determinants is too complicated to state here. However, we can find the inverse of an $n \times n$ matrix **A**, if it exists, using row operations.

Elementary Row Operations

Gaussian elimination is a most useful computational technique where only elementary row operations are performed. There are three such operations:

1. Interchanging two rows of a matrix.
2. Multiplying a row by an invertible scalar.
3. Replacing a row by itself plus a multiple of another row.

These row operations correspond to things that can be done to sets of simultaneous equations without affecting the solutions of these equations. If **A** is an $m \times n$ matrix with elements from a field R, then using operation (3) it is possible to introduce zero entries into the matrix and transform it into its *row reduced echelon form*. The following are examples of matrices in this form:

$$\begin{bmatrix} 1 & 0 & 2 \\ 0 & 1 & 2 \\ 0 & 0 & 0 \end{bmatrix}, \quad \begin{bmatrix} 1 & 0 & 0 \\ 0 & 1 & 0 \\ 0 & 0 & 1 \end{bmatrix}, \quad \begin{bmatrix} 0 & 0 & 1 & 0 \\ 0 & 0 & 0 & 1 \\ 0 & 0 & 0 & 0 \end{bmatrix},$$

$$\begin{bmatrix} 1 & 2 & 0 & 0 & 4 & 0 \\ 0 & 0 & 1 & -1 & 0 & 0 \\ 0 & 0 & 0 & 0 & 0 & 1 \end{bmatrix}$$

We can now find the inverse of an $n \times n$ matrix **A**, if it exists, by solving for $\mathbf{x}^T$ in the equation

$$\mathbf{A}\mathbf{x}^T = \mathbf{y}^T = \mathbf{I}\mathbf{y}^T \quad \text{or} \quad (\mathbf{A}, \mathbf{I})(\mathbf{x}^T, -\mathbf{y}^T)^T = \mathbf{0}^T$$

where $(\mathbf{A}, \mathbf{I})$ is the $n \times 2n$ matrix whose left side is $\mathbf{A}$ and whose right side is the $n \times n$ identity matrix $\mathbf{I}$, and where $\mathbf{x}, \mathbf{y}$ are vectors of unknowns. If the row-reduced echelon form for the $n \times 2n$ matrix $(\mathbf{A}, \mathbf{I})$ is in the form $(\mathbf{I}, \mathbf{B})$, that is, if the leftmost n columns of the row-reduced echelon form of the $n \times 2n$ matrix $(\mathbf{A}, \mathbf{I})$ form the $n \times n$ identity matrix $\mathbf{I}$, then

$$(\mathbf{A}, \mathbf{I}) \begin{bmatrix} \mathbf{x}^T \\ -\mathbf{y}^T \end{bmatrix} = \mathbf{0}^T \quad \text{iff} \quad (\mathbf{I}, \mathbf{B}) \begin{bmatrix} \mathbf{x}^T \\ -\mathbf{y}^T \end{bmatrix} = \mathbf{0}^T \quad \text{iff} \quad \mathbf{x}^T = \mathbf{B}\mathbf{y}^T$$

iff $\mathbf{B}$ is the inverse of $\mathbf{A}$. Therefore, in order to compute $\mathbf{A}^{-1}$, the inverse of $\mathbf{A}$, write side by side $(\mathbf{A}, \mathbf{I})$ and do row operations on the whole matrix to try and reduce $\mathbf{A}$ to $\mathbf{I}$. If this succeeds, then by the same operations $\mathbf{I}$ is transformed into $\mathbf{A}^{-1}$.

Example. Let us compute the inverse of

$$\mathbf{A} = \begin{bmatrix} 1 & 2 \\ 3 & 4 \end{bmatrix}$$

Actually, for this matrix we have a formula to obtain $\mathbf{A}^{-1}$, but the same result is obtained by the following sequence of row operations

$$\begin{bmatrix} 1 & 2 & 1 & 0 \\ 3 & 4 & 0 & 1 \end{bmatrix} \to \begin{bmatrix} 1 & 2 & 1 & 0 \\ 0 & -2 & -3 & 1 \end{bmatrix} \to \begin{bmatrix} 1 & 2 & 1 & 0 \\ 0 & 1 & 3/2 & -1/2 \end{bmatrix} \to$$

$$\begin{bmatrix} 1 & 0 & -2 & 1 \\ 0 & 1 & 3/2 & -1/2 \end{bmatrix}$$

that is, we have

$$\begin{bmatrix} -2 & 1 \\ 3/2 & -1/2 \end{bmatrix} \begin{bmatrix} 1 & 2 \\ 3 & 4 \end{bmatrix} = \begin{bmatrix} 1 & 0 \\ 0 & 1 \end{bmatrix}$$

We next present certain theorems that show the effect of row operations on determinants; their proofs are quite unenlightening and are omitted. (They can be found in most books on Linear Algebra)

Theorem A.3. If any two rows (or columns) of a determinant are interchanged, the determinant changes sign. To see the validity of this theorem, consider

$$|\mathbf{A}| = \det \begin{bmatrix} a & b \\ c & d \end{bmatrix} \quad \text{and} \quad |\mathbf{B}| = \det \begin{bmatrix} c & d \\ a & b \end{bmatrix}.$$

Then $|\mathbf{A}| = ad - bc$ and $|\mathbf{B}| = bc - ad = -(ad - bc)$. Do the same for columns.

Theorem A.4. If the elements of a row (or column) of a determinant are multiplied by k, the determinant is multiplied by k. Compare

$$|\mathbf{A}| = \det\begin{bmatrix} a & b \\ c & d \end{bmatrix} \quad \text{and} \quad |\mathbf{B}| = \det\begin{bmatrix} ka & kb \\ c & d \end{bmatrix}$$

Clearly, $|\mathbf{A}| = ad - dc$, whereas $|\mathbf{B}| = kad - kbc = k(ad - bc) = k|\mathbf{A}|$. Likewise for columns.

Theorem A.5. If two rows (or columns) of a determinant are equal, the determinant is 0. Consider

$$|\mathbf{A}| = \det\begin{bmatrix} a & b \\ a & b \end{bmatrix} = ab - ab = 0$$

Theorem A.6. If a multiple of one row (or column) is added to another row (or column) of a determinant, the value is unchanged. Consider, for example

$$|\mathbf{A}| = \det\begin{bmatrix} a & b \\ c & d \end{bmatrix} \quad \text{and} \quad |\mathbf{B}| = \det\begin{bmatrix} a+kc & b+kd \\ c & d \end{bmatrix}$$

Then $|\mathbf{A}| = ad - bc$ and $|\mathbf{B}| = (a + kc)d - (b + kd)c = ad + kcd - bc - kcd = ad - bc = |\mathbf{A}|$. Repeat the same for columns.

Finally, we have the following.

Theorem A.7 (Interchange of Rows and Columns). If the rows of a determinant become columns and the columns become rows (in the same order), the determinant does not change. (In other words, the determinant of the transpose of a matrix **A** is the same as the determinant of **A**.) That is

$$|\mathbf{A}| = \det\begin{bmatrix} a & b \\ c & d \end{bmatrix} = \det\begin{bmatrix} a & c \\ b & d \end{bmatrix} = ad - bc$$

Expanding a Determinant according to the Elements of a Row or Column

From the previous example we saw that we can very easily compute the value of a 2×2 determinant. But what about a 3×3 or, in general, an $n \times n$ determinant? In what follows we shall given an answer to this question.

Definition A.8. Given an $n \times n$ determinant $|\mathbf{A}|$, and an element a_{mk} of it, we call the *minor* determinant, or *cofactor*, of the element a_{mk}, the $(n-1) \times (n-1)$ determinant obtained by deleting from **A** the mth row and the kth column. The minor is denoted by $|\mathbf{A}_{mk}|$.

The following is a major result.

Theorem A.9. Given an $n \times n$ determinant $|\mathbf{A}|$, we have

$$|\mathbf{A}| = (-1)^{m+1} a_{m1} |\mathbf{A}_{m1}| + (-1)^{m+2} a_{m2} |\mathbf{A}_{m2}| + \cdots + (-1)^{m+n} a_{mn} |\mathbf{A}_{mn}|$$

Example. To compute the value of

$$|\mathbf{A}| = \det \begin{bmatrix} a_{11} & a_{12} & a_{13} \\ a_{21} & a_{22} & a_{23} \\ a_{31} & a_{32} & a_{33} \end{bmatrix}$$

we use Theorem A.9, and expanding according to the elements of the first row, we obtain

$$|\mathbf{A}| = (-1)^2 a_{11} \det \begin{bmatrix} a_{22} & a_{23} \\ a_{32} & a_{33} \end{bmatrix} + (-1)^3 a_{12} \det \begin{bmatrix} a_{21} & a_{23} \\ a_{31} & a_{33} \end{bmatrix}$$
$$+ (-1)^4 a_{13} \det \begin{bmatrix} a_{21} & a_{22} \\ a_{31} & a_{32} \end{bmatrix}$$

from which we easily obtain the six terms that we saw before. We can also expand according to the elements of a column.

Subspaces, Bases, and Dimension

For V a field, we denote by $\mathbf{V}_n$ the set of all n-tuples $(v_1, \ldots, v_n)$ with entries from V. In $\mathbf{V}_n$ the operations of addition and scalar multiplication are defined as follows:

$$(a_1, \ldots, a_n) + (b_1, \ldots, b_n) = (a_1 + b_1, \ldots, a_n + b_n),$$
$$s(v_1, \ldots, v_n) = (sv_1, \ldots, sv_n)$$

for s in V. These operations make $\mathbf{V}_n$ into what is called a *vector space*.

A subset $\mathbf{S}$ of $\mathbf{V}_n$ is called a *subspace* if whenever $\mathbf{x}, \mathbf{y}$ are in $\mathbf{S}$, so is their linear combination, that is, so is $r\mathbf{x} + s\mathbf{y}$ for any r, s in V. In this book, the relevant examples of subspaces are null spaces and row spaces of $m \times n$ matrices.

Let $\mathbf{A}$ be an $m \times n$ matrix. The *null space* of $\mathbf{A}$ is the set $\mathbf{S}$ of all $\mathbf{x}$ in $\mathbf{V}_n$ such that $\mathbf{A}\mathbf{x}^T = \mathbf{0}^T$. If we consider $\mathbf{A}$ as a matrix of coefficients of a set of homogeneous equations, then the null space of $\mathbf{A}$ is the set of solutions to the corresponding equations. The set $\mathbf{S}$ is a subspace of $\mathbf{V}_n$, because if $\mathbf{x}, \mathbf{y}$ are solutions, so is any linear combination of them: If r, s are any elements in V, then

$$\mathbf{A}(r\mathbf{x}^T + s\mathbf{y}^T) = r(\mathbf{A}\mathbf{x}^T) + s(\mathbf{A}\mathbf{y}^T) = r\mathbf{0}^T + s\mathbf{0}^T = \mathbf{0}^T$$

(Observe that the set of $\mathbf{x}$ such that $\mathbf{Ax}^T = \mathbf{b}^T$, for fixed $\mathbf{b} \neq \mathbf{0}$, is not a subspace, because if $\mathbf{Ax}^T = \mathbf{b}^T$, then $\mathbf{A}(\mathbf{x}^T + \mathbf{x}^T) = 2\mathbf{b}^T$, not $\mathbf{b}^T$.)

A *basis of a vector space* $\mathbf{S}$ is a set $\{\mathbf{x}_1, \ldots, \mathbf{x}_r\}$ of vectors in $\mathbf{S}$ such that any vector $\mathbf{a}$ in $\mathbf{S}$ can be written as a linear combination of $\mathbf{x}_1, \ldots, \mathbf{x}_r$ in exactly one way. There is an equivalent way of defining a basis: we say that a set $\{\mathbf{x}_1, \ldots, \mathbf{x}_r\}$ of vectors of $\mathbf{S}$ *spans* $\mathbf{S}$ if any vector in $\mathbf{S}$ can be written as a linear combination of $\mathbf{x}_1, \ldots, \mathbf{x}_r$ (not necessarily in one way), and we say that the vectors $\mathbf{x}_1, \ldots, \mathbf{x}_r$ are *linearly independent* if the only solution to the equation

$$c_1 \mathbf{x}_1 + c_2 \mathbf{x}_2 + \cdots + c_r \mathbf{x}_r = 0$$

is the solution $c_1 = 0, c_2 = 0, \ldots, c_r = 0$. Then a basis of $\mathbf{S}$ is a set of linearly independent vectors that spans $\mathbf{S}$.

If $\mathbf{S}$ is the null space of a matrix $\mathbf{A}$, then, as we have seen in Section 6.2.4, the basis of $\mathbf{S}$ is found by putting $\mathbf{A}$ into row-reduced echelon form with the help of algorithm NS in Section 6.2.4.

In general, a space has many different bases. However, a well-known theorem informs us that all bases of a space $\mathbf{S}$ contain the same number of vectors. The *dimension* of a space is equal to the number of vectors in any basis of the space. Thus the dimension of $\mathbf{V}_n$ is n.

The *row space* of an $m \times n$ matrix $\mathbf{A}$ is the set of all n-tuples that are linear combinations of the rows of $\mathbf{A}$. It is easy to see that if we do an elementary row operation to $\mathbf{A}$, the row space of the new matrix is the same as the row space of $\mathbf{A}$. Therefore, since $\mathbf{A}_e$, the row-reduced echelon form of $\mathbf{A}$, is obtained by a sequence of elementary row operations, the row space of $\mathbf{A}_e$ is the same as the row space of $\mathbf{A}$.

It can be easily verified that the nonzero rows of the row-reduced echelon form of a matrix $\mathbf{A}$ constitute a basis of the row space of $\mathbf{A}$. Therefore, the number of nonzero rows in the row-reduced echelon form of a matrix $\mathbf{A}$ is the dimension of the row space of $\mathbf{A}$. This dimension is called the (row) *rank* of $\mathbf{A}$, and a well-known theorem informs us that for an $m \times n$ matrix $\mathbf{A}$, the rank of $\mathbf{A}$ plus the dimension of the null space of $\mathbf{A}$ equals n, the number of columns of $\mathbf{A}$.

Index

A

Abel, N. H., 335, 397
Abelian group, 67
Abnormal PRS, *see* Incomplete PRS
Absolute pseudoprime(s), 66, 80, 98
Additive group, 32
Adleman, L. M., 77, 79, 98, 202, 214
Afrati, F., 146, 154, 158, 213
Akritas, A. G., 152, 154, 213, 216, 217, 227, 254, 265, 282, 334, 336, 350, 357, 364, 366, 381, 382, 394, 395, 396, 397, 398
Albert, A. A., 138, 154
Algebraic computations, *see* Computer algebra
Algebraic number, 138
Algorithm(s):
    analysis, 9–10
    complexity of, 9
    definition, 9
    deterministic, 77
    exponential-time, 10
    polynomial-time, 10
    probabilistic, 77
    running-time, 9
Allotrious factor(s), 227, 261
Analysis of algorithms, 9–10
Anderson, G., 267, 282
Approximation of real roots,
    definition, 335
    using bisection, 382–383
    using continued fractions, 383–388
Arago, F., 336
Arithmetic:
    long integer, 11–14
    modular, 60–61
    polynomial, 15–16, 102–103
Arnon, D., 301, 330
Asmuth, C., 76
Associate polynomials, 122
Associative operation, 29
Asymmetric cryptosystem(s), 190, 201–203
    knapsack, 203–206
    RSA, 206–208
Auto key cipher(s), 194
Available space list, 8
Average-case bound, 10

B

Bareiss, E. H., 255, 264, 283
Bareiss's algorithm, 255–256
Barnett, S., 220, 283
Base vector, 87
Basis of a space, 412
BCH code, 174–188
Bender, E. A., 296, 330
Berlekamp, E. R., 138, 140, 154, 158, 213, 243, 283, 290, 293, 296, 300, 330
Berlekamp's algorithm, 304–316
Bertrand's postulate, 93
Bigradient(s), 235
Bijection, 28
Binary:
    codes, 159
    exponentiation method, 64–65
    operation, 29
    relation, 26

Binary: *(Continued)*
   symmetric channel, 159
Blake, I. F., 158, 213
Blakley, G. R., 208, 213
Block cipher(s), 191
   digraphs, 196
   polygraphs, 196
   trigraphs, 196
Blocks of a Partition, 25
Bloom, J., 76
B-Numbers, 82
Bocher, M., 230, 283, 336, 397
Borosh, I., 213
Bose, R. C., 158, 174, 188, 213
Bound(s):
   average-case, 10
   on minimum distance between any two roots, 353–357
   on number of codewords, 163
   on number of positive roots of polynomials, 340
   on number of real roots in interval, 338–339, 357–358
   on values of coefficients of factors of polynomial, 317–319
   on values of positive roots of polynomials, 349–353
   worst-case, 10
Bradley, G. H., 42, 98
Brillhart, J., 82, 99
Brown, W. S., 220, 225, 268, 282, 283
Bruno, F. Faa di, 233, 234, 256–260
Bruno's form of the resultant, 233
Bubble pivot, 256, 268
Budan, F. D., 334, 336, 357
Budan's theorem, 357–358
Burnside, W. S., 335, 397
Butler, M. C. R., 292, 330

C

Caesar's cipher, 192
Cajori, F., 107, 154, 363, 382, 396, 397
Calmet, J., 292, 330
Camion, P., 292, 330
Cantor, D. G., 301, 315, 330, 395, 397
Cantor, M., 286, 330
Cardano, G., 335
Cardano–Descartes rule of signs, 340

Cardinality procedure, 327
Carmichael numbers, 66, 80, 98
Cartesian product of sets, 25
Casting out nines, 61–62
Cauchy's rule, 350, 395
Ceiling function, 9
Cell, 7
Characteristic of a field, 34
Chaudhuri, D. K., 158, 174, 213
Check digit(s), 159
Chen, J., 395, 398
Childs, L., 34, 98, 127, 138, 154, 188, 213, 296, 329, 330
Chinese remainder theorem/Algorithm, *see* Greek–Chinese remainder theorem; Greek–Chinese remainder algorithm; Nikomachos of Gerasa
Cipher, 189
   auto key, 194
   block, 191
   Caesar's, 192
   Hill, 196
   modular, 191
   one-time pads, 195
   permutation, 190
   product, 200
   substitution, 190
   substitution/permutation, 190
   stream, 191
   transposition, 190
   Vigenere, 193, 195
Ciphertext, 190
Class representative, 26
Closed set under an operation, 33
Code(s):
   BCH, 174–188
   binary, 159
   cyclic, 174
   dual, 169
   equivalent, 167
   error-correcting, 158–188
   Hamming, 164–174
   linear, 166
   minimum distance of, 162
   (n,k), 159
   ordered, 169
   orthogonal, 169
   perfect, 164

repetition, 165
single-parity-check, 165
Codeword(s), 160
  bound on the number of, 163
Codominance, 9
Cofactor, 410
Cohen, H., 77, 79, 98
Collins, G. E., 42, 98, 220, 282, 283, 381, 394, 396, 398
Collins-Akritas method, 381, 394, 396
Commutative:
  operation, 29
  Ring, 33
Complete PRS, 230
Complete residue system, 60
Complexity of algorithms, 9
Complex numbers set, 25
Composite integer, 52
Composition of functions, 30
Computer algebra:
  definition, 4
  publications on, 16
  systems, 16-18
Computing time function, 10
Congruence:
  equations, 70-77
  relation, 60
Content of polynomial, 216
Continued fractions, 44-52
Continued fractions expansion:
  convergent, 47
  of irrational numbers, 46-47
  partial quotients of, 45
  of polynomials, 153
  of rational numbers, 45-46
Continued fractions method of 1978 for isolation of real roots, 373-380
Convergent, 47
Coset(s), 171
Coset leader, 172
c-primitive polynomial, 144, 184
Cryptanalysis, 188
Cryptogram, 190
Cryptography, 188-209
Cryptosystem(s), 189
  asymmetric, 190, 201-203
  knapsack, 203-206
  monoalphabetic, 191

polyalphabetic, 191
polygraphic, 196
product-cipher, 200
public key, 190, 201-203
RSA, 206-208
single key, 190
symmetric, 190
Cyclic code(s), 174
Cyclic group(s), 68

D

Danielopoulos, S. D., 152, 154, 364, 396
Data encryption system, 200
Data structures, 7-9
Decoding:
  maximum-likelihood, 160
  nearest-neighbor, 161
Decomposition (prime-power) of integers, 54
Decryption by iteration, 208
Degree:
  of a polynomial, 102
  procedure, 327
Descartes, R., 25, 335
Descartes' rule of signs, 340
Determinant of a matrix, 407
  expansion of, 410
Deterministic algorithms, 77
Dickson, L. E., 52, 95, 98, 335, 398
Difference of sets, 25
Differentiation method, 381, 394
Diffie, W., 201, 213
Digits:
  check, 159
  information, 159
Dimension of a space, 412
Diophantine equations, 35-36
Dirichlet, P. G. L., 42, 98
Discriminant of a polynomial, 243
Disjoint sets, 25
Distinct degree (partial) factorization, 299-304
Divisibility of integers, 35
Division:
  long integer, 12-14
  polynomial, 103-104
  synthetic, 109
Division property:
  of integers, 35
  of polynomials, 103

Divisor:
  of an integer, 35
  of a polynomial, 103
Dixon, J. D., 77, 82, 98
Domain:
  Euclidean, 121
  integral, 33
  unique factorization, 54, 129
Dominance, 9
Dual code(s), 169
Dudley, U., 56, 98

E

Echelon form of a matrix, 408
Ecker, A., 201, 213
Eisenstein's irreducibility criterion, 131
Eliminant, *see* Resultant of Polynomials
Empty set, 24
Equal:
  functions, 30
  sets, 24
Equivalence:
  class, 26
  class representative, 26
  relation, 26
Equivalent code(s), 167
Eratosthenes, 56
Erdoes, P., 57, 98, 99
Error-correcting code(s), 158–188
Error-locator polynomial, 188
Error word (vector), 160
Euclidean algorithm:
  extended, 42–44, 124
  for integers, 39
  for polynomials, 122, 218–219
Euclidean domain, 121
Euclidean norm, 15
Euclidean property:
  of integers, 35
  of polynomials, 103
Euclidean PRS Algorithm, 219–220
Euclid's theorem, 55
Euler's phi function, 55
Euler's theorem, 63
Evaluation of polynomials, 107–110
Expansion:
  continued fractions, 44–52, 153
  of a determinant, 410

Exponential-time algorithm, 10
Extended Euclidean algorithm,
  for integers, 42–44
  for polynomials, 124
Extension of a Field, 117, 136

F

Factor base B, 82
Factor of an integer, 52
·Factor of a polynomial, 103
Factor set, 26
Factor theorem for functions, 31
Factorization domains, 54, 129
Factorization of integers, 52–83
Factorization of polynomials, 285–332,
    292–316
  distinct degree, 299–304
  lifting of, 316–326
  over a finite field, 292–316
  over the integers, 285–332
  squarefree, 133–134, 293–296
Feistel, H., 198, 213
Fermat numbers, 83
Fermat, P., 63, 83, 98
Fermat's "little" theorem, 63
Ferrari, L., 335
Fibonacci, 98
Fibonacci sequence, 22
Field(s):
  characteristic of, 34
  definition, 34
  extensions, 117
  finite, 34, 135–151
  Galois, 34, 135–151
  splitting, 138
Finite field(s), 34, 135–151
  construction of, 144–146
  representation of, 144–146
Fitch, J., 20
Fixed radix representation of integers, 89
Flanders, H., 14, 20
Floating point numbers, 4–6
  normalized, 5
Floor function, 9
Forsythe, G. E., 5, 20
Fourier, J. B. J., 334, 336, 338
Fourier's sequence, 338

Fourier's theorem, 338–339
Friedman, G., 195
Frobenius, G., 291
Fryer, W. D., 282, 283
Function(s):
 bijective, 28
 ceiling, 9
 composition, 30
 computing-time, 10
 definition, 28
 equal, 30
 Euler's phi, 55
 floor, 9
 identity, 28
 injective, 28
 inverse, 30
 Moebius, 296
 multiplicative, 296
 one-way, 202–203
 successor, 29
 surjective, 28
 surjective quotient, 29
 trapdoor, 202–203
Fundamental theorem of algebra, 127
Fundamental theorem of arithmetic, 53

G

Galois Fields [GF(p)], 34, 135–151
 construction, 144–146
 representation, 144–146
Galyean, P. H., 397
Gass, F., 195, 213
Gauss, C. F., 59, 98, 129, 130
Gaussian elimination, 254–256, 408–409
Gelfond, A. O., 226
Generator (primitive root or element) modulo m, 68
Generating substitutions, 360
 effect, 362
 implementation of, 363
 inversion, 360
 stretching, 360
 translation, 360
Generator of a group, 68
Generator matrix of a code, 166
Generator polynomial, 176
Goldman, J. R., 330

Goldstein, L. J., 57, 99
Greatest common divisor (*gcd*) of integers:
 computation, 38–44
 definition, 36
 notation, 36
Greatest common divisor (*gcd*) of polynomials:
 computation of (over a Field), 122–124
 computation of (over the integers), 215–284
 definition, 121
 modular algorithms for, 224–227
Greek–Chinese interpolation, 113–114, 224–227
Greek–Chinese remainder algorithm, 74–75
Greek–Chinese remainder theorem:
 for integers, 71–72
 for polynomials, 304–305
Greek–Chinese representation of integers, 73
Gregory, R. T., 83, 99
Group(s):
 abelian, 67
 additive, 32
 commutative, 67
 cyclic, 68
 definition, 32
 generator, 68, 137
 multiplicative, 32
 order of an element in, 68
 primitive root, 68, 137
 symmetric, 32
 of units, 67
Gunji, H., 301, 330
Guy, R. K., 82, 99

H

Habicht, W., 216, 217, 248, 249, 252, 268–278, 282, 283
Habicht's subresultant PRS algorithm, 221–222, 243–253
Hadamard (inequality), 98, 253, 267, 353
Hamming, R. W., 213
Hamming:
 code(s), 164–174
 distance, 160
 weight, 160

Heindel, L. E., 395, 398
Hellman, M. E., 201, 204, 213, 214
Hensel, K., 290, 291, 321, 330
Hensel's Lemma, 321
Herlestam, T., 208, 213
Hill, L. S., 196, 213
Hill cipher, 196
Hocquenghem, A., 158, 174, 213
Homogeneous system of linear equations, 404
Homomorphism, 28
Horner's method, *see* Ruffini–Horner method
Horowitz, E., 6, 20
Householder, A. S., 236, 283
Hurwitz, A., 337, 395, 398

I

Ideal, 174
Identity function, 28
Image, 28
Incomplete PRS, 230
Index set, 25
Information:
  digits, 159
  rate, 159
Injective function, 28
Inner product of vectors, 146
Integer(s):
  arithmetic algorithms, 11–14
  β-length of, 9
  composite, 52
  congruence of, 60
  divisibility of, 35
  division property of, 35
  divisor, 35
  Euclidean algorithm for, 39
  Euclidean property of, 35
  extended Euclidean algorithm for, 44
  factorization of, 52–83
  fixed-radix representation of, 89
  gcd computation, 38–44
  gcd definition, 36
  gcd notation, 36
  Greek–Chinese representation of, 73
  irreducible, 52
  least common multiple, 37
  length, 9
  list representation, 7
  long, 7
  mixed-radix representation of, 89
  modulo m, 59–77
  precision of, 7
  prime, 53
  prime-power decomposition of, 54
  quotient, 35
  reducible, 52
  relatively prime, 37
  remainder, 35
  set of, 24
  small, 7
  squarefree, 73
Integral:
  domain, 33
  part of a number, 45
Interpolation, 110–114
  Greek–Chinese, 113–114
  Lagrange, 110–111
Intersection of sets, 25
Inverse:
  function, 30
  of a matrix, 405
  multiplicative, 62–66
  relation, 26
Inversion, 360
  effect on roots, 362
Involutory Matrix, 196
Irrational Number(s), 46–47
  continued fractions expansion of, 46–47
Irreducibility Tests, 298
Irreducible:
  factors of polynomials, 126–132
  integer, 52
  polynomial, 114, 130
Isolation of real roots,
  Collins–Akritas method for, 381, 394, 396
  continued fractions method of 1978 for, 334, 373–380
  definition, 335
  differentiation method for, 381, 394
  "modified Uspensky's" method for, 394, 396. *See also* Collins–Akritas method
  Sturm's method for, 334, 341–349
  "Uspensky's" method for, 396. *See also* Vincent's continued fractions method

Vincent's continued fractions method of 1836
 for, 334, 373–374
Isomorphism, 28

K

Kahn, D., 189, 191, 195, 213
Kaltofen, E., 330
Kasiski, F. W., 194
Knapsack Problem, 203
 multiplicative, 206
Knuth, D., 14, 20, 42, 82, 83, 99, 153, 154, 268, 282, 283, 291, 313, 330
Krishnamurthy, E. V., 197, 213
Kronecker, L., 286
Kronecker's Factorization Method, 286–288

L

Lagrange, J. L., 95, 99, 383, 384, 396, 398
Lagrange Interpolation, 110–111
Laidacker, M. A., 257, 283
Lamé, G., 39, 42, 99
Lamé's theorem, 39–41
Landau, E., 330
Lang, S., 51, 99
Lauer, M., 292, 331
Lazard, D., 288, 331
Leading coefficient of polynomial, 102
Least absolute value residue system, 60
Least common multiple of integers, 37
Legendre, A. M., 82, 98, 99
Lehman, R. S., 82, 99
Leibniz, G. W., 66
Lempel, A., 198, 209, 213
Length of an integer, 9
Lenstra, A. K., 286, 300, 326, 331
Lenstra, H. W. Jr., 79, 98, 286, 331
Leonardo of Pisa, 98
LeVeque, W. J., 69, 99
Levinson, N., 57, 99, 158, 214
Lewis, H. R., 198, 214
Lidl, R., 197, 214, 313, 331
Lifting of a (mod-p) factorization, 316–326
 linear, 321
 quadratic, 325
Linear code(s), 166
Linear fractional substitution(s), 358

Linear independence, 412
Linear lifting of a (mod-p) factorization, 321
Lipson, J. D., 42, 99, 230, 283
List(s):
 available space, 8
 definition, 6
 operations on, 6
 representation of integers, 7
 representation of polynomials, 14–15
Lloyd, E. K., 366, 398
Long integer(s), 7
 arithmetic, 11–14
Loos, R., 282, 283, 330, 381, 394
Lovasz, L., 286, 331
Lower bound(s):
 on the minimum distance between any two roots, 353–357
 on the values of the positive roots of polynomials, 349–353
Lucas, E., 77, 78, 92, 99
LUCIFER, 200

M

Mackiw, G., 188, 214
McEliece, R. J., 330, 331
MacWilliams, F. J., 158, 188, 214
Mahler, K., 349, 353, 398
Malcolm, M. E., 5, 20
Map, or Mapping, see Function(s)
Marcus, M., 353, 398
Matrix:
 arithmetic, 401–404
 determinant of, 407
 generator, 166
 involutory, 196
 null space of, 309, 411
 parity check, 166
 rank of, 309, 412
 row reduced echelon form of, 408
 transpose, 401–402
Matrix-Triangularization Subresultant PRS algorithm, 222–224, 260–267
Mauborgne, O., 195
Maximum-likelihood decoding, 160
Max-norm, 15
Merkle, R. C., 204, 209, 214

Mersenne number(s), 83
Metric, 160
Mignotte, M., 319, 330, 331, 353, 381, 398
Mills, W. H., 56, 99
Minc, H., 353, 398
Minimal polynomial, 139
Minimum distance of a code, 162
Minimum root separation, 346
Miola, A., 226, 283, 326, 331
Mixed-radix representation of integers, 89
"Modified Uspensky's" method, 394, 396.
  See also Isolation of real roots
Modular arithmetic, 60–61
Modular cipher, 191
Modular GCD algorithms for polynomials, 224–227
Modulus, 59
Moebius function, 296
Moebius Inversion Formula, 297
Moebius substitution(s), 358
Moenck, R. T., 288, 331
Moler, C. B., 5, 20
Monic polynomial, 102
Monoalphabetic Cryptosystem, 191
Morrison, M. A., 82, 99
Moses, J., 226, 283
Motzkin, T. S., 42, 99
Muir, T., 234, 235, 262, 283
Multiple residue arithmetic, 88
Multiplicative function, 296
Multiplicative group, 32
Multiplicative inverse (modulo m), 62–66
Multiplicative knapsack problem, 206
Multiplicity of a root, 106
Musser, D. R., 289, 331

N

Natural numbers set, 24
Nearest neighbor decoding, 161
Netto, E., 120, 154, 230, 283
Newton, I., 286
Ng, K. H., 382, 398
Nikomachos of Gerasa, 94, 98
Niven, I., 93, 99
Node, 7
Nonhomogeneous system of linear equations, 404

Nonnegative residue system(s), 60
Nontrivial ring, 33
Nordgaard, M. A., 382, 398
Norm, 160
Norm of polynomials:
  Euclidean, 15
  max, 15
  sub-infinity, 15
  sub-one, 15
  sum, 15
Normal PRS, see Complete PRS
NP-complete problems, 198
NP-problems, 198
Null space, 309, 411
Null space algorithm, 313

O

Obreschkoff, N., 337, 395, 398
Olds, C. D., 44, 99
One-to-one function, 28
One-time pads, 195
One way function, 202–203
Onto function, 28
Operation:
  associative, 29
  commutative, 29
Order of a group element, 68
Ordered code(s), 169
Orthogonal:
  code(s), 169
  polynomials, 176

P

Panton, A. W., 335, 397
Papadimitriou, C. H., 198, 214
Parity:
  bit, 165
  check matrix, 166
  check polynomial, 178
Partial factorization of polynomials, 299–304
Partial quotient, 45
Partition(s):
  block, 25
  representative, 25
  of a set, 25

Pavelle, R., 4, 20
Perfect code(s), 164
Permutation:
  box, 199
  cipher, 190
Peterson, W. W., 158, 188, 214
Petr, K., 292, 331
Petricle, S. R., 16, 20
Phi function, 55
Pilz, G., 197, 214, 331
Pivot, 256, 268
Plaintext, 189
Principle subresultant coefficient, 239, 241
Poggendorff, J. C., 366, 398
Pohlig, S. C., 209, 214
Polyalphabetic cryptosystem(s), 191
Polygraphic cryptosystem(s), 196
Polynomial(s):
  arithmetic, 15–16, 102–103
  associates, 122
  content of, 216
  continued fractions expansion of, 153
  c-primitive, 144, 184
  degree of, 102
  discriminant of, 243
  division property of, 103
  divisor of, 103
  error-locator, 188
  Euclidean algorithm for, 122, 218
  Euclidean property of, 103
  evaluation of, 107–110
  extended Euclidean algorithm for, 124
  factor of, 103
  factorization of, 292–316 (over finite Fields) and 285–332 (over the integers)
  gcd computation 120–135 (over a Field) and 215–284 (over the integers)
  gcd definition, 121
  generator, 176
  ideal, 174
  interpolation, 110–114
  irreducible, 114, 130
  irreducible factors of, 126–132
  leading coefficient of, 102
  list representation of, 14–15
  minimal, 139
  modular gcd algorithm, 224–227
  monic, 102
  norms of, 15
  orthogonal, 176
  parity check, 178
  partial factorization of, 299–304
  prime, 114
  prime factorization theorem for, 128
  primitive, 130
  primitive part of, 216
  pseudodivision of, 218
  pseudoquotient, 218
  pseudoremainder, 218
  quotient, 104
  relatively prime, 123
  remainder, 104
  remainder sequence, 122
  resultant of, 230–243
  root of, 105
  Schubert–Kronecker factorization of, 286–288
  squarefree, 132
  squarefree factorization of, 133–134, 293–296
  synthetic division of, 109
  trace, 303
  unit, 114–115
  zero of, 105
Polynomial Remainder Sequence (PRS):
  abnormal, 230
  complete, 230
  incomplete, 230
  normal, 230
Polynomial Remainder Sequence (PRS) algorithm(s):
  Euclidean, 219–220
  Habicht's subresultant, 221–222, 243–253
  matrix-triangularization subresultant, 222–224, 260–267
  primitive, 220
  "reduced", see Sylvester's reduced (subresultant) PRS algorithm
  "subresultant", see Habicht's subresultant PRS algorithm
  Sylvester's reduced (subresultant), 221–222, 227–230
Polynomial-time algorithm(s), 10
Pomerance, C., 77, 79, 98, 99
Power set, 25

Pratt, V. R., 77, 99
Primality testing, 77–83
Prime(s):
  integer, 53
  residue system, 63
  unlucky, 224
Prime Factorization Theorem for Polynomials, 128
Prime number theorem, 57, 98
Prime polynomial, 114
Prime-power decomposition, 54
Primitive element (or root), 68, 137
Primitive element theorem, 137
Primitive part of a polynomial, 216
Primitive polynomial, 130
Primitive PRS algorithm, 220
Principle subresultant coefficient, 239, 241
Probabilistic algorithm(s), 77
Problems:
  P, 198
  NP, 198
  NP-complete, 198
  (un)solvable, 198
Product-cipher systems, 200
Pseudodivision of polynomials, 218
Pseudoprimality test(s), 79
Pseudoprime(s), 66
Pseudoquotient, 218
Pseudoremainder, 218
Public key cryptosystem(s), 190, 201–203
  knapsack, 203–206
  RSA, 206–208

Q

Quadratic irrational number, 47
Quadratic lifting of a (mod p) factorization, 325
Quotient(s),
  integral, 35
  partial, 45
  polynomial, 104
  set, 26
  trial, 12

R

Rabin, M. O., 77, 82, 99, 292, 331
Radix:
  fixed, 89
  mixed, 89
Ramachandran, V., 197, 213
Rank of a matrix, 309, 412
Rational numbers set, 24–25
  continued fractions expansion of, 45
Real numbers set, 25
Real root(s), see Root(s)
Real root approximation, see Approximation of real roots
Real root isolation (methods), see Isolation of real roots
Real root approximation, see Isolation of real roots
Reduced PRS algorithm, see Sylvester's reduced (subresultant) PRS algorithm
Reducible integer(s), 52
Relation(s):
  binary, 26
  congruence, 60
  equivalence, 26
  inverse, 26
Relatively prime:
  integers, 37
  polynomials, 123
Remainder:
  arithmetic, 60
  integral, 35
  modulo m, 60
  polynomial, 104
  sequence, 122
Repetition code(s), 165
Representative(s):
  of an equivalence class, 26
  of a partition, 25
Residue system(s):
  complete, 60
  least absolute value, 60
  nonnegative, 60
  prime, 63
  symmetric, 60
Resultant of polynomials, 230–243
  Bruno's form of, 233, 256–260
  definition, 231
  properties of, 231–232, 237–239
  representation of, 232–233
  Sylvester's form of, 222, 260
  Trudi's form of, 235, 256–260

Rice, J., 381, 394, 398
Richards, I., 44, 99
Ring(s):
    commutative, 33
    definition of, 32
    nontrivial, 33
Rivest, R. L., 202, 214
Root(s):
    approximation, 335
    approximation methods, *see* Approximation of real roots
    isolation, 335
    isolation methods, *see* Isolation of real roots
    multiplicity, 106, 129
    of polynomial, 105
    primitive, 68, 137
    separation, 346
    simple, 129
Rothstein, M., 20
Round-off error(s), 5
Row space, 412
Ruffini, P., 335, 396, 398
Ruffini–Horner method, 107–110
Rule of signs, 340
Rumely, R. S., 79, 98
Rump, S. M., 349, 398
Running time of algorithm(s), 9
Russell, B., 97
Russell's paradox, 97
Russian peasant multiplication method, 65

S

Sahni, S., 20
Sarrus, P. F., 66
Schmid, B., 214
Schroeder, M. R., 42, 100
Schubert, F., 286
Schubert–Kronecker factorization method, 286–288
Schwarz, S., 301, 331
Scott, N. R., 83, 100
Separation of roots, 346
Set(s), 24–26
    Cartesian product, 25
    closed under operation, 33
    of complex numbers, 25
    difference, 25
    disjoint, 25
    empty, 24
    equal, 24
    factor, 26
    finite, 29
    image of, 28
    index, 25
    of integers, 24
    intersection, 25
    of natural numbers, 24
    notation of, 24
    paradox, 97
    partition of, 25
    product of, 25
    power set of, 25
    quotient, 26
    of rational numbers, 24–25
    of real numbers, 25
    singleton, 24
    subset, 24
    union, 25
Shamir, A., 202, 214
Shannon, C. E., 158, 200, 214
Sieve of Eratosthenes, 56
Sign variation, 338
Simple field extension, 117, 136
Simple root, 129
Simmons, G. J., 201, 214, 298, 331
Sims, C. C., 24, 100, 129, 154, 301, 331
Single key cryptosystem(s), 190
Single-parity-check code(s), 165
Singleton, 24
Sinkov, A., 191, 196, 214
Slepian table, 172
Sloane, N. J. A., 214
Smith, S., 301, 332
Solovay, R., 77, 100
Solvable problems, 198
Specht, W., 318, 319, 330, 332, 398
Splitting field, 138
Squarefree:
    factorization, 133–134, 293–296
    factors, 133
    integers, 73
    polynomial, 132
Standard residue digits, 87
Stickelberger's theorem, 329
Strassen, V., 100

Stream cipher(s), 191
Stretching, 360
  effect of, 362
Sturm, C., 227, 261, 263, 334, 336, 342, 344, 396, 399
Sturm's bisection method of 1829, 341–349
Sturm's sequence, 341
  properties of, 342
Sturm's theorem, 342–343
Sub-infinity norm, 15
Sub-one norm, 15
Subresultant(s), 233–237
  coefficient(s), 239, 241
Subresultant PRS algorithm, *see* Habicht's subresultant PRS algorithm
Subresultant PRS method(s), 221–224, 227–253, 260–267. *See also* Polynomial Remainder Sequence (PRS) algorithm
  matrix-triangularization, 222–224, 260–267
  Sylvester–Habicht pseudodivisions, 221–222, 227–253
Subset, 24
Subset problem, 203
Substitution(s):
  effect of, 362
  generating, 360
  implementation of, 363
  Moebius, 358
Substitution box, 199
Substitution cipher(s), 190
Successor:
  field, 7–8
  function, 29
Sum-norm, 15
Superincreasing sequence, 203
Surjection, 28
Surjective quotient function, 29
Swinnerton-Dyer, H. P. F., 316, 332
Sylvester, J. J., 216, 217, 227, 229, 268–278, 282, 283
Sylvester–Habicht pseudodivisions subresultant PRS method, 221–222, 227–253
Sylvester's form of the resultant, 222, 260
Sylvester's reduced (subresultant) PRS algorithm, 221–222, 227–230

Symbolic and algebraic computations, *see* Computer algebra
Symmetric cryptosystem(s), 190
Symmetric group, 32
Symmetric residue system, 60
Systems of computer algebra, 16–18
Syndrome, 171
Synthetic division, 109

T

Tartaglia, N., 335
Taub, H., 146, 154
Taylor's expansion formula, 358
Terminal node, 375
Todhunter, I., 335, 399
Totient function, 55
Trace polynomial, 303
Trager, B. T., 135, 154
Transendental number, 138
Translation, 360
  effect of, 362
Transpose:
  vector, 401
  matrix, 401–402
Transposition cipher, 190
Trapdoor function, 202–203
Traub, J. F., 283
Trial quotient, 12
Trotter, H., 51, 99
Trudi, N., 235, 236, 256–260
Trudi's form of the resultant, 235, 256–260
Truncation, 5
Tuples, 25
Turnbull, H. W., 335, 399

U

Union of sets, 25
Unique factorization domain, 54, 129
Unit(s):
  in a Ring, 33
  group of, 67
  polynomial, 114–115
Unlucky prime, 224
Unsolvable problems, 198
Upper bound(s),
  on the number of codewords, 163

on the number of positive roots of
   polynomials, 340
on the number of real roots in an interval,
   338–339, 357–358
on the values of the coefficients of factors of
   polynomials, 317–319
on the values of the positive roots of
   polynomials, 349–353
Uspensky, J. V., 395, 396, 399
"Uspensky's Method", 396. *See also* Isolation
   of real roots

V

Vandermonde Determinant, 354
van der Sluis, A., 395, 399
van der Waerden, B. L., 335, 399
Van Vleck, E. B., 222, 254, 260, 263, 264,
   271, 282, 283
Variation, *see* Sign variation
Vector(s):
   inner product of, 146
   space, 146, 411
   transpose, 401
Vector Space(s), 146, 411
   basis of, 309, 412
   dimension of, 309, 412
Verbaeten, P., 382, 399
Vernam, G. S., 195
Vigenere cipher, 193, 195
Vincent, A. J. H., 334, 336, 357, 366, 396, 399
Vincent's Continued Fractions Method of 1836,
   373–374
Vincent's theorem, 366

W

Wakerly, J., 158, 214
Wang, P., 135, 154, 288, 332
Wang's theorem, 395
Wang, X., 395
Weisner, L., 335, 399
Weldon, E. J., 214
Well-ordering principle, 35
Wilf, H. S., 41, 82, 100
Williams, H. C., 77, 82, 100, 208, 214
Wilson's theorem, 66
Worst-case bound, 10
Wunderlich, M., 82, 100

Y

Yun, D. Y. Y., 135, 154, 226, 283, 324, 326,
   331, 332

Z

Zassenhaus, H., 301, 315, 325, 326, 332
Zerodivisor (in a ring), 33
Zero of polynomial, 105: *See also* Root
Zimmer, H. G., 397
Zuckerman, H. S., 93, 99